THE
BENCHTOP
ELECTRONICS
REFERENCE
MANUAL
2nd Edition

Dedication

To my beautiful wife, Joyce,
who has been my inspiration and strength for over forty years,
and to those I hold dear:
Jackie, Gill, Phil, Ray, Jaimie, Nicola, Lisa, Katie, Peter, Paul, and Paula
also in memory of
Charles H. and Margaret A. Veley

THE
BENCHTOP
ELECTRONICS
REFERENCE
MANUAL
2nd Edition

Victor F. C. Veley

TAB **TAB BOOKS**

Blue Ridge Summit, PA

FIRST EDITION
FIRST PRINTING

© 1990 by **TAB BOOKS**
TAB BOOKS is a division of McGraw-Hill, Inc.

Library of Congress Cataloging-in-Publication Data

Veley, Victor F. C.
    The benchtop electronics reference manual / by Victor F.C. Veley.
—2nd ed.
        p.   cm.
    Includes index.
    ISBN 0-8306-7414-4   ISBN 0-8306-3414-2 (pbk.)
    1. Electronics—Handbooks, manuals, etc.   I. Title.
TK7825.V45   1990
621.381—dc20   90-37038
                                                CIP

TAB BOOKS offers software for sale. For information and a catalog, please contact TAB Software Department, Blue Ridge Summit, PA 17294-0850.

Questions regarding the content of this book should be addressed to:

Reader Inquiry Branch
TAB BOOKS
Blue Ridge Summit, PA 17294-0214

Acquisitions Editor: Roland S. Phelps
Book Editor: David Gauthier
Production: Katherine Brown
Book Design: Jaclyn J. Boone

Contents

PART 2   Alternating Current Principles

Acknowledgments

I wish to acknowledge a considerable debt to my wife, Joyce, for preparing the manuscript. Without her help and inspiration the creation of this new edition would not have been possible. In addition I also wish to thank Ms. Marta E. Ruiz who skillfully and diligently made many corrections on my "pet" computer, Lazarus, which was built from scrap parts and therefore "rose from the dead."

Introduction

To INCREASE THE BREADTH AND DEPTH OFFERED IN THE ORIGINAL TEXT, A further 40 chapters have been added to this second edition. The new material is divided into two parts and covers the following subject areas:

1. Mathematics for Electronics. Twenty new chapters are devoted to such topics as the measurement of angles; trigonometry including multiple angles; the binomial series; exponential, logarithmic and hyperbolic functions; differential and integral calculus.
2. Digital Principles. Twenty new chapters cover the topics of number systems, logic gates, Boolean algebra and the Karnaugh Map.

Like the first edition, the real strength of this book lies in its unique format. Two hundred of the more common electronics topics have been selected from the subject areas of dc, ac, solid-state and tube circuits, communications including microwave, applied mathematics and digital principles. Each topic is treated separately and is explored in three stages.

1. The basic principles are discussed in a "friendly" style which invites the reader to think in terms of electronics and communications.
2. As many relevant equations as possible appear in the mathematical derivations. All these equations use the modern SI system of metric units.
3. One or more examples are used to illustrate the use of the various equations. Wherever possible practical values are used in the examples. It follows that many of the circuits can be constructed in the laboratory; the calculated results can then be compared with the measured values.

The result of using this format is to create an excellent reference guide for the electronics and communications technicians already working in industry or in the armed services. This book is also a valuable aid to students who are studying electronics in the community colleges, private vocational schools, and senior high schools.

One hundred of the topics are devoted to the subject of direct and alternating current. The book can therefore serve as the text for introductory courses in electronics. When all topics are taken into account, the material covers most of the requirements for the *FCC Commercial General Radiotelephone Operator's License.*

Conventions used in this book are as follows: Small capitals or numbers following a large capital refer to the component (e.g., R1, RE). Subscripted numbers or letters refer to the unit of measure (e.g., capacitance C_2, resistance R_1).

1
PART

Direct current principles

Chapter 1
The international system (SI) of units

AS FAR AS ELECTRICAL UNITS ARE CONCERNED, WE ARE FORTUNATE TO LIVE IN an age in which one unified system has been adopted. Prior to about 1960 there were really three systems of measurement units. To start with, we had the practical everyday units, some of which you may already know: the ampere, the volt, and the watt. The second system of units was based on magnetism and the third on electrostatics. The last two were referred to as cgs systems because they used the centimeter as the unit of length, the gram for mass, and the second for time. Frankly, it was a complete mess! Not only were there three possible units in which an electrical quantity could be measured, but between the three systems there were horrible conversion factors that were virtually impossible to memorize.

In 1901, Professor Giorgi of Italy proposed a new system founded on the meter (100 centimeters, or slightly greater than 1 yard), the kilogram (1000 grams, or about 2.2 pounds), and the second—as the units of length, mass, and time. For electricity it was necessary to define a fourth fundamental unit and then build on this foundation to establish other units. In 1948, the ampere, which measures electrical current, was internationally adopted as a fourth unit. We therefore have the MKSA (meter, kilogram, second, ampere) system, which is also referred to as the international system or SI (Systéme Internationale d'Unites). The three previous systems have therefore been replaced by a single system, with the added attraction that the old practical units are part of the new system.

In any study of electricity and electronics we often need to use mechanical units to measure such quantities as force and energy. We are therefore going to use this section to establish the mechanical SI units, and we shall also see that some of these units are directly transferable to the electrical system. For example, electrical energy and mechanical energy are each measured by the same unit.

UNIT OF FORCE

Isaac Newton stated that when a force is applied to an object or mass (Fig. 1-1A), the object or mass accelerates so that its speed or velocity increases. In the SI system the velocity is measured in meters per second. We shall use the abbreviations of m for meter and s for second, so that meters per second is written as m/s. Acceleration will be expressed as meters per second per second. For example, if a mass starts from rest with zero velocity and is given an acceleration of 3 meters per second per second, its velocity after 1 second is 3 meters per second, after 2 s is 6 m/s, after 3 s is 9 m/s and so on.

If the force applied to a particular mass is increased, the acceleration will be greater. However, if the force is kept the same but the mass is greater, the acceleration will be less. The unit of force in the SI system is the *newton* (N), which will give a mass of 1 kilogram an acceleration of 1 meter per second per second in the direction of the force.

When a mass is falling under the force of the earth's gravity, its acceleration is 9.81 m/s^2. Therefore, the gravitational force on the mass of 1 kg is 9.81 N. This force is sometimes referred to as 1-kilogram weight. Of course, on the moon the force of gravity exerted on 1 kg would be less than 9.81 N.

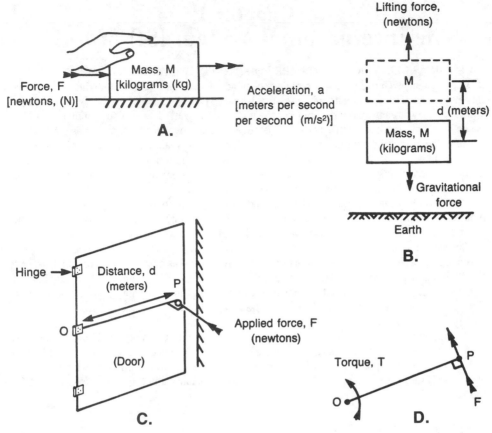

Figure 1-1

UNIT OF ENERGY OR WORK

Energy is the capacity for doing work, and therefore both these quantities are measured by the same unit. When a force is applied through a certain distance in the direction of the force, the energy must be supplied so that work may be performed. A good example is lifting a mass at a constant speed against the force of gravity (Fig. 1-1B). When the mass has been raised through a given height or distance, you have expended some energy in performing a certain amount of work. The larger the force and the longer the distance through which the force is applied, the greater is the amount of work that must be done. The value of the work performed is then found by multiplying the force in newtons by the distance in meters. The result of multiplying any two quantities together is called their product. We can therefore say that work is the product of force and distance. In the SI system, the unit of mechanical energy or work is the *joule (J)*. For example, if a force of 5 N is applied through a distance of 2 m, the work done in joules is 2 m × 5 N = 10 J. One joule can therefore be thought of as 1 meter-newton, as it is the result of multiplying 1 meter by 1 newton.

Notice what we are doing. The newton was defined in terms of our fundamental units of mass (kilogram), length (meter), and time (second). The joule is derived from the newton and the meter; in other words, each new unit is defined in terms of its predecessors. This is the logical manner in which a system of units is established.

4

UNIT OF TORQUE

A torque produces a twisting effect, which we use every day when we open a door by either pushing or pulling its handle. The force that we apply is most effective when its direction is at right angles (90°) to the line joining the handle to the door hinge (Fig. 1-1C). This is clearly so because if the force and the line were in the same direction, the door would not open at all. If the handle were positioned close to the hinge, it would be difficult to open the door. We are therefore led to the conclusion that the value of the torque must be equal to the result of multiplying the applied force, F, by the distance, d, which is at right angles (90°) to the force's direction. This distance is the length between the point, P, at which the force is applied and the pivot, O, about which the twisting effect or the rotation occurs (Fig. 1-1D). Torque is equal to the product of force and distance, but with one very important difference. In the case of work, the directions of the force and the distance are the same, but with torque, the directions are 90° apart. Consequently, torque is not measured in joules but in *newton-meters*.

UNIT OF POWER

There is often much confusion over the distinction between *work* and *power*. *Power* is the rate at which work is performed or energy is expended. As soon as you see the word *rate,* you must realize that time is involved. Let us take a mechanical example. Suppose that you have a heavy weight and you ask a powerful adult to lift it up a certain height against the force of gravity. The adult will be able to perform this task quickly (in a short time) because of his or her power. However, you could have a complex pulley system attached to the weight and at the end of the system there might be a wheel. A small child could be capable of turning the handle on the wheel and the weight would slowly rise to the same height achieved by the powerful adult. Neglecting the weight of the pulley system and its friction, the total work performed by the adult and the child is the same, but the child took a much longer time because he or she is much less powerful than the adult. An old British unit is the horsepower, which is equivalent to 550 foot-pounds per second. This means that a motor whose mechanical output is 1 horsepower (hp) would be capable of lifting a mass of 550 pounds through a distance of 1 foot against the force of gravity, and do it in a time of 1 second.

The SI unit of power is the *watt,* whose unit symbol is W. The power is 1 watt if 1 joule of energy is created or used every second. For example, when you switch on a 60-W electric light bulb, 60 J of energy are released from the bulb every second, mostly in the form of heat but a small amount as light. Therefore, watts are equivalent to joules per second, or joules are the same as watts × seconds, which may be written as *watts-seconds*. Since the horsepower and the watt both measure the same quantity, the two must be related and, in fact, 1 horsepower is equivalent to 746 watts.

If the 60-W light bulb is left on for 1 h (3600 s), the energy consumed is 60 W × 3600 s = 216000 J, and this consumption would appear on the electricity bill. It is clear that, for everyday purposes, the joule is too small a unit. In fact, it takes about 8 J of energy to lift a 2-lb book through a distance of 1 yard and about half a million joules to boil a kettle of water. A larger unit would be the *watthour* (Wh), which is the energy consumed when a power of 1 watt is operated for a time of 1 hour. Since 1 h is the same as 3600 s, 1 watthour = 1 W × 3600 s = 3600 J. The unit on the electricity bill is still larger; it is the *kilowatt-hour* (kWh), which will be equal to 1000 Wh or 3,600,000 J.

MATHEMATICAL DERIVATIONS
Unit of force

Acceleration,

$$a = \frac{F}{m} \text{ meters per second per second} \tag{1-1}$$

Force,

$$F = m \times a \text{ newtons} \tag{1-2}$$

Velocity,

$$v = a \times t \text{ meters per second} \tag{1-3}$$

Acceleration,

$$a = \frac{v}{t} \text{ meters per second per second} \tag{1-4}$$

where: F = force (newtons, N)
  m = mass (kilograms, kg)
  a = acceleration (meters per second per second, m/s^2)
  v = velocity (meters per second, m/s)

Unit of energy or work

Work,

$$W = d \times F \text{ joules or meter-newtons} \tag{1-5}$$

where: W = work done (joules, J)
  F = force (newtons, N)
  d = distance (meters, m)

Unit of torque

Torque,

$$T = F \times d \text{ newton-meters} \tag{1-6}$$

where: T = torque (newton-meters, N m)
  F = applied force (newtons, N)
  d = distance (meters, m)

Unit of power

Power,

$$P = \frac{W}{t} \tag{1-7}$$

Work,

$$W = P \times t \text{ joules} \tag{1-8}$$

From Equation 1-5, $W = d \times F$ so that

Power,

$$P = \frac{W}{t} = \frac{d}{t} \times F = v \times F \text{ watts} \tag{1-9}$$

where: P = power (watts, W)

W = work (joules, J)

d = distance (meters, m)

v = velocity (meters per second, m/s)

Example 1-1

A force of 150 N is continuously applied to a 30 kg mass that is initially at rest. Calculate the values of the acceleration and velocity after 8 s. When the mass has been moved through a distance of 2 km, how much is the total work performed?

Solution

Acceleration,

$$a = \frac{F}{N} \qquad (1\text{-}1)$$

$$= \frac{150 \text{ N}}{30 \text{ kg}}$$

$$= 5 \text{ m/s}^2$$

Velocity after 8 s,

$$v = a \times t \qquad (1\text{-}3)$$
$$= 5 \text{ m/s}^2 \times 8 \text{ s}$$
$$= 40 \text{ m/s}$$

Work done,

$$W = d \times F \qquad (1\text{-}5)$$
$$= 2 \times 10^3 \text{ m} \times 150 \text{ N}$$
$$= 3 \times 10^5 \text{ J}$$

Example 1-2

A metal block whose mass is 250 gm is given an acceleration of 15 cm/s^2. Calculate the value of the accelerating force.

Solution

Accelerating force,

$$F = m \times a \qquad (1\text{-}2)$$
$$= 250 \times 10^{-3} \text{ kg} \times 15 \times 10^{-2} \text{ m/s}^2$$
$$= 3750 \times 10^{-5} \text{ N}$$
$$= 0.0375 \text{ N}$$

Example 1-3

A mass of 1500 kg is lifted vertically with a velocity of 180 m/min. Calculate the value of the required power in kilowatts.

Solution

Force,

$$F = m \times a \qquad \text{(1-2)}$$
$$= 1500 \times 9.81$$
$$= 14715 \text{ N}$$

Distance through which the mass is lifted in 1 s is 180/60 = 3 m. Therefore the work done in 1 s is:

Work,

$$W = d \times F \qquad \text{(1-5)}$$
$$= 3 \text{ m} \times 14715 \text{ N}$$
$$= 44145 \text{ J}$$

Power,

$$P = \frac{W}{t} \qquad \text{(1-7)}$$

$$= \frac{44145 \text{ J}}{1 \text{ s}}$$

$$= 44.145 \text{ kW}$$

Example 1-4

A perpendicular force of 250 N is used to create a torque about an axis of rotation. If the distance from the axis to the application point of the force is 8 cm, calculate the value of the torque.

Solution

Torque,

$$T = F \times d \qquad \text{(1-6)}$$
$$= 250 \text{ N} \times 8 \times 10^{-2} \text{ m}$$
$$= 20 \text{ Nm}$$

Chapter 2
Unit of charge

THE WORD *CHARGE* MEANS A QUANTITY OF ELECTRICITY AND ITS LETTER symbol is Q. The SI unit is the coulomb (Charles A. Coulomb, 1736–1806) which was originally defined from a series of experiments performed in the 1830s by Michael Faraday (1791–1867). These experiments involved the flow of electron current through a chemical solution and demonstrated the phenomenon of electrolytic conduction.

In one experiment a bar of silver and a nickel plate (referred to as the electrodes) are immersed in a silver nitrate solution (Fig. 2-1), which acts as the electrolyte. The silver bar is called the *anode* and is connected to one terminal of the battery while the nickel plate or *cathode* is

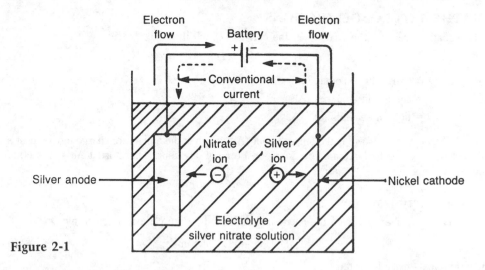

Figure 2-1

connected to the other terminal. As time passed, Faraday observed that silver was lost from the anode and an equal amount was deposited on the cathode. In addition, the silver was being transferred at a constant rate. At the time of this experiment the existence of the electron was unknown and therefore Faraday (wrongly) considered that the movement of the electricity through the circuit was carrying the silver from the anode to the cathode. It was then assumed that the anode was connected to the battery's positive terminal while the cathode was joined to the negative terminal so that the direction of the electrical flow or current was from positive to negative. In fact there is a movement of positive silver ions from the anode to the cathode while negative nitrate ions are traveling in the opposite direction through the electrolyte. The electron flow occurs in the external copper connecting wires with the electrons leaving the battery's negative terminal and entering the positive terminal. Faraday's assumption is often referred to as the conventional or mathematical current flow, which is considered to flow from the battery's positive terminal to its negative terminal and is therefore in the opposite direction to the *actual* physical electron flow.

From his experiments Faraday stated his law of electrolysis. *"The mass of silver leaving the anode and deposited on the cathode is directly proportional to the quantity of electricity or charge passing through the electrolyte."*

For a silver nitrate solution the coulomb is that quantity of electricity that causes 1.118×10^{-6} kg of silver to be deposited on the cathode. This quantity, 1.118×10^{-6} kg/C, is called silver's electrochemical equivalent. Other electrochemical equivalent values are: copper 3.294×10^{-7}, nickel 3.04×10^{-7} and zinc 3.38×10^{-7} kg/C.

From a study of atomic structure, it follows that the practical charge of 1 coulomb must be equivalent to the negative charge associated with a certain number of electrons. In fact, the coulomb represents the charge carried by 6.24×10^{18} electrons so that the charge, e, possessed by a single electron is only $1/(6.24 \times 10^{18}) = 1.602 \times 10^{-19}$C.

The *ampere* is the *fundamental* SI unit for measuring the current and has already been referred to in Chapter 1. We can derive the coulomb from the ampere by including the unit of time, which is the second. When a current of one ampere flows for a time of one second, a charge of one coulomb passes a particular point in an electrical circuit. In other words amperes are equivalent to coulombs per second or coulombs are the same as amperes × seconds.

MATHEMATICAL DERIVATIONS

The equations relating the charge, Q, the current, I, and the time, t, are:

$$Q = I \times t, \qquad I = Q/t, \qquad t = Q/I \tag{2-1}$$

where: Q = charge in coulombs (C)

I = current in amperes (A)

t = time in seconds (s)

A larger unit of charge is the ampere-hour (Ah) which is the amount of charge moved past a point when a current of one ampere flows for a time of one hour. Therefore 1 Ah = $1 \times 60 \times 60$ = 3600 C.

Law of electrolysis

Total mass, M, liberated from the anode and deposited on the cathode is given by:

$$M = zQ = zIt \text{ kilograms} \tag{2-2}$$

where: M = mass liberated (kg)

z = electrochemical equivalent (kg/C)

Q = charge in coulombs (C)

I = current in amperes (A)

t = time in seconds (s)

Example 2-1

If a steady current of 2.5 A exists at a given point for a period of 15 min, calculate the amount of charge flowing past that point in coulombs and ampere-hours.

Solution

Charge,

$$Q = I \times t \tag{2-1}$$
$$= 2.5 \text{ A} \times 15 \times 60 \text{ s}$$
$$= 2250 \text{ C}$$
$$= \frac{2250}{3600} \text{ Ah}$$
$$= 0.625 \text{ Ah}$$

Example 2-2

A constant current of 4.5 A flows through a copper sulfate solution for a time of 1.7 h. What is the value of the mass liberated from the copper anode?

Solution

Mass liberated,

$$m = zIt \tag{2-2}$$
$$= 3294 \times 10^{-7} \text{ kg/C} \times 4.5 \text{ A} \times 1.7 \times 60 \times 60 \text{ s}$$
$$= 9.07 \times 10^{-3} \text{ kg}$$

Chapter 3
Unit of electromotive force (EMF)

IN ESTABLISHING THE INTERNATIONAL SYSTEM OF ELECTRICAL UNITS, WE BEGAN in Chapter 1 by defining the ampere as the fourth fundamental unit of the MKSA or SI system. By combining the ampere with a unit of time (the second) we were able to derive the coulomb as the unit of charge or quantity of electricity. In Chapter 1 we also defined the joule as the SI unit of energy or work. The same unit can be applied to electrical energy because the various forms of energy (mechanical, electrical, heat, etc.) are interchangeable.

In obtaining the mechanical SI units of Chapter 1 we saw that a force was necessary in order to accelerate the mass of an object and cause the object to move. In a similar way the electrical equivalent of this force imparts a velocity to the free electrons so that a current is able to flow. This "electron moving force" is the *electromotive force* (EMF) whose letter symbol is E. The EMF is the force that gives electricity its motion (Fig. 3-1) and its unit is the *volt*, named after the Italian Count Alessandro Volta (1745–1827) who invented the first chemical cell able to generate electricity. The unit symbol for the volt is the letter V.

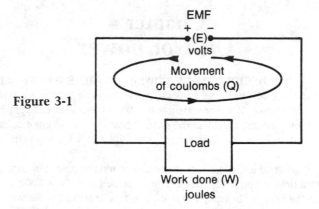

Figure 3-1

In order to drive the charge through a circuit, work must be done and it is the electrical source (for example, a battery) that must be capable of providing the necessary energy. The volt may therefore be defined in terms of the coulomb and the joule. The EMF is 1 volt if, when 1 coulomb is driven around an electrical circuit, the work done is 1 joule. If we increase either the EMF and/or the charge, the amount of work done will be greater.

A much smaller unit of work is the electronvolt (eV) which is the work done when the charge carried by an electron is driven around a circuit by an EMF of one volt. Since one coulomb is equivalent to the charge carried by 6.24×10^{18} electrons, one joule must be equal to 6.24×10^{18} electron-volts.

MATHEMATICAL DERIVATIONS
The equations relating the work done, the charge and the electromotive force are:

$$W = Q \times E, \ Q = \frac{W}{E} \tag{3-1}$$

$$E = \frac{W}{Q}$$

where: W = work done in joules (J)

Q = charge in coulombs (C)

E = EMF in volts (V)

Example 3-1

What is the work done if an EMF of 6 V drives a charge of 3 C around an electrical circuit?

Solution

Work done,

$$W = Q \times E$$
$$= 3 \text{ C} \times 6 \text{ V}$$
$$= 18 \text{ J}$$

Chapter 4
Unit of power

IN CHAPTER 1 WE DECIDED THAT POWER IS THE RATE AT WHICH WORK IS performed or the rate at which energy is created. The letter symbol for power is P while its unit is the watt (unit symbol, W). For example a 100-watt incandescent light bulb releases 100 joules of energy every second, mostly in the form of heat but a small amount as light. Therefore watts are equivalent to joules per second or joules are the same as watts × seconds (watt-seconds).

The energy consumed is the product of the power and the time. For example, if a 100-watt bulb is left on for a time of one hour, the energy consumed is 60 W × 3600 s = 216,000 joules. For everyday use the joule is obviously too small a unit. A larger unit is the watt-hour (Wh) which is the energy consumed when a power of one watt is operated for a time of one hour. Therefore one watt-hour is the same as 1 W × 3600 s = 3600 J. However, the unit on the electricity bill is still larger. It is the kilowatt-hour (kWh) which is equivalent to 1000 Wh or 3,600,000 J.

We know that the power in watts is equal to the same number of joules per second while the

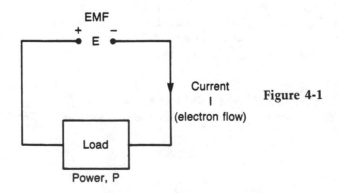

Figure 4-1

work in joules is the product of the EMF, E, in volts and the charge, Q, in coulombs (equation 3-1). The power is therefore the product of the EMF, E, and the current, I (Fig. 4-1), because the current in amperes is equivalent to the same number of coulombs per second (equation 2-1).

MATHEMATICAL DERIVATIONS

Power,

$$P = \frac{W}{t} \text{ watts} \tag{4-1}$$

Work,

$$W = P \times t = E \times Q \text{ joules} \tag{4-2}$$

where:　P = power in watts (W)
　　　　W = work in joules (J)
　　　　E = EMF in volts (E)
　　　　Q = charge in coulombs (C)

Power,

$$P = \frac{W}{t} \tag{4-3}$$

$$= \frac{E \times Q}{t}$$

$$= E \times \frac{Q}{t}$$

$$= E \times I \text{ watts}$$

where:　I = current in amperes (A)

Current,

$$I = \frac{P}{E} \text{ amperes} \tag{4-4}$$

EMF,

$$E = \frac{P}{I} \text{ volts} \tag{4-5}$$

Example 4-1

For a period of 3 h, a current of 4 A is drawn from an electrical source whose EMF is 120 V. Calculate the values of the power and the total energy consumption.

Solution

Power,

$$P = E \times I \tag{4-3}$$

$$= 120 \text{ V} \times 4 \text{ A}$$

$$= 480 \text{ W}$$

Energy consumption,

$$W = P \times t \tag{4-2}$$
$$= 480 \text{ W} \times 3 \text{ h}$$
$$= 1440 \text{ Wh}$$
$$= \frac{1440}{1000} \text{ kWh}$$
$$= 1.44 \text{ kWh}$$

Example 4-2

A 60 W electric light bulb is operated from a 120 V source. What is the value of the current supplied to the bulb?

Solution

Current,

$$I = \frac{P}{E} \tag{4-4}$$
$$= \frac{60 \text{ W}}{120 \text{ V}} = 0.5 \text{ A}$$

Chapter 5
Ohm's law, resistance, and conductance

IN CHAPTER 3 WE SAW THAT THE ELECTROMOTIVE FORCE, E, IS RESPONSIBLE for creating the current, I. Consequently, there should be some sort of relationship between E and I. For example, it is only logical to assume that if we increase the EMF applied to an electrical circuit, the current will also increase. However, in 1827 George Simon Ohm stated that the exact relationship between E and I was linear. Under the law which bears Ohm's name, the current flowing through a conductor (such as a very long length of thin wire made from silver, copper, aluminum, etc.) is *directly proportional* to the EMF applied across the conductor. This occurs under constant physical conditions of temperature, humidity, and pressure. This means that if we triple the voltage, the current will also be tripled and if we halve the voltage, the current will also be divided by 2. Whatever we do to the voltage, the same will happen to the current. However, please do not think that Ohm's law is obviously true. Most electronic components *within their operating range* obey Ohm's law, but others do not. For example, if you double the forward voltage across a semiconductor diode, the current will increase, but will not double. In general, solid-state devices and tubes do not obey Ohm's law.

In Fig. 5-1 a voltage source whose EMF can be varied, is connected across a long length of copper wire, which acts as the conductor. An ammeter is placed in the path of the current, I, and will record the current value in amperes. A voltmeter is connected across the battery and will read the value of the EMF, E. To start with, it is obvious that if the applied voltage is zero, the current is also zero. We next assume that when the initial EMF is 12 V, the recorded current is

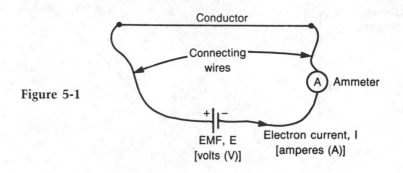

Figure 5-1

2 A. If this EMF is now doubled to 24 V, the current will also double to 4 A. When the voltage is again doubled to 48 V, the new current is 8 A. If the EMF is multiplied by 10 to a value of 10×12 V = 120 V, the accompanying current is 10×2 A = 20 A. Finally, if the initial EMF is halved to 12 V/2 = 6 V, the current drops to 2 A/2 = 1 A. These corresponding values of E and I are illustrated in Table 5-1.

Table 5-1. Example of constant ratio of E/I.

EMF (E) volts	Current (I) amperes	Ratio E/I
6 V	1 A	6
12 V	2 A	6
24 V	4 A	6
48 V	8 A	6
120 V	20 A	6

In Table 5-1, the ratio of $E:I$ is calculated for each corresponding voltage and current. In every case the answer is the same (6) which is a constant for the circuit of Fig. 5-1. Ohm's law may therefore be restated as: *"Under constant physical conditions, the ratio of the voltage applied across a conductor to the current flowing through the conductor is a constant."* This constant measures the conductor's opposition to current flow and is called its resistance. The letter symbol for resistance is R while its unit is the *ohm,* which is denoted by the Greek capital letter, Omega Ω. Therefore in Fig. 5-1 the conductor's resistance is 6 Ω.

Resistance is that property of an electrical circuit that opposes or limits the flow of current. The component possessing this property is called a resistor whose schematic symbol is ⋀⋀⋀. Practical examples of resistors are described in Chapter 8.

If the graph of E versus I is plotted for the values of Table 5-1, the result is the straight line illustrated in Fig. 5-2. This means that there is a linear relationship between E and I and the straight line is the graphical way of showing that the current is directly proportional to the voltage. By contrast, when a component (such as a transistor) does not obey Ohm's law, its voltage/current relationship is some form of *curve* and not a straight line.

POTENTIAL DIFFERENCE ACROSS A RESISTOR

In Fig. 5-3 a source whose EMF is 12 V is connected across a resistor and the measured (electron flow) current is 6 mA. Therefore, the value of the resistor is $E:I$ = 12 V:6 mA = 2 kΩ. The electron flow from X to Y develops a voltage, V_R, across the resistor. This voltage is called

15

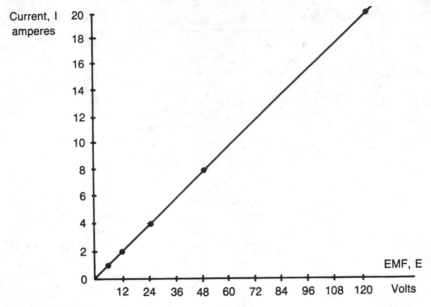

Figure 5-2

the potential difference (PD) or difference of potential (DP) and exactly balances the EMF of the source. Using a water analogy, the back pressure of a pipe always balances the forward pressure of the pump.

When current flows through a resistor, the component becomes hot as the result of friction between the free electrons and the atoms of the material(s) from which the resistor is made. This heat represents an energy *loss* that must be supplied from the voltage source. The resistor dissipates (lost) power in the form of heat and has a power or wattage rating that should not be exceeded.

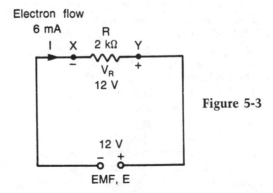

Figure 5-3

CONDUCTANCE

We have learned that resistance is a measure of the *opposition* to current flow. However, it is just as valid to introduce an electrical property that measures the *ease* with which current is allowed to

16

flow. Such a property is called the conductance, whose letter symbol is G and whose SI unit is the siemens (S).

Because a high resistance is obviously equivalent to a low conductance, G and R are reciprocals and are inversely related. It follows that because the ratio $E:I$ is the same as the resistance, the ratio $I:E$ must be equal to the conductance.

MATHEMATICAL DERIVATIONS

Ohm's law

$$\frac{E}{I} = R \text{ ohms} \tag{5-1}$$

$$E = I \times R \text{ volts} \tag{5-2}$$

$$I = \frac{E}{R} \text{ amperes} \tag{5-3}$$

provided the resistance, R, is a constant over the range of operating conditions.

From Equation 4-3:

Power dissipated,

$$P = E \times I \tag{5-4}$$
$$= (I \times R) \times I$$
$$= I^2R \text{ watts}$$

It follows that

$$I = \sqrt{\frac{P}{R}} \text{ amperes} \tag{5-5}$$

and

$$R = \frac{P}{I^2} \text{ ohms} \tag{5-6}$$

Also,

Power dissipated,

$$P = E \times I \tag{5-7}$$
$$= E \times \frac{E}{R}$$
$$= \frac{E^2}{R} \text{ watts}$$

This yields,

$$E = \sqrt{P \times R} \text{ volts} \tag{5-8}$$

and

$$R = \frac{E^2}{P} \text{ ohms} \tag{5-9}$$

Conductance,

$$G = \frac{1}{R} = \frac{I}{E} \text{ siemens} \tag{5-10}$$

The equations 5-1 through 5-9 together with the relationships $P = E \times I$, $I = P/E$, $E = P/I$, enable any two of the four quantities (E, I, R, P) to be found if the other two are given. Of special importance are the two equations $I = \sqrt{P/R}$ and $E = \sqrt{P \times R}$. If we know a resistor's wattage rating, P, and its value, R, we can use these two equations to obtain the highest voltage and current values that will not allow the resistor to overheat.

Example 5-1

All the following questions are related to Fig. 5-1.

(a) If $E = 24$ V and $I = 3$ A, find the values of P, R, and G.
(b) If $E = 80$ V and $P = 320$ W, find the values of I, R, and G.
(c) If $E = 30$ V and $R = 5$ kΩ, find the values of I and P.
(d) If $I = 4$ mA and $P = 48$ mW, find the values of E, R, and G.
(e) If $I = 11$ mA and $R = 3$ kΩ, find the values of E and P.
(f) If $P = 28$ mW and $R = 7$ kΩ, find the values of I and E.
(g) If $P = 12$ W and $R = 3$ Ω, find the values of E and I.
(h) If $G = 0.25$ mS and $I = 8$ mA, find the values of E and P.

Solution

(a) Power, $P = E \times I = 24$ V $\times$ 3 A $= 72$ W.

Resistance, $R = \dfrac{E}{I} = \dfrac{24 \text{ V}}{3 \text{ A}} = 8 \ \Omega.$

Conductance, $G = \dfrac{1}{R} = \dfrac{1}{8 \ \Omega} = 0.125$ S.

(b) Current, $I = \dfrac{P}{E} = \dfrac{320 \text{ W}}{80 \text{ V}} = 4$ A.

Resistance, $R = \dfrac{E}{I} = \dfrac{80 \text{ V}}{4 \text{ A}} = 20 \ \Omega.$

Conductance, $G = \dfrac{1}{R} = \dfrac{1}{20 \ \Omega} = 0.05$ S.

(c) Current, $I = \dfrac{E}{R} = \dfrac{30 \text{ V}}{5 \text{ k}\Omega} = 6$ mA.

Power, $P = E \times I = 30$ V $\times$ 6 mA $= 180$ mW.

(d) Voltage, $E = \dfrac{P}{I} = \dfrac{48 \text{ mW}}{4 \text{ mA}} = 12$ V.

Resistance, $R = \dfrac{E}{I} = \dfrac{12 \text{ V}}{4 \text{ mA}} = 3$ kΩ.

Conductance, $G = \dfrac{1}{R} = \dfrac{1}{3 \text{ k}\Omega} = 0.33$ mS.

(e) Voltage, $E = I \times R = 11$ mA $\times$ 3 kΩ $= 33$ V.

Power, $P = E \times I = 33$ V $\times 11$ mA $= 363$ mW.

(f) Current, $I = \sqrt{P/R} = \sqrt{28 \text{ mW}/7 \text{ k}\Omega} = 2$ mA.

Voltage, $E = I \times R = 2$ mA $\times 7$ k$\Omega = 14$ V.

(g) Voltage, $E = \sqrt{P \times R} = \sqrt{12 \ \mu\text{W} \times 3 \text{ M}\Omega} = 6$ V.

Current, $I = \dfrac{6 \text{ V}}{3 \text{ M}\Omega} = 2 \ \mu$A.

(h) Voltage, $E = \dfrac{I}{G} = \dfrac{8 \text{ mA}}{0.25 \text{ mS}} = 32$ V.

Power, $P = E \times I = 32$ V $\times 8$ mA $= 256$ mW.

Chapter 6
Resistance of a cylindrical conductor

NORMALLY A LOAD IS JOINED TO ITS VOLTAGE BY MEANS OF COPPER CONNECTING wires. Ideally these wires should have zero resistance but in practice their resistance, although small, is not negligible. Consequently, there will be some voltage drop across the connecting wires that will also dissipate power in the form of heat. To keep these effects to a minimum, the wires must have adequate thickness since the thickness determines the current-carrying capacity.

A length of a metal wire with a circular cross section is an example of a cylindrical conductor. If two identical lengths are joined end-to-end in series, the resistance will be doubled. The conductor's resistance is therefore directly proportional to its length (Fig. 6-1). However, if the conductor is thickened by doubling its cross-sectional area, it is equivalent to connecting two identical lengths in parallel and the resistance is halved. To summarize, the resistance of a cylindrical conductor is directly proportional to its length, is inversely proportional to its cross-sectional area, and also depends on the material from which the conductor is made.

The specific resistance or resistivity is the factor that allows us to compare the resistances of different materials. The letter symbol for the specific resistance is the Greek lowercase letter rho, ρ, while its SI unit is the ohm-meter (Ω-m). A conductor made from a particular material will have a resistance of ρ ohms if its length is one meter and its cross-sectional area is one square

Figure 6-1

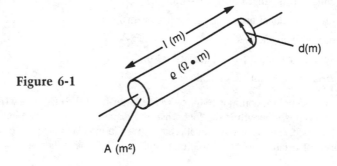

meter. The values of specific resistance for various materials are shown in Table 6-1. Since the specific resistance varies with the temperature, the values quoted in the table are for 20° C.

Although copper is excellent for connecting wires, it is far too good a conductor for a load such as a toaster. For example the element of a 600 W/120 V toaster has a resistance of $E^2/P = (120 \text{ V})^2/600 \text{ W} = 24 \text{ }\Omega$, which could not be obtained from any practical length of copper wire with adequate current-carrying capacity. Consequently the toaster element is made from nichrome (an alloy of nickel and chromium) wire, which has a much higher specific resistance than copper (see Table 6-1).

Table 6-1. Specific resistance (ρ) values for various materials.

Material	SI unit (specific resistance in $\Omega \cdot$ m at 20° C)	British unit (specific resistance in $\Omega \cdot$ cmil/ft at 20° C)
Silver	1.46×10^{-8}	9.86
Annealed copper	1.724×10^{-8}	10.37
Aluminum	2.83×10^{-8}	17.02
Tungsten	5.5×10^{-8}	33.08
Nickel	7.8×10^{-8}	46.9
Pure iron	1.2×10^{-7}	72.2
Constantan	4.9×10^{-7}	294.7
Nichrome	1.1×10^{-6}	660.0
Germanium (semiconductor)	0.55	3.3×10^8
Silicon (semiconductor)	550	3.3×10^{11}
Mica (insulator)	2×10^{10}	12.0×10^{18}

MATHEMATICAL DERIVATIONS

Resistance of a cylindrical conductor,

$$R = \frac{\rho L}{A} \tag{6-1}$$

$$= \frac{\rho L}{\pi d^2/4}$$

$$= \frac{4 \rho L}{\pi d^2} \text{ ohms}$$

where: R = conductor's resistance in ohms (Ω)

L = conductor's length in meters (m)

A = cross-sectional area in square meters (m^2)

d = diameter in meters (m)

ρ = specific resistance or resistivity (Ω-m)

π = circular constant, 3.1415926...

Unfortunately, it is still more common to quote values in terms of the customary (British) system of units. Here, the chosen length is the foot while the diameter is measured in mils where one linear mil is 0.001 inch. The cross-sectional area is given in circular mils where one circular mil (cmil) is the area of a circle whose diameter is one mil. This simplifies the calculation of the cross-sectional area because, for example, if the diameter is 4 mils, the area is $4^2 = 16$ cmil. In this way the complication of using π is avoided.

In customary units the resistance of a cylindrical conductor,

$$R = \frac{\rho L}{A} \tag{6-2}$$

$$= \frac{\rho L}{d^2} \text{ ohms}$$

where: R = conductor's resistance in ohms (Ω)

L = length in feet (ft)

A = cross-sectional area in circular mils (cmil)

d = diameter in linear mils (mil)

ρ = specific resistance (Ω-cmil/ft)

The conversion factor between the customary and SI unit system is $1 \, \Omega\text{-m} = 6.015 \times 10^8 \, \Omega\text{-cmil/ft}$.

The American Wire Gauge (AWG) number is based on the customary units. For example, household wiring is normally AWG #14, which has a diameter of 64 mils, a cross-sectional area of 4110 cmil, and a resistance of 2.58 ohms per 1000 ft at 25° C. The higher the gauge number, the thinner the wire and the less is its current-carrying capacity.

Example 6-1

What is the resistance of a cylindrical aluminum conductor which is 5 m long and has a cross-sectional area of 6 mm^2?

Solution

Resistance of aluminum conductor,

$$R = \frac{\rho L}{A} \tag{6-1}$$

$$= \frac{2.83 \times 10^{-8} \, \Omega\text{-m} \times 5 \text{ m}}{6 \times 10^{-6} \text{ m}^2}$$

$$= 2.36 \times 10^{-2} \, \Omega$$

Example 6-2

A tungsten filament has a length of 5 inches and a diameter of 2 mil. What is the filament's resistance at 20° C?

Solution

Resistance of tungsten filament,

$$R = \frac{\rho L}{d^2} \tag{6-2}$$

$$= \frac{33.08 \, \Omega\text{-cmil/ft} \times 5/12 \text{ ft}}{(2 \text{ mil})^2}$$

$$= 3.45 \, \Omega$$

Chapter 7
Temperature coefficient of resistance

WHEN A CONDUCTOR IS HEATED, THE RISE IN TEMPERATURE CAUSES THE random motion of the free electrons to increase. If a voltage is then applied to the conductor, it is more difficult to move the electrons along the wire to create the flow of current and the resistance therefore increases. For good conductors such as silver, copper, and aluminum, the increase in the resistance rises in a linear manner with the increase in the temperature so that their resistance versus temperature graphs are straight lines.

By contrast with the good conductors, there are alloys, for example, constantan and manganin, whose resistances are practically independent of temperature. Such alloys are used in the manufacture of precision wirewound resistors.

Semiconductors such as silicone, germanium and carbon have very few free electrons and holes at room temperature. However, if the temperature is increased more electrons are moved from the valence band to the conduction band and the resistance falls.

The change in the resistance with the change in the temperature is measured by the temperature coefficient of resistance whose symbol is the Greek lowercase letter alpha, α. If the two changes are in the same direction (for example, the resistance increases when the temperature increases) the coefficient will be positive. Therefore, conductors have positive coefficients while semiconductors have negative coefficients.

The value of a temperature coefficient is normally referred to a temperature of 20° C and is defined as the change in ohms for every ohm of resistance at 20° C for every one degree Celsius rise above 20° C. For example the value of α for silver is $+ 0.0038$ $\Omega/\Omega/°$ C. Consequently, if a length of silver wire has a resistance of 1 Ω at 20° C, its resistance will *increase* by 0.0038 Ω for every degree Celsius *rise* in temperature above 20° C. The corresponding graph of resistance versus temperature is shown in Fig. 7-1.

Table 7-1 shows the values of the resistance temperature coefficients for various materials.

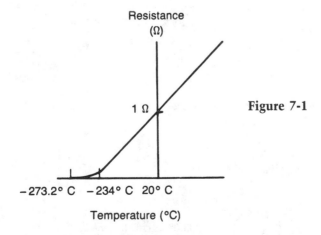

Figure 7-1

**Table 7-1. Temperature coefficient
of resistance for various materials.**

Material	Temperature coefficient of resistance, α, at 20° C ($\Omega/\Omega/°$ C)
Silver	+0.0038
Copper	+0.00393
Aluminum	+0.0039
Tungsten	+0.0045
Iron	+0.0055
Nickel	+0.006
Constantan	+0.0000008
Carbon	−0.0005

MATHEMATICAL DERIVATIONS

If the temperature is increased to $T°$ C,

$$\text{Increase in the resistance} = \alpha \times R_{20°\,C} \times (T°\,C - 20°\,C) \text{ ohms} \qquad (7\text{-}1)$$

Resistance at $T°$ C,

$$R_{T°\,C} = R_{20°\,C} + \alpha \times R_{20°\,C} \times (T°\,C - 20°\,C) \qquad (7\text{-}2)$$
$$= R_{20°\,C}\,[1 + \alpha\,(T°\,C - 20°\,C)] \text{ ohms}$$

If the silver conductor is cooled its resistance will decrease by 0.0038 Ω for every 1° C drop in temperature. Consequently when the temperature drops by 1 $\Omega \times$ 1° C/0.0038 Ω = 263° C, the resistance of the conductor should theoretically be zero. This would occur at a temperature of +20° C − 263° C = −243° C. However, the resistance/temperature graph is non-linear at low temperatures (Fig. 7-1) so that the resistance does not virtually disappear until we approach a temperature which is close to −273.2° C. This temperature is the ultimate limit of absolute zero or 0° K (Kelvin Scale) which cannot be achieved in practice.

Example 7-1

At 20° C a length of silver wire has a resistance of 12 Ω. What is its resistance at (a) 0° C and (b) 100° C?

Solution

(a) Resistance at 0° C, $R_{0°\,C} = R_{20°\,C}[1 + (0°\,C - 20°\,C)]$ (7-1)
$$= 12\,\Omega\,[1 - 20 \times 0.0038]$$
$$= 11.09\,\Omega$$

(b) Resistance at 100° C, $R_{100°\,C} = R_{20°\,C}\,[1 + (100°\,C - 20°\,C)]$ (7-1)
$$= 12\,\Omega\,[1 + 80 \times 0.0038]$$
$$= 15.65\,\Omega$$

23

Chapter 8
Composition resistors—the color code

A COMPOSITION RESISTOR IS THE MOST COMMON TYPE OF RESISTOR USED IN electronics. These resistors are inexpensive and are manufactured in standard resistance values which normally range from 2.7 Ω to 22 MΩ. The resistance values remains reasonably constant over a limited temperature range so that composition resistors are linear components and obey Ohm's law. However, such resistors can only dissipate a limited amount of power so that their normal wattage ratings are only 1/8 W, 1/4 W, 1/2 W, 1 W, and 2 W. These ratings are judged from the resistor's physical size (Fig. 8-1A). Knowing the values of the power rating *(P)* and the resistance *(R),* the formulas $I = \sqrt{P/R}$ amperes and $E = \sqrt{P \times R}$ volts will enable you to calculate the maximum values of the voltage and the current that will not allow the resistor's power rating to be exceeded.

Composition resistors are normally manufactured from powdered carbon with its specific resistance of more than 3.325×10^{-5} Ωm. The carbon is mixed with an insulating substance (for example, talc) and a binding material such as resin. The proportions of carbon and talc are then varied to produce a wide range of resistances. The resistor is finally coated with an insulating material that is baked to a hard finish. This coating provides protection against moisture and mechanical damage (Fig. 8-1B). At either end of the resistor a tinned copper "pigtail" wire is deeply embedded and provides an adequate contact area for making a good connection with the carbon/talc mixture.

The main disadvantage of composition resistors is that they are not manufactured to precise values but are normally sorted into three groups. The first group has a 5% tolerance which means that their measured resistances are within ±5% of their rated values. The second group has a tolerance between ±5% and ±10%, while the tolerance of the third and final group lies between ±10% and ±20%. Any resistor whose tolerance is greater than ±20%, is discarded.

As an example, let us consider a 10% tolerance resistor with a rated value of 10 kΩ. The actual resistance value must lie between 10 kΩ + 10/100 × 10 kΩ = 11 kΩ and 10 kΩ − 10/100 × 10 kΩ = 9 kΩ. Consequently there would be no point in manufacturing 10.5 kΩ or 9.5 kΩ resistors since both these values lie within the tolerance of the rated 10 kΩ resistor. It follows that composition resistors are only made in certain standard values. Between 10 Ω and

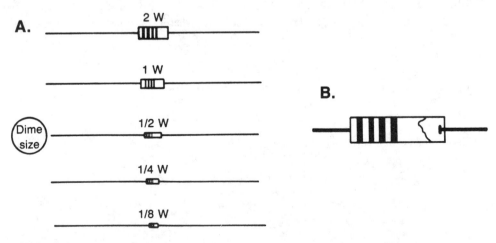

Figure 8-1

100 Ω in the 20% range, the only values manufactured are 10 Ω, 15 Ω, 22 Ω, 33 Ω, 47 Ω, 68 Ω and 100 Ω. These values are chosen so that the upper tolerance limit of one resistor equals (or slightly overlaps) the lower limit of the next highest value resistor. For example, the upper limit of a 22 Ω resistor is $22 + 20/100 \times 22 = 26.4$ Ω, while the lower limit of a 33 Ω resistor is $33 - 20/100 \times 33 = 26.4$ Ω. The ±10% and ±5% range are then formed by including additional standard values in the gaps of the 20% range (Table 8-1).

Table 8-1. Standard resistor values.

5% tolerance	10% tolerance	20% tolerance
10	10	10
11		
12	12	
13		
15	15	15
16		
18	18	
20		
22	22	22
24		
27	27	
30		
33	33	33
36		
39	39	
43		
47	47	47
51		
56	56	
62		
68	68	68
75		
82	82	
91		

The higher resistances are formed by adding zeros to the values shown in Table 8-1. The least costly method of indicating the rated resistance value is to use a four-band color code. The first two bands represent the significant figures of the rated value, the third band is the multiplier, which indicates the *number* of zeros, while the fourth band (if present) indicates the tolerance (Fig. 8-2). For example:

33 ± 5%	Orange, orange, black (no zeros) gold.
560 ± 10%	Green, blue, brown, silver.
2200 ± 20%	Red, red, red (no fourth band for ±20% tolerance).
39000 ± 10%	Orange, white, orange, silver.
150000 ± 20%	Brown, green, yellow.

Black is not allowed in the first band so that if the rated value is less than 10 Ω, the four band system cannot be used. Gold and silver are then placed in the third band as multipliers of 0.1 and 0.01 respectively. For example, a 5.6 Ω resistor could be color-coded green, blue, gold.

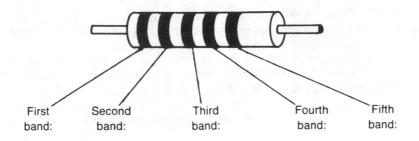

| First band: | Second band: | Third band: | Fourth band: | Fifth band: |

Color code	First significant figure	Second significant figure	Multiplier	Tolerance (%)	Fail-rate percent per 1000 hr
Black		0	$\times\ 1\ =\ \times\ 10^0$		
Brown	1	1	$\times\ 10\ =\ \times\ 10^1$		1.0
Red	2	2	$\times\ 100\ =\ \times\ 10^2$		0.1
Orange	3	3	$\times\ 1000\ =\ \times\ 10^3$		0.01
Yellow	4	4	$\times\ 10000\ =\ \times\ 10^4$		0.001
Green	5	5	$\times\ 100000\ =\ \times\ 10^5$		
Blue	6	6	$\times\ 1000000\ =\ \times\ 10^6$		
Violet	7	7	$\times\ 10000000\ =\ \times\ 10^7$		
Gray	8	8			
White	9	9			
Gold			$\times\ 0.1$	± 5	
Silver			$\times\ 0.01$	± 10	
No color				± 20	

Figure 8-2

Wirewound resistors are manufactured to more precise values than the composition type and are available in high wattage ratings (2 W and upwards). They are commonly made from constantan wire with its low temperature coefficient.

Example 8-1

A 1/2 W composition resistor is color-coded brown, green, brown, gold. What are the permitted upper and lower limits of its resistance value?

Solution

The rated value of the resistor is 150 Ω with a $\pm5\%$ tolerance. The permitted upper and lower limits are therefore:

$$150 \pm \frac{5}{100} \times 150\ \Omega = 150 \pm 7.5\ \Omega = 157.5\ \Omega \text{ and } 142.5\ \Omega.$$

Example 8-2

The color bands of a 2 W composition resistor are red, violet, brown, silver. What is the maximum value of the current that can flow through the resistor without exceeding its power rating?

Solution

$$\text{Maximum current, } I = \sqrt{\frac{P}{R}}$$

$$= \sqrt{\frac{2\text{ W}}{270\text{ }\Omega}} = 0.086\text{ A}$$

$$= 86\text{ mA}$$

Note:

$$\text{Maximum voltage, } E = I \times R$$
$$= 0.086\text{ A} \times 270\text{ }\Omega$$
$$= 23.2\text{ V}$$

Chapter 9
Resistors in the series arrangement

RESISTORS IN SERIES ARE JOINED END TO END SO THAT THERE IS ONLY A SINGLE path for the current. Starting at the negative battery terminal (where a surplus of electrons exists) there is an electron flow through the connecting wires (which are assumed to possess zero resistance) and the three resistors. Finally this flow reaches the positive terminal where there is a deficit of electrons. The battery through its chemical energy is then responsible for maintaining the surplus of electrons at the negative terminal and the deficit at the positive terminal. It follows from the above discussion that the current is the same throughout the circuit and that the ammeters (Fig. 9-1) $A1$, $A2$, $A3$, and $A4$ all have the same reading.

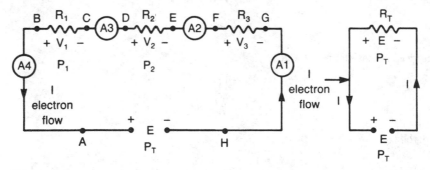

Figure 9-1

Across each resistor is developed a difference of potential (DP) [sometimes referred to as the potential difference (PD)]; in this sense "potential" is another word for "voltage." For example, V_1 is the amount of voltage that is required to drive the current, I, through the resistor, $R1$. Since $V_1 = IR_1$, the voltage is often called the "IR drop" where the meaning of the word drop is illustrated in Example 9-2.

The sum of the voltages across the resistors must exactly balance the source voltage, E; this is an example of Kirchhoff's Voltage Law (KVL) which we will fully explore in Chapter 26. Since the current through each resistor is the same, the highest value resistor will develop the greatest voltage drop and the lowest value resistor will have the smallest voltage drop. In the extreme case of the connecting wires, which theoretically have zero resistance, there will be no voltage drop so that if, for example, a voltmeter were connected between points A and B (Fig. 9-1), its reading would be zero.

Notice that other voltages exist in the circuit apart from V_1, V_2, V_3, and E. If a voltmeter were connected between points B and E, its reading will be the sum of the voltages V_1 and V_2, while between the points D and G the voltage is equal to $V_2 + V_3$. It is often said that the order in which series resistors are connected is immaterial. This is true as far as the current and the individual voltage drops are concerned, but not in terms of the other voltages that exist in the circuit (see Example 9-1).

In each resistor a certain amount of power is dissipated in the form of heat. (Care must be taken that the amount of power dissipated does not exceed the resistor's wattage rating.) The sum of the individual powers dissipated is equal to the total power derived from the source.

Finally, what is the purpose of connecting resistors in series? One obvious reason is to increase the total equivalent resistance, R_T, and thereby limit the current to a safe value; however this could also be achieved by using a higher value single resistor. More importantly, adding resistors enables you to increase the overall wattage rating and to obtain nonstandard resistance values by using standard components (Chapter 8). In addition, series resistors may be connected across a source voltage to provide a voltage divider circuit (Chapter 11).

MATHEMATICAL DERIVATIONS

Voltage drop across the resistor $R1$,

$$V_1 = V_{AB} = I \times R_1, \tag{9-1}$$

$$I = \frac{V_1}{R_1}, \ R_1 = \frac{V_1}{I}$$

Voltage drop across the resistor $R2$,

$$V_2 = V_{CD} = I \times R_2, \tag{9-2}$$

$$I = \frac{V_2}{R_2}, \ R_2 = \frac{V_2}{I}$$

Voltage drop across the resistor $R3$,

$$V_3 = V_{FG} = I \times R_3, \tag{9-3}$$

$$I = \frac{V_3}{R_3}, \ R_3 = \frac{V_3}{I}$$

Source voltage,

$$E = V_1 + V_2 + V_3 \tag{9-4}$$
$$= I \times R_1 + I \times R_2 + I \times R_3$$
$$= I \times (R_1 + R_2 + R_3) \text{ volts} \tag{9-5}$$

If the total equivalent resistance is R_T,
Source Voltage,

$$E = I \times R_T \text{ volts} \tag{9-6}$$

Comparing equations 9-4 and 9-5,
Total equivalent resistance,

$$R_T = R_1 + R_2 + R_3 \text{ ohms} \tag{9-7}$$

If N resistors are connected in series,
Total equivalent resistance,

$$R_T = R_1 + R_2 + R_3 \ldots + R_N \text{ ohms} \tag{9-8}$$

If all N resistors are of equal value R,
Total equivalent resistance,

$$R_T = NR \text{ ohms} \tag{9-9}$$

Power dissipated in the $R1$ resistor,

$$P_1 = I \times V_1 = I^2 R_1 = V_1^2/R_1 \text{ watts} \tag{9-10}$$

Power dissipated in the $R2$ resistor,

$$P_2 = I \times V_2 = I^2 R_2 = V_2^2/R_2 \text{ watts} \tag{9-11}$$

Power dissipated in the $R3$ resistor,

$$P_3 = I \times V_3 = I^2 R_3 = V_3^2/R_3 \text{ watts} \tag{9-12}$$

Total power dissipated,

$$P_T = P_1 + P_2 + P_3 \text{ watts} \tag{9-13}$$

Power derived from the source voltage,

$$P_T = I \times E = I^2 R_T = E^2/R_T \text{ watts} \tag{9-14}$$

Example 9-1

In Fig. 9-2, calculate the values of R_T, I_1, V_1, V_2, V_3, V_4, P_1, P_2, P_3, P_4, P_T. What are the voltages between the points PA, PB, PC, PD, PE, PF, PG, PH, PN, AB, AC, AD, AE, AF, AG, AH, AN, BC, BD, BE, BF, BG, BH, BN, CD, CE, CF, CG, CH, CN, DE, DF, DG, DH, DN, EF, EG, EH, EN, FG, FH, FN, GH, GN, HN?

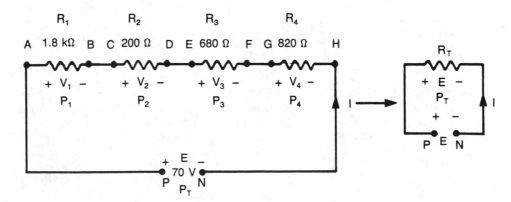

Figure 9-2

Solution

We must first convert the 1.8 kΩ into ohms so that 1.8 kΩ = 1800 Ω
Total equivalent resistance,

$$R_T = R_1 + R_2 + R_3 + R_4 \qquad\qquad (9\text{-}7)$$
$$= 1800 + 200 + 680 + 820 = 3500 \ \Omega = 3.5 \ \text{k}\Omega$$

Current, $I = \dfrac{E}{R_T} = \dfrac{70 \ \text{V}}{3.5 \ \text{k}\Omega} = 20 \ \text{mA}$ $\qquad\qquad (9\text{-}7)$

Voltage drop, $V_1 = V_{AB} = I \times R_1 = 20 \ \text{mA} \times 1.8 \ \text{k}\Omega = 36 \ \text{V}$ $\qquad (9\text{-}1)$

Voltage drop, $V_2 = V_{CD} = I \times R_2 = 20 \ \text{mA} \times 200 \ \Omega = 4 \ \text{V}$ $\qquad (9\text{-}1)$

Voltage drop, $V_3 = V_{EF} = I \times R_3 = 20 \ \text{mA} \times 680 \ \Omega = 13.6 \ \text{V}$ $\qquad (9\text{-}1)$

Voltage drop, $V_4 = V_{GH} = I \times R_4 = 20 \ \text{mA} \times 820 \ \Omega = 16.4 \ \text{V}$ $\qquad (9\text{-}1)$

Notice that the highest value resistor (R_1 = 1800 Ω) carries the greatest voltage drop (V_1 = 36 V) while the lowest value resistor (R_2 = 200 Ω) has the least voltage drop (V_2 = 4 V).

Voltage check:
Source voltage, $E = V_1 + V_2 + V_3 + V_4 = 36 + 4 + 13.6 + 16.4 = 70 \ \text{V}$ $\qquad (9\text{-}4)$

Power dissipated in the $R1$ resistor,

$$P_1 = I \times V_1 = 20 \ \text{mA} \times 36 \ \Omega \qquad\qquad (9\text{-}10)$$
$$= 720 \ \text{mW}$$

Power dissipated in the $R2$ resistor,

$$P_2 = I \times V_2 = 20 \ \text{mA} \times 4 \ \text{V} \qquad\qquad (9\text{-}11)$$
$$= 80 \ \text{mW}$$

Power dissipated in the $R3$ resistor,

$$P_3 = I \times V_3 = 20 \ \text{mA} \times 13.6 \ \text{V} \qquad\qquad (9\text{-}12)$$
$$= 272 \ \text{mW}$$

Power dissipated in the $R4$ resistor,

$$P_4 = I \times V_4 = 20 \ \text{mA} \times 16.4 \ \text{V} \qquad\qquad (9\text{-}13)$$
$$= 328 \ \text{mW}$$

Total power dissipated,

$$P_T = P_1 + P_2 + P_3 + P_4 \qquad\qquad (9\text{-}14)$$
$$= 720 + 80 + 272 + 328$$
$$= 1400 \ \text{mW}$$

The highest value resistor ($R1$ = 1800 Ω) dissipates the greatest power (P_1 = 720 mW) while the lowest value resistor ($R2$ = 200 Ω) has the least power dissipation (P_2 = 80 mW).
Notice that a 1/2-watt resistor would be adequate for $R2$ and $R3$. By contrast, $R1$ would need to be a 1-watt resistor, while $R4$ would require a 2-watt resistor.

Total power derived from the source,

$$P_T = I \times E = 20 \ \text{mA} \times 70 \ \text{V} \qquad\qquad (9\text{-}15)$$
$$= 1400 \ \text{mW}$$

Voltage between the points PA, $V_{PA} = 0$ V since PA is a connecting wire of zero resistance.

$V_{PB} = V_{PA} + V_{AB} = 0$ V $+ 36$ V $= 36$ V.
$V_{PC} = V_{PA} + V_{AB} + V_{BC} = 0$ V $+ 36$ V $+ 0$ V $= 36$ V.
$V_{PD} = V_{PA} + V_{AB} + V_{BC} + V_{CD} = 0$ V $+ 36$ V $+ 0$ V $+ 4$ V $= 40$ V.
$V_{PE} = V_{PA} + V_{AB} + V_{BC} + V_{CD} + V_{DE} = 0$ V $+ 36$ V $+ 0$ V $+ 4$ V $+ 0$ V $= 40$ V.
$V_{PF} = V_{PA} + V_{AB} + V_{BC} + V_{CD} + V_{DE} + V_{EF} = 0$ V $+ 36$ V $+ 0$ V $+ 4$ V $+ 0$ V $+ 13.6$ V
    $= 53.6$ V.
$V_{PG} = V_{PA} + V_{AB} + V_{BC} + V_{CD} + V_{DE} + V_{EF} + V_{FG}$
    $= 0$ V $+ 36$ V $+ 0$ V $+ 4$ V $+ 0$ V $+ 13.6$ V $+ 0$ V $= 53.6$ V.
$V_{PH} = V_{PA} + V_{AB} + V_{BC} + V_{CD} + V_{DE} + V_{EF} + V_{FG} + V_{GH}$
    $= 0$ V $+ 36$ V $+ 0$ V $+ 4$ V $+ 0$ V $+ 13.6$ V $+ 0$ V $+ 16.4$ V $= 70$ V.
$V_{PN} = V_{PA} + V_{AB} + V_{BC} + V_{CD} + V_{DE} + V_{EF} + V_{FG} + V_{GH} + V_{HN}$
    $= 0$ V $+ 36$ V $+ 0$ V $+ 4$ V $+ 13.6$ V $+ 0$ V $+ 16.4$ V $+ 0$ V $= 70$ V.

Notice that if for example, the resistors $R2$ and $R3$ were interchanged, V_{PD} would become 36 V $+ 13.6$ V $= 49.6$ V. Since P and A are only joined by a connecting wire of zero resistance, $V_{AB} = V_{PB}$, $V_{AC} = V_{PC}$, etc.

$V_{BC} = 0$ V.
$V_{BD} = V_{BC} + V_{CD} = 0$ V $+ 4$ V $= 4$ V.
$V_{BE} = V_{BC} + V_{CD} + V_{DE} = 0$ V $+ 4$ V $+ 0$ V $= 4$ V.
$V_{BF} = V_{BC} + V_{CD} + V_{DE} + V_{DF} = 0$ V $+ 4$ V $+ 0$ V $+ 13.6$ V $= 17.6$ V.
$V_{BG} = V_{BC} + V_{CD} + V_{DE} + V_{DF} + V_{FG} = 0$ V $+ 4$ V $+ 0$ V $+ 13.6$ V $+ 0$ V $= 17.6$ V.
$V_{BH} = V_{BC} + V_{CD} + V_{DE} + V_{DF} + V_{FG} + V_{GH}$
    $= 0$ V $+ 4$ V $+ 0$ V $+ 13.6$ V $+ 16.4$ V $= 34.0$ V.
$V_{BN} = V_{BC} + V_{CD} + V_{DE} + V_{DF} + V_{FG} + V_{GH} + V_{HN}$
    $= 0$ V $+ 4$ V $+ 0$ V $+ 13.6$ V $+ 0$ V $+ 16.4$ V $+ 0$ V $= 34.0$ V.

Between B and C there is a connecting wire of zero resistance so that $V_{BD} = V_{CD}$, $V_{BE} = V_{CE}$, etc.

$V_{DE} = 0$ V.
$V_{DF} = V_{DE} + V_{EF} = 0$ V $+ 13.6$ V $= 13.6$ V.
$V_{DG} = V_{DE} + V_{EF} + V_{FG} = 0$ V $+ 13.6$ V $+ 0$ V $= 13.6$ V.
$V_{DH} = V_{DE} + V_{EF} + V_{FG} + V_{GH} = 0$ V $+ 13.6$ V $+ 16.4$ V $= 30$ V.
$V_{DN} = V_{DE} + V_{EF} + V_{FG} + V_{GH} + V_{HN} = 0$ V $+ 13.6$ V $+ 16.4$ V $+ 0$ V $= 30$ V.

Between D and E there is a connecting wire of zero resistance so that $V_{DF} = V_{EF}$, $V_{DG} = V_{EG}$, etc.

$V_{FG} = 0$ V.
$V_{FH} = V_{FG} + V_{GH} = 0$ V $+ 16.4$ V $= 16.4$ V.
$V_{FN} = V_{FG} + V_{GH} + V_{HN} = 0$ V $+ 16.4$ V $+ 0$ V $= 16.4$ V.
$V_{HN} = 0$ V.

Example 9-2

In Fig. 9-3 the negative probe of the voltmeter is connected permanently to the point G. What are the readings of the voltmeter when the positive probe is in turn connected to the points D, E, F, and G?

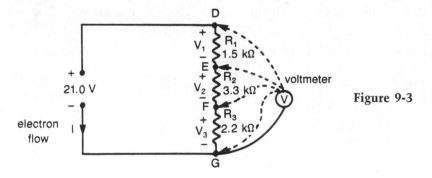

Figure 9-3

Solution

Total equivalent resistance, $R_T = R_1 + R_2 + R_3 = 1.5 + 3.3 + 2.2 = 7.0$ kΩ

$$\text{Current, } I = \frac{E}{R_T} = \frac{21.0 \text{ V}}{7 \text{ k}\Omega} = 3 \text{ mA}.$$

Voltage drop, $V_1 = I \times R_1 = 3$ mA $\times$ 1.5 kΩ = 4.5 V.

Voltage drop, $V_2 = I \times R_2 = 3$ mA $\times$ 3.3 kΩ = 9.9 V.

Voltage drop, $V_3 = I \times R_3 = 3$ mA $\times$ 2.2 kΩ = 6.6 V.

Voltage check: Source voltage, $E = V_1 + V_2 + V_3 = 4.5 + 9.9 + 6.6 = 21.0$ V.

Positive probe connected to the point D: Reading of voltmeter = source voltage = 21.0 V.

Positive probe connected to the point E: Reading of the voltmeter = $V_2 + V_3$ = 9.9 + 6.6 = 16.5 V.

We can therefore say that there has been a *voltage drop* of $21.0 - 16.5 = 4.5$ V between the points D and E. This voltage drop is, of course, equal to the value of V_1.

Positive probe connected to the point F: Reading of the voltmeter = V_3 = 6.6 V.

There has been a voltage drop of $16.5 - 6.6 = 9.9$ V between points E, F and a voltage drop of $21.0 - 6.6 = 14.4$ V between the points D, F.

Positive probe connected to the point G: The two probes are joined together so that the voltmeter reading is zero. The voltage drops between points D and G, E and G, F and G are respectively 21.0 V, 16.5 V, and 6.6 V.

Example 9-3

A series circuit consists of four 1 W resistors whose values are 120 Ω, 150 Ω, 270 Ω, 330 Ω. Calculate their maximum power dissipation without exceeding the wattage rating of any of the resistors.

Solution

The current in each resistor to produce its 1 W rating is different; using the relationship $I = \sqrt{P/R}$, these currents are:

$$I_{120 \, \Omega} = \sqrt{\frac{1 \text{ W}}{120 \, \Omega}} = 0.0913 \text{ A}.$$

$$I_{150 \, \Omega} = \sqrt{\frac{1 \text{ W}}{150 \, \Omega}} = 0.0816 \text{ A}.$$

$$I_{270\,\Omega} = \sqrt{\frac{1\text{ W}}{270\ \Omega}} = 0.0609 \text{ A.}$$

$$I_{330\,\Omega} = \sqrt{\frac{1\text{ W}}{330\ \Omega}} = 0.0550 \text{ A.}$$

We must select the smallest of these currents, 0.0550 A, because if the current were allowed to exceed this value, the power dissipation in the 330 Ω resistor would be exceeded. Using the relationship $P = I^2 R_T$ the maximum power dissipation of this series string is:

$$(0.0550 \text{ A})^2 \times (120 \ \Omega + 150 \ \Omega + 270 \ \Omega + 330 \ \Omega) = 2.27 \text{ W.}$$

The maximum power dissipation is therefore only 2.27 W and *not* 4 W.

Chapter 10
Ground—voltage reference level

GROUND MAY BE REGARDED AS ANY LARGE MASS OF CONDUCTING MATERIAL IN which there is essentially *zero* resistance between any two ground points. Examples of ground are the metal chassis of a transmitter, the aluminum chassis of a receiver or a wide strip of copper plating on a printed-circuit board.

The main reason for using a ground system is to simplify the circuitry by saving on the amount of wiring. This is done by using ground as the return paths for many circuits so that at any time there are a number of currents flowing through the ground system.

Figures 10-1 A and B show two circuits that contain four connecting wires. However, if a common ground (alternative symbols , usually to signify earth ground, and ⟋⊓ to signify chassis ground) is used for the return path as in Fig. 10-1C, only two connecting wires are required; the schematic equivalent of Fig. 10-1C is shown in Fig. 10-1D. Here is where the concept of zero resistance is vital. If the ground resistance were not zero, the current flow in one circuit would develop a voltage between the ground points X and Y, and this would cause a current to flow in the other circuit. In other words the two circuits would interfere with each other. It is therefore important to regard all ground points in a schematic as being electrically joined together as a single path. This zero resistance property of ground is essential in order to achieve isolation between all those circuits that use the same ground.

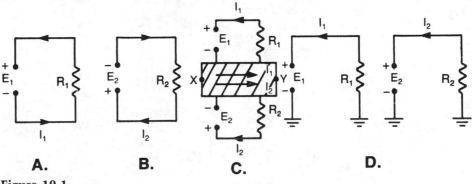

A.   **B.**   **C.**   **D.**

Figure 10-1

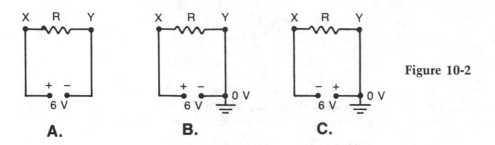

Figure 10-2

In our previous discussions we have regarded a voltage drop, potential difference or difference of potential as existing between *two* points; to have referred to the voltage at a single point would have been meaningless. However, if we create a common reference level to represent the voltage at one point, we can quote the voltages at other points in relation to the reference level. Ground is therefore chosen as a reference level of zero volts and the voltage or *potential* at any point is then measured relative to ground. This is the manner in which the expected potentials at various points are indicated on a schematic.

In Fig. 10-2A we could either say that the point X is 6 V positive with respect to the point Y or that the point Y is 6 V negative with respect to the point X. However, if the negative battery terminal is grounded at zero volts (Fig. 10-2B), point X carries a potential of positive 6 V (+6 V) with respect to ground. By contrast if the battery is reversed and its positive terminal is grounded (Fig. 10-2C), the potential at X is now negative 6 V (−6 V) with respect to ground. When troubleshooting a circuit, it is normal practice to connect the black or common probe of a voltmeter permanently to ground and to measure the potentials at various points using the red probe. The voltmeter's function switch has two positions marked "+DC" and "−DC"; these enable either positive or negative potentials to be measured without removing the common probe from ground. Remember that a potential with respect to ground must always be indicated as either "+" or "−."

In any particular circuit the potentials at the various points will depend on which point is grounded. However, the potential differences between any two points must always remain the same.

It is worth mentioning that both positive and negative potentials are used in the operation of electronic circuits. For example, pnp transistor amplifiers normally need negative potentials while npn amplifiers require positive potentials.

Example 10-1

In Fig. 10-3 ground in turn the points A, B, C, and D. In each case calculate the values of the potentials at the various points in the circuit.

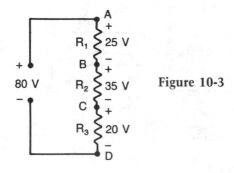

Figure 10-3

Solution

Point A Grounded

Potential at $A = 0$ V
Potential at $B = -25$ V
Potential at $C = -(25 + 35) = -60$ V
Potential at $D = -(25 + 35 + 20) = -80$ V

Point B Grounded

Potential at $A = +25$ V
Potential at $B = 0$ V
Potential at $C = -35$ V
Potential at $D = -(35 + 20) = -55$ V

Point C Grounded

Potential at $A = +(25 + 35) = +60$ V
Potential at $B = +35$ V
Potential at $C = 0$ V
Potential at $D = -20$ V

Grounding at a point such as B or C would be regarded as an *intermediate* ground.

Point D Grounded

Potential at $A = +(25 + 35 + 20) = +80$ V
Potential at $B = +(35 + 20) = +55$ V
Potential at $C = +20$ V
Potential at $D = 0$ V

Note that in this circuit we must not ground two separate points simultaneously. For example, if points B and C were grounded at the same time, it would be equivalent to placing a short circuit of zero resistance across the resistor $R2$.

Chapter 11
Voltage division rule—
voltage divider circuit

WE HAVE ALREADY LEARNED THAT, IN A SERIES STRING OF RESISTORS, THE highest voltage drop is developed across the largest value resistor while the lowest voltage drop appears across the smallest value resistor. In fact the voltage drops across all the resistors are directly proportional to the resistor values so that we may use the method of proportion to solve a problem in which, for example, we are given the values of R_1, R_2, V_1 and then asked to find the value of V_2 (Fig. 11-1). The same rule of proportion can be applied to the powers dissipated in the individual resistors.

The source voltage is divided between the series resistors in a proportional manner that is determined by the resistor values. The fraction of the source voltage developed across a particular resistor is equal to the ratio of the resistor's value to the circuit's total resistance. This result is

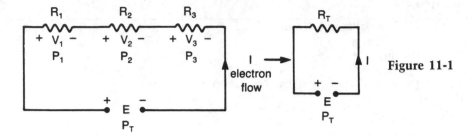

Figure 11-1

known as the *voltage division rule* (VDR). In the same manner the total power derived from the source divides between the resistors to produce the individual powers dissipated.

MATHEMATICAL DERIVATIONS

Voltage drop,

$$V_1 = IR_1 \text{ volts} \tag{11-1}$$

Voltage drop,

$$V_2 = IR_2 \text{ volts} \tag{11-2}$$

Therefore: Current,

$$I = \frac{V_1}{R_1} = \frac{V_2}{R_2} \text{ amperes} \tag{11-3}$$

and

$$\frac{V_1}{V_2} = \frac{R_1}{R_2} \tag{11-4}$$

This yields
Voltage drop,

$$V_1 = V_2 \times \frac{R_1}{R_2}, \; V_2 = V_1 \times \frac{R_2}{R_1} \text{ volts} \tag{11-5}$$

Power dissipated,

$$P_1 = I^2 R_1 \text{ watts} \tag{11-6}$$

Power dissipated,

$$P_2 = I^2 R_2 \text{ watts} \tag{11-7}$$

Then,

$$I^2 = \frac{P_1}{R_1} = \frac{P_2}{R_2} \tag{11-8}$$

and

$$\frac{P_1}{P_2} = \frac{R_1}{R_2} \tag{11-9}$$

This yields:
Power dissipated,

$$P_1 = P_2 \times \frac{R_1}{R_2}, \ P_2 = P_1 \times \frac{R_2}{R_1} \text{ watts} \tag{11-10}$$

These results apply to any number of resistors in series.
Voltage Division Rule (VDR): Current,

$$I = \frac{E_T}{R_T} \text{ amperes} \tag{11-11}$$

Voltage drop, $V_1 = I R_1$ volts

It follows that: Voltage drop,

$$V_1 = \frac{E_T}{R_T} \times R_1 = E_T \times \frac{R_1}{R_T} \text{ volts (Voltage Division Rule)} \tag{11-12}$$

Then
Voltage drop,

$$V_2 = E_T \times \frac{R_2}{R_T} \text{ and } V_3 = E_T \times \frac{R_3}{R_T} \text{ volts} \tag{11-13}$$

Also,

$$\frac{V_1}{E_T} = \frac{R_1}{R_T} \text{ and } E_T = V_1 \times \frac{R_T}{R_1} = V_2 \times \frac{R_T}{R_2} = V_3 \times \frac{R_T}{R_3} \text{ volts} \tag{11-14}$$

For *two* resistors in series:
Voltage drop,

$$V_1 = E_T \times \frac{R_1}{R_T} = E_T \times \frac{R_1}{R_1 + R_2} \text{ volts} \tag{11-15}$$

Voltage drop,

$$V_2 = E_T \times \frac{R_2}{R_1 + R_2} \text{ volts} \tag{11-16}$$

For *N equal* value resistors:
Voltage drops,

$$V_1 = V_2 = V_3 \ldots = V_N = \frac{E}{N} \text{ volts} \tag{11-17}$$

Power division: Total power delivered from the source,

$$P_T = I^2 R_T \text{ watts} \tag{11-18}$$

Power dissipated in the $R1$ resistor,

$$P_1 = I^2 R_1 \text{ watts} \tag{11-19}$$

Therefore:

$$I^2 = \frac{P_T}{R_T} = \frac{P_1}{R_1} \tag{11-20}$$

This yields:

$$P_1 = P_T \times \frac{R_1}{R_T}, \; P_2 = P_T \times \frac{R_2}{R_T}, \; P_3 = P_T \times \frac{R_3}{R_T} \text{ watts} \qquad (11\text{-}21)$$

The voltage divider circuit is the practical outcome of the voltage division rule. By its use we can obtain a number of different voltages from a single voltage source. These voltages will then be available as the supplies for various loads.

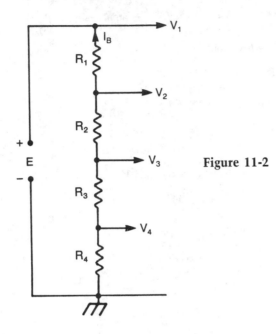

Figure 11-2

Referring to Fig. 11-2, V_1 is the voltage drop across the series combination of R_1, R_2, R_3, R_4 and is therefore equal to the source voltage, $+E$. V_2 is dropped across R_2, R_3 and R_4 in series. Therefore, using the voltage division rule,

$$V_2 = + E \times \frac{(R_2 + R_3 + R_4)}{R_1 + R_2 + R_3 + R_4} \text{ volts} \qquad (11\text{-}22)$$

Similarly,

$$V_3 = + E \times \frac{(R_3 + R_4)}{R_1 + R_2 + R_3 + R_4} \text{ volts} \qquad (11\text{-}23)$$

and

$$V_4 = + E \times \frac{R_4}{R_1 + R_2 + R_3 + R_4} \text{ volts} \qquad (11\text{-}24)$$

Notice that V_1, V_2, V_3, and V_4 are all positive potentials.

To develop the voltages V_1, V_2, V_3, V_4 it is necessary for a so-called "bleeder current", I_B, to flow through the resistors. The provision of the different output voltages is therefore achieved at the expense of the power dissipated in the series string.

The voltage divider described so far would be regarded as operating under no load conditions. As soon as loads are connected across the voltages V_1, V_2, V_3, V_4, the circuit becomes a more complex series-parallel arrangement and the equations 11-22, 11-23, 11-24 no longer apply.

Example 11-1

In Fig. 11-3, calculate the values of V_1, V_3, and the potential at the point X.

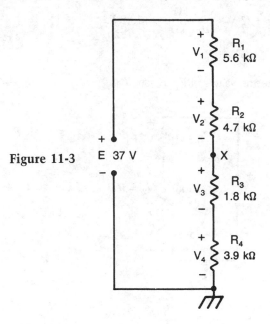

Figure 11-3

Solution

Voltage drop,

$$V_1 = E \times \frac{R_1}{R_T} \tag{11-12}$$

$$= 37 \text{ V} \times \frac{5.6 \text{ k}\Omega}{5.6 \text{ k}\Omega + 4.7 \text{ k}\Omega + 1.8 \text{ k}\Omega + 3.9 \text{ k}\Omega}$$

$$= 37 \text{ V} \times \frac{5.6 \text{ k}\Omega}{16.0 \text{ k}\Omega}$$

$$= 12.95 \text{ V}.$$

Voltage drop,

$$V_3 = V_1 \times \frac{R_3}{R_1} \tag{11-5}$$

$$= 12.95 \text{ V} \times \frac{1.8 \text{ k}\Omega}{5.6 \text{ k}\Omega} = 4.16 \text{ V}.$$

Voltage check:

Voltage drop,

$$V_3 = E \times \frac{R_3}{R_T} = 37 \text{ V} \times \frac{1.8 \text{ k}\Omega}{16.0 \text{ k}\Omega} = 4.16 \text{ V} \qquad (11\text{-}12)$$

$$\text{Potential at } X = +37 \text{ V} \times \frac{(1.8 \text{ k}\Omega + 3.9 \text{ k}\Omega)}{5.6 \text{ k}\Omega + 4.7 \text{ k}\Omega + 1.8 \text{ k}\Omega + 3.9 \text{ k}\Omega}$$

$$= +37 \text{ V} \times \frac{5.7 \text{ k}\Omega}{16.0 \text{ k}\Omega} = +13.2 \text{ V}.$$

Example 11-2

In Fig. 11-4, calculate the values of V_1, E, and P_3.

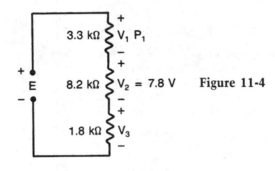

Figure 11-4

Solution

Voltage drop,

$$V_1 = V_2 \times \frac{R_1}{R_2} \qquad (11\text{-}5)$$

$$= 7.8 \text{ V} \times \frac{3.3 \text{ k}\Omega}{8.2 \text{ k}\Omega}$$

$$= 3.14 \text{ V}$$

Source voltage,

$$E = V_2 \times \frac{R_T}{R_2} \qquad (11\text{-}14)$$

$$= 7.8 \text{ V} \times \frac{(3.3 \text{ k}\Omega + 8.2 \text{ k}\Omega + 1.8 \text{ k}\Omega)}{8.2 \text{ k}\Omega}$$

$$= 7.2 \text{ V} \times \frac{13.3 \text{ k}\Omega}{8.2 \text{ k}\Omega}$$

$$= 12.65 \text{ V}.$$

Total power,

$$P_T = \frac{E^2}{R_T} \qquad (11\text{-}18)$$

$$= \frac{(12.65 \text{ V})^2}{13.3 \text{ k}\Omega}$$

$$= 12.03 \text{ mW}$$

Power dissipated,

$$P_3 = P_T \times \frac{R_3}{R_T} \tag{11-2}$$

$$= 12.03 \text{ mW} \times \frac{1.8 \text{ k}\Omega}{13.3 \text{ k}\Omega}$$

$$= 1.63 \text{ mW}$$

Example 11-3

In Fig. 11-5, calculate the values of the potentials V_1, V_2, V_3, and V_4.

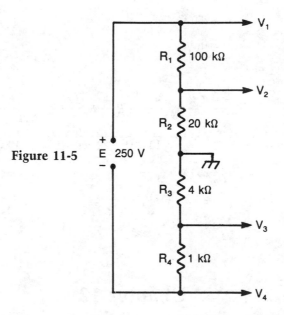

Figure 11-5

Solution

Potential,

$$V_1 = +E \times \frac{(R_1 + R_2)}{R_1 + R_2 + R_3 + R_4} \tag{11-22}$$

$$= +250 \text{ V} \times \frac{(100 \text{ k}\Omega + 20 \text{ k}\Omega)}{100 \text{ k}\Omega + 20 \text{ k}\Omega + 4 \text{ k}\Omega + 1 \text{ k}\Omega}$$

$$= +250 \text{ V} \times \frac{120 \text{ k}\Omega}{125 \text{ k}\Omega}$$

$$= +240 \text{ V}$$

Potential,

$$V_2 = +E \times \frac{R_2}{R_1 + R_2 + R_3 + R_4} \tag{11-23}$$

$$= +250 \text{ V} \times \frac{20 \text{ k}\Omega}{100 \text{ k}\Omega + 20 \text{ k}\Omega + 4 \text{ k}\Omega + 1 \text{ k}\Omega}$$

$$= +250 \text{ V} \times \frac{20 \text{ k}\Omega}{125 \text{ k}\Omega}$$

$$= +40 \text{ V}$$

Potential,

$$V_3 = -E \times \frac{(R_3 + R_4)}{R_1 + R_2 + R_3 + R_4} \tag{11-24}$$

$$= -250 \text{ V} \times \frac{(4 \text{ k}\Omega + 1 \text{ k}\Omega)}{100 \text{ k}\Omega + 20 \text{ k}\Omega + 4 \text{ k}\Omega + 1 \text{ k}\Omega}$$

$$= -250 \text{ V} \times \frac{5 \text{ k}\Omega}{125 \text{ k}\Omega}$$

$$= -10 \text{ V}$$

Potential,

$$V_4 = -E \times \frac{R_4}{R_1 + R_2 + R_3 + R_4} \tag{11-24}$$

$$= -250 \text{ V} \times \frac{1 \text{ k}\Omega}{100 \text{ k}\Omega + 20 \text{ k}\Omega + 4 \text{ k}\Omega + 1 \text{ k}\Omega}$$

$$= -250 \text{ V} \times \frac{1 \text{ k}\Omega}{125 \text{ k}\Omega}$$

$$= -2 \text{ V.}$$

Chapter 12
Sources connected in series-aiding and in series-opposing

TWO SOURCES ARE IN SERIES-AIDING WHEN THEIR POLARITIES ARE SUCH AS TO drive the current in the same direction around the circuit. This is illustrated in Fig. 12-1A where the negative terminal of E_1 is directly connected to the positive terminal of E_2. Both voltages will then drive the electrons around the circuit in the counterclockwise direction. If both voltage sources are reversed, they will still be connected in series-aiding but the electron flow will now be in the clockwise direction.

The normal purpose of connecting sources in series-aiding is to increase the amount of voltage applied to a circuit. A good example is the insertion of two 1½-V "D" cells into a flashlight to create a total of $2 \times 1½$ V = 3 V. In the case of the flashlight the positive center terminal of

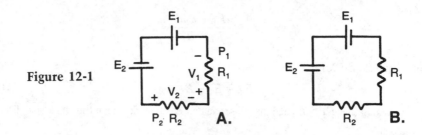

Figure 12-1

one cell must be in contact with the negative casing of the other cell to provide the series-aiding connection.

If the E_1 cell is reversed as in Fig. 12-1B the polarities of the sources will be such as to drive currents in opposite direction around the circuit. The connection is therefore series-opposing and the total voltage available is the *difference* of the individual EMFs. The greater of the two EMFs will then determine the actual direction of the current flow. In the particular case when two identical voltage sources are connected in series-opposing, the total voltage is zero. Consequently the current is also zero, as are the individual voltage drops and the powers dissipated in the resistors.

MATHEMATICAL DERIVATIONS

In Fig. 12-1A:

Total equivalent EMF,

$$E_T = E_1 + E_2 \text{ volts} \tag{12-1}$$

Total equivalent resistance,

$$R_T = R_1 + R_2 \text{ ohms} \tag{12-2}$$

Current,

$$I = \frac{E_1 + E_2}{R_1 + R_2} \text{ amperes} \tag{12-3}$$

Voltage drop,

$$V_1 = I\,R_1 \text{ volts} \tag{12-4}$$

Voltage drop,

$$V_2 = I\,R_2 \text{ volts} \tag{12-5}$$

Power dissipated,

$$P_1 = I^2 R_1 = V_1{}^2/R_1 = I \times V_1 \text{ watts} \tag{12-6}$$

Power dissipated,

$$P_2 = I^2 R_2 = V_2{}^2/R_2 = I \times V_2 \text{ watts} \tag{12-7}$$

Total power delivered from the source,

$$
\begin{aligned}
P_T &= P_1 + P_2 \\
&= I \times E_T \\
&= I \times (E_1 + E_2) \\
&= I \times (V_1 + V_2) \\
&= I^2 \times (R_1 + R_2)
\end{aligned}
$$

$$= \frac{(E_1 + E_2)^2}{R_1 + R_2} \text{ watts} \tag{12-8}$$

In Fig. 12-1B:
Total equivalent EMF,

$$E_T = E_1 - E_2 \text{ volts} \tag{12-9}$$

This assumes that $E_1 > E_2$ and that the resultant electron flow is in the clockwise direction.

Current,

$$I = \frac{E_1 - E_2}{R_1 + R_2} \text{ amperes} \tag{12-10}$$

Total power,

$$P_T = I \times (E_1 - E_2) = \frac{(E_1 - E_2)^2}{R_1 + R_2} \text{ watts} \tag{12-11}$$

The other equations are the same as those for the series-aiding connection.

Example 12-1

In Fig. 12-2 calculate the values of E_T, R_T, I, V_1, V_2, P_1, P_2 and P_T. What is the value of the potential at the point X? If the E_2 source is reversed, recalculate the values of E_T, I, V_1, V_2, P_1, P_2 and P_T.

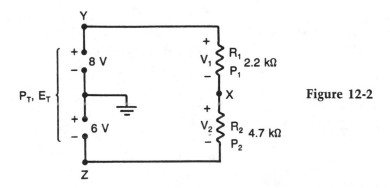

Figure 12-2

Solution

Total equivalent EMF, $E_T = E_1 + E_2 = 8 + 6 = 14$ V $\tag{12-1}$

Total equivalent resistance, $R_T = R_1 + R_2 = 2.2 + 4.7 = 6.9$ kΩ $\tag{12-2}$

$$\text{Current, } I = \frac{E_T}{R_T} = \frac{14 \text{ V}}{6.9 \text{ kΩ}} = 2.03 \text{ mA}$$

and is in the counterclockwise direction (electron flow) $\tag{12-3}$

Voltage drop, $V_1 = I \times R_1 = 2.03$ mA $\times 2.2$ kΩ $= 4.46$ V $\tag{12-4}$

Voltage drop, $V_2 = I \times R_2 = 2.03$ mA $\times 4.7$ kΩ $= 9.54$ V $\tag{12-5}$

Voltage check:

$$E_T = V_1 + V_2 = 4.46 + 9.54 = 14 \text{ V}$$

Power dissipated, $P_1 = I \times V_1 = 2.03$ mA $\times$ 4.44 V = 9.01 mW   (12-6)
Power dissipated, $P_2 = I \times V_2 = 2.03$ mA $\times$ 9.54 V = 19.4 mW   (12-7)
Total power, $P_T = I \times E_T = 2.03$ mA $\times$ 14 V = 28.4 mW   (12-8)

Power check:
$P_T = P_1 + P_2 = 9.01 + 19.4 = 28.41$ mW.
Potential at $Y = +8$ V.
Potential at $X = (+8$ V$) + (-4.46$ V$) = +3.54$ V.

Potential check:
Potential at $Z = -6$ V.
Potential at $X = (-6$ V$) + (+9.54$ V$) = +3.54$ V.

E_2 source reversed:
Total equivalent EMF, $E_T = E_1 - E_2 = 8$ V $- 6$ V $= 2$ V.   (12-9)

$$\text{Current, } I = \frac{E_T}{R_T} = \frac{2 \text{ V}}{6.9 \text{ k}\Omega} = 0.29 \text{ mA} \qquad (12\text{-}10)$$

and is in the counterclockwise direction (electron flow).

Voltage drop, $V_1 = I \times R_1 = 0.29$ mA $\times$ 2.2 kΩ = 0.64 V.
Voltage drop, $V_2 = I \times R_2 = 0.29$ mA $\times$ 4.7 kΩ = 1.36 V.

Voltage check:

$$E_T = V_1 + V_2 = 0.64 + 1.36 = 2.0 \text{ V.}$$

Power dissipated, $P_1 = I \times V_1 = 0.29$ mA $\times$ 0.64 V = 0.186 mW.
Power dissipated, $P_2 = I \times V_2 = 0.29$ mA $\times$ 1.36 V = 0.394 mW.
Total power $P_T = I \times E_T = 0.29$ mA $\times$ 2 V = 0.58 mW.

Power check:

$$P_T = P_1 + P_2 = 0.186 + 0.394 = 0.58 \text{ mW.}$$

Example 12-2

In Fig. 12-3, calculate the values of I and the voltmeter readings V_1 and V_2.

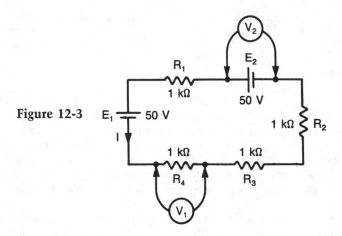

Figure 12-3

45

Solution

The two sources are connected in series-opposing.

Total equivalent EMF, $E_T = E_1 - E_2 = 50 - 50 = 0$ V. (12-9)

Current, $I = 0$ A.

Voltage drop, $V_1 = 0$ V.

With respect to the voltmeter V_2, one probe is effectively connected to both positive terminals while the other probe is joined to the two negative terminals. Therefore, the reading of V_2 is 50 V.

Chapter 13
The potentiometer and the rheostat

THE POTENTIOMETER (OR *POT*) IS ANOTHER PRACTICAL APPLICATION OF THE voltage division rule. Essentially it consists of a length of resistance wire (for example, nichrome wire) or a thin carbon track along which a moving contact or slider may be set at any point (Fig. 13-1A). The wire is wound on an insulating base that may either be straight or, more conveniently, formed into a circle.

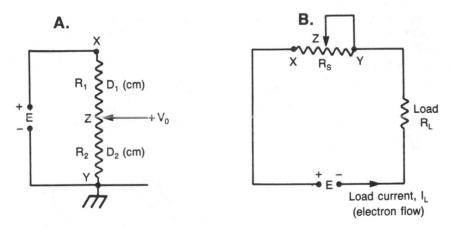

Figure 13-1

The purpose of the potentiometer is to obtain an output voltage, V_o, which may be varied from zero up to the full value of the source voltage, E. There are two terminals, X and Y, at the ends of the potentiometer while the slider is connected to the third terminal, Z. The source voltage is then applied between the end terminals, X and Y, while the output voltage appears between the slider terminal, Z, and the end terminal, Y, which is commonly grounded. A practical example of the potentiometer is a receiver's volume control. In this case the source is an audio signal but the principles of the potentiometer apply equally well to both dc and ac. In either case care must be taken not to exceed the potentiometer's power rating.

While a potentiometer varies an output *voltage,* the rheostat is used for controlling the *current* in a series circuit (Fig. 13-1B). The construction of a rheostat is similar to that of a potentiometer except that the rheostat needs only one end terminal, X, and the terminal, Z, which is connected to the moving contact. The third terminal, Y, is then directly joined to the terminal, Z. A practical example of a rheostat is the dimmer control for the lights on a car's dashboard.

MATHEMATICAL DERIVATIONS

In Fig. 13-1A, the total resistance of the potentiometer is $R_1 + R_2$, while the resistance between Z and Y is R_2. Then by the voltage division rule (VDR),

Output voltage,

$$V_0 = E \times \frac{R_2}{R_1 + R_2} \text{ volts} \tag{13-1}$$

If the slider is moved to the terminal, Y,

$$R_2 = 0 \ \Omega \text{ and } V_0 = 0 \text{ V} \tag{13-2}$$

When the slider is moved to the terminal, X,

$$R_1 = 0 \ \Omega \text{ and } V_0 = E \text{ volts} \tag{13-3}$$

If the terminal, Z, is at the center position,

$$R_1 = R_2 \text{ and } V_0 = E/2 \text{ volts} \tag{13-4}$$

The results of Equations 13-1, 13-2, 13-3, and 13-4 are only true if no load is connected across the output voltage. Once a load is connected, there would be an additional current flow through R_1, increasing its voltage drop so that the output voltage would fall.

If the ratio of R_2 to the total resistance of the potentiometer is equal to the ratio of the slider's travel, D_2 (from its starting position), to the total travel available, $D_1 + D_2$, the potentiometer is said to be "linear." However, if the resistance and the travel ratios are connected by a logarithmic relationship, we have a "log pot." Whether a potentiometer is linear or logarithmic depends on the winding of the resistance wire or the construction of the carbon track.

For a linear potentiometer,

Output voltage,

$$V_0 = E \times \frac{D_2}{D_1 + D_2} \text{ volts} \tag{13-5}$$

where D_1 and D_2 may be conveniently measured in centimeters.

In Fig. 13-1B, the slider is moved to terminal X. None of the rheostat's resistance is now included in the circuit so that the current has its maximum value given by

Maximum current,

$$I_{\max} = \frac{E}{R_L} \text{ amperes} \tag{13-6}$$

When the slider is subsequently moved to the terminal, Y, the whole of the rheostat's resistance, R_S, is in series with the load, R_L, so that a minimum current will flow.

Minimum current,

$$I_{\min} = \frac{E}{R_S + R_L} \text{ amperes} \tag{13-7}$$

The rheostat can therefore control any level of current between the maximum and minimum values.

Example 13-1

In Fig. 13-1A, $E = 20$ V, $D_1 = 25$ cm, $D_2 = 35$ cm. If the potentiometer is linear and has a total resistance of 500 Ω, calculate the values of the output voltage and the power dissipation.

Solution

Output voltage,

$$V_0 = +E \times \frac{D_1}{D_1 + D_2} \qquad (13\text{-}5)$$

$$= +20 \text{ V} \times \frac{25 \text{ cm}}{25 \text{ cm} + 35 \text{ cm}}$$

$$= +20 \text{ V} \times \frac{25 \text{ cm}}{60 \text{ cm}}$$

$$= +8.33 \text{ V}.$$

Power dissipation,

$$P = \frac{E^2}{R_1 + R_2} = \frac{(20 \text{ V})^2}{500 \text{ }\Omega} = 0.8 \text{ W}.$$

Example 13-2

In Fig. 13-1B, $E = 9$ V, $R_S = 6$ Ω and $R_L = 4$ Ω. Calculate the maximum and minimum values of the circuit current. Assuming that the rheostat is linear, what is the value of the load current when the slider is in its center position?

Solution

Maximum load current,

$$I_{max} = \frac{E}{R_L} \qquad (13\text{-}6)$$

$$= \frac{9 \text{ V}}{4 \text{ }\Omega} = 2.25 \text{ A}$$

Minimum load current,

$$I_{min} = \frac{E}{R_L + R_S} \qquad (13\text{-}7)$$

$$= \frac{9 \text{ V}}{4 \text{ }\Omega + 6 \text{ }\Omega} = 0.9 \text{ A}$$

With the slider in the center position,

$$R_S = \frac{6 \text{ }\Omega}{2} = 3 \text{ }\Omega.$$

The load current is then

$$I_L = \frac{E_L}{R_L + R_S} = \frac{9 \text{ V}}{4 \text{ }\Omega + 3 \text{ }\Omega} = 1.29 \text{ A}.$$

Chapter 14
The series voltage-dropping resistor

EVERY LOAD NORMALLY HAS A VOLTAGE/CURRENT RATING, AS WELL AS A WATTAGE rating. For example, a 75 W bulb is also rated at 120 V. If the available source voltage is greater than the required load voltage, it is possible to insert a series dropping resistor so that the load is still operated correctly. Figure 14-1 illustrates a load such as a transistor amplifier, which needs a supply voltage of 9 V with a corresponding current of 8 mA. If the available source voltage is 15 V, the series resistor is inserted to drop the voltage down to the 9 V required by the load. The voltage across R_D is therefore 15 V − 9 V = 6 V. This advantage of using a dropping resistor is achieved at the expense of its cost and the power dissipation.

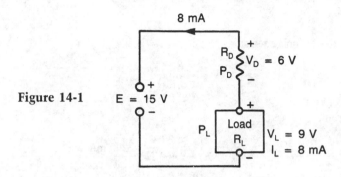

Figure 14-1

Care must be taken to calculate the required ohmic value of the dropping resistor, R_D, as well as its required power dissipation because the resistor would overheat if its power rating were too low (see Example 14-1).

MATHEMATICAL DERIVATIONS

Voltage across the dropping resistor,

$$V_D = E - V_L \tag{14-1}$$
$$= E - I_L R_L \text{ volts}$$

Ohmic value of the dropping resistor,

$$R_D = \frac{V_D}{I_L} \text{ ohms} \tag{14-2}$$

Load resistance,

$$R_L = \frac{V_L}{I_L} \text{ ohms} \tag{14-3}$$

Power dissipation of the dropping resistor,

$$P_D = I_L \times V_D \text{ watts} \tag{14-4}$$

Load power,

$$P_L = I_L \times V_L \text{ watts} \tag{14-5}$$

Total power delivered from the source,

$$P_T = P_L + P_D \tag{14-6}$$
$$= I_L \times (V_L + V_D)$$
$$= I_L{}^2(R_L + R_D)$$
$$= \frac{E^2}{(R_L + R_D)} \text{ watts}$$

Example 14-1

A relay whose coil resistance is 400 Ω is designed to operate when the voltage across the relay is 80 V. If the relay is operated from a 120 V supply, calculate the ohmic value of the required series dropping resistor and its power dissipation.

Solution

Voltage across the dropping resistor,

$$V_D = E - V_L \tag{14-1}$$
$$= 120 \text{ V} - 80 \text{ V}$$
$$= 40 \text{ V}$$

Relay current,

$$I_L = \frac{V_L}{R_L} = \frac{80 \text{ V}}{400 \text{ }\Omega} = 200 \text{ mA} \tag{14-2}$$

Ohmic value of the dropping resistor,

$$R_D = \frac{V_D}{I_L} \tag{14-3}$$
$$= \frac{40 \text{ V}}{200 \text{ mA}}$$
$$= 200 \text{ }\Omega$$

Power dissipation of the dropping resistor,

$$P_D = I_L \times V_D \tag{14-4}$$
$$= 200 \text{ mA} \times 40 \text{ V}$$
$$= 8 \text{ W}$$

A 200 Ω, 10 W resistor would be a suitable choice.

Chapter 15
Resistors in parallel

ALL LOADS (LIGHTS, VACUUM CLEANER, TOASTER, ETC.) IN YOUR HOME ARE normally stamped with a 120 V rating. It follows that all these loads must be connected directly across the household supply because although each load has its own power rating, its

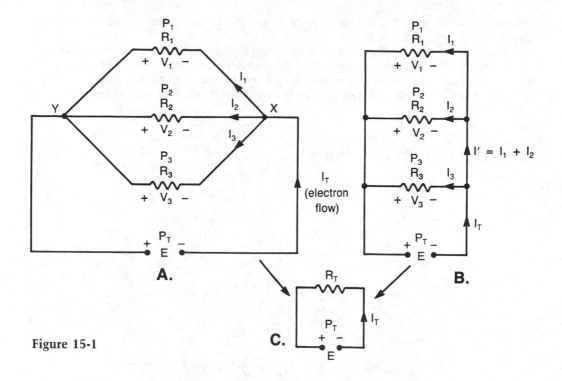

Figure 15-1

required voltage value in every case is the same. This type of arrangement is known as parallel and is illustrated in Fig. 15-1 A and B in which the three resistors are connected between two common points X, Y (Fig. 15-1A) or two common lines (Fig. 15-1B). The points or lines are then directly connected to the voltage source. It follows that the voltage drop across each parallel resistor is the same and equal to the source voltage. This is in contrast with the series arrangement where voltage division occurred.

Each of the resistors in a parallel circuit is a path for current flow and each path is called a branch. Because the voltage across each branch is the same, each branch will carry its own current and the individual branch currents will all be different (unless the branch resistances are the same in which case their currents would also be equal). Notice that the branch currents are independent of one another. In other words, if one branch current is switched off, the other branch currents are unaffected. This would be equivalent to saying that, if you switch off the light in the kitchen, the TV set in the lounge continues to operate.

The total source current, I_T, splits at the point X into individual branch currents. The current in a particular branch is inversely proportional to the value of the resistance in that branch. Consequently the lowest-value resistor will carry the greatest current, while the smallest current will flow through the highest-value resistor. In terms of electron flow the branch currents, I_1, I_2, I_3 are all leaving the junction point, X, while the total current I_T is entering the same junction point. Recombination of the branch currents then occurs at the other junction point, Y. It follows that the total current is the *sum* of the individual branch currents. This is an expression of *Kirchhoff's Current Law* (KCL), which is further explored in Chapter 26.

The total equivalent resistance, R_T, is defined as the ratio of the source voltage, E, to the source current, I_T. If an additional branch is added to a parallel arrangement of resistors, the total current must *increase* by the amount of the new branch current and therefore the total equivalent

51

resistance must *decrease.* For parallel resistors the value of R_T must always be less than the lowest-value resistor in the arrangement.

Each parallel resistor dissipates its own power and the sum of the individual powers dissipated is equal to the total power delivered from the source. The greatest power is dissipated by the lowest-value resistor, which carries the greatest branch current. For example, the resistance of a nickel-iron 1000 W heater element is only one-tenth that of the tungsten filament for a 100 W electric light bulb.

MATHEMATICAL DERIVATIONS

Source voltage,

$$E = V_1 = V_2 = V_3 \text{ volts} \tag{15-1}$$

Branch currents,

$$I_1 = \frac{V_1}{R_1} = \frac{E}{R_1}, I_2 = \frac{V_2}{R_2} = \frac{E}{R_2}, I_3 = \frac{V_3}{R_3} = \frac{E}{R_3} \tag{15-2}$$

Total source current,

$$I_T = I_1 + I_2 + I_3 \tag{15-3}$$

$$= \frac{E}{R_1} + \frac{E}{R_2} + \frac{E}{R_3}$$

$$= E \times \left(\frac{1}{R_1} + \frac{1}{R_2} + \frac{1}{R_3} \right) \text{ amperes}$$

Notice that the current I', which exists in the circuit of Fig. 15-1B, cannot be measured in the circuit of Fig. 15-1A.

Total equivalent resistance,

$$R_T = \frac{E}{R_T} = E \times \frac{1}{R_T} \text{ ohms} \tag{15-4}$$

Comparing equations 15-3, 15-4:

$$\frac{1}{R_T} = \frac{1}{R_1} + \frac{1}{R_2} + \frac{1}{R_3} \tag{15-5}$$

For N resistors in parallel,

$$I_T = I_1 + I_2 + I_3 + \dots I_N \tag{15-6}$$

$$\frac{1}{R_T} = \frac{1}{R_1} + \frac{1}{R_2} + \frac{1}{R_3} + \dots \frac{1}{R_N} \tag{15-7}$$

This is known as the "reciprocal formula" since $1/R_T$, $1/R_1$, etc., are the reciprocals of R_T, R_1, etc.

Then

$$R_T = \frac{1}{\dfrac{1}{R_1} + \dfrac{1}{R_2} + \dfrac{1}{R_3} + \dots \dfrac{1}{R_N}} \text{ ohms} \tag{15-8}$$

If all N resistors are of equal value R,

$$R_T = \frac{R}{N} \text{ ohms} \qquad (15\text{-}9)$$

Conductance,

$$G_1 = \frac{1}{R_1}, \ G_2 = \frac{1}{R_2}, \ G_3 = \frac{1}{R_3} \ \ldots \ G_N = \frac{1}{R_N} \text{ siemens} \qquad (15\text{-}10)$$

Combining equations 15-7 and 15-8,
Total conductance,

$$G_T = \frac{1}{R_T} = G_1 + G_2 + G_3 + \ldots \ G_N \text{ siemens} \qquad (15\text{-}11)$$

For *two* resistors in parallel:

$$R_T = \frac{R_1 \times R_2}{R_1 + R_2}, \ R_1 = \frac{R_2 \times R_T}{R_2 - R_T}, \ R_2 = \frac{R_1 \times R_T}{R_1 - R_T} \text{ ohms} \qquad (15\text{-}12)$$

The equation $R_T = (R_1 \times R_2)/(R_1 + R_2)$ is commonly referred to as the *product-over-sum* formula. If more than two parallel resistors are involved, the formula may be repeated indefinitely by considering pairs of resistors but the method then becomes far more cumbersome than the use of the reciprocal formula to which the electronic calculator is particularly suited.

Power dissipated in the $R1$ resistor,

$$P_1 = I_1 \times E = I_1^2 R_1 = E^2/R_1 \text{ watts} \qquad (15\text{-}13)$$

Power dissipated in the $R2$ resistor,

$$P_2 = I_2 \times E = I_2^2 R_2 = E^2/R_2 \text{ watts} \qquad (15\text{-}14)$$

Power dissipated in the $R3$ resistor,

$$P_3 = I_3 \times E = I_3^2 R_3 = E^2/R_3 \text{ watts} \qquad (15\text{-}15)$$

Total power delivered from the source,

$$
\begin{aligned}
P_T &= P_1 + P_2 + P_3 \qquad (15\text{-}16)\\
&= I_T \times E\\
&= I_T^2 \times R_T\\
&= E^2/R_T \text{ watts}
\end{aligned}
$$

Example 15-1

In Fig. 15-2 calculate the values of I_1, I_2, I_3, I_4, I', I'', I_T, R_T, P_1, P_2, P_3, P_4, and P_T.

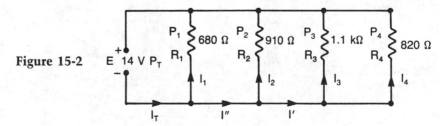

Figure 15-2

Solution

$$\text{Branch current, } I_1 = \frac{E}{R_1} \tag{15-2}$$

$$= \frac{14 \text{ V}}{680 \text{ }\Omega}$$

$$= 20.6 \text{ mA}$$

$$\text{Branch current, } I_2 = \frac{E}{R_2} \tag{15-4}$$

$$= \frac{14 \text{ V}}{910 \text{ }\Omega}$$

$$= 15.4 \text{ mA}$$

$$\text{Branch current, } I_3 = \frac{E}{R_3} \tag{15-2}$$

$$= \frac{14 \text{ V}}{1.1 \text{ k}\Omega}$$

$$= 12.7 \text{ mA}$$

$$\text{Branch current, } I_4 = \frac{E}{R_4} \tag{15-2}$$

$$= \frac{14 \text{ V}}{820 \text{ }\Omega}$$

$$= 17.1 \text{ mA}$$

$$\text{Current } I' = I_3 + I_4$$

$$= 12.7 + 17.1$$

$$= 29.8 \text{ mA.}$$

$$\text{Current } I'' = I_2 + I'$$

$$= 15.4 + 29.8$$

$$= 45.2 \text{ mA.}$$

$$\text{Total current, } I_T = I_1 + I_2 + I_3 + I_4 \tag{15-6}$$

$$= 20.6 + 15.4 + 12.7 + 17.1$$

$$= 65.8 \text{ mA}$$

$$\text{Total resistance, } R_T = \frac{E}{I} \tag{15-4}$$

$$= \frac{14 \text{ V}}{65.8 \text{ mA}}$$

$$= 213 \text{ }\Omega.$$

Total resistance check:
Total resistance,

$$R_T = \frac{1}{\dfrac{1}{R_1} + \dfrac{1}{R_2} + \dfrac{1}{R_3} + \dfrac{1}{R_4}} \tag{15-8}$$

$$= \cfrac{1}{\cfrac{1}{680} + \cfrac{1}{910} + \cfrac{1}{1100} + \cfrac{1}{820}}$$

$$= 213 \ \Omega.$$

Note that the value of R_T (213 Ω) is less than the lowest-value resistor (680 Ω) in the parallel arrangement.

Power dissipated in $R1$ resistor,

$$\begin{aligned} P_1 &= I_1 \times E \\ &= 20.6 \text{ mA} \times 14 \text{ V} \\ &= 288 \text{ mW} \end{aligned} \tag{15-13}$$

Power dissipated in $R2$ resistor,

$$\begin{aligned} P_2 &= I_2 \times E \\ &= 15.4 \text{ mA} \times 14 \text{ V} \\ &= 216 \text{ mW} \end{aligned} \tag{15-14}$$

Power dissipated in $R3$ resistor,

$$\begin{aligned} P_3 &= I_3 \times E \\ &= 12.7 \text{ mA} \times 14 \text{ V} \\ &= 178 \text{ mW} \end{aligned} \tag{15-15}$$

Power dissipated in $R4$ resistor,

$$\begin{aligned} P_4 &= I_4 \times E \\ &= 17.1 \text{ mA} \times 14 \text{ V} \\ &= 239 \text{ mW} \end{aligned}$$

Total power dissipated,

$$\begin{aligned} P_T &= P_1 + P_2 + P_3 + P_4 \\ &= 288 + 216 + 178 + 239 \\ &= 921 \text{ mW} \end{aligned} \tag{15-16}$$

Power check:
Total power delivered from the source,

$$\begin{aligned} P_T &= I_T \times E \\ &= 65.8 \text{ mA} \times 14 \text{ V} \\ &= 921 \text{ mW} \end{aligned} \tag{15-16}$$

Example 15-2

A parallel circuit consists of four 1-watt resistors whose values are 120 Ω, 150 Ω, 270 Ω, and 330 Ω. Calculate their maximum power dissipation without exceeding the wattage rating of any of the resistors.

Solution

The voltage drop across each resistor to produce its 1W rating is different; using the relationship $E = \sqrt{P \times R}$, these voltages are

$$E_{120 \ \Omega} = \sqrt{1W \times 120 \ \Omega} = 10.95 \text{ V}$$
$$E_{150 \ \Omega} = \sqrt{1W \times 150 \ \Omega} = 12.25 \text{ V}$$

$$E_{270\,\Omega} = \sqrt{1W \times 270\ \Omega} = 16.43\ V$$
$$E_{330\,\Omega} = \sqrt{1W \times 330\ \Omega} = 18.17\ V$$

We must select the smallest of these voltages, 10.95 V, because if the current were allowed to exceed this value, the maximum dissipation in the 120 Ω resistor would be exceeded. Using the relationship $P_T = E^2 G_T$, the maximum power dissipation of this network is:

$$(10.95\ V)^2 \times \left(\frac{1}{120\ \Omega} + \frac{1}{150\ \Omega} + \frac{1}{270\ \Omega} + \frac{1}{330\ \Omega} \right) = 2.61\ W$$

This result should be compared with the solution to Example 9-3.

∽∾

Chapter 16
Open and short circuits

AN OPEN CIRCUIT IS REGARDED AS A BREAK THAT OCCURS IN A CIRCUIT. FOR example a resistor might burn out (Fig. 16-1A) due to excessive power dissipation or become disconnected from an adjacent resistor. In either case the break would have theoretically infinite resistance so that no current can exist in an open circuit.

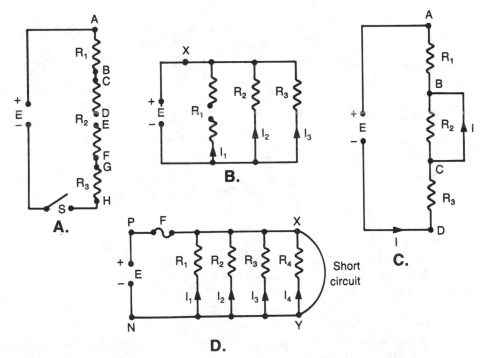

Figure 16-1

THE OPEN CIRCUIT

In the series string of Fig. 16-1A, the resistor $R2$ has burned out so that the total equivalent resistance is theoretically infinite. There is zero current throughout the circuit and no voltage drops across the resistors $R1$, $R3$. Consequently the source voltage appears across the open circuit and could be measured by a voltmeter connected between the points D and E.

To locate the position of the break, the switch, S, is opened to remove power from the circuit. Using an ohmmeter on a high resistance range, the common probe is connected to the point H. With the red probe at point A, the ohmmeter reading is virtually infinite because the measured resistance contains the open circuit. The reading will not change as the red probe is moved in turn to the points B, C, and D. However, as soon as the red probe is moved to the point E, the resistance falls to the relatively low value of $R3$. This change in the reading indicates that the open circuit exists between the points D and E. The ohmmeter has been used to conduct a *continuity test,* which has located a *discontinuity*—the open circuit.

If the resistor $R1$ burns out in the parallel circuit of Fig. 16-1B, the current I_1 is zero but the currents I_2 and I_3 remain the same. The total current I_T is reduced and the total equivalent resistance is greater. However, if a break occurs at the point X, all current ceases and the total equivalent resistance is theoretically infinite.

THE SHORT CIRCUIT

A short circuit has theoretically zero resistance so that a current can flow through a short circuit without developing any voltage drop. Although resistors rarely become shorted, a short circuit can occur as the result of the bare connecting wires coming into contact or being joined together by a stray drop of solder. The short circuit is therefore a zero-resistance path in parallel with (shunting across) the resistor.

An example of a short circuit is shown in the series circuit of Fig. 16-1C. No current will flow through the resistor $R2$ and there will be no voltage drop between the points B and C. The total equivalent resistance will be given by $R_T = R_1 + R_3$ and the current, I, will increase as a result of the short circuit.

In the parallel circuit of Fig. 16-1D, a short circuit has developed across the resistor $R4$. The same short circuit also exists in parallel with the resistors $R1$, $R2$, $R3$ so that the currents I_1, I_2, I_3, I_4 are all zero. However, the path $NYXP$ has an extremely low resistance so that a high current would flow and the protection fuse, F, would melt. This would prevent possible damage to the connecting wires and the source.

Example 16-1

In Fig. 16-2, an open circuit occurs between the points B and C. What are the potential values at the points A, B, C, D?

Solution

The current throughout the circuit is zero and there are zero voltage drops across the resistors. Consequently, the potentials at the points A, B are both +11 V, while the potentials at the points C, D are *zero*.

Example 16-2

In Fig. 16-3 the 180 Ω resistor burns out. What are the new values of I_1, I_2, I_3, and I_T?

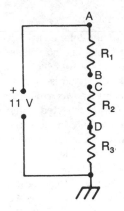

Figure 16-2

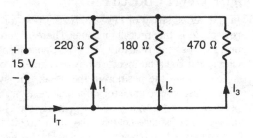

Figure 16-3

Solution

$$\text{Current, } I_1 = \frac{15 \text{ V}}{220 \text{ }\Omega}$$

$$= 68.2 \text{ mA.}$$

Current, $I_2 = 0$.

$$\text{Current, } I_3 = \frac{15 \text{ V}}{470 \text{ }\Omega}$$

$$= 31.9 \text{ mA.}$$

$$\text{Current, } I_T = I_1 + I_2 + I_3$$

$$= 68.2 + 0 + 31.9$$

$$= 100.1 \text{ mA.}$$

Example 16-3

In Fig. 16-4, a short develops across the resistor $R2$. Calculate the new potential values at the points A, B, C, D.

Solution

By the voltage division rule (VDR),
Potential at A,

$$V_A = +23\text{V} \times \frac{(2.7 \text{ k}\Omega + 3.3 \text{ k}\Omega)}{2.7 \text{ k}\Omega + 3.3 \text{ k}\Omega + 1.8 \text{ k}\Omega}$$

$$= +23 \text{ V} \times \frac{6.0 \text{ k}\Omega}{7.8 \text{ k}\Omega}$$

$$= +17.69 \text{ V.}$$

Potentials at B, C,

$$V_B = V_C$$

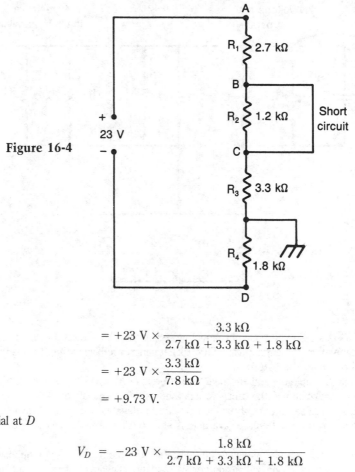

Figure 16-4

$$= +23 \text{ V} \times \frac{3.3 \text{ k}\Omega}{2.7 \text{ k}\Omega + 3.3 \text{ k}\Omega + 1.8 \text{ k}\Omega}$$

$$= +23 \text{ V} \times \frac{3.3 \text{ k}\Omega}{7.8 \text{ k}\Omega}$$

$$= +9.73 \text{ V}.$$

Potential at D

$$V_D = -23 \text{ V} \times \frac{1.8 \text{ k}\Omega}{2.7 \text{ k}\Omega + 3.3 \text{ k}\Omega + 1.8 \text{ k}\Omega}$$

$$= -23 \text{ V} \times \frac{1.8 \text{ k}\Omega}{7.8 \text{ k}\Omega}$$

$$= -5.31 \text{ V}.$$

Chapter 17
Voltage sources in parallel

FIGURE 17-1 ILLUSTRATES N IDENTICAL CELLS, EACH OF EMF E, WHICH ARE parallel-connected across a number (M) of resistive loads. All positive terminals are joined together so that any one of these terminals may be chosen as the positive output terminal; the negative terminals are likewise connected. If a voltmeter V is connected across any one of the cells, it must also be in parallel with all the other cells, so that the voltmeter reading is only E volts

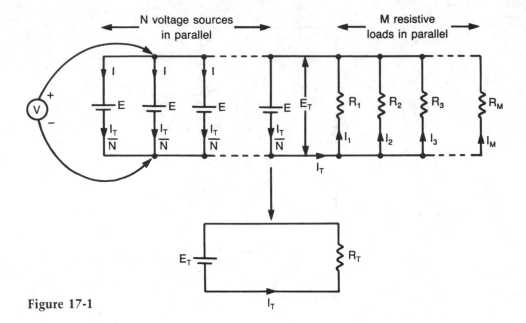

Figure 17-1

(the EMF of one cell). The purpose of connecting cells in parallel is therefore *not* to increase the total voltage. However, the current capability has been increased since the total load current, I_T, will be shared between the batteries.

An everyday example of parallel-connected batteries occurs when you give someone a "jump start" for his/her car. You use the cables to join the positive terminals of the two 12 V batteries together and make the same type of connection with the negative terminals. However, with this parallel arrangement the total voltage available is still only equal to 12 V (the EMF of one battery).

Notice that in the loops containing the cells, the arrangement is series-opposing, so that there is no loading effect of one cell on another. With identical cells, no parallel-opposing arrangement is possible with respect to the loads, because the cells would be in series-aiding around the loops. A very high circulating current would flow between the cells but no current would be supplied to the loads.

MATHEMATICAL DERIVATIONS

For N identical cells in parallel:

Total voltage of the parallel combination,

$$E_T = E \text{ volts} \tag{17-1}$$

Total equivalent resistance of the parallel loads,

$$R_T = \cfrac{1}{\cfrac{1}{R_1} + \cfrac{1}{R_2} + \cfrac{1}{R_3} + \cdots \cfrac{1}{R_N}} \tag{17-2}$$

Total load current,

$$I_T = \frac{E_T}{R_T} \text{ amperes} \tag{17-3}$$

Current supplied by each cell

$$I = \frac{I_T}{N} \text{ amperes} \qquad (17\text{-}4)$$

Example 17-1

Four identical 6 V cells are connected in parallel across two loads whose resistances are 20 Ω and 5 Ω. Calculate the value of the current supplied by each cell.

Solution

Total voltage applied to the parallel loads,

$$E_T = 6 \text{ V (EMF of one cell)} \qquad (17\text{-}1)$$

Total load,

$$R_T = \frac{20 \times 5}{20 + 5}$$

$$= 4 \ \Omega$$

Total load current,

$$I_T = \frac{E_T}{R_T} = \frac{6 \text{ V}}{4 \ \Omega} \qquad (17\text{-}3)$$

$$= 1.5 \text{ A}$$

Current supplied by each cell

$$I = \frac{I_T}{N} = \frac{1.5 \text{ A}}{4} \qquad (17\text{-}4)$$

$$= 0.375 \text{ A}$$

Chapter 18
The current division rule

THE *CURRENT DIVISION RULE* IS USED IF YOU ARE GIVEN THE VALUES OF THE parallel resistors $R1$, $R2$, $R3$ . . . R_N (Fig. 18-1) and the value of the total current, I_T, and are then asked to determine the individual branch currents I_1, I_2, I_3 . . . I_N. Because the voltage across each branch is the same, the individual branch currents are in inverse proportion to the values of the resistors. In other words, the branch with the lowest resistance will carry the highest current while the smallest branch current will flow in the branch with the greatest resistance.

Because the circuit's total resistance, R_T, is equal to the source voltage, E_T, divided by the total current, I_T, the fraction of the source current flowing through a particular branch is equal to

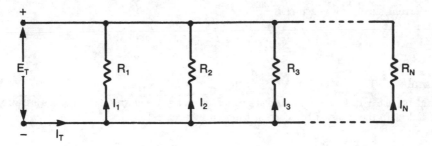

Figure 18-1

the ratio of the total equivalent resistance to the branch resistance. This relationship is called the current division rule.

MATHEMATICAL DERIVATIONS

Source voltage, $E_T = I_1 R_1 = I_2 R_2 = I_3 R_3 \ldots = I_N R_N = I_T R_T$ volts

Then

$$\frac{I_1}{I_2} = \frac{R_2}{R_1} \text{ etc.}$$

Therefore

$$I_1 = I_2 \times \frac{R_2}{R_1}, \; I_2 = I_1 \times \frac{R_1}{R_2} \ldots \text{ etc., amperes} \tag{18-1}$$

Also

$$\frac{I_1}{I_T} = \frac{R_T}{R_1}$$

This yields

$$I_1 = I_T \times \frac{R_T}{R_1}, \; I_2 = I_T \times \frac{R_T}{R_2} \text{ etc., amperes} \tag{18-2}$$

and

$$I_T = I_1 \times \frac{R_1}{R_T} = I_2 \times \frac{R_2}{R_T} \text{ amperes} \tag{18-3}$$

Equation 18-2 represents the *current division rule*.

If there are only two resistors in parallel,

$$R_T = \frac{R_1 R_2}{R_1 + R_2} \text{ ohms}$$

Then

$$I_1 = I_T \times \frac{R_1 R_2}{R_1(R_1 + R_2)} = I_T \times \frac{R_2}{R_1 + R_2} \text{ amperes} \tag{18-4}$$

and

62

$$I_2 = I_T \times \frac{R_1 R_2}{R_2(R_1 + R_2)} = I_T \times \frac{R_1}{R_1 + R_2} \text{ amperes} \tag{18-5}$$

To calculate one of the branch currents for two resistors in parallel, you must first multiply the source current by the resistance *in the other branch* and then divide by the sum of the two resistances.

Example 18-1

In Fig. 18-2, find the value of the current, I_2.

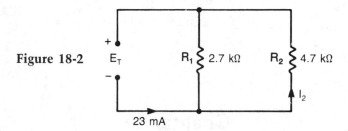

Figure 18-2

Solution

Current,

$$I_2 = I_T \times \frac{R_1}{R_1 + R_2} \tag{18-5}$$

$$= 23 \text{ mA} \times \frac{2.7 \text{ k}\Omega}{2.7 \text{ k}\Omega + 4.7 \text{ k}\Omega}$$

$$= \frac{23 \times 2.7}{7.4}$$

$$= 8.4 \text{ mA, rounded off.}$$

Example 18-2

In Fig. 18-3, find the value of the total current, I_T.

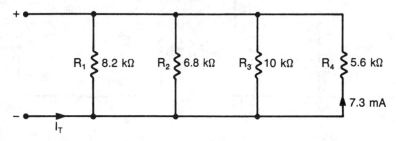

Figure 18-3

Solution

Total equivalent resistance,

$$R_T = \cfrac{1}{\cfrac{1}{8.2} + \cfrac{1}{6.8} + \cfrac{1}{10} + \cfrac{1}{5.6}} \tag{18-3}$$

$$= 1.83 \ k\Omega.$$

Total current,

$$I_T = I_4 \times \frac{R_4}{R_T} \tag{18-3}$$

$$= 7.3 \ mA \times \frac{5.6 \ k\Omega}{1.83 \ k\Omega}$$

$$= 22.3 \ mA, \ \text{rounded off.}$$

Chapter 19
Series-parallel arrangements of resistors

IN CHAPTER 9 WE LEARNED ABOUT THE PROPERTIES RELATED TO SERIES STRINGS of resistors while in Chapter 15 we studied the results of connecting resistors in parallel banks. Briefly summarizing, the single current was the same throughout the series circuit and the source voltage was divided among the resistors in proportion to their resistances. By contrast, the (same) source voltage was applied across each resistor in parallel and the (total) source current was divided between the various branches in inverse proportion to their resistances. However, series strings and parallel banks may be combined to create so-called *series-parallel* circuits of which the two simplest are shown in Fig. 19-1A and B.

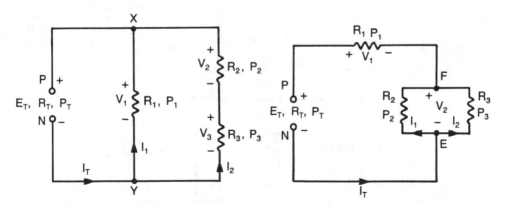

Figure 19-1

In Fig. 19-1A, the resistors $R2$ and $R3$ are joined end-to-end so that the same current must flow through each of these resistors. Consequently, the resistors $R2$ and $R3$ are in series. However, the (electron flow) current splits at the point Y and recombines at the point X. Therefore $R1$ is in parallel with the series combination of $R2$ and $R3$.

The (electron flow) current splits at the point, E, in Fig. 19-1B and recombines at the point, F, so that $R2$ and $R3$ are in parallel. However, this parallel combination is joined to one end of $R1$ so that this resistor is in series with $R_{2,3}$. ($R_{2,3}$ is the equivalent resistance of the $R2$, $R3$ parallel combination.)

The equations for the circuits of Fig. 19-1A, B are fully developed in the mathematical derivations. However, there is an infinite variety of complex series-parallel networks of resistors. For example, we might be faced with the circuit of Fig. 19-2A and be asked to find the total equivalent resistance presented to the source. This problem is solved in a number of steps.

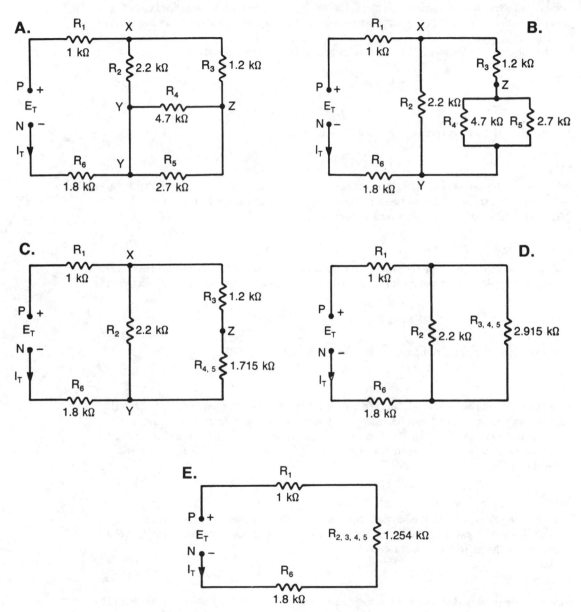

Figure 19-2

Step 1

Identify all points that are electrically different in the circuit. These are the points P, N, X, Y, and Z. The circuit can then be redrawn in the more conventional manner of Fig. 19-2B.

Corresponding points are labeled in the two drawings and you must make certain that the same resistors are connected between any two such points.

Step 2

Combine all obvious series strings and parallel banks. Redraw the resulting simplified circuit. The general practice is to start furthest away from the source and then work your way towards the source.

In our example $R4$ and $R5$ are in parallel and their equivalent resistance,

$$R_{4,5} = \frac{R_4 \times R_5}{R_4 + R_5} = \frac{4.7 \times 2.7}{4.7 + 2.7}$$
$$= 1.715 \text{ k}\Omega$$

The circuit may then be redrawn as shown in Fig. 19-2C.

Step 3

Add together the series equivalent resistances and again draw the circuit. In Fig. 19-2C, $R3$ is in series with $R_{4,5}$ so that the equivalent resistance of the combination is $R_{3,4,5} = 1.2 + 1.715 = 2.915$ kΩ. The circuit when redrawn is shown in Fig. 19-2D.

Step 4

Use the product-over-sum or reciprocal formula, to combine the equivalent parallel resistances. In our example $R2$ is in parallel with $R_{3,4,5}$ and the equivalent resistance of this combination is

$$R_{2,3,4,5} = \frac{2.915 \times 2.2}{2.915 + 2.2} = 1.254 \text{ k}\Omega.$$

When the circuit is again redrawn, its appearance is as in Fig. 19-2E.

Step 5

Repeat steps 3 and 4 alternatively and redraw the circuit as many times as necessary (note that, with experience, it is possible to eliminate most, if not all, of the intermediate drawings).

The redrawn circuit of Fig. 19-2E is clearly a simple series arrangement and therefore the total equivalent resistance presented to the source is $R_T = R_1 + R_{2,3,4,5} + R_6 = 1 + 1.254 + 1.8 = 4.054$ kΩ.

Step 6

Knowing the value of the total equivalent resistance you can determine the total source current, I_T, and then work away from the source and calculate in turn the branch currents and the voltage drops across the individual resistors.

Step 7

Calculate the powers dissipated in the individual resistors and the total power drawn from the source.

MATHEMATICAL DERIVATIONS
In Fig. 19-1A

Equivalent resistance of $R1$ and $R2$ in series,

$$R_{1,2} = R_1 + R_2 \text{ ohms} \tag{19-1}$$

Total equivalent resistance,

$$R_T = R_{1,2}/R_3 \tag{19-2}$$

$$= \frac{(R_1 + R_2) \times R_3}{R_1 + R_2 + R_3} \text{ ohms}$$

Total current,

$$I_T = \frac{E_T}{R_T} \text{ amperes} \tag{19-3}$$

Branch current,

$$I_1 = \frac{E_T}{R_1} \text{ amperes} \tag{19-4}$$

By Kirchhoff's current law (KCL):

$$I_T = I_1 + I_2, \ I_1 = I_T - I_2, \ I_2 = I_T - I_1 \text{ amperes} \tag{19-5}$$

By the Voltage Division Rule (VDR)

$$V_2 = I_2 R_2 = E_T \times \frac{R_2}{R_2 + R_3}, \ V_3 = I_2 R_3 = E_T \times \frac{R_3}{R_2 + R_3} \text{ volts} \tag{19-6}$$

$$E_T = V_1 = V_2 + V_3 \text{ volts} \tag{19-7}$$

Total power delivered from the source,

$$P_T = E_T I_T \tag{19-8}$$

$$= P_1 + P_2 + P_3 \text{ watts}$$

In Fig. 19-1B

Equivalent resistance of $R2$ and $R3$ in parallel

$$R_{2,3} = \frac{R_2 \times R_3}{R_2 + R_3} \text{ ohms}$$

Total equivalent resistance,

$$R_T = R_1 + R_{2,3} \tag{19-9}$$

$$= R_1 + \frac{R_2 \times R_3}{R_2 + R_3}$$

$$= \frac{R_1 R_2 + R_2 R_3 + R_3 R_1}{R_2 + R_3} \text{ ohms}$$

Total source current,

$$I_T = \frac{E_T}{R_T} = \frac{E_T \times (R_2 + R_3)}{R_1 R_2 + R_2 R_3 + R_3 R_1} \text{ amperes} \tag{19-10}$$

Using the CDR,
Branch current,

$$I_1 = \frac{V_2}{R_2} = I_T \times \frac{R_3}{R_2 + R_3} \tag{19-11}$$

$$= \frac{E_T \times R_3}{R_1 R_2 + R_2 R_3 + R_3 R_1} \text{ amperes}$$

Branch current,

$$I_2 = \frac{V_2}{R_3} = I_T \times \frac{R_2}{R_2 + R_3} \tag{19-12}$$

$$= \frac{E_T \times R_2}{R_1 R_2 + R_2 R_3 + R_3 R_1} \text{ amperes}$$

Also

$$I_T = I_1 + I_2 = \frac{V_1}{R_1}, \ I_1 = \frac{V_2}{R_2} = I_T - I_2, \ I_2 = \frac{V_2}{R_3} = I_T - I_1 \text{ amperes} \tag{19-13}$$

Voltage drop across the resistor $R1$,

$$V_1 = I_T R_1 \text{ volts} \tag{19-14}$$

By KVL,

$$E_T = V_1 + V_2, \ V_1 = E_T - V_2, \ V_2 = E_T - V_1 \text{ volts} \tag{19-15}$$

Total power drawn from the source,

$$P_T = E_T I_T$$
$$= P_1 + P_2 + P_3 \text{ watts}$$

Example 19-1

In Fig. 19-1A, $R_1 = 3.9 \text{ k}\Omega$, $R_2 = 1 \text{ k}\Omega$, $R_3 = 1.5 \text{ k}\Omega$. If $E_T = 9 \ V$, calculate the values of R_T, I_T, I_1, I_2, V_1, V_2.

Solution

Equivalent resistance of $R2$ and $R3$ in series,

$$R_{2,3} = R_2 + R_3 \tag{19-1}$$
$$= 1 + 1.5$$
$$= 2.5 \text{ k}\Omega.$$

Total equivalent resistance,

$$R_T = \frac{R_1 \times R_{2,3}}{R_1 + R_{2,3}} \tag{19-2}$$

$$= \frac{3.9 \times 2.5}{3.9 + 2.5}$$

$$= 1.52 \text{ k}\Omega.$$

Total source current,

$$I_T = \frac{E_T}{R_T} \tag{19-3}$$

$$= \frac{9 \text{ V}}{1.52 \text{ k}\Omega}$$

$$= 5.91 \text{ mA}.$$

Branch current,

$$I_1 = \frac{E_T}{R_1} \tag{19-4}$$

$$= \frac{9 \text{ V}}{3.9 \text{ k}\Omega}$$

$$= 2.31 \text{ mA}.$$

Branch current,

$$I_2 = I_T - I_1 \tag{19-5}$$
$$= 5.91 - 2.31$$
$$= 3.6 \text{ mA}.$$

Voltage drop,

$$V_1 = I_2 R_2 \tag{19-6}$$
$$= 3.6 \text{ mA} \times 1 \text{ k}\Omega$$
$$= 3.6 \text{ V}.$$

Voltage drop,

$$V_2 = I_2 R_3 \tag{19-6}$$
$$= 3.6 \text{ mA} \times 1.5 \text{ k}\Omega$$
$$= 5.4 \text{ V}.$$

Voltage check:
Source voltage,

$$E_T = V_1 + V_2 = 3.6 \text{ V} + 5.4 \text{ V} = 9 \text{ V} \tag{19-7}$$

Example 19-2

In Fig. 19-1B, $R_1 = 1.5$ kΩ, $R_2 = 3.3$ kΩ. $R_3 = 5.6$ kΩ and $E_T = 7$ V. Calculate the values of R_T, I_T, I_1, I_2, V, V_2.

Solution

Equivalent resistance of $R2$ and $R3$ in parallel,

$$R_{2,3} = \frac{R_2 \times R_3}{R_2 + R_3}$$

$$= \frac{3.3 \times 5.6}{3.3 + 5.6}$$

$$= 2.076 \text{ k}\Omega.$$

Total equivalent resistance,

$$R_T = R_1 + R_{2,3} \qquad (19\text{-}9)$$
$$= 1.5 + 2.076$$
$$= 3.576 \text{ k}\Omega.$$

Total source current,

$$I_T = \frac{E_T}{R_T} \qquad (19\text{-}10)$$
$$= \frac{7 \text{ V}}{3.576 \text{ k}\Omega}$$
$$= 1.96 \text{ mA}.$$

Voltage drop,

$$V_1 = I_T \times R_1 \qquad (19\text{-}13)$$
$$= 1.96 \text{ mA} \times 1.5 \text{ k}\Omega$$
$$= 2.94 \text{ V}.$$

Voltage drop,

$$V_2 = E - V_1 \qquad (19\text{-}15)$$
$$= 7 - 2.94$$
$$= 4.06 \text{ V}.$$

Branch current,

$$I_1 = \frac{V_2}{R_2} \qquad (19\text{-}11)$$
$$= \frac{4.06 \text{ V}}{3.3 \text{ k}\Omega}$$
$$= 1.23 \text{ mA}.$$

Branch current,

$$I_2 = \frac{V_2}{R_3} \qquad (19\text{-}12)$$
$$= \frac{4.06 \text{ V}}{5.6 \text{ k}\Omega}$$
$$= 0.73 \text{ mA}.$$

Current check: Total current, $I_T = I_1 + I_2 = 1.23 + 0.73 = 1.96$ mA $\qquad$ (19-13)

Example 19-3

In Fig. 19-3A, calculate the value of the total equivalent resistance presented to the source.

Solution

The circuit is redrawn in Fig. 19-3B.

Total resistance of $R3$ and $R4$ in series,

$$R_{3,4} = 1 + 2.2$$
$$= 3.2 \text{ k}\Omega.$$

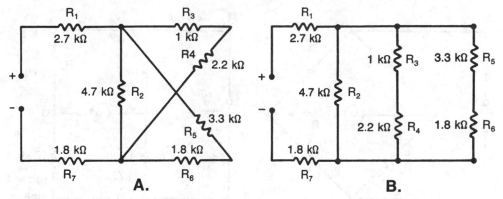

Figure 19-3

Total resistance of $R5$ and $R6$ in series,

$$R_{5,6} = 3.3 + 1.8$$
$$= 5.1 \text{ k}\Omega.$$

Total resistance of $R2$, $R3$, 4, $R5,6$ in parallel,

$$R_{2,3,4,5,6} = \cfrac{1}{\cfrac{1}{4.7} + \cfrac{1}{3.2} + \cfrac{1}{5.1}}$$
$$= 1.386 \text{ k}\Omega.$$

Total equivalent resistance,

$$R_T = R_1 + R_{2,3,4,5,6} + R_7$$
$$= 2.7 + 1.386 + 1.8$$
$$= 5.886 \text{ k}\Omega.$$

Chapter 20
The Wheatstone bridge circuit

THE WHEATSTONE BRIDGE CIRCUIT IS USED TO OBTAIN AN ACCURATE measurement of an unknown resistance, R_x. Conventionally the bridge circuit consists of four resistor arms $R1$, $R2$, $R3$, and R_x (Fig. 20-1A) with a center link ("bridge") that contains a sensitive current indicating device or galvanometer, G. However, the circuit may be redrawn as in Fig. 20-1B and we then observe that the bridge circuit is really an example of a series-parallel resistor arrangement.

The whole of the source voltage must be dropped across $R1$, $R2$, and also, $R3$, R_x. However, unless there is a special relationship between the four resistance values, the current through the resistor $R1$ will not be equal to the current through the resistor $R2$ and the same applies to the currents through the resistors $R3$, R_x.

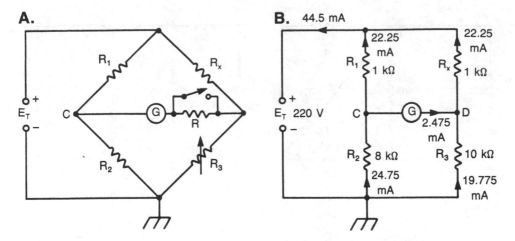

Figure 20-1

As an example, consider the circuit of Fig. 20-1B. If the low resistance of the sensitive galvanometer is neglected, the resistor $R1$ is in parallel with the resistor R_x while the resistor $R3$ is in parallel with the resistor $R2$. The two parallel combinations are then in series. The total equivalent resistance presented to the source is $1 \text{ k}\Omega \parallel 1 \text{ k}\Omega + 8 \text{ k}\Omega \parallel 10 \text{ k}\Omega = 500 \ \Omega +$ $4444 \ \Omega = 4944 \ \Omega$. The total current, $I_T = 220 \text{ V}/4944 \ \Omega = 44.5 \text{ mA}$, which divides equally between the two 1 kΩ resistors. The voltage drop across each of these resistors is 44.5/2 mA $\times$ 1 kΩ = 22.25 V. The voltage drop across each of the 8 kΩ, 10 kΩ resistors is 220 V $-$ 22.25 V = 197.75 V. Consequently, the current in the 8 kΩ resistor is 197.75 V/8 kΩ = 24.72 mA, while the 10 kΩ resistor carries a current of 197.75 V/10 kΩ = 19.775 mA. The current in the link is then 22.25 $-$ 19.775 (or 24.72 $-$ 22.25) = 2.475 mA, with the electron flow in the direction from left to right (Fig. 20-1B). However, if the 8 kΩ resistor was replaced by a 10 kΩ resistor, the currents in the four resistors would be all the same and the current in the link (as recorded by the galvanometer) would be zero. Under such conditions the bridge is said to be *balanced* although the situation in which $R_1 = R_x$ and $R_2 = R_3$ is a special case and is not the general condition for a balance to occur.

There will be no voltage difference between the points C and D if the ratios $R_1:R_2$ and $R_x:R_3$ are equal. The reading of the galvanometer is zero and the bridge is balanced. The value of the unknown resistor, R_x is then equal to $R_3 \times R_1/R_2$ (see mathematical derivations). Notice that in this condition the products, $R_2 R_x$ and R_1, of the opposing resistors are equal in the conventional bridge circuit of Fig. 20-1A.

To measure the value of R_x, $R3$ is an accurately calibrated variable resistor, which is used to balance the bridge. A protection resistor, R, is commonly included to prevent the galvanometer from being subjected to excessive current. With R included, a rough balance is first obtained. Afterwards the resistor is shorted out by the switch and the value of $R3$ is finally adjusted for an accurate balance.

Normally the high-quality fixed resistors $R1$ and $R2$ have possible values of 1 Ω, 10 Ω, 100 Ω or 1000 Ω. By using switches to change the values of the resistors, the ratio $R_1:R_2$ can be set for 1000, 100, 10, 1, 0.1, 0.01 and 0.001. The measured reading of R_x can therefore range from 1/1000 of the lowest value of $R3$ to 1000 times the highest value of $R3$.

MATHEMATICAL DERIVATIONS

For a balanced bridge,

$$\frac{R_x}{R_3} = \frac{R_1}{R_2} \tag{20-1}$$

or

$$R_x R_2 = R_1 R_3 \tag{20-2}$$

or

$$R_x = R_3 \times \frac{R_1}{R_2} \text{ ohms} \tag{20-3}$$

Alternatively, using the VDR:

Potential at C, $V_C = E \times \dfrac{R_2}{R_1 + R_2}$

Potential at D, $V_D = E \times \dfrac{R_3}{R_3 + R_x}$

If the bridge is balanced, $V_C = V_D$. Therefore,

$$E \times \frac{R_2}{R_1 + R_2} = E \times \frac{R_3}{R_3 + R_x} \tag{20-4}$$

$$R_2 R_3 + R_2 R_x = R_1 R_3 + R_2 R_3$$

$$R_2 R_x = R_1 R_3$$

$$R_x = R_3 \times \frac{R_1}{R_2} \text{ ohms}$$

Example 20-1

In Fig. 20-1A, $R1 = 10 \ \Omega$, $R2 = 1000 \ \Omega$. When the bridge is balanced, the value of $R3$ is adjusted to 5.7 Ω. What is the value of the unknown resistor, R_x?

Solution

For a balanced bridge,

$$R_x = R_3 \times \frac{R_1}{R_2} \tag{20-3}$$

$$= 5.7 \ \Omega \times \frac{10 \ \Omega}{1000 \ \Omega}$$

$$= 0.057 \ \Omega.$$

Chapter 21
The loaded voltage divider circuit

IN CHAPTER 11 WE DESCRIBED THE UNLOADED VOLTAGE DIVIDER CIRCUIT, WHICH was a series string of resistors, capable of producing a number of different voltages from a single voltage source. In such a circuit there was only the single bleeder current that flowed through the series resistors. The values of those resistors were then calculated to provide the necessary voltages. However, as soon as we try to make use of the divider circuit by connecting loads across the various output points X, Y, Z, the arrangement becomes an application of a series-parallel network (Fig. 21-1). Knowing the values of the load currents, it is necessary to recalculate the required values of $R1$, $R2$, $R3$, in order to provide the correct output voltages, V_{L1}, V_{L2} and V_{L3}.

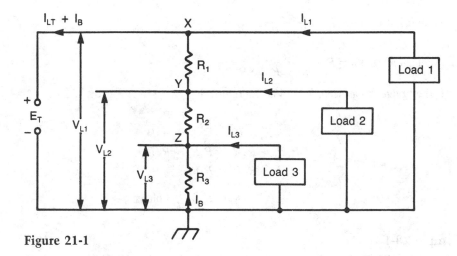

Figure 21-1

MATHEMATICAL DERIVATIONS

The bleeder current, I_B, is normally about 10% of the total load current, I_{LT}, which is the sum of I_{L1}, I_{L2} and I_{L3}. Therefore

$$I_{LT} = I_{L1} + I_{L2} + I_{L3} \text{ amperes} \tag{21-1}$$

The voltage drop across $R3$ must equal the load voltage, V_{L3}. It follows that:

$$V_{L3} = I_B R_3, \; R_3 = \frac{V_{L3}}{I_B} \tag{21-2}$$

The sum of the currents I_{L3} and I_B must flow through $R2$. When V_{L3} is added to the voltage drop across $R2$, the result must equal the value of V_{L2}. In equation form:

$$V_{L2} = V_{L3} + (I_{L3} + I_B) R_2 \text{ volts} \tag{21-3}$$

This yields

$$R_2 = \frac{V_{L2} - V_{L3}}{I_{L3} + I_B} \text{ ohms} \tag{21-4}$$

The sum of the currents I_{L2}, I_{L3} and I_B flows through $R1$. The sum of this resistor's voltage drop and V_{L2} is equal to V_{L1}, which is the same as the source voltage, E. Then

$$V_{L1} = E = R_1(I_{L2} + I_{L3} + I_B) + V_{L2} \text{ volts} \qquad (21\text{-}5)$$

so that

$$R_1 = \frac{V_{L1} - V_{L2}}{I_{L2} + I_{L3} + I_B} \text{ ohms} \qquad (21\text{-}6)$$

Example 21-1

In Fig. 21-1, $E = 60$ V, $V_{L2} = 35$ V, $V_{L3} = 10$ V, $I_{L1} = 170$ mA, $I_{L2} = 70$ mA, $I_{L3} = 40$ mA. Suggest suitable values for $R1$, $R2$, and $R3$ (assume that the bleeder current is 10% of the total load current).

Solution

Total load current,

$$I_{LT} = I_{L1} + I_{L2} + I_{L3} \qquad (21\text{-}1)$$
$$= 170 + 70 + 40$$
$$= 280 \text{ mA}.$$

Bleeder current, $I_B = 280 \times 10/100 = 28$ mA.
The required voltage divider resistances are:

$$R_3 = \frac{V_{L3}}{I_B} \qquad (21\text{-}2)$$

$$= \frac{10 \text{ V}}{28 \text{ mA}}$$

$$= 357 \ \Omega.$$

$$R_2 = \frac{V_{L2} - V_{L3}}{I_{L3} + I_B} \qquad (21\text{-}4)$$

$$= \frac{35 \text{ V} - 10 \text{ V}}{40 \text{ mA} + 28 \text{ mA}} = \frac{25 \text{ V}}{68 \text{ mA}}$$

$$= 368 \ \Omega$$

$$R_1 = \frac{V_{L1} - V_{L2}}{I_{L2} + I_{L3} + I_B} \qquad (21\text{-}6)$$

$$= \frac{60 \text{ V} - 35 \text{ V}}{70 \text{ mA} + 40 \text{ mA} + 28 \text{ mA}}$$

$$= \frac{25 \text{ V}}{138 \text{ mA}}$$

$$= 181 \ \Omega$$

The power dissipations of these resistors are:

$$P_{R1} = 25 \text{ V} \times 138 \text{ mA} = 3450 \text{ mW}.$$

$$P_{R2} = 25 \text{ V} \times 68 \text{ mA} = 1700 \text{ mW}.$$
$$P_{R3} = 10 \text{ V} \times 28 \text{ mA} = 280 \text{ mW}.$$

Suitable resistors would be $R1:180 \ \Omega$, 5 W, $R2:390 \ \Omega$, 2 W, and $R3:330 \ \Omega$, 1/2 W.

Chapter 22
Internal resistance—voltage regulation

THE SOURCES WE DISCUSSED IN CHAPTERS 12 AND 17, WERE IN A CERTAIN SENSE, idealized because we assumed that the source voltage remained constant and independent of the load current. This is not true in practice as we know from our experience with a car battery. When a large load current is drawn from such a battery, the terminal voltage falls and with a "bad" battery, the terminal voltage is so low that the car will not start. The explanation for this effect lies within the voltage source itself and is due to its internal resistance.

All sources possess internal resistance to some extent. For example, a primary (nonrechargeable) 1½ V "D" cell has an internal resistance on the order of 1 ohm. This value of resistance depends on the size of the electrodes, their separation, and the nature of the electrolyte. By contrast the basic lead acid secondary (rechargeable) cell has a negative electrode of sponge lead, a positive electrode of lead peroxide and an electrolyte of sulfuric acid. The large size of the electrodes results in a very low internal resistance which is typically less than 10 mΩ. This is the major difference between a 12 V car battery and a 12 V cell for operating a transistor radio. The cell for the transistor radio has a much higher internal resistance and is therefore quite incapable of starting a car (whose initial current requirement is about 100 A or more!).

When a load current is drawn from a voltage source, the terminal voltage, V_L, falls as the load current, I_L, is increased. A model which explains this effect, is illustrated in Fig. 22-1. This model assumes that the source contains a constant voltage (EMF or E) which is in series with the

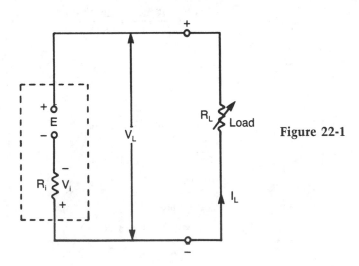

Figure 22-1

internal resistance, R_i. Then if the load resistance, R_L, is decreased, the load current, I_L, will rise. This will cause an increase in the voltage drop, V_i, across the internal resistance, R_i, and therefore the terminal voltage, V_L, will drop.

Having established the model, we must have some means of finding the values of E and R_i for any electrical source. The value of E is best measured by disconnecting the load, R_L, and then placing a voltmeter across the terminals. Assuming that the voltmeter draws virtually no current from the source, the voltage drop across the internal resistance is negligible. The voltmeter reading will then be equal to the value of E, which is often referred to as the open-circuit or no-load terminal voltage.

Having determined the value of E, we then connect an ammeter of negligible resistance between the terminals so that only the internal resistance will now limit the current. The value of this current is therefore E/R_i amperes and because there is practically zero resistance between the terminals, this level is normally referred to as the *short-circuit terminal* current. The internal resistance may then be calculated from the ratio of the open-circuit terminal voltage to the short circuit terminal current.

It may not be practical to measure the short-circuit current because such a large current may possibly damage the electrical source. A superior method of finding the internal resistance is to vary the value of the load resistance until the terminal load voltage, V_L, is equal to $E/2$. V_i will also be $E/2$ and the measured value of R_L will then equal the internal resistance, R_i.

The degree to which the load voltage, V_L, varies with changes in the load current is measured by the percentage voltage regulation. Ideal regulation means that the load voltage remains constant and independent of any changes in the load current.

MATHEMATICAL DERIVATIONS

In Fig. 22-1,

$$\text{Open-circuit or no-load terminal voltage} = E \text{ volts} \tag{22-1}$$

Load current,

$$I_L = \frac{V_L}{R_L} = \frac{V_i}{R_i} = \frac{E}{R_i + R_L} \text{ amperes} \tag{22-2}$$

Terminal load voltage,

$$V_L = I_L R_L \tag{22-3}$$
$$= E - I_L R_i$$
$$= E - \frac{E R_i}{R_i + R_L} \text{ volts}$$

This yields
Load voltage,

$$V_L = E \times \frac{R_L}{R_i + R_L} \text{ volts} \tag{22-4}$$

Typical graphs of V_L vs. R_L and I_L vs. R_L are illustrated in Fig. 22-2.

Equation 22-4 is an example of the voltage division rule (VDR) because the source voltage, E, is divided between the load resistance, R_L, and the internal resistance, R_i. To obtain a high voltage across the load, R_L must typically be at least 5 to 10 times the value of R_i.

$$\text{Short-circuit terminal current} = \frac{E}{R_i} \text{ amperes} \tag{22-5}$$

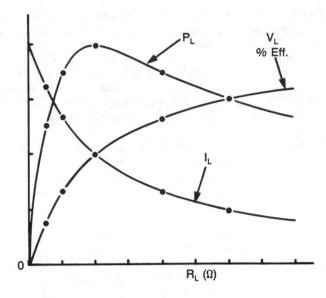

Figure 22-2

Internal resistance,

$$R_i = \frac{\text{open-circuit terminal voltage, } E}{\text{short-circuit terminal current, } E/R_i} \quad \text{ohms} \qquad (22\text{-}6)$$

$$\text{Percentage voltage regulation} = \frac{V_{NL} - V_{FL}}{V_{FL}} \times 100\% \qquad (22\text{-}7)$$

where: V_{NL} = terminal voltage under no-load or minimum load conditions.

V_{FL} = terminal voltage under full-load (current) conditions when the load resistance is R_{FL}.

No-load voltage,

$$V_{NL} = V_{FL}\left(1 + \frac{\% \text{ Regulation}}{100}\right) \quad \text{volts} \qquad (22\text{-}8)$$

Full-load voltage,

$$V_{FL} = \frac{V_{NL}}{\left(1 + \dfrac{\% \text{ Regulation}}{100}\right)} \quad \text{volts} \qquad (22\text{-}9)$$

For an ideal source, $V_{FL} = E$ and the percentage regulation is zero.
If the no-load voltage is E volts,

$$\text{Percentage regulation} = \frac{R_i}{R_{FL}} \times 100\% \qquad (22\text{-}10)$$

Example 22-1

An electrical source has an open-circuit terminal voltage of 48 V and a short circuit current of 32 A. If the full-load resistance is 12 Ω, calculate the values of the full-load voltage and the percentage of regulation.

Solution

Internal resistance,

$$R_i = \frac{\text{open-circuit voltage, } E}{\text{short-circuit current, } E/R_i} \qquad (22\text{-}6)$$

$$= \frac{48 \text{ V}}{32 \text{ A}}$$

$$= 1.5 \ \Omega.$$

Full-load voltage,

$$V_{FL} = \frac{E \times R_{FL}}{R_{FL} + R_i} \qquad (22\text{-}4)$$

$$= \frac{48 \text{ V} \times 12 \ \Omega}{12 \ \Omega + 1.5 \ \Omega}$$

$$= 42.7 \text{ V}$$

$$\text{Percentage regulation} = \frac{V_{NL} - V_{FL}}{V_{FL}} \times 100\% \qquad (22\text{-}7)$$

$$= \frac{48 - 42.7}{42.7} \times 100$$

$$= 12.5\%$$

Percentage regulation check:

$$\text{Percentage regulation} = \frac{R_i}{R_{FL}} \times 100\% \qquad (22\text{-}10)$$

$$= \frac{1.5}{12} \times 100$$

$$= 12.5\%$$

Example 22-2

An electrical source has a no-load voltage of 250 V and a percentage regulation of 3.5%. What is the value of the full-load voltage?

Solution

Full-load voltage,

$$V_{FL} = \frac{V_{NL}}{\left(1 + \dfrac{\% \text{ Regulation}}{100}\right)} \qquad (22\text{-}9)$$

$$= \frac{250}{\left(1 + \dfrac{3.5}{100}\right)}$$

$$= \frac{250}{1.035}$$

$$= 241.5 \text{ V}$$

$$\backsim\!\!\!\sim\!\!\!\sim$$

Chapter 23
Maximum power transfer—
percentage efficiency

IN CHAPTER 22 WE EXPLORED THE CONDITION FOR A HIGH VOLTAGE ACROSS THE load. We decided that the load resistance, R_L, should be many times greater than the internal resistance, R_i, so that when the constant EMF, E, divided between these two resistances, very little voltage was dropped across R_i and most appeared across the load.

Because there is both a load voltage, V_L, and a load current, I_L, it follows that there is a power, P_L, developed in the load (Fig. 23-1A). However, the condition for maximum power in the load cannot be the same as the condition for a high load voltage because for open- and short-circuit loads, the load power in both cases is zero. Consequently, the load power must occur between these two extremes and this is well illustrated in the circuit of Fig. 23-1B. Here we choose particular load resistances, R_L, and calculate the corresponding values of I_L, V_L, and P_L. The results are shown in Table 23-1.

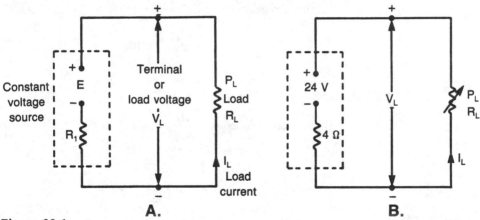

Figure 23-1

As an example let us adjust the value of R_L to be 8 Ω. The total resistance is $R_i + R_L = 4 \, \Omega + 8 \, \Omega = 12 \, \Omega$. The load current is $E/(R_i + R_L) = 24 \text{ V}/12 \, \Omega = 2$ A and the load voltage is $V_L = I_L \times R_L = 2 \text{ A} \times 8 \, \Omega = 16$ V. The corresponding load power is $P_L = I_L \times V_L = 2 \text{ A} \times 16 \text{ V} = 32$ W. The graphs of V_L, P_L, percentage efficiency versus R_L and I_L are illustrated in Fig. 23-2A and B.

From the graph of P_L versus R_L it is clear that the load power reaches its maximum value of 36 W when the load resistance is 4 Ω. It is no numerical coincidence that this value of load resistance is the same as the internal resistance of the source.

In the mathematical derivations it will be proved that there is maximum power transfer to the load when the load resistance is *matched (made equal)* to the internal resistance. This result is of more than academic significance. In the case of a radio transmitter the antenna represents a load whose ohmic value must be matched to the load required by the final stage of the transmitter; only then can rf (radio frequency) power be effectively transferred from the transmitter to the antenna. In another example the load of a loudspeaker must be matched to the load required by its receiver's output audio stage.

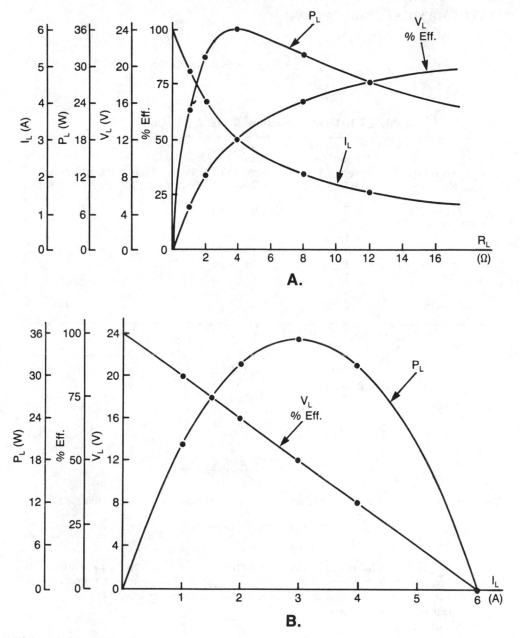

Figure 23-2

The percentage efficiency is determined from the ratio of the *load* power to the *total* power drawn from the constant voltage EMF. When there is maximum power transfer to the load, the percentage efficiency is only 50% and it may therefore be necessary to compromise between the values of load power and percentage efficiency.

MATHEMATICAL DERIVATIONS

Load current,

$$I_L = \frac{V_L}{R_L} = \frac{E}{R_i + R_L} \text{ amperes} \tag{23-1}$$

Table 23-1. Load Resistance R_L vs. I_L, V_L, and P_L.

Load Resistance (Ω) R_L	Load Current (A) $I_L = E/(R_i + R_L)$	Load Voltage (V) $V_L = I_L \times R_L$	Load Power (W) $P_L = I_L \times V_L$	% Efficiency $P_L/P_T \times 100\%$
Zero (short-circuit)	6	0	0	0
2	4	8	32	33⅓
4	3	12	36	50
8	2	16	32	66⅔
20	1	20	20	83⅓
Infinite (open-circuit)	0	24	0	100

Load voltage,

$$V_L = I_L \times R_L = \frac{ER_L}{R_i + R_L} \text{ volts} \tag{23-2}$$

Load power,

$$P_L = I_L \times V_L = \frac{E}{R_i + R_L} \times \frac{ER_L}{R_i + R_L} \tag{23-3}$$

$$= \frac{E^2 R_L}{(R_i + R_L)^2} \text{ watts}$$

The load power reaches its maximum value when its reciprocal, $1/P_L$, is at its minimum level.

$$\frac{1}{P_L} = \frac{(R_i + R_L)^2}{E^2 R_L} = \frac{1}{E^2}\left[\frac{(R_i - R_L)^2}{R_L} + 4R_i\right] \tag{23-4}$$

Because the lowest value of a square term such as $(R_i - R_L)^2$ is zero, $1/P_L$ will reach its minimum level when $(R_i - R_L)^2 = 0$ or $R_L = R_i$. This proves that there is maximum power transfer to the load when the load and the internal resistances are matched.

Alternatively:

Load power,

$$P_L = \frac{E^2 R_L}{(R_i + R_L)^2}$$

Differentiating P_L with respect to R_L,

$$\frac{dP_L}{dR_L} = E^2 \times \left[\frac{(R_i + R_L)^2 - 2R_L(R_i + R_L)}{(R_i + R_L)^4}\right]$$

$$= E^2 \times \left[\frac{(R_i + R_L) - 2R_L}{(R_i + R_L)^3} \right]$$

$$= E^2 \left[\frac{R_i - R_L}{(R_i + R_L)^3} \right]$$

For the maximum value of P_L, $dP_L/dR_L = 0$.

Therefore

$$R_i = R_L \text{ ohms} \tag{23-5}$$

Maximum power developed in the load,

$$P_{Lmax} = \frac{E^2 R_i}{(R_i + R_L)^2} \tag{23-6}$$

$$= \frac{E^2}{4R_i}$$

$$= \frac{E^2}{4R_L} \text{ watts}$$

$$\text{Percentage Efficiency} = \frac{P_L}{P_T} \times 100 \tag{23-7}$$

$$= \frac{I_L^2 R_L}{I_L^2 (R_i + R_L)} \times 100$$

$$= \frac{R_L}{R_i + R_L} \times 100\%$$

Because $V_L = E \times R_L/(R_i + R_L)$, the graphs of percentage efficiency and V_L versus R_L will be similar in appearance (Fig. 23-2A).

Example 23-1

In Fig. 23-3 what is the value of R_L that will permit maximum power transfer to the load? Calculate the value of the maximum load power involved. What would be the new values of the load power if the chosen value of R_L were (a) halved and (b) doubled? In each case calculate the value of the percentage efficiency.

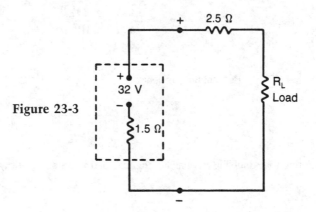

Figure 23-3

Solution

For the maximum power transfer to the load, the value of R_L must be matched to the total of all the resistances that are not associated with the load. Therefore, the required value for R_L is $1.5 + 2.5 = 4 \ \Omega$. Then:

$$\text{Maximum load power} = \frac{E^2}{4R_L} \tag{23-6}$$

$$= \frac{(32 \text{ V})^2}{4 \times 4 \ \Omega}$$

$$= 64 \text{ W}$$

$$\text{Percentage efficiency} = \frac{R_L}{R_i + R_L} \times 100 \tag{23-7}$$

$$= \frac{R_L}{2R_L} \times 100$$

$$= 50\%$$

(a) When R_L is doubled to $2 \times 4 \ \Omega = 8 \ \Omega$,

$$\text{Load power} = \frac{E^2 R_L}{(R_i + R_L)^2} \tag{23-3}$$

$$= \frac{(32 \text{ V})^2 \times 8 \ \Omega}{(4 \ \Omega + 8 \ \Omega)^2}$$

$$= 56.9 \text{ W}$$

$$\text{Percentage efficiency} = \frac{R_L}{R_i + R_L} \times 100 \tag{23-7}$$

$$= \frac{8 \ \Omega \times 100}{4 \ \Omega + 8 \ \Omega}$$

$$= 66.7\%$$

(b) When R_L is halved to $4 \ \Omega/2 = 2 \ \Omega$,

$$\text{Load power} = \frac{E^2 R_L}{(R_i + R_L)^2} \tag{23-3}$$

$$= \frac{(32 \text{ V})^2 \times 2 \ \Omega}{(4 \ \Omega + 2 \ \Omega)^2}$$

$$= 56.9 \text{ W}$$

$$\text{Percentage efficiency} = \frac{R_L}{R_i + R_L} \times 100 \tag{23-7}$$

$$= \frac{2 \ \Omega \times 100}{4 \ \Omega + 2 \ \Omega}$$

$$= 33.3\%$$

It is worth noting that doubling and halving the matched value of R_L always provides equal load powers.

Chapter 24
The constant current source

IN CHAPTER 22 WE PROPOSED A *CONSTANT VOLTAGE* MODEL TO REPRESENT AN electrical source. This consisted of the constant EMF, *E*, in series with the internal resistance, R_i (Fig. 24-1A). But is this the only model we can devise? As far as the load is concerned, it is also possible to have a *constant current* source that is capable of generating a variable EMF. The value of the constant current is the short-circuit terminal value previously discussed in Chapter 22. Across this current generator is a parallel internal resistance that has the same value as the series internal resistance of the constant voltage model (Fig. 24-1B).

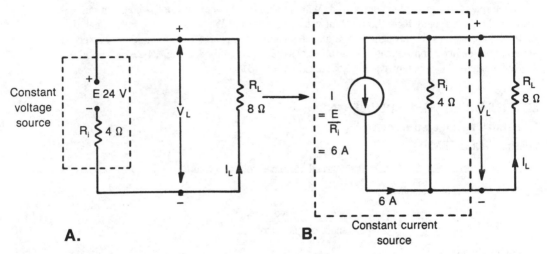

A.   **B.**

Figure 24-1

Let us compare the two models by an example in which an electrical source has an open-circuit voltage of 24 V and a short circuit current of 6 A. Then the internal resistance is 24 V/ 6 A = 4 Ω. The two comparable models are illustrated in Fig. 24-1A and B and the arrow convention in the constant current source indicates the same direction of electron flow as produced by the constant voltage source.

If we now connect an 8-ohm load across the terminals of each source, the load current for the constant voltage generator is:

$$I_L = \frac{24 \text{ V}}{4 \text{ } \Omega + 8 \text{ } \Omega} = 2 \text{ A.}$$

For the constant current model the load current by the CDR rule is:

$$I_L = 6 \text{ A} \times \frac{4 \text{ } \Omega}{4 \text{ } \Omega + 8 \text{ } \Omega} = 2 \text{ A.}$$

Consequently, as far as the load is concerned, both models are equally valid. However, the generators themselves are not exact equivalents because under open-circuit load conditions, there is power dissipated in the internal resistance of the current generator but not in the internal resistance of the voltage generator.

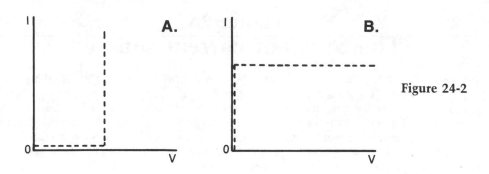

A.

B.

Figure 24-2

One question has not been answered. Why do we need the concept of the constant current source when we already have the constant voltage generator? We shall see that in the analytical methods of Chapters 26 through 32, one model is sometimes preferable to the other. Moreover some active devices approximate more to the constant voltage model (Fig. 24-2A) while others tend to behave like constant current sources (Fig. 24-2B).

MATHEMATICAL DERIVATIONS
Constant voltage source

Constant voltage EMF,

$$E = \text{open-circuit terminal voltage (volts)} \tag{24-1}$$

Series internal resistance,

$$R_i = \frac{\text{open-circuit voltage, } E}{\text{short-circuit current, } I} \text{ ohms} \tag{24-2}$$

Load current,

$$I_L = \frac{E}{R_i + R_L} \text{ amperes} \tag{24-3}$$

Load voltage,

$$V_L = I_L R_L = \frac{E R_L}{R_i + R_L} \text{ volts} \tag{24-4}$$

Constant current source

Constant current,

$$I_L = \text{short-circuit terminal current (amperes)} \tag{24-5}$$

Parallel internal resistance,

$$R_i = \frac{\text{open-circuit voltage, } E}{\text{short-circuit current, } I} \text{ ohms} \tag{24-6}$$

Load current,

$$I_L = I \times \frac{R_i}{R_i + R_L} = \frac{E}{R_i + R_L} \text{ amperes} \tag{24-7}$$

Load voltage,

$$V_L = I_L R_L = I \times \frac{R_i R_L}{R_i + R_L} = \frac{E R_L}{R_i + R_L} \text{ volts} \qquad (24\text{-}8)$$

Example 24-1

An electrical source has an open circuit voltage of 32 V and a short circuit current of 16 A. Draw the schematic of its constant source model. If a load of 2.5 Ω is connected across the terminals of the source, calculate the values of the load current and the load voltage. See Fig. 24-3.

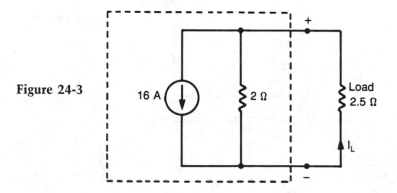

Figure 24-3

Constant current source for Example 24-1.

Solution

Internal resistance,

$$R_i = \frac{E}{I} \qquad (24\text{-}6)$$

$$= \frac{32 \text{ V}}{16 \text{ A}}$$

$$= 2 \ \Omega$$

Load current,

$$I_L = I \times \frac{R_i}{R_i + R_L} \qquad (24\text{-}7)$$

$$= 16 \text{ A} \times \frac{2 \ \Omega}{2 \ \Omega + 2.4 \ \Omega}$$

$$= 7.11 \text{ A}$$

Load voltage,

$$V_L = I_L R_L \qquad (24\text{-}8)$$

$$= 7.11 \text{ A} \times 2.5 \ \Omega$$

$$= 17.8 \text{ V}$$

Example 24-2

An electrical source has an open circuit voltage of 15 V and a short circuit current of 25 A. Draw the schematic of its constant voltage model. If a load of 4 Ω is connected across the terminals of the source, calculate the values of the load current and the load voltage. See Fig. 24-4.

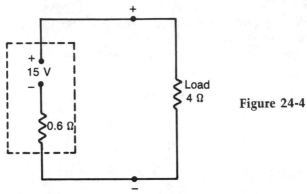

Figure 24-4

Constant voltage source for Example 24-2.

Solution

Internal resistance,

$$R_i = \frac{E}{I} = \frac{15 \text{ V}}{25 \text{ A}} = 0.6 \text{ Ω} \tag{24-2}$$

Load current,

$$I_L = \frac{E}{R_i + R_L} \tag{24-3}$$

$$= \frac{15 \text{ V}}{0.6 \text{ Ω} + 4 \text{ Ω}} = \frac{15 \text{ V}}{4.6 \text{ Ω}}$$

$$= 3.26 \text{ A}$$

Load voltage,

$$V_L = I_L R_L$$
$$= 3.26 \text{ A} \times 4 \text{ Ω}$$
$$= 13.0 \text{ V}$$

Chapter 25
Practical sources in series, parallel, and series-parallel

WHEN THE CELLS OF CHAPTER 12 WERE CONNECTED IN SERIES-AIDING, THE TOTAL EMF available was the sum of the individual EMFs. However, these cells were idealized in the sense that we ignored their internal resistances.

PRACTICAL SOURCES IN SERIES

Figure 25-1A illustrates N practical sources connected in series-aiding. The total no-load terminal voltage will still be the sum of the individual EMFs but the internal resistances are also in series and must be added to obtain the equivalent internal resistance. As a result a series-aiding combination of sources will increase the total EMF available, but this advantage is only achieved at the expense of the higher total internal resistance.

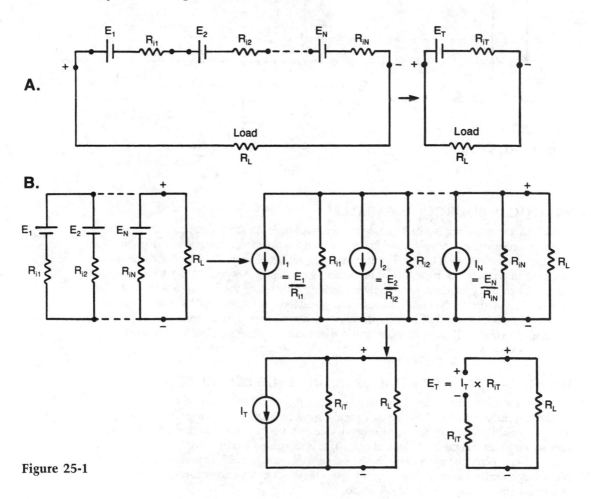

Figure 25-1

C.

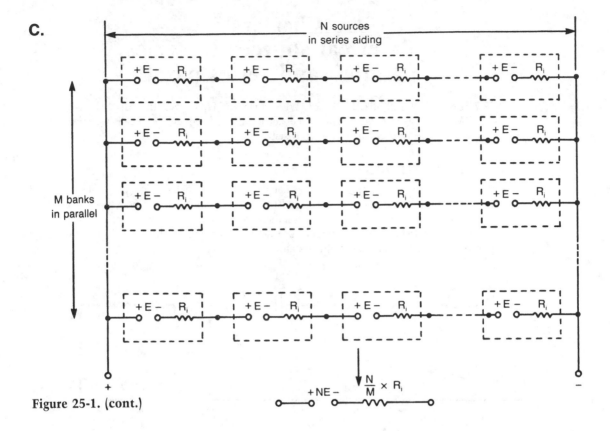

Figure 25-1. (cont.)

PRACTICAL SOURCES IN PARALLEL

If practical parallel sources are expressed in terms of their constant voltage equivalents, the resulting circuit cannot be analyzed without the need for simultaneous equations (Fig. 25-1B). However, if each source is replaced by its constant current generator, the individual currents can be added to produce the total current of the final equivalent generator. At the same time the individual internal resistances are directly in parallel and therefore the reciprocal formula can be used to obtain the relatively low total equivalent resistance. It follows that the main purpose of connecting cells in parallel is to reduce the effective internal resistance as opposed to increasing the available voltage. If necessary, the final constant current generator can be converted to its equivalent voltage model.

IDENTICAL (PRACTICAL) SOURCES IN SERIES-PARALLEL

We have learned that a series connection increases the available voltage at the expense of a greater internal resistance. By contrast a parallel connection reduces the internal resistance but does not raise the voltage. It is obviously possible to have the best of both worlds by using a series-parallel arrangement of identical cells. This is illustrated in Fig. 25-1C where N cells in series are connected to form a single "bank" and raise the voltage while there are M such banks in parallel to lower the internal resistance. However, these advantages are achieved at the expense of a large, cumbersome and costly dc source.

MATHEMATICAL DERIVATIONS
Cells in series (Fig. 25-1A)
Total EMF,

$$E_T = E_1 + E_2 + E_3 + \ldots + E_N \text{ volts} \qquad (25\text{-}1)$$

Total internal resistance,

$$R_{iT} = R_{i1} + R_{i2} + R_{i3} + \ldots + R_{iN} \text{ ohms} \qquad (25\text{-}2)$$

If the sources are identical,
Total EMF,

$$E_T = NE \text{ volts} \qquad (25\text{-}3)$$

Total internal resistance,

$$R_{iT} = NR_i \text{ ohms} \qquad (25\text{-}4)$$

Cells in parallel (Fig. 25-1B)
Equivalent constant currents,

$$I_1 = \frac{E_1}{R_{i1}}, \; I_2 = \frac{E_2}{R_{i2}}, \qquad (25\text{-}5)$$

$$I_3 = \frac{E_3}{R_{i3}} \cdots I_N = \frac{E_N}{R_{iN}} \text{ amperes}$$

Total equivalent constant current,

$$I_T = I_1 + I_2 + I_3 \ldots + I_N \text{ amperes} \qquad (25\text{-}6)$$

Total equivalent internal resistance,

$$R_{iT} = \cfrac{1}{\cfrac{1}{R_{i1}} + \cfrac{1}{R_{i2}} + \cfrac{1}{R_{i3}} + \ldots + \cfrac{1}{R_{iN}}} \text{ ohms} \qquad (25\text{-}7)$$

Total equivalent EMF,

$$E_T = I_T \times R_{iT} \text{ volts} \qquad (25\text{-}8)$$

If the sources are identical,
Total equivalent constant current,

$$I_T = NI \text{ amperes} \qquad (25\text{-}9)$$

Total equivalent internal resistance,

$$R_{iT} = \frac{R_i}{N} \text{ ohms} \qquad (25\text{-}10)$$

$$\text{Total equivalent EMF} = E \text{ volts} \qquad (25\text{-}11)$$

Cells in series-parallel (Fig. 25-1C)

$$\text{Total EMF of one bank} = NE \text{ volts} \qquad (25\text{-}12)$$

$$\text{Total internal resistance of one bank} = NR_i \text{ ohms} \qquad (25\text{-}13)$$

$$\text{Total EMF of } M \text{ banks} = NE \text{ volts} \qquad (25\text{-}14)$$

$$\text{Total internal resistance of } M \text{ banks} = \frac{NR_i}{M} \text{ ohms} \qquad (25\text{-}15)$$

Example 25-1

A voltage source has an EMF of 1.8 V and an internal resistance of 0.3 Ω. If twelve such sources are connected in series-aiding across a 25 Ω load, what is the value of the load voltage?

Solution

Total constant voltage EMF,

$$E_T = 12 \times 1.8 \text{ V} = 21.6 \text{ V} \qquad (25\text{-}3)$$

Total internal resistance,

$$R_{iT} = 12 \times 0.3 \text{ Ω} = 3.6 \text{ Ω} \qquad (25\text{-}4)$$

Using the voltage division rule,

Load voltage,

$$V_L = E_T \times \frac{R_L}{R_{iT} + R_L}$$

$$= 21.6 \text{ V} \times \frac{25 \text{ Ω}}{3.6 \text{ Ω} + 25 \text{ Ω}}$$

$$= 21.6 \text{ V} \times \frac{25 \text{ Ω}}{28.6 \text{ Ω}}$$

$$= 18.9 \text{ V}$$

Example 25-2

In Fig. 25-2A, calculate the values of I_L and V_L.

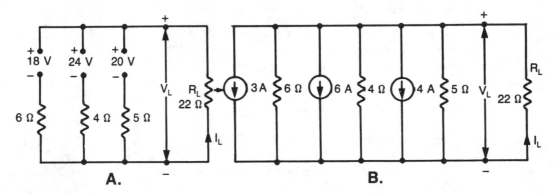

A.

B.

Figure 25-2

Solution

Convert the voltage sources of Fig. 25-2A into the current sources of Fig. 25-2B. The values of the equivalent current generators are 18 V/6 Ω = 3 A, 24 V/4 Ω = 6 A and 20 V/5 Ω = 4 A. The total equivalent current is therefore,

$$I_T = 3 \text{ A} + 6 \text{ A} + 4 \text{ A} \qquad (25\text{-}6)$$
$$= 13 \text{ A}$$

Total equivalent resistance,

$$R_{iT} = \cfrac{1}{\cfrac{1}{6} + \cfrac{1}{4} + \cfrac{1}{5}} \qquad (25\text{-}7)$$
$$= 1.62 \ \Omega$$

Load current,

$$I_L = 13 \text{ A} \times \frac{1.62 \ \Omega}{1.62 \ \Omega + 22 \ \Omega}$$
$$= 0.93 \text{ A}$$

Load voltage,

$$V_L = 0.93 \text{ A} \times 22 \ \Omega$$
$$= 20.5 \text{ V}$$

Example 25-3

Twelve cells, each with an EMF of 1.5 V and an internal resistance of 0.1 Ω are connected in series-aiding. Eight such banks are joined in parallel so that the total number of individual cells is 12 × 8 = 96 cells. Calculate the open-circuit terminal voltage available and the corresponding value of the total internal resistance.

Solution

Total EMF,

$$E_T = NE \qquad (25\text{-}12)$$
$$= 12 \times 1.5 \text{ V}$$
$$= 18 \text{ V}$$

Total internal resistance,

$$R_{iT} = \frac{NR_i}{M} \qquad (25\text{-}15)$$
$$= \frac{12 \times 0.1 \ \Omega}{8}$$
$$= 0.15 \ \Omega$$

Chapter 26
Kirchhoff's voltage and current laws

IN PREVIOUS SECTIONS WE HAVE LOOSELY REFERRED TO *KIRCHHOFF'S VOLTAGE law* (KVL) as the requirement for a voltage balance around any closed circuit or loop while *Kirchhoff's current law* (KCL) fulfilled the need for the current balance that must exist at any junction joint. These laws are commonly used to analyze circuits that are too difficult to be solved by Ohm's law alone.

A more formal KVL statement would be: "The *algebraic* sum of the constant voltage EMFs and the voltage drops around *any closed* electrical loop is *always* zero."

Because the algebraic sum is involved, we must have a convention that distinguishes between positive and negative voltages. Normally we start at any point in a loop and move around in the clockwise direction. A voltage is then positive if the negative polarity of that voltage is first encountered. In the circuit of Fig. 26-1 there are three loops (*XYZX*, *XWYX*, and *XWYZX*) so that three KVL equations can be obtained.

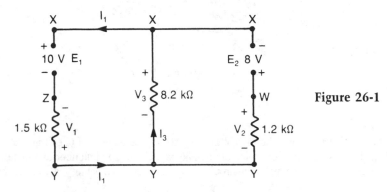

Figure 26-1

Initially the separate currents (electron flow) must be specified in the circuit. Sometimes the direction of a particular current is doubtful, such is the case with I_3 because the voltages E_1 and E_2 are tending to drive currents in opposite directions through the 8.2 kΩ resistor. However, this is not a serious problem because if you choose the wrong direction, the mathematical sign of the current will be revealed as negative.

MATHEMATICAL DERIVATIONS

If the currents are measured in milliamperes:

Loop XYZX (starting at X)

The first KVL equation is:

$$+ (-V_3) + (-V_1) + (+E_1) = 0 \tag{26-1}$$
$$-I_3 \times 8.2 \text{ k}\Omega - I_1 \times 1.5 \text{ k}\Omega + 12 \text{ V} = 0$$
$$1.5 \, I_1 + 8.2 \, I_3 = 12$$

Loop XWYX (starting at X)

The second KVL equation is:

$$+ (+E_2) + (-V_2) + (+V_3) = 0 \tag{26-2}$$
$$8\ \text{V} - I_2 \times 1.2\ \text{k}\Omega + I_3 \times 8.2\ \text{k}\Omega = 0$$
$$+1.2\ I_2 - 8.2\ I_3 = 8$$

Loop XWYZX (starting at X)

The third KVL equation is:

$$+(+E_2) + (-V_2) + (-V_1) + (+E_1) = 0 \tag{26-3}$$
$$+8\ \text{V} - I_2 \times 1.2\ \text{k}\Omega - I_1 \times 1.5\ \text{k}\Omega + 12\ \text{V} = 0$$
$$1.5\ I_1 + 1.2\ I_2 = 8 + 12 = 20$$

Although we have three equations, they are not independent because the third KVL equation is the result of adding together the first and second equations. Consequently, there are three unknowns (I_1, I_2 and I_3) and we still need a third equation, which is of the KCL type.

Formally stated, Kirchhoff's current law is: "The *algebraic* sum of the currents existing at any junction point is zero."

Conventionally we assume that the electron flow currents entering a junction point are positive while those leaving are negative. Therefore at the junction point, Y:

$$+(+I_1) + (-I_2) + (-I_3) = 0 \tag{26-4}$$
$$I_1 = I_2 + I_3$$

Combining equations 26-1 and 26-4:

$$1.5\ (I_2 + I_3) + 8.2\ I_3 = 12 \tag{26-5}$$
$$1.5\ I_2 + 9.7\ I_3 = 12$$

Multiplying equation 26-2 by 1.5 and equation 26-5 by 1.2:

$$1.8\ I_2 - 12.3\ I_3 = 12 \tag{26-6}$$

$$1.8\ I_2 + 11.64\ I_3 = 14.4 \tag{26-7}$$

Subtracting equation 26-6 from equation 26-7:

$$23.9\ I_3 = 2.4$$
$$I_3 \approx +0.1\ \text{mA}.$$

The positive sign for I_3 indicates that the chosen direction for this current was correct. Substituting this value of I_3 in equation 26-1,

$$1.5\ I_1 + 0.823 = 12$$

$$I_1 = \frac{11.177}{1.5} = 7.45\ \text{mA}.$$

From equation 26-4,

$$I_2 = I_1 - I_3 = 7.35\ \text{mA}.$$

Checking for loop $XYZX$:

$$-0.1\ \text{mA} \times 8.2\ \text{k}\Omega - 7.45\ \text{mA} \times 1.5\ \text{k}\Omega + 12\ \text{V}$$
$$= -0.82 - 11.18 + 12 = 0\ \text{V}.$$

The individual voltage drops and powers dissipated can be obtained from the normal Ohm's law equations.

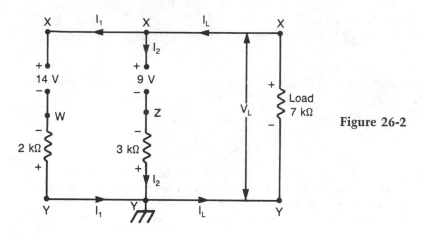

Figure 26-2

Example 26-1

In Fig. 26-2 calculate the value of the load current, I_L, and the potential at the junction X.

Solution

The (electron flow) currents in milliamperes are as assigned.

The KVL equation for the loop $XYWX$ (starting at X) is:

$$+(-I_L \times 7 \text{ k}\Omega) + (-I_1 \times 2 \text{ k}\Omega) + 14 \text{ V} = 0.$$
$$2 I_1 + 7 I_L = 14.$$

For the loop $XYZX$ (starting at X):

$$+(-I_L \times 7 \text{ k}\Omega) + (-I_2 \times 3 \text{ k}\Omega) + 9 \text{ V} = 0.$$
$$3 I_2 + 7 I_L = 9$$

The KCL equation for the X junction is:

$$I_1 + I_2 = I_L$$

Therefore

$$2 I_1 + 7 (I_1 + I_2) = 9 I_1 + 7 I_2 = 14 \qquad (26\text{-}1\text{-}1)$$

and

$$3 I_2 + 7 (I_1 + I_2) = 7 I_1 + 10 I_2 = 9 \qquad (26\text{-}1\text{-}2)$$

Multiplying the equation 26-1-1 by 7 and the equation 26-1-2 by 9,

$$63 I_1 + 49 I_2 = 98 \qquad (26\text{-}1\text{-}3)$$
$$63 I_1 + 90 I_2 = 81 \qquad (26\text{-}1\text{-}4)$$

Subtracting the equation 26-1-3 from the equation 26-1-4,

$$(90 - 49) I_2 = 41 I_2 = 81 - 98 = -17$$

$$I_2 = -\frac{17}{41} = -0.417 \text{ mA}.$$

The negative sign indicates that the electron flow direction is from the junction Y to the junction X (and not from X to Y as indicated). This means that the 14 V source is, in fact, charging the 9 V source.

Substituting the value of I_2 in the equation 26-1-3,

$$63\, I_1 = 98 + (49 \times 0.417) = 118.4$$

$$I_1 = \frac{118.4}{63} = 1.88 \text{ mA.}$$

Therefore

$$I_L = 1.88 - 0.417 = 1.463 \text{ mA.}$$

Potential at the junction X, $V_L = +I_L \times R_L = +(1.463 \text{ mA} \times 7 \text{ k}\Omega) = +10.24$ V. Checking the loop $XYWX$:

$$2 \text{ k}\Omega \times 1.88 \text{ mA} + 7 \text{ k}\Omega \times 1.463 \text{ mA} - 14 \text{ V} = 0$$

$$3.76 + 10.24 - 14 = 0.$$

Chapter 27
Mesh analysis

A CLOSED VOLTAGE LOOP MAY ALSO BE REFERRED TO AS A MESH. IN MESH analysis each of the mesh currents i_1, i_2, i_3 (Fig. 27-1) flow around a complete loop although an individual resistor may carry one or more mesh currents. For example, the mesh currents i_1 and i_2 flow in opposite directions through the resistor $R1$.

The normal convention is to assign clockwise mesh (electron flow) currents in each of the loops and then write down the KVL equation for each loop. The main advantages of this method are (1) we will not require any KCL equations and (2) the KVL equations can be written by inspection and do not require precise voltage conventions.

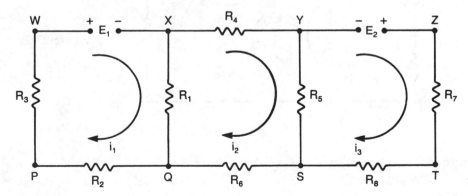

Figure 27-1

MATHEMATICAL DERIVATIONS

In Fig. 27-1

KVL equation for the mesh $PWXQP$ (starting at P):

$$i_1R_3 - E_1 + i_1R_1 + i_1R_2 - i_2R_1 = 0 \tag{27-1}$$
$$-i_2R_1 + i_1(R_1 + R_2 + R_3) = E_1$$

KVL equation for the mesh $QXYSQ$ (starting at Q):

$$i_2R_1 - i_1R_1 + i_2R_4 + i_2R_5 - i_3R_5 + i_2R_6 = 0 \tag{27-2}$$
$$-i_1R_1 - i_3R_5 + i_2(R_1 + R_4 + R_5 + R_6) = 0$$

KVL equation for the mesh $SYZTS$ (starting at S):

$$i_3R_5 - i_2R_5 + E_2 + i_3R_7 + i_3R_8 = 0 \tag{27-3}$$
$$-i_2R_5 + i_3(R_5 + R_7 + R_8) = -E_2$$

From the pattern of these results it follows that the KVL equations can be written down from inspection if we observe the following rules.

1. Add together all the resistances in the loop and multiply the results by that loop's mesh current. This applies to the terms $i_1(R_1 + R_2 + R_3)$, $i_2(R_1 + R_4 + R_5 + R_6)$, and $i_3(R_5 + R_7 + R_8)$.
2. If a resistor in a particular mesh carries a current assigned to another mesh, the corresponding IR drop is given a negative sign. This refers to the terms $-i_2R_1$, $-i_1R_1$, $-i_3R_5$, and $-i_2R_5$.
3. The signs of the voltage sources are determined by the normal KVL convention.

Example 27-1

In Fig. 27-2, calculate the value of the potential at the point X. Note: This is the same problem that was previously solved by Kirchhoff's laws in Example 26-1.

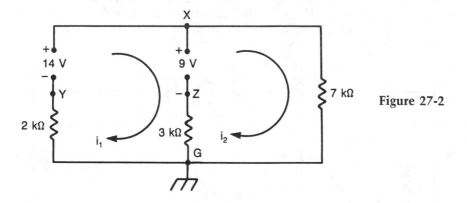

Figure 27-2

Solution

For mesh $XZGYX$:

$$i_1(2\text{ k}\Omega + 3\text{ k}\Omega) - i_2 \times 3\text{ k}\Omega - 9\text{ V} + 14\text{ V} = 0 \tag{27-1-1}$$
$$5i_1 - 3i_2 = -5$$

For mesh $XGZX$:

$$i_1 \times 3 \text{ k}\Omega + i_2(3 \text{ k}\Omega + 7 \text{ k}\Omega) + 9 \text{ V} = 0 \qquad (27\text{-}1\text{-}2)$$

$$-3 \, i_1 + 10 \, i_2 = -9$$

Multiply Equation 27-1-1 by 3 and Equation 27-1-2 by 5,

$$15 \, i_1 - 9 \, i_2 = -15$$

$$-15 \, i_1 + 50 \, i_2 = -45$$

Therefore

$$41 \, i_2 = -60$$

$$i_2 = \frac{-60}{41} \text{ mA.}$$

Potential at point X is $i_2 R_L = -(-60/41 \text{ mA}) \times 7 \text{ k}\Omega = +10.24 \text{ V.}$

Notice that the mesh analysis solution is shorter than that obtained from Kirchhoff's laws.

~

Chapter 28
Nodal analysis

NODAL ANALYSIS IS ANOTHER METHOD FOR SOLVING CIRCUITS, BUT THIS TIME, only current sources are involved. By contrast, those circuits that were solved by mesh analysis or Kirchhoff's laws contained voltage sources and required the use of simultaneous equations. However, on many occasions it is possible to convert the voltage sources into their equivalent current sources and subsequently solve the problem with a single nodal equation rather than with two or more simultaneous equations.

A node is another term for a junction point at which two or more electrical currents exist. Therefore the junction N is a node (Fig. 28-1) and the purpose of the analysis is to find the value of the potential at N. As a convention we shall assume that the potential at N is negative. The true polarity will be shown by the sign of the value for V_L. If the potential at N is negative, the electrons through the resistors must flow from the nodal point to ground. The generators that force (electron flow) currents into the node are then positive while those driving currents out of the node are negative.

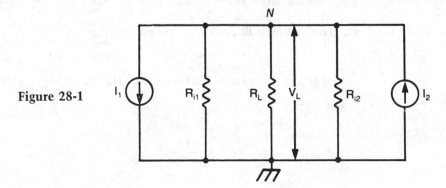

Figure 28-1

MATHEMATICAL DERIVATIONS
In Fig. 28-1

Total electron flow currents leaving node N through resistors $= \dfrac{V_L}{R_{i1}} + \dfrac{V_L}{R_L} + \dfrac{V_L}{R_{i2}}$ amperes

$$\text{Total of the generator currents} = (-I_1) + (+I_2) \text{ amperes} \qquad (28\text{-}2)$$

Consequently the nodal equation is:

$$(-I_1) + (+I_2) = \frac{V_L}{R_{i1}} + \frac{V_L}{R_L} + \frac{V_L}{R_{i2}} \qquad (28\text{-}3)$$

Example 28-1

In Fig. 28-2A, calculate the value of the potential at the point N. Note: This is the same problem that was previously solved by three simultaneous equations in Example 26-1 (Kirchhoff's laws) and by two simultaneous equations in Example 27-1 (Mesh Analysis).

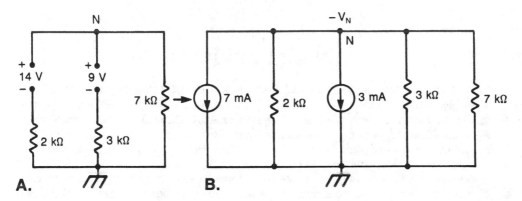

A.   **B.**

Figure 28-2

Solution

Convert the voltage sources of Fig. 28-2A into their constant current equivalents (Fig. 28-2B).

Total electron flow currents leaving node N through resistors $= \dfrac{V_N}{2\text{ k}\Omega} + \dfrac{V_N}{3\text{ k}\Omega} + \dfrac{V_N}{7\text{ k}\Omega}$

$$= 0.9762\ V_N \text{ mA} \qquad (28\text{-}1)$$

$$\text{Total of generator currents} = (-7) + (-3) = -10 \text{ mA} \qquad (28\text{-}2)$$

Therefore,

$$0.9762\ V_N = -10 \text{ mA}$$

$$V_N = -\frac{10}{0.9762}$$

$$= -10.24 \text{ V.}$$

The potential at N is therefore $+10.24$ V.

Note that the nodal solution involved only a *single* equation.

Chapter 29
The superposition theorem

THE PRINCIPLE OF SUPERPOSITION MAY BE USED TO SOLVE A NUMBER OF problems that contain more than one source and only linear resistances. The method requires you to consider the currents and voltages created by each source in turn and then finally combine (superimpose) the results from all the sources. A formal statement of this theorem follows.

*In a network of **linear** resistances containing more than one source, the resultant current flowing at any one point is the algebraic sum of the currents that would flow at that point if each source is considered separately while all other sources are replaced by their equivalent internal resistances. This last step is carried out by short-circuiting all sources of constant voltage and open-circuiting all sources of constant current.*

This theorem has the advantage of allowing each source to be considered separately so that only Ohm's law equations are required in the solution. However, each time one of the sources is applied to the circuit, a different voltage will normally appear across a particular resistor. The superposition theorem therefore requires the resistance to be linear so that its resistance does not change with the amount of the voltage drop but remains at a constant value. In other words, the voltage drop is always directly proportional to the current.

If a large number of sources are involved, use of the superposition theorem is not recommended because the circuit has to be solved separately for each source before combining the results. The final analysis would probably be far more tedious than if we used either the mesh or nodal methods.

MATHEMATICAL DERIVATIONS

In Fig. 29-1A, let us assume that our purpose is to find the value of the current, I_3.

Short out the constant voltage source, E_1, to produce the circuit of Fig. 29-1B.

Total equivalent resistance, R_T, presented to the E_2 source is

$$R_T = R_2 + R_4 \parallel (R_1 + R_3) \text{ ohms} \qquad (29\text{-}1)$$

Then

$$I_2' = \frac{E_2}{R_T} \text{ and } I_3' = I_2' \times \frac{R_1 + R_3}{R_1 + R_3 + R_4} \text{ amperes} \qquad (29\text{-}2)$$

Short out the constant voltage source, E_2, to produce the circuit of Fig. 29-1C.

Total equivalent resistance, R_T, presented to the E_1 source is

$$R_T = R_1 + R_3 + R_2 \parallel R_4 \text{ ohms} \qquad (29\text{-}3)$$

Then

$$I_1'' = \frac{E_1}{R_T} \text{ and } I_3'' = I_1 \times \frac{R_2}{R_2 + R_4} \text{ amperes} \qquad (29\text{-}4)$$

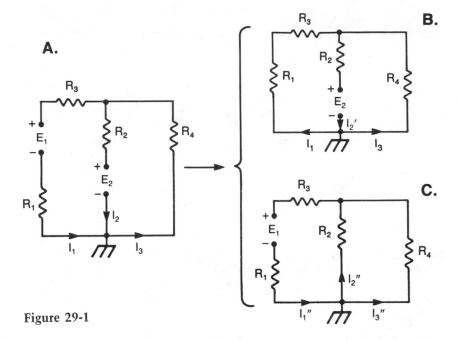

Figure 29-1

Superimposing these results,

$$I_3 = I_3' + I_3'' \text{ amperes} \qquad (29\text{-}5)$$

Similarly,

$$I_1 = I_1' - I_1'' \text{ and } I_2 = I_2' - I_2'' \text{ amperes} \qquad (29\text{-}6)$$

The signs of the answers for I_1 and I_2 will indicate the actual directions of the electron flows.

Example 29-1

In Fig. 29-2A, calculate the value of the potential at the point X. Note: This is the same problem that was previously solved by the methods of Kirchhoff's laws, Mesh Analysis, and Nodal Analysis.

Solution

Short out the 14 V constant voltage source to produce the circuit of Fig. 29-2B. Total resistance presented to the 9 V source,

$$R_T = 3 \text{ k}\Omega + 2 \text{ k}\Omega \parallel 7 \text{ k}\Omega$$

$$= 3 + \frac{2 \times 7}{2 + 7} = 3 + \frac{14}{9} = \frac{41}{9}$$

$$= 4.56 \text{ k}\Omega.$$

$$\text{Current, } I_2' = \frac{9 \text{ V}}{4.56 \text{ k}\Omega}$$

$$= 1.97 \text{ mA.}$$

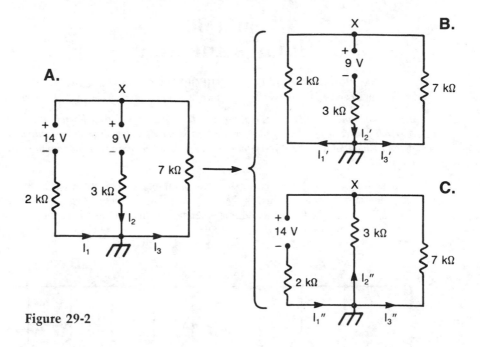

Figure 29-2

$$\text{Current, } I_3' = 1.97 \text{ mA} \times \frac{2 \text{ k}\Omega}{2 \text{ k}\Omega + 7 \text{ k}\Omega}$$

$$= 0.44 \text{ mA (CDR rule)}.$$

Short out the 9 V constant source to produce the circuit of Fig. 29-2C. Total resistance presented to the 14 V source,

$$R_T = 2 \text{ k}\Omega + 3 \text{ k}\Omega \parallel 7 \text{ k}\Omega = 2 + \frac{3 \times 7}{3 + 7}$$

$$= 4.1 \text{ k}\Omega.$$

$$\text{Current } I_1'' = \frac{14 \text{ V}}{4.1 \text{ k}\Omega}$$

$$= 3.41 \text{ mA}.$$

$$\text{Current } I_3'' = 3.41 \text{ mA} \times \frac{3 \text{ k}\Omega}{3 \text{ k}\Omega + 7 \text{ k}\Omega}$$

$$= 1.023 \text{ mA}.$$

When the results are superimposed,

$$I_3 = I_3' + I_3'' = 0.440 + 1.023 = 1.463 \text{ mA} \tag{29-5}$$

$$\text{Potential at } X = +(1.463 \text{ mA} \times 7 \text{ k}\Omega)$$

$$= +10.24 \text{ V}.$$

Chapter 30
Millman's theorem

IN FACT WE HAVE ALREADY USED MILLMAN'S THEOREM WHEN ANALYZING PRACTICAL parallel sources in Chapter 25. However, the formal statement of Millman's theorem is:

Any number of constant current sources that are directly connected in parallel can be converted into a single current source whose total generator current is the algebraic sum of the individual source currents and whose total internal resistance is the result of combining the individual source resistances in parallel.

The theorem actually refers to only current sources, but it may also be applied to a mixture of parallel voltage and current sources (Fig. 30-1A) by converting all voltage sources into their constant current equivalents. The final generator may then be shown either as a constant current or as a constant voltage source.

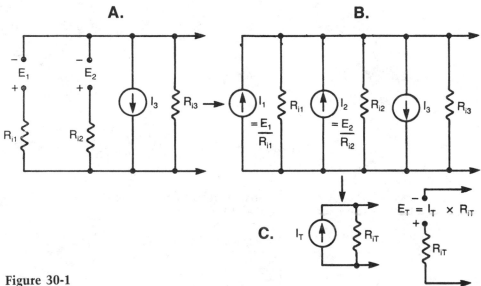

Figure 30-1

MATHEMATICAL DERIVATIONS
In Fig. 30-1B

Total generator current,

$$I_T = +(+I_1) + (+I_2) + (-I_3) \text{ amperes} \tag{30-1}$$

Total internal resistance,

$$R_{iT} = \frac{1}{\dfrac{1}{R_{i1}} + \dfrac{1}{R_{i2}} + \dfrac{1}{R_{i3}}} \text{ ohms} \tag{30-2}$$

Total generator voltage,

$$E_T = I_T \times R_{iT} \text{ volts} \tag{30-3}$$

Example 30-1

In Fig. 30-2A, calculate the value of the potential at the point N. Note: This is the same problem that we previously solved by three simultaneous equations in Example 26-1 (Kirchhoff's law's), by two simultaneous equations in Example 27-1 (Mesh Analysis) and by one equation in Example 28-1 (Nodal Analysis).

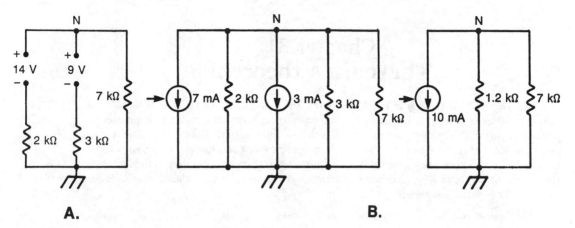

A.   **B.**

Figure 30-2

Solution

Convert the two constant voltage sources into their equivalent constant current sources (see Fig. 30-2B).
Total generator current,

$$I_T = +(+7 \text{ mA}) + (+3 \text{ mA}) \tag{30-1}$$
$$= 10 \text{ mA}$$

Total internal resistance,

$$R_{iT} = \cfrac{1}{\cfrac{1}{2} + \cfrac{1}{3}} \tag{30-2}$$
$$= 1.2 \text{ k}\Omega$$

Total resistance of the circuit,

$$1.2 \text{ k}\Omega \parallel 7 \text{ k}\Omega = \frac{1.2 \times 7}{1.2 + 7}$$
$$= \frac{8.4}{8.2}$$
$$= 1.024 \text{ k}\Omega$$

Potential at the point N,

$$= +1.024 \text{ k}\Omega \times 10 \text{ mA}$$
$$= +10.24 \text{ V.}$$

Notice that the use of Millman's theorem does not even require the solution of one linear equation.

Chapter 31
Thévenin's theorem

*T*HÉVENIN'S THEOREM IS THE MOST VALUABLE ANALYTICAL TOOL WHEN DEALING with complex networks. It enables you to focus your attention on a particular component or any circuit part that is connected between two terminals and is regarded as the load. The remainder of the circuit is then represented by a single generator with a constant voltage, E_{TH}, which is in series with an internal resistance, R_{TH}. The purpose of Thévenin's theorem is then to obtain the values of E_{TH} and R_{TH} from the components and sources in the original circuit. Stated formally, Thévenin's theorem is:

The current in a load connected between two output terminals, X and Y, of a network of resistors and electrical sources (Fig. 31-1A) is no different than if that same load were connected across a simple constant voltage generator (Fig. 31-1B) whose EMF, E_{TH}, is the open circuit voltage measured between X and Y (Fig. 31-1C) and whose series internal resistance, R_{TH}, is the resistance of the network looking back into the terminals X and Y with all sources replaced by resistances equal to their internal resistances (Fig. 31-1D). This last step involves short-circuiting all sources of constant voltage and open-circuiting all sources of constant current.

Example of Thévenin's Theorem.

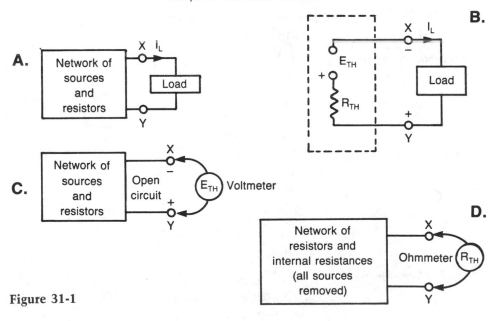

Figure 31-1

The process of *Thévenizing* the circuit is illustrated in Fig. 31-1. As an example let us Thévenize the same circuit (Fig. 31-2A) which we have already explored with Kirchhoff's laws, Mesh Analysis, Nodal Analysis, and Millman's Theorem.

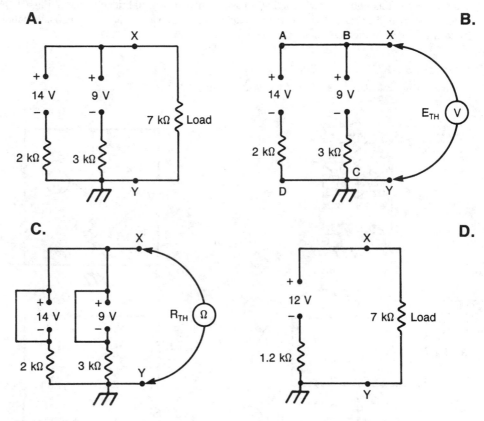

Figure 31-2

The 7 kΩ resistor is regarded as the load. When this load is removed, we consider that a voltmeter is connected between X and Y to record the value of E_{TH}. In the loop $ABCD$ the sources are in series-opposing and the current, I, is

$$\frac{14\ V - 9\ V}{2\ k\Omega + 3\ k\Omega} = 1\ mA$$

in the direction shown.

The value of E_{TH} can be thought of as either 14 V − (1 mA × 2 kΩ) = 12 V or 9 V + (1 mA × 3 kΩ) = 12 V (Fig. 31-2B).

To find the value of R_{TH}, the 14-V and 9-V constant voltage sources are shorted out and we visualize that an ohmmeter is connected between X and Y (Fig. 31-2C). The ohmmeter reading will then be 2 kΩ ‖ 3 kΩ = (2 × 3)/(2 + 3) = 1.2 kΩ. The load is now replaced across the Thévenin equivalent generator (Fig. 31-2D) and by the VDR, the potential at X will be +12 V × 7 kΩ/(1.2 kΩ + 7 kΩ) = +10.24 V.

You are probably not impressed with the performance of Thévenin's theorem in solving this particular problem. The solution certainly seemed much more involved than the Millman treatment. However, with other types of problems, Thévenin's theorem has a number of advantages and nowhere is this better shown than in the case of the unbalanced bridge circuit of Fig. 31-3A. Using mesh or nodal analysis such a circuit would require three algebraic simultaneous equations while the Thévenin method does not require algebra at all.

Example 31-1

In Fig. 31-3A, calculate the value of the load current, I_L.

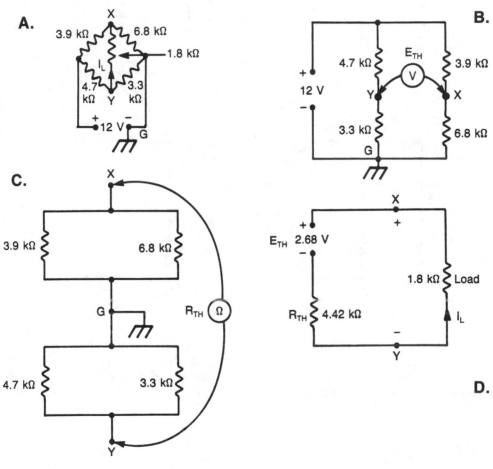

Figure 31-3

Solution

Regard the 1.8 kΩ resistor as the load. When this load is removed and replaced by a voltmeter, the remaining circuit is as shown in Fig. 31-3B. By the voltage division rule: The potential at Y is:

$$+12 \text{ V} \times \frac{3.3 \text{ k}\Omega}{3.3 \text{ k}\Omega + 4.7 \text{ k}\Omega} = +4.95 \text{ V}$$

while the potential at X is:

$$+12 \text{ V} \times \frac{6.8 \text{ k}\Omega}{6.8 \text{ k}\Omega + 3.9 \text{ k}\Omega} = +7.63 \text{ V}.$$

Consequently, the value of E_{TH} is $+(7.63 - 4.95) = +2.68$ V (terminal X is positive with respect to the terminal Y).

When a short is placed across the 12-V source, the equivalent circuit is shown in Fig. 31-3C. The value of R_{TH} is the series combination of 3.9 kΩ ‖ 6.8 kΩ and 3.3 kΩ ‖ 4.7 kΩ. Using the "product-over-sum" formula,

$$R_{TH} = \frac{3.9 \times 6.8}{3.9 + 6.8} + \frac{3.3 \times 4.7}{3.3 + 4.7} = 2.48 + 1.94$$

$$= 4.42 \text{ k}\Omega$$

The 1.8 kΩ load resistor is now replaced across the equivalent Thévenin generator (Fig. 31-3D). The load current, $I_L = 2.68$ V/(4.42 kΩ + 1.8 kΩ) = 0.43 mA.

Chapter 32
Norton's theorem

A FORMAL STATEMENT OF NORTON'S THEOREM IS: *The Current in a Load connected between two output terminals, X and Y, of a complex network containing electrical sources and resistors is no different than if that same load were connected to a* **constant current** *source, whose generator current, I_N, is equal to the* **short-circuit** *current measured between X and Y. This constant current generator is placed in parallel with a resistance, R_N, which is equal to the resistance of the network looking back into the terminals X and Y with all sources replaced by resistances equal to their internal resistances.*

This last step would involve short-circuiting all sources of constant voltage and open-circuiting all sources of constant current. Figure 32-1A, B, C, and D illustrates the various steps that are taken in Nortonizing a circuit.

It is clear that Norton's Theorem involves a constant current generator while Thévenin's theorem reduces to a constant voltage source. We can readily convert between the two generators by using the results of Chapter 24. For example, the series Thévenin resistance and the parallel Norton resistance have the same value.

We will now Nortonize the same circuit to which we have applied all other analytical methods. The 7 kΩ resistor is regarded as the load. When the load is removed, we consider that an ammeter is connected between X and Y. This ammeter will record the Norton current, I_N, which is equal to 14 V/2 kΩ + 9 V/3 kΩ = 10 mA (Fig. 32-2B). The Norton resistance is $(2 \times 3)/(2 + 3)$ = 1.2 kΩ (Fig. 32-2C) and therefore the equivalent constant current generator is as shown in Fig. 32-2D. Using the current division rule, the load current:

$$I_L = 10 \text{ mA} \times \frac{1.2 \text{ k}\Omega}{1.2 \text{ k}\Omega + 7 \text{ k}\Omega} = 1.4634 \text{ mA}$$

and the potential at X is $+ (1.4634 \text{ mA} \times 7 \text{ k}\Omega) = +10.24$ V.

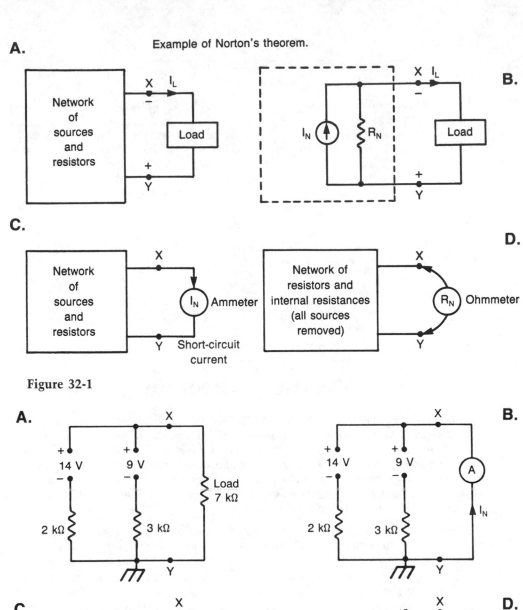

Example of Norton's theorem.

Figure 32-1

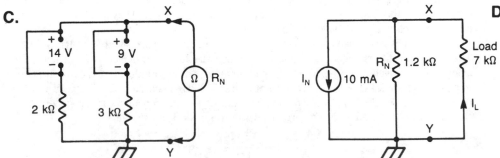

Figure 32-2

For this type of problem there was little to choose between the Thévenin and Norton methods of analysis.

Example 32-1

In Fig. 32-3 calculate the value of the load current, I_L.

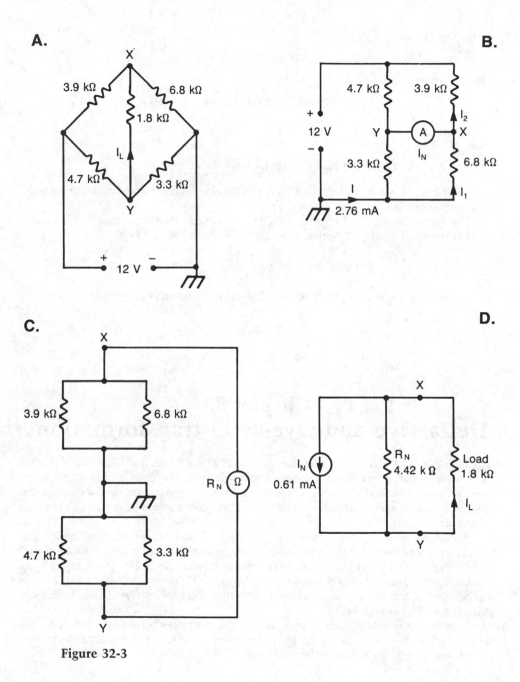

Figure 32-3

Solution

Regard the 1.8 kΩ resistor as the load. When this load is removed and replaced by an ammeter, the remaining circuit as shown in Fig. 32-3B. The total resistance presented to the 12 V source is:

$$4.7 \text{ k}\Omega \parallel 3.9 \text{ k}\Omega + 3.3 \text{ k}\Omega \parallel 6.8 \text{ k}\Omega = 2.13 + 2.22$$
$$= 4.35 \text{ k}\Omega,$$

so that the current I is:

$$12 \text{ V}/4.35 \text{ k}\Omega = 2.76 \text{ mA}.$$

Using the current division rule,

$$I_1 = 2.76 \text{ mA} \times 3.3 \text{ } k\Omega/(3.3 \text{ k}\Omega + 6.8 \text{ k}\Omega)$$
$$= 0.90 \text{ mA},$$

while,

$$I_2 = 2.76 \text{ mA} \times 4.7 \text{ k}\Omega/(4.7 \text{ k}\Omega + 3.9 \text{ k}\Omega) = 1.51 \text{ mA}.$$

Therefore, $I_N = 1.51 - 0.90 = 0.61$ mA with the electron flow in the direction of the terminal Y to the terminal X.

The Norton resistance, R_N, is the same as the Thévenin resistance, R_{TH}, and is equal to 4.42 kΩ. With the equivalent Norton generator as shown in Fig. 32-3D, the load current

$$I_L = 0.61 \text{ mA} \times \frac{4.42 \text{ k}\Omega}{1.8 \text{ k}\Omega + 4.42 \text{ k}\Omega} = 0.43 \text{ mA}.$$

This is the same result as we obtained with Thévenin's Theorem in Chapter 31.

Chapter 33
Delta-wye and wye-delta transformations

WE HAVE ALREADY EXAMINED THE UNBALANCED BRIDGE CIRCUIT, WHICH CANNOT be regarded as a series-parallel arrangement because the center arm bears no simple series or parallel relationship to the other four resistors. However, on many occasions it is possible to convert such circuits into simpler arrangements by means of a delta-wye (or a wye-delta) transformation.

A triangular arrangement of resistors (Fig. 33-1A) is known as a delta (Δ, the Greek capital letter) connection. However, the same arrangement can be changed to resemble another Greek capital letter, pi, Π (Fig. 33-1B) so that the two names are interchangeable. The same applies to a second configuration, which is referred to as a Wye (letter Y) connection (Fig. 33-2A). This can also be modified to look like the letter tee (T), so that either tee or wye is used to describe identical networks of resistors.

If the delta-wye networks are to be equivalent, the resistances between the points, X, Y, Z in both configurations must be the same. The required equations for the two transformations ($\Delta \rightarrow Y$ and $Y \rightarrow \Delta$) are shown in the following:

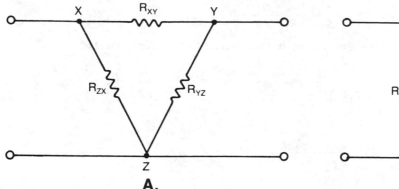

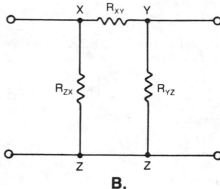

A. B.

Figure 33-1

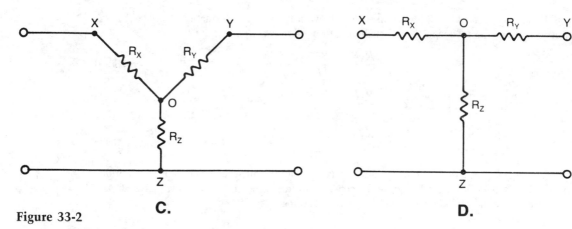

C. D.

Figure 33-2

MATHEMATICAL DERIVATIONS
In Fig. 33-1A:

$$\text{Resistance between the points } X \text{ and } Y = \frac{R_{XY} \times (R_{YZ} + R_{ZX})}{R_{XY} + (R_{YZ} + R_{ZX})} \qquad (33\text{-}1)$$

In Fig. 33-2A:

$$\text{Resistance between the points } X \text{ and } Y = R_X + R_Y \qquad (33\text{-}2)$$

Therefore,

$$R_X + R_Y = \frac{R_{XY}(R_{YZ} + R_{ZX})}{R_{XY} + R_{YZ} + R_{ZX}} \qquad (33\text{-}3)$$

By symmetry,

$$R_Y + R_Z = \frac{R_{YZ}(R_{ZX} + R_{XY})}{R_{XY} + R_{YZ} + R_{ZX}} \qquad (33\text{-}4)$$

113

$$R_Z + R_X = \frac{R_{ZX}(R_{XY} + R_{YZ})}{R_{XY} + R_{YZ} + R_{ZX}} \tag{33-5}$$

Add Equations 33-3 and 33-5. From their sum subtract Equation 33-4 and divide the result by 2. This yields

$$R_X = \frac{R_{XY}R_{ZX}}{R_{XY} + R_{YZ} + R_{ZX}} \tag{33-6}$$

Similarly,

$$R_Y = \frac{R_{YZ}R_{XY}}{R_{XY} + R_{YZ} + R_{ZX}} \tag{33-7}$$

and

$$R_Z = \frac{R_{ZX}R_{YZ}}{R_{XY} + R_{YZ} + R_{ZX}} \tag{33-8}$$

Equations 33-6, 33-7, 33-8 are used to achieve a $\Delta \rightarrow Y$ transformation. For the $Y \rightarrow \Delta$ transformation, the following three equations can be derived:

$$R_{XY} = \frac{R_X R_Y + R_Y R_Z + R_Z R_X}{R_Z} \tag{33-9}$$

$$R_{YZ} = \frac{R_X R_Y + R_Y R_Z + R_Z R_X}{R_X} \tag{33-10}$$

$$R_{ZX} = \frac{R_X R_Y + R_Y R_Z + R_Z R_X}{R_Y} \tag{33-11}$$

These transformations can be used to advantage with the analysis of the unbalanced bridge circuit. In the following we will choose for convenience the same circuit which we have already solved with the aid of Thévenin's and Norton's theorems.

Example 33-1

In Fig. 33-3A, calculate the value of the total equivalent resistance presented to the 12 V source.

Solution

In Fig. 33-3A, regard X, Y, Z as the corners of a delta formation of resistors and then convert this arrangement into its wye equivalent. Then,

$$R_X = \frac{R_{XY}R_{ZX}}{R_{XY} + R_{YZ} + R_{ZX}} \tag{33-6}$$

$$= \frac{1.8 \times 6.8}{3.3 + 6.8 + 1.8}$$

$$= 1.03 \ k\Omega$$

$$R_Y = \frac{R_{XY}R_{YZ}}{R_{XY} + R_{YZ} + R_{ZY}} \tag{33-7}$$

$$= \frac{1.8 \times 3.3}{3.3 + 6.8 + 1.8}$$

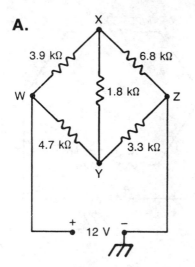

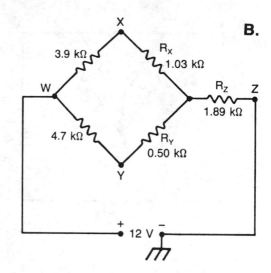

Figure 33-3

$$= 0.50 \; k\Omega$$

$$R_Z = \frac{R_{XZ}R_{YZ}}{R_{XY} + R_{YZ} + R_{ZX}} \tag{33-8}$$

$$= \frac{6.8 \times 3.3}{3.3 + 6.8 + 1.8}$$

$$= 1.89 \; k\Omega$$

The circuit has now been transformed to the relatively simple series-parallel arrangement of Fig. 33-3B.

The total resistance, R_T presented to the 12 V source is:

$$R_T = 1.89 + \frac{(3.9 + 1.03) \times (4.7 + 0.5)}{(3.9 + 1.03) + (4.7 + 0.5)}$$

$$= 1.89 + \frac{4.93 \times 5.2}{10.13}$$

$$= 4.42 \; k\Omega$$

An alternative solution is to regard X as the center of a wye formation and convert this formation to its delta equivalent (Fig. 33-4A).

Then

$$R_{WZ} = \frac{3.9 \times 6.8 + 6.8 \times 1.8 + 1.8 \times 3.9}{1.8} = \frac{45.78}{1.8}$$

$$= 25.43 \; k\Omega$$

$$R_{WY} = \frac{3.9 \times 6.8 + 6.8 \times 1.8 + 1.8 \times 3.9}{6.8} = \frac{45.78}{6.8}$$

$$= 6.73 \; k\Omega$$

115

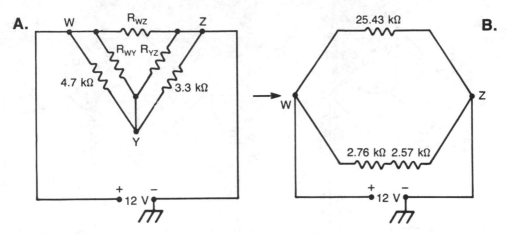

Figure 33-4

$$R_{YZ} = \frac{3.9 \times 6.8 + 6.8 \times 1.8 + 1.8 \times 3.9}{3.9} = \frac{45.78}{3.9}$$

$$= 11.74 \text{ k}\Omega$$

The circuit then reduces to the schematic of Fig. 33-4B.
Total equivalent resistance, R_T, presented to the 12 V source is:

$$R_T = \frac{25.43 \times (2.76 \times 2.57)}{25.43 + (2.76 + 2.57)}$$

$$= 4.41 \text{ k}\Omega$$

Chapter 34
Magnetic flux—magnetic flux density

FIGURE 34-1 SHOWS A TOROIDAL SOFT IRON RING ON WHICH IS WOUND A COIL OF N turns. A current of I amperes flows through the coil so that the ring is magnetized and a magnetic flux is established in the iron.

The flux is represented by magnetic lines of force. These are the directions in which isolated north poles would travel in a magnetic field. Such lines form closed loops, cannot intersect, and mutually repel one another. The degree of concentration of the lines of force indicates the strength of the magnetic field.

In a solid magnetized ring the flux lines would be continuous and no magnetic poles would exist. However, if a radial cut is made in the ring to create an air gap (Fig. 34-1), a north pole exists on one side of the gap while a south pole exists on the other side.

In a magnetic field the total number of lines is referred to as the magnetic flux whose letter symbol is the Greek lower case letter, phi (ϕ). In the SI system the unit of magnetic flux is the

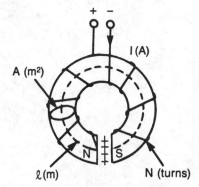

Figure 34-1

Demonstration of magnetic flux

Weber, whose unit symbol is Wb. This unit is defined through the phenomenon of electromagnetic induction (Chapter 38). If a conductor cuts a flux of 1 weber in one second, the voltage induced in the conductor is 1 volt. In practice the weber is too large a unit for measuring a typical magnetic flux so that the more common units are the microweber (μWb) and the milliweber (mWb).

To indicate the actual strength of a magnetic field we must have a unit for measuring the concentration of the lines of force. Referring to Fig. 34-1, the concentration in the air gap will be directly proportional to the magnetic flux, ϕ, but inversely proportional to the circular cross-sectional area, A, of the radial cut. The ratio of ϕ (Wb) to A (m^2) is called the flux density whose letter symbol is B and whose unit is the tesla (T).

MATHEMATICAL DERIVATIONS

$$\text{Flux Density, } B = \frac{\phi}{A} \text{ teslas} \tag{34-1}$$

$$\text{Flux, } \phi = BA \text{ webers} \tag{34-2}$$

Example 34-1

A flux density of 0.6 T is constant over a total area of 2.5 cm^2. Calculate the value of the total flux.

Solution

Area, A = 3.5 cm^2 = 3.5 × 10^{-4} m^2

$$\text{Total flux, } \phi = B \times A = 0.6 \text{ T} \times 2.5 \times 10^{-4} m^2 \tag{34-2}$$
$$= 1.5 \times 10^{-4} \text{ Wb}$$
$$= 150 \ \mu\text{Wb}.$$

Example 34-2

A magnetic field has a flux of 0.37 mWb, which is uniformly spread over an area of 4.3 cm^2. Calculate the value of the flux density.

Solution

$$\text{Area, } A = 4.3 \text{ cm}^2 \tag{34-1}$$
$$= 4.3 \times 10^{-4} \text{m}^2$$

$$\text{Flux density, } B = \frac{\phi}{A} = \frac{0.37 \times 10^{-3} \text{ Wb}}{4.3 \times 10^{-4} \text{m}^2}$$
$$= 0.86 \ T.$$

Chapter 35
The motor effect

LET AN EXTERNAL VOLTAGE, E, DRIVE AN ELECTRON FLOW, I, THROUGH A conductor that is situated in a uniform magnetic field. In other words, electrical energy is being supplied so that a force may be exerted on the conductor. The result is the so-called *motor effect* whereby electrical energy is converted into mechanical energy. The reverse effect occurs in a generator that changes mechanical energy into electrical energy.

Because it is carrying a direct current, the conductor will have its own surrounding magnetic field whose flux pattern consists of concentric circles with the conductor at the center. The two fields will therefore interact to produce a resultant flux pattern. Below the conductor (Fig. 35-1A) the magnetic fields have the same direction and will therefore combine to produce a concentration of the flux lines. However, the fields above the conductor have opposite directions and therefore the flux lines are spread far apart. The resultant flux pattern is illustrated in Fig. 35-1B and because the flux lines mutually repel one another, there will be an upward force exerted on the conductor. If the conductor is free to accelerate, the result would be to equalize the distribution of the flux lines in the resultant field.

The direction of the force may be found by the right hand or motor rule, which is illustrated in Fig. 35-1C. The *first* finger indicates the direction of the uniform magnetic *field*, whose flux density is B teslas. The *second* finger is the direction of the *current* (electron flow) while the thu*mb* shows the direction of the *motion* due to the force.

The principle we have illustrated is used in electrical motors and the moving coil instrument; it is also the means by which an electrical beam is deflected in one type of cathode-ray tube (CRT).

MATHEMATICAL DERIVATIONS

The magnitude on the force, F, exerted on the conductor is directly proportional to the flux density, B, teslas, of the uniform magnetic field, the current, I, amperes, flowing through the conductor and its length, l meters. The equation is therefore:

$$\text{Force, } F = B \times I \times l \text{ newtons} \tag{35-1}$$

EXAMPLE 35-1

A conductor of length 20 cm is placed at 90° to the direction of a uniform magnetic field whose flux

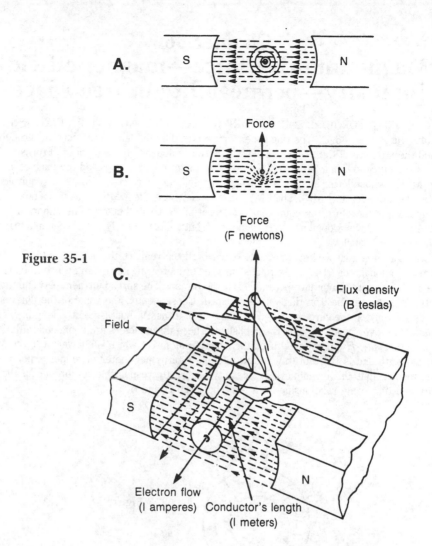

A.

B.

Force

Force
(F newtons)

Figure 35-1

C.

Flux density
(B teslas)

Field

Electron flow
(I amperes)   Conductor's length
(I meters)

density is 0.9 T. If the conductor is carrying a current of 4 A, what is the magnitude of the force exerted on the conductor?

Solution

$$B = 0.9 \ T, \ I = 4 \ A, \ l = 20 \ cm = 0.2 \ m.$$

Force exerted on the conductor,

$$F = B \times I \times l \tag{35-1}$$
$$= 0.9 \times 4 \times 0.2$$
$$= 0.72 \ N$$

Chapter 36
Magnetomotive force—magnetic field intensity—permeability of free space

CONSIDER A TOROIDAL SOFT-IRON RING THAT IS WOUND WITH AN EXCITING COIL of N turns (Fig. 36-1). The ring is being magnetized by the flow of the current through the coil and the strength of the magnetic field will therefore depend on the number of turns and the value of the current. Their product, IN ampere-turns or amperes, is called the magnetomotive force (MMF) whose letter symbol is $\mathcal{F}$. The magnetomotive force is the magnetic equivalent of the electromotive force (EMF). The EMF *creates* a current I, the MMF *establishes* a flux ϕ. Flux in magnetism is therefore the equivalent of current in electricity but with one important difference. A current flows around an electrical network but a flux is merely established in a magnetic circuit and does not flow.

Let us assume that we keep the same values of the current (I), the number of turns (N) and the cross-sectional area (A), but the ring is enlarged by increasing its circumference (the average of the inner and outer circumferences). Clearly the turns will be spread further apart and the flux density, B teslas, in the iron ring will be reduced. Consequently, the value of the flux density depends not only on the current, I, and the number of turns, N, but also on the length, l, of the magnetic path over which the flux is established. These three factors are contained in the *magnetic field intensity, H* (sometimes called the magnetizing force) which is equivalent to the MMF per unit length and is therefore measured in ampere-turns per meter or amperes per meter.

It follows that the magnetic field intensity is directly responsible for producing the flux density and that some relationship must exist between H and B.

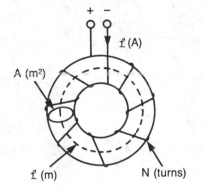

Figure 36-1

MATHEMATICAL DERIVATIONS

Magnetomotive force (MMF),

$$\mathcal{F} = IN = H \times l \text{ ampere-turns (AT) or amperes (A)} \qquad (36\text{-}1)$$

Magnetizing force,

$$H = \frac{IN}{l} = \frac{\mathcal{F}}{l} \text{ ampere-turns per meter (AT/m) or amperes per meter (A/m)} \qquad (36\text{-}2)$$

Figure 36-2 shows the cross section of a long straight conductor X, which is situated in a vacuum and carries an electron current of 1 A exiting from the paper. One of the circular lines of

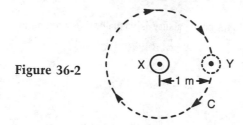

Figure 36-2

force has a radius of 1 m so that the length of its magnetic path is 2π meters. The MMF associated with this path is 1 A and the corresponding field intensity, H is $1/2\pi$ ampere per meter.

Y is a second conductor that is parallel to and 1 m away from the first conductor. Assuming that the second conductor also carries a current of 1 A, we will use the "motor rule" of Equation 35-1 to find the force exerted on one meter length of the second conductor. This force is:

$$F \text{ (newton)} = B \times I \times l \tag{36-3}$$
$$= B \times 1 \text{ A} \times 1 \text{ m}$$

where B is the flux density in teslas at the position of the second conductor.

From the definition of the ampere, this force must also be 2×10^{-7} N. Therefore,

$$F = B \times 1 \text{ A} \times 1 \text{ m}$$
$$= 2 \times 10^{-7} N$$

This yields

$$B = 2 \times 10^{-7} T \tag{36-4}$$

Consequently, a field intensity, H, of $1/2\pi$ ampere per meter is responsible for a flux density of $2 \times 10^{-7}T$. Therefore, in a vacuum (free space), the ratio of B to H is:

$$\mu_o = \frac{B}{H} = \frac{2 \times 10^{-7} \ T}{1/(2\pi)\text{A/m}} = 4\pi \times 10^{-7} \tag{36-5}$$
$$= 12.57 \times 10^{-7} \text{ SI units}$$

In a vacuum the ratio of B to H is called the permeability of free space whose letter symbol is μ_o and whose unit is the henry per meter. The term *permeability* measures the ability of a medium to permit the establishment of a flux density, B, by applying a magnetic field intensity, H. For free space only,

$$\mu_o = \frac{B}{H}, \tag{36-6}$$
$$B = \mu_o H, \ H$$
$$= \frac{B}{\mu_o}$$

Most materials are nonmagnetic and have virtually the same permeability as that of free space.

Example 36-1

An air gap has a length of 0.75 mm and a cross-sectional area of 2.5 cm². What is the value of the MMF required to establish a magnetic flux of 0.45 mWb in the air gap?

Solution

Because air is nonmagnetic, its permeability is virtually the same as in free space.

$$A = 2.5 \times 10^{-4} \text{ m}^2$$
$$l = 0.75 \times 10^{-3} \text{ m}$$
$$\phi = 0.45 \times 10^{-3} \text{ Wb.}$$

Flux density,

$$B = \frac{\phi}{A} = \frac{0.45 \times 10^{-3} \text{ Wb}}{2.5 \times 10^{-4} \text{ m}^2}$$
$$= 1.8 \ T.$$

Magnetic field intensity,

$$H = \frac{B}{\mu_o} = \frac{1.8 \ T}{12.57 \times 10^{-7} \text{ H/m}} \tag{36-6}$$
$$\mathscr{F} = 1.43 \times 10^6 \text{ AT/m.}$$

Magnetomotive force,

$$\mathscr{F} = H \times l \tag{36-1}$$
$$= 1.43 \times 10^6 \text{ AT/m} \times 0.75 \times 10^{-3} \text{ m}$$
$$= 1072.5 \text{ AT.}$$

Chapter 37
Relative permeability
—Rowland's law—reluctance

A CIRCULAR COIL THAT CONSISTS OF N TURNS AND CARRIES A CURRENT OF I amperes, is placed in a vacuum (free space). As the result of the magnetic field intensity, H, a certain flux density $B(B = \mu_o H$ teslas) will be established. If the same coil is now wound on a ferromagnetic toroidal ring (Fig. 37-1), the flux density for the same magnetic field intensity will be greatly increased. The factor by which the flux density is increased, is called the relative permeability, μ_r, which has no units and is just a number. However, the value of μ_r is not a constant but depends on the magnetic field intensity (Fig. 37-2). This is a result of the ferromagnetic material's atomic structure.

Because the MMF of a magnetic circuit can be compared with the EMF of an electrical circuit while the flux is regarded as similar to the current, it follows that the ratio of MMF: flux must be the magnetic equivalent of the resistance. This ratio is equal to a factor called the reluctance so that MMF, $\mathscr{F}$/Flux, ϕ = Reluctance, $\mathscr{R}$; this equation is sometimes referred to as Rowland's law. The reluctance is measured in ampere-turns per weber and is directly proportional to the length of the magnetic circuit but is inversely proportional to the relative permeability and the cross-

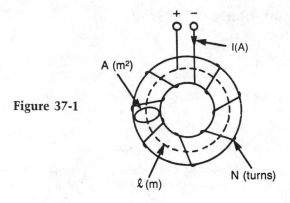

Figure 37-1

sectional area. This relationship recalls the equation for the resistance, R, of a cylindrical conductor, namely

$$R = \frac{\text{length, } \ell}{\text{Conductivity, } \sigma \times \text{cross-sectional area, } A} \text{ ohms.}$$

The principles outlined in the analyses of series and parallel resistor circuits can also be applied to magnetic circuits. For example, if a radial cut is made in the iron ring and we assume that the same flux is established in both the air gap and the iron, this is equivalent to saying that the current is the same throughout a series circuit. A certain number of ampere turns would be necessary to establish the flux in the air gap and a different number would create the same flux in the iron. The total required MMF would then be obtained by adding together the individual numbers of ampere turns. This is the same as finding the source voltage by adding together the individual voltage drops across two resistors in series.

Finally, we must mention that the concept of power is another important difference between magnetic circuits and electrical circuits. For an electrical circuit containing resistors, energy must be continuously supplied from the source, but when a flux is established in a magnetic circuit, no further energy is required.

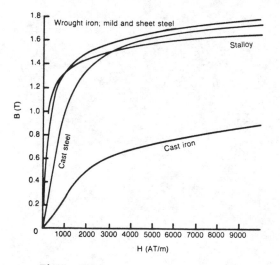

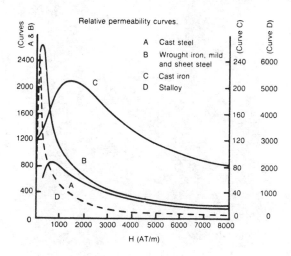

Figure 37-2

MATHEMATICAL DERIVATIONS

Relative permeability,

$$\mu_r = \frac{\text{Flux density with a magnetic core}}{\text{Flux density with a free space core}}$$

The flux density for a free space core is $\mu_o H$.

It follows that:

Flux density with a magnetic core,

$$B = \mu_o \mu_r H = \mu H \text{ teslas} \tag{37-2}$$

where $\mu = \mu_o \mu_r$ and is the absolute permeability which is measured in henrys per meter.

Remembering that the total flux, $\phi = B \times A$ Wb and the magnetomotive force, $\mathscr{F} = H \times \ell$ ampere turns,

$$\frac{\text{MMF}, \mathscr{F}}{\text{Flux}, \phi} = \frac{H \times \ell}{B \times A} = \frac{H \times \ell}{\mu_o \mu_r H \times A} = \frac{\ell}{\mu_o \mu_r A} \tag{37-3}$$

$$= \frac{\ell}{\mu A}$$

Therefore Rowland's law is

$$\text{Reluctance}, \mathfrak{R} = \frac{\mathscr{F}}{\phi} \tag{37-4}$$

$$= \frac{\ell}{\mu A} \text{ ampere-turns per weber}$$

Example 37-1

An air gap has a length of 2.5 mm and a cross-sectional area of 7 cm². What are the values of the gap's reluctance and the MMF required to establish a flux of 0.9 mWB across the gap?

Solution

$$\mu_r = 1, \ A = 7 \times 10^{-4} \text{m}^2, \ \ell = 2.5 \times 10^{-3} \text{ m}.$$

Reluctance of the air gap,

$$\mathfrak{R} = \frac{\ell}{\mu A} = \frac{\ell}{\mu_o \mu_r A} \tag{37-3}$$

$$= \frac{2.5 \times 10^{-3}}{12.57 \times 10^{-7} \times 7 \times 10^{-4}}$$

$$= 2.841 \times 10^6 \text{ AT/Wb}$$

$$\text{MMF}, \mathscr{F} = \mathfrak{R} \times \phi$$

$$= 2.841 \times 10^6 \times 0.9 \times 10^{-3}$$

$$= 2557 \text{ AT}$$

Alternatively:

Flux density,

$$B = \frac{\phi}{A} = \frac{0.9 \times 10^{-3}}{7 \times 10^{-4}}$$
$$= 1.286 \text{ T.}$$

Magnetic field intensity,

$$H = \frac{B}{\mu_o} = \frac{1.286}{12.57 \times 10^{-7}}$$
$$= 1.023 \times 10^6 \text{ AT/m.}$$
$$\text{MMF, } \mathscr{F} = H \times \ell = 1.023 \times 10^6 \times 2.5 \times 10^{-3}$$
$$= 2557 \text{ AT.}$$

Example 37-2

A toroidal ring of cast steel has a mean circumference of 45 cm and a cross-sectional area of 4 cm². A radial cut creates an air gap with a thickness of 1.6 mm. Neglecting any flux leakage, what is the total MMF required to establish a flux density of 1.32 T in the air gap?

Solution

The air gap and the cast steel ring must be regarded as in series. From Fig. 37-2 a flux density of 1.32 T corresponds to a magnetic field intensity of 2000 AT/m. MMF required to establish the flux in the steel ring is

$$\mathscr{F}_1 = H_1 \times \ell_1 = 2000 \times 45 \times 10^{-2}$$
$$= 900 \text{ AT.}$$

For the air gap the magnetic field intensity is

$$H_2 = \frac{B}{\mu_o} = \frac{1.32}{12.57 \times 10^{-7}}$$
$$= 1.05 \times 10^6 \text{ AT/m.}$$

MMF required to establish the flux in the air gap is

$$\mathscr{F}_2 = H_2 \times \ell_2 = 1.05 \times 10^6 \times 1.6 \times 10^{-3}$$
$$= 2520 \text{ AT.}$$

Total MMF required,

$$\mathscr{F}_T = \mathscr{F}_1 + \mathscr{F}_2 = 900 + 2520$$
$$= 3420 \text{ AT.}$$

Chapter 38
Electromagnetic induction—
Faraday's law—Lenz's law

ELECTROMAGNETIC INDUCTION IS THE LINK BETWEEN MECHANICAL ENERGY, magnetism, and electrical energy. In other words we are discussing the principles behind the electrical generator. In 1831 Faraday discovered that when a conductor cuts or is cut by (mechanical energy) a magnetic field (magnetism), and EMF (electricity) is *induced* in the conductor. Notice that is only necessary for there to be relative motion between the conductor and the flux lines of the magnetic field.

Faraday's law is illustrated in Fig. 38-1A. This shows a conductor that is mechanically moved to cut the flux lines of a magnetic field at right angles (90°). The ends of the conductor are connected to a sensitive voltmeter that records the value of the EMF. When the conductor is stationary, the free electrons are moving in random directions. However, when the conductor cuts the magnetic flux, the electrons are given a movement in a particular direction. Through the motor effect, forces are then exerted on the moving charges so that the electrons are driven along the conductor. One end is then negatively charged and the other end is positively charged so that the EMF is induced in the conductor. The magnitude of this EMF is directly proportional to the flux density, the length of the conductor, and its velocity at right angles to the lines of force. These are the results of Faraday's law, which states that the magnitude of the EMF depends on the rate of the cutting of the magnetic flux.

The directions of the current (electron flow), the magnetic flux and the conductor's motion are mutually perpendicular. This may be found by the *left-hand* "generator" rule (as opposed to the right-hand "motor" rule of Chapter 35). The *first* finger indicates the direction of the *flux*, the

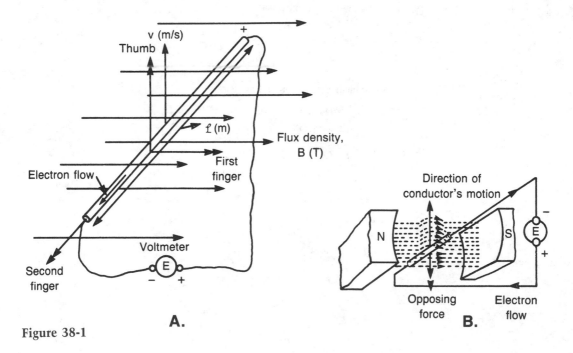

Figure 38-1

second finger is the direction of the electron flow through the conductor and the thumb shows the direction of the conductor's *movement* (Fig. 38-1A).

The output electrical energy is created from the work done in moving the conductor. The conductor itself carries a current and is therefore surrounded by its own flux, which reacts with the uniform magnetic field. Consequently, the conductor experiences a force whose direction opposes the motion of the conductor. This may be verified by applying the right-hand "motor" rule. Mechanical work must therefore be done in order to move the conductor against the opposing force (Fig. 38-1B). This is all summarized in Lenz's law, which states that *the direction of the induced EMF is such as to oppose the change that originally created the EMF.* Such a law sounds more like a riddle so let us examine the statement in some detail. The "original change" is the conductor's motion that creates the induced EMF by cutting the uniform flux. This EMF produces a current flow in a particular direction so that a second magnetic field appears around the conductor. The reaction between the two fields causes a force to be exerted on the conductor and the direction of this force opposes the conductor's original motion. Lenz's law is really saying that you cannot obtain the electrical energy without supplying the mechanical work.

MATHEMATICAL DERIVATIONS

Faraday's law states that the magnitude of the EMF depends on the amount of the magnetic flux, ϕ, cut by the conductor in one second. Therefore

$$\text{Induced EMF, } E = -\frac{\Delta\phi}{\Delta T} \text{ volts} \tag{38-1}$$

The minus sign illustrates the meaning of Lenz's law.

If a flux of one weber is cut in a time of one second, the induced EMF is one volt. This enables the weber unit to be defined.

Let us assume the uniform flux density is B teslas, the conductor's length is l meters and the conductor is moving upwards at 90° to the magnetic field with a velocity of v m/s. In 1 second (ΔT) the conductor covers a vertical distance of v meters and sweeps out an area, $\Delta A = v \times l \, \text{m}^2$. Then:

$$\text{Flux cut by the conductor in 1 s} = B \times \Delta A = Blv \text{ webers} \tag{38-2}$$

Value of the induced EMF,

$$E = \frac{\Delta\phi}{\Delta T} \tag{38-3}$$

$$= \frac{Blv \text{ (Wb)}}{1 \text{ s}}$$

$$= Blv \text{ volts}$$

Example 38-1

A conductor cuts a flux of 650 μWb in a time of 0.4 ms. What is the value of the induced EMF?

Solution

$$\text{Flux change, } \Delta\phi = 650 \times 10^{-6} \text{ Wb.}$$

$$\text{Time change, } \Delta T = 0.4 \times 10^{-3} \text{ s.}$$

Value of the induced EMF,

$$E = \frac{\Delta\phi}{\Delta T} \qquad (38\text{-}1)$$

$$= \frac{650 \times 10^{-6} \text{ Wb}}{0.4 \times 10^{-3} \text{ s}}$$

$$= 1.625 \text{ V.}$$

Example 36-2

A conductor of length 30 cm is moving at 90° to the direction of a uniform magnetic field with a flux density of 0.7 T. If the conductor's velocity is 130 m/s, what is the value of the induced EMF?

Solution

$B = 0.7$ T, $l = 30 \times 10^{-2}$ m, $v = 130$ m/s

Value of the induced EMF,

$$E = Blv \qquad (38\text{-}3)$$

$$= 0.7 \text{ T} \times 30 \times 10^{-2} \text{ m} \times 130 \text{ m/s}$$

$$= 27.3 \text{ V.}$$

Chapter 39
Self-inductance

AN ELECTRICAL CIRCUIT CAN ONLY POSSESS THREE PASSIVE QUANTITIES— resistance, inductance, and capacitance. In previous chapters we have already learned that resistance opposes and therefore limits the current. By contrast, inductance is defined as that electrical property that prevents any sudden or abrupt *change* of current and also limits the rate of the *change* in the current. This means that, if an electrical circuit contains inductance, the current of that circuit can neither rise nor fall instantaneously.

Although a straight length of wire possesses some inductance, the property is most marked in a coil which is called an inductor and whose circuit symbol is . When a current increases or decreases in a coil, the surrounding magnetic field will expand or collapse and cut the turns of the coil itself. On the principle of electromagnetic induction (Chapter 38), a voltage or counter EMF will be induced into the coil but this voltage will depend on the *rate* at which the current is *changing* rather than on the value of the current itself. Due to the *change* in the current, the coil creates its own *moving* flux that cuts its own turns and induces the voltage across the coil. For these reasons the property of the coil is referred to as its self-inductance whose letter symbol is L.

In Fig. 39-1A we are comparing the properties of resistance and inductance. When the switch is closed in position 1, the current in the resistor immediately rises from zero to a constant value of 10 V/5 Ω = 2 A (Fig. 39-1B). The voltage drop across the resistor (10 V) then exactly balances (or opposes) the source voltage (10 V).

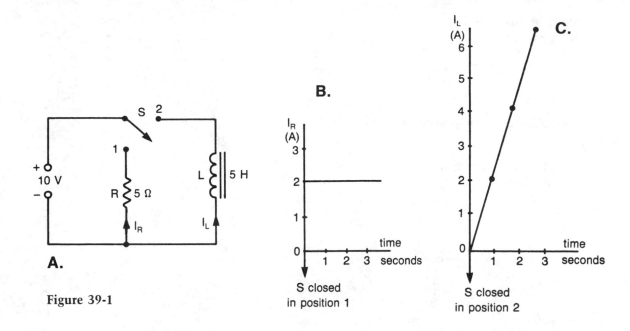

Figure 39-1

B.

S closed
in position 1

C.

S closed
in position 2

If the switch is now moved to position 2, the resistor current immediately drops to zero but the inductor current starts to grow in the coil. This creates a magnetic flux which, as it expands outwards, cuts the turns of the coil and induces the counter EMF, which must exactly balance the source voltage (assuming that the coil only possesses inductance and has zero resistance). Because the source voltage is fixed, it follows that the current must start at zero and must then grow at a constant rate that is measured in amperes per second (Fig. 39-1C).

Although the counter EMF is proportional to the rate of the change of the current, its value must also depend on some factor of the coil. This factor is called the self-inductance, L, which is determined by such physical quantities as the number of turns, the length, the cross-sectional area and the nature of the core on which the coil is wound. The unit of inductance is the henry (H) which is defined as follows. The inductance is one henry if, when the current is changing at the rate of one ampere per second, the induced EMF is one volt. Consequently, in Fig. 39-1A where the source voltage is 10 V and the inductance is 5 H, the rate of the growth of the current is 10 V/5 H = 2 A/s.

MATHEMATICAL DERIVATIONS
Switch S in position 1

Voltage drop across the resistor,

$$V_R = -I_R \times R \text{ volts} \tag{39-1}$$

Using Kirchhoff's voltage law (KVL):

$$E + V_R = 0 \tag{39-2}$$

Equations 39-1, 39-2 yield

$$E = I_R \times R \text{ volts} \tag{39-3}$$

Switch S in position 2

Counter EMF,

$$V_L = -L \times \frac{\Delta I_L}{\Delta T} \text{ volts} \tag{39-4}$$

where: $\Delta I_L / \Delta T$ – rate of the change of the current in amperes per second $\tag{39-5}$

Using Kirchhoff's voltage law (KVL),

$$E + V_L = 0 \tag{39-6}$$

From Equations 39-4, 39-5
Source voltage,

$$E = L \times \frac{\Delta I_L}{\Delta T} \text{ volts} \tag{39-7}$$

Therefore,
Rate of the change of the current,

$$\frac{\Delta I_L}{\Delta T} = \frac{E}{L} \text{ amperes per second} \tag{39-8}$$

Self-inductance,

$$L = \frac{E}{\Delta I_L / \Delta T} \text{ henrys} \tag{39-9}$$

Current after T seconds,

$$I_L = \frac{\Delta I_L}{\Delta T} \times T \text{ amperes} \tag{39-10}$$

Example 39-1

A 12-V dc source is suddenly switched across a 3-H inductor of negligible resistance. What are the values of the initial current, the rate of the growth of the current, the counter EMF, and the current after 0.6 second?

Solution

Initial current is *zero*.
Rate of the growth of the current,

$$\frac{\Delta I_L}{\Delta T} = \frac{E}{L} \tag{39-8}$$

$$= \frac{12 \text{ V}}{3 \text{ H}}$$

$$= 4 \text{ A/s.}$$

Counter EMF,

$$V_L = (-)E = 12 \text{ V.} \tag{39-6}$$

Current after 0.6 s,

$$I_L = \frac{\Delta I_L}{\Delta T} \times T \tag{39-10}$$

$$= 4 \text{ A/s} \times 0.6 \text{ s} = 2.4 \text{ A.}$$

Example 39-2

The counter EMF induced in a coil is 30 V when the current changes by 12 mA in a time of 5 μs. What is the value of the coil's self-inductance?

Solution

Rate of the change of the current,

$$\frac{\Delta I_L}{\Delta T} = \frac{12 \times 10^{-3} \text{ A}}{5 \times 10^{-6} \text{ s}}$$

$$= 2400 \text{ A/s}$$

Self-inductance,

$$L = \frac{E}{\Delta I_L / \Delta T} = \frac{30 \text{ V}}{2400 \text{ A/s}} \tag{39-9}$$

$$= 0.00125 \text{ H}$$

$$= 1.25 \text{ mH.}$$

Chapter 40
Factors determining
a coil's self-inductance

FOR A COIL A HIGH VALUE OF INDUCTANCE REQUIRES THAT A LARGE COUNTER EMF is induced for a given rate of change in the current. Clearly the number of turns is one of the factors involved, since each turn can be regarded as a conductor and the magnetic field intensity is directly proportional to the number of turns (Fig. 40-1). In fact the value of the inductance is directly proportional to the *square* of the number of turns (as will be proved in the mathematical derivations that follow).

The size of each turn is another factor that determines the value of the inductance. If the cross-sectional area of each turn is increased, the coil's reluctance [$\Re = l/(\mu_o \mu_r A)$] will be less and the flux density will be greater. Looking at it another way, if we continue to reduce the cross-sectional area, the coil would ultimately vanish and the inductance would obviously be zero.

If the number of turns and their size are kept constant, but the coil's length is increased, the turns are spread farther apart and there is a greater reluctance. Consequently, the flux density and the inductance are reduced.

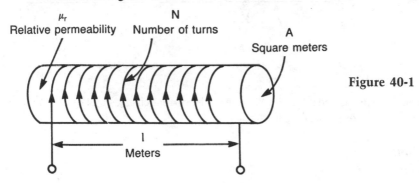

Factors affecting the value of the self-inductance

μ_r
Relative permeability

N
Number of turns

A
Square meters

l
Meters

Figure 40-1

Finally, we can replace a nonmagnetic core with one that is made from ferromagnetic material with a high relative permeability. This greatly lowers the amount of the reluctance and therefore increases the flux density. The result is a much higher inductance value.

Summarizing, the inductance value is directly proportional to the square of the number of turns, their cross-sectional area and the coil's permeability, but is inversely proportional to the coil's length.

The values of practical inductors range from henrys to microhenrys. Those with a value of several henrys are referred to as chokes (circuit symbol ⌒⌒) which are made with a large number of copper turns wound on an iron core. The presence of the core is indicated by the two lines drawn to one side of the circuit symbol. Such a core is laminated (cut into slices that are insulated from each other) to reduce the loss owing to eddy currents. Because the relative permeability varies with the amount of the direct current flowing through the coil, the inductance value is normally quoted for a particular dc current level.

A coil with a value of a few microhenrys only consists of a few copper turns that are wound on a nonmagnetic core (circuit symbol ⌒). Between these two extremes are millihenry inductors with cores of either iron dust (circuit symbol ⌒⌒) or a ferrite material.

MATHEMATICAL DERIVATIONS

Regarding each turn as a conductor and using Faraday's law,

$$\text{EMF, } E = N \frac{\Delta\phi}{\Delta T} \tag{40-1}$$

$$= L \frac{\Delta I}{\Delta T} \text{ volts}$$

This yields
Self-Inductance,

$$L = N \frac{\Delta\phi}{\Delta I} \tag{40-2}$$

$$= \frac{\text{change in flux linkages}}{\text{change in current}} \text{ henrys}$$

The "flux linkages" is the product of the flux and the number of turns with which the flux is linked. The inductance is one henry if a current change of one ampere creates a change in the flux linkages of 1 Wb turn.

132

Self-Inductance,

$$L = N \frac{\Delta\phi}{\Delta I} \tag{40-3}$$

$$= N \frac{\Delta(BA)}{\Delta I}$$

$$= NA \frac{\Delta B}{\Delta I} \text{ henrys}$$

From $B = \mu_o\mu_r H$ teslas (Equation 37-2),

$$L = N \times \frac{\Delta B}{\Delta I} \tag{40-4}$$

$$= NA \times \mu_o\mu_r \frac{\Delta H}{\Delta I} \text{ henrys}$$

Using $H = IN/l$ amperes per meter,

$$L = NA \times \mu_o\mu_r \frac{\Delta(IN/l)}{\Delta I}$$

Self-Inductance,

$$L = \frac{\mu_o\mu_r N^2 A}{l} \text{ henrys} \tag{40-5}$$

where μ_o, permeability of free space $= 4\pi \times 10^{-7}$ henry per meter. Equation 40-5 demonstrates mathematically that the inductance value is directly proportional to the square of the number of turns, the cross-sectional area and the core's relative permeability, but is inversely proportional to the coil's length.

For the nonmagnetic core, $\mu_r = 1$, and therefore,
Self-Inductance,

$$L = \frac{\mu_o N^2 A}{l} \text{ henrys} \tag{40-6}$$

Example 40-1

A coil is wound with 800 turns on a ferromagnetic core with a relative permeability of 1100. If the coil's length is 5.5 cm and its cross-sectional area is 3.75 cm², calculate the value of the coil's self-inductance.

Solution

Self-Inductance,

$$L = \frac{\mu_o\mu_r N^2 A}{l} \text{ henrys} \tag{40-5}$$

$$= \frac{4 \times \pi \times 10^{-7} \times 1100 \times 800^2 \times 3.75 \times 10^{-4}}{5.5 \times 10^{-2}}$$

$$= 3.02 \text{ H.}$$

If the same coil were wound on a nonmagnetic core, the self-inductance would only be 3.02/1100 H = 2.75 mH.

Chapter 41
Energy stored in the
magnetic field of an inductor

THE INDUCTOR IS ASSUMED TO BE IDEAL IN THE SENSE THAT THERE ARE NO resistance losses associated with the component. When S is closed, the growth rate of the current is E/L amperes per second so that the counter EMF induced in the inductor exactly balances the source voltage (Fig. 41-1A). After a period of t seconds the current is equal to I amperes, and because the growth of the current is linear with respect to time, the average current during the period is $I/2$ amperes (Fig. 41-1B). Around the inductor is established a magnetic field that represents the energy derived from the source. If the switch is then opened, the magnetic field collapses and the energy is dissipated in the form of an arc, which occurs between the contacts of the switch.

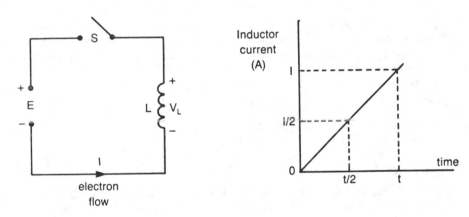

Figure 41-1

MATHEMATICAL DERIVATIONS

Source voltage,

$$E = L \times \text{rate of change of current} \qquad (41\text{-}1)$$

$$= \frac{L \times I}{t} \text{ volts}$$

Average power at the source over the period of t seconds

$$P = E \times \frac{I}{2} = L \times \frac{I}{t} \times \frac{I}{2} \qquad (41\text{-}2)$$

$$= \frac{1}{2} \times \frac{LI^2}{t} \text{ watts}$$

Total energy stored in the magnetic field = average power × time

$$= \frac{1}{t} \times \frac{LI^2}{2} \times t \qquad (41\text{-}3)$$

$$= \frac{1}{2} \times LI^2 \text{ joules}$$

Alternatively

Let a current I be flowing through an inductor, L. If the current increases by an amount, δI, in a time of δT, the induced voltage is $L \times \delta I/\delta t$ and the additional energy delivered to the inductor is $(L \times \delta I)/\delta t \times I \times \delta t = LI\delta I$ joules. Integrating over the period of t seconds during which the current increased from zero to I amperes:

Total energy stored in the magnetic field

$$W = \int_{o}^{I} LI \, dI \qquad (41\text{-}4)$$

$$= \frac{1}{2} LI^2 \text{ joules}$$

Example 41-1

In Fig. 41-1, $E = 60$ V and $L = 8$ H. When the switch is closed, what is the rate of growth of the current? Calculate the amount of energy stored in the magnetic field after a time of 5 seconds.

Solution

Rate of growth of the current = E/L = 60 V/8 H = 7.5 amperes per second.

After 5 seconds, current I = 7.5 A/s × 5 s = 37.5 amperes.

Energy stored in magnetic field

$$W = \frac{1}{2} LI^2 = \frac{1}{2} \times 8 \text{ H} \times (37.5 \text{ A})^2 \qquad (41\text{-}3)$$

$$= 5625 \text{ J}$$

Chapter 42
Inductors in series

IT IS ASSUMED THAT ALL THREE INDUCTORS OF FIG. 42-1 ARE IDEAL IN THE SENSE that they have no losses and they only possess the property of inductance. Because these components are joined end-to-end, the same *rate of change* in the current will be associated with all of the inductors. In each inductor the expanding magnetic field will induce a counter EMF (V_{L1}, V_{L2}, V_{L3}) and the sum of these EMFs must exactly balance the source voltage.

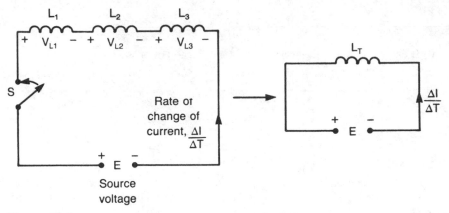

Figure 42-1

When the switch, S, is closed there will be a constant rate of growth in the current, and the magnetic field surrounding each inductor will expand. After a period of time there will be a certain amount of energy stored in each of the magnetic fields, and the sum of these energies will equal the total energy derived from the source voltage. Because the inductors are regarded as ideal components, there is no power *dissipated* in the circuit as heat.

From the mathematical derivations we shall see that the total inductance is the sum of the individual inductances. Consequently, one use of connecting inductors in series is to create a nonstandard total inductance value from standard components in the same way a series arrangement of small inductors can avoid the use of a physically larger inductor. However, in such an arrangement it is essential that the currents ratings of the series inductors are compatible.

MATHEMATICAL DERIVATIONS

$$\text{Rate of growth in the current} = -\frac{\Delta I}{\Delta T} \text{ amperes per second} \qquad (42\text{-}1)$$

ΔI represents "a small change in the current" while ΔT is the corresponding "small change in the time"

Counter EMF,

$$V_{L1} = -L_1 \frac{\Delta I}{\Delta T}, \ V_{L2} = -L_2 \frac{\Delta I}{\Delta T}, \ V_{L3} = -L_3 \frac{\Delta I}{\Delta T} \text{ volts} \qquad (42\text{-}2)$$

The negative sign for each counter EMF and the rate of growth in the current indicates that the counter EMF opposes (balances) the source voltage. It would be equally correct to write $V_R = -I \times R$ for the voltage drop across the resistor although this is rarely done.

Kirchhoff's voltage law (KVL):

$$E + V_{L1} + V_{L2} + V_{L3} = 0 \qquad (42\text{-}3)$$

Combining Equations 42-2, 42-3:

$$E = L_1 \frac{\Delta I}{\Delta T} + L_2 \frac{\Delta I}{\Delta T} + L_3 \frac{\Delta I}{\Delta T} \qquad (42\text{-}4)$$

$$= \frac{\Delta I}{\Delta T} \times (L_1 + L_2 + L_3) \text{ volts}$$

If L_T is the total equivalent inductance,

$$E = L_T \times \frac{\Delta I}{\Delta T} = \frac{\Delta I}{\Delta T} \times L_T \text{ volts} \qquad (42\text{-}5)$$

Comparing Equations 42-4, 42-5:

$$L_T = L_1 + L_2 + L_3 \text{ henrys} \qquad (42\text{-}6)$$

For N inductors connected in series,

$$L_T = L_1 + L_2 + L_3 + \cdots + L_N \text{ henrys} \qquad (42\text{-}7)$$

If the N inductors are all of equal value, L,

$$L_T = NL \text{ henrys} \qquad (42\text{-}8)$$

Voltage division:

$$V_{L1} = -L_1 \times \frac{\Delta I}{\Delta T} = L_1 \times \frac{E}{L_T} = E \times \frac{L_1}{L_T} \text{ volts} \qquad (42\text{-}9)$$

$$V_{L2} = E \times \frac{L_2}{L_T} \text{ volts} \qquad (42\text{-}10)$$

$$V_{L3} = E \times \frac{L_3}{L_T} \text{ volts} \qquad (42\text{-}11)$$

Consequently, the equations for inductors in *series* are comparable with the equations for resistors in *series*.

Energy stored in the magnetic field surrounding the $L1$ inductor,

$$W_1 = \frac{1}{2} L_1 I^2 \text{ joules} \qquad (42\text{-}12)$$

Energy stored in the magnetic field surrounding the $L2$ inductor,

$$W_2 = \frac{1}{2} L_2 I^2 \text{ joules} \qquad (42\text{-}13)$$

Energy stored in the magnetic field surrounding the $L3$ inductor,

$$W_3 = \frac{1}{2} L_3 I^2 \text{ joules} \qquad (42\text{-}14)$$

Total energy derived from the source,

$$W_T = W_1 + W_2 + W_3 \qquad (42\text{-}15)$$

$$= \frac{1}{2} L_T I^2 \text{ joules}$$

Example 42-1

In Fig. 42-1, $L1 = 4$ H, $L2 = 8$ H, $L3 = 20$ H, and $E = 43$ V. Calculate the values of L_T, and the rate of growth of the current, V_1, V_2, and V_3. If the switch is closed for a period of four seconds, what is the energy stored in the magnetic field surrounding each inductor and what is the total energy derived from the source voltage?

Solution

Total inductance

$$L_T = L_1 + L_2 + L_3 = 4 + 8 + 20 \qquad (42\text{-}7)$$
$$= 32 \ H$$

Rate of growth in the current,

$$\frac{\Delta I}{\Delta T} = \frac{E}{L_T} = \frac{43 \ V}{32 \ H} = 1.344 \ \text{A/s} \qquad (42\text{-}5)$$

Voltage across $L1$ inductor,

$$V_{L1} = L_1 \times \frac{E}{L_T} = 4 \ H \times \frac{43 \ V}{32 \ H} = 5.375 \ V \qquad (42\text{-}9)$$

Voltage across $L2$ inductor,

$$V_{L2} = L_2 \times \frac{E}{L_T} = 8 \ H \times \frac{43 \ V}{32 \ H} = 10.75 \ V \qquad (42\text{-}10)$$

Voltage across $L3$ inductor,

$$V_{L3} = L_3 \times \frac{E}{L_T} = 20 \ H \times \frac{43 \ V}{32 \ H} = 26.875 \ V \qquad (42\text{-}11)$$

Voltage check:
Source voltage,

$$E = V_{L1} + V_{L2} + V_{L3} = 5.375 + 10.75 + 26.875 = 43 \ V \qquad (42\text{-}3)$$

After a period of four seconds,
Current,

$$I = 1.344 \ \text{A/s} \times 4 \ \text{s} = 5.376 \ \text{A}.$$

Energy stored in $L1$ inductor,

$$W_1 = \frac{1}{2} L_1 I^2 = \frac{1}{2} \times 4 \ H \times (5.376 \ \text{A})^2 = 57.8 \ \text{joules} \qquad (42\text{-}12)$$

Energy stored in $L2$ inductor,

$$W_2 = \frac{1}{2} L_2 I^2 = \frac{1}{2} \times 8 \ H \times (5.376 \ \text{A})^2 = 115.6 \ \text{joules} \qquad (42\text{-}13)$$

Energy stored in $L3$ inductor,

$$W_3 = \frac{1}{2} L_3 I^2 = \frac{1}{2} \times 20 \ H \times (5.376 \ \text{A})^2 = 289.0 \ \text{joules} \qquad (42\text{-}15)$$

Total energy stored,

$$W_T = 57.8 + 115.6 + 289.0 = 462.4 \ \text{joules} \qquad (42\text{-}15)$$

Energy check:
Total energy derived from the source,

$$W_T = \frac{1}{2} L_T I^2 \qquad \text{(42-15)}$$

$$= \frac{1}{2} \times 32 \text{ H} \times (5.376 \text{ A})^2$$

$$= 462.4 \text{ joules}$$

❧

Chapter 43
The L/R time constant

FROM CHAPTER 39 WE KNOW THAT THE PROPERTY OF SELF-INDUCTANCE prevents any sudden change of the current and also limits the rate at which the current can change. Consequently, when the switch S, is closed in position 1 (Fig. 43-1A), the current must take a certain time before reaching a final steady level, which is determined by the value of the resistor, R. We are therefore going to discuss the factors that control the duration of the so-called *transient* or *changing* state. This is the interval between the time at which the switch, S, is closed in position 1, and the time when the final or steady-state conditions are reached.

Immediately after the switch is closed in position 1, the current must initially be zero so that the voltage drop across the resistor, R, must also be zero. Consequently to satisfy Kirchhoff's Voltage law, the counter EMF created in the inductor, L, must equal the source voltage, E. The initial rate of the current growth is therefore at its maximum value of E/L amperes per second.

As the current increases and the magnetic field expands outwards, the voltage drop across the resistor rises and the counter EMF in the inductor falls because at all times the sum of these two voltages must exactly balance the source voltage, E. When the counter EMF falls, the rate of the current's growth is correspondingly less, so there is a situation in which the greater the value of the current, the less is its rate of growth. Theoretically, it would take infinite time for the

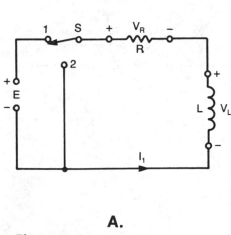

A.

Figure 43-1

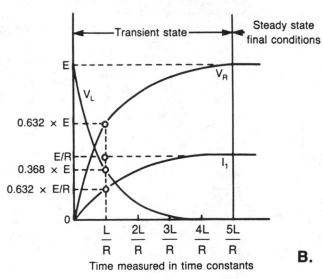

B.

Time measured in time constants

current to reach its final steady value of E/R amperes, but as we shall see in the mathematical derivations, the current reaches its final value to within 1% after a finite time interval that is determined by the values of the inductance and the resistance.

Remember that the initial rate of current growth is E/L amperes per second while the final current value is E/R amperes. Consequently, *if* the initial rate of current growth were maintained, the current would reach its final value after an interval of

$$\frac{E/R \text{ A}}{E/L \text{ A/s}} = \frac{L}{R} \text{ seconds}$$

which is referred to as the *time constant* of the circuit. However, because the rate of current growth falls off, the current only reaches 63.2% of its final value after one time constant. It requires *five* time constants before the current can rise to within 1% of E/R amperes (Fig. 43-1B) and we can assume that the transient growth period has been completed.

Notice that the conditions at the start of the transient state can be predicted by regarding the inductor as an open circuit while the final steady-state values can be obtained by considering the inductor to be a short circuit.

Assuming that the steady-state conditions have been reached, let us now move the switch, S, to position 2. The current cannot change instantaneously so that I_2 must still be E/R amperes and the voltage across the resistor must remain at E volts. By the KVL rule the inductor's counter EMF will also be E volts but its polarity is reversed because the magnetic field is starting to collapse rather than expand (Fig. 43-2A).

As the current decays, V_R and V_L must fall together in order to maintain the voltage balance around the closed loop. When V_L decreases, the rate of current decay is less, so that theoretically it would take infinite time for the current to reach zero. However, after one time constant the

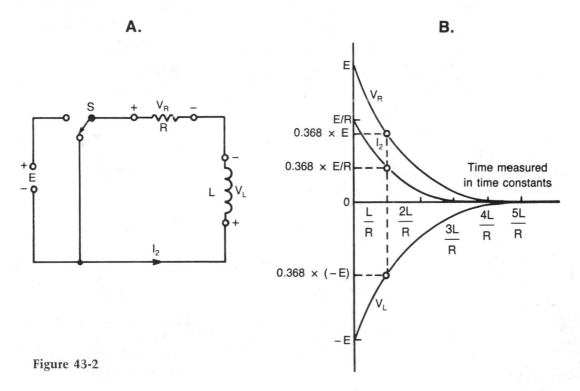

A.

B.

Figure 43-2

current has *lost* 63.2% of its initial value and has therefore dropped to 36.8% of E/R amperes. After five time constants we assume that the transient decay period has been concluded in the sense that I_2, V_R and V_L have all fallen to less than 1% of their original values (Fig. 43-2B).

MATHEMATICAL DERIVATIONS

In Fig. 43-1A, the switch, S, is closed in position 1. The initial conditions are:
$I_1 = 0$ amperes, $V_R = 0$ volts, $V_L = E$ volts, initial rate of current growth = E/L amperes per second.

At all times:

$$V_R + V_L = E \tag{43-1}$$

$$I_1 \times R + L\frac{dI_1}{dt} = E$$

Integrating the equation throughout,

$$\int dt = \int \frac{L}{E - I_1R}\, dI_1$$

$$t + k = -\frac{L}{R}\, ln\, (E - I_1R)$$

When $t = 0$, $I_1 = 0$ and $k = -L/R\, ln\, E$.

Therefore

$$t = -\frac{L}{R}\, ln\, \left(\frac{E - I_1R}{E}\right) \tag{43-2}$$

$$I_1 = \frac{E}{R}\, (1 - e^{-Rt/L})\ \text{amperes}$$

where $e = 2.7183 \ldots$ and is the base of the natural logarithms.

Voltage drop across the resistor,

$$V_R = I_1 \times R = E(1 - e^{-Rt/L})\ \text{volts} \tag{43-3}$$

The expressions for I_1 and V_R represent the equations of exponential growth.

When $t = L/R$ seconds, $I_1 = E/R\, (1 - e^{-1}) = 0.632\, E/R$ amperes (63.2% of the final steady-state current). After $t = 5L/R$ seconds, $I_1 = E/R\, (1 - e^{-5}) = 0.993\, E/R$ amperes so that the current has reached a level which is within 1% of its ultimate value.

Counter EMF induced in the coil,

$$V_L = E - V_R = E\, e^{-Rt/L} \tag{43-4}$$

The expression for V_L represents an equation for exponential decay.

In Fig. 43-2A, the switch, S, is moved to position 2. The initial conditions are:
$I_2 = E/R$ amperes, $V_R = E$ volts, $V_L = -E$ volts and the initial rate of the current's *decay*, $d\, I_2/dt = -E/L$ amperes per second.

At all times:

$$V_R + V_L = 0.$$

$$I_2R + L\frac{dI_2}{dt} = 0.$$

$$\int dt = -\frac{L}{R} \int \frac{1}{I_2} dI_2$$

$$t + k = -\frac{L}{R} \ln I_2$$

When $t = 0$, $I_2 = E/R$ amperes and $k = -L/R \ln E/R$. Therefore

$$t = -\frac{L}{R} \ln \left(\frac{R \, I_2}{E}\right) \tag{43-5}$$

$$I_2 = \frac{E}{R} \, e^{-Rt/L} \text{ amperes}$$

$$V_R = I_2 \times R \tag{43-6}$$
$$= E \, e^{-Rt/L} \text{ volts}$$

$$V_L = -E \, e^{-Rt/L} \text{ volts} \tag{43-7}$$

The expressions for I_2, V_R, V_L all represent equations of exponential decay (Fig. 43-2B).

When $t = L/R$ seconds, $I_2 = 0.368 \, E/R$ amperes, $V_R = 0.368 \, E$ volts and $V_L = -0.368 \, E$ volts. After $t = 5 \, L/R$ seconds, the values of I_2, V_R, V_L have all dropped to 0.7% of their original levels. The transient decay period is therefore virtually finished.

Example 43-1

In Fig. 43-1A, $R = 3.9$ kΩ, $L = 25$ mH and $E = 80$ V. When S is closed in position 1, what are the initial values of I_1, V_R, V_L and their values after time intervals of 4.49, 6.41, 12.82, 19.23, 25.64, 32.05 and 64.10 microseconds?

Solution

Time constant, $L/R = 25$ mH/3.9 kΩ = 6.41 μs. Initial values of I_1, V_R, V_L are respectively zero amperes, zero volts, and 80 volts.

The time interval of 4.49 μs is equal to $4.49/6.41 = 0.7 \times$ time constant. Final current, E/R, is 80 V/3.9 kΩ = 20.5 mA. Then,

$$I_1 = E/R \, (1 - e^{-Rt/L}) = 20.5 \, (1 - e^{-0.7}) = 10.25 \text{ mA}.$$

Note that the current has reached 50% of its final value in the interval of $0.7 \times$ time constant.

$$V_R = 80 \times 0.5 = 40 \text{ V}.$$
$$V_L = 80 - 40 = 40 \text{ V}.$$

The time interval of 6.41 μs is equal to one time constant and therefore

$$I_1 = 20.5 \, (1 - e^{-1}) = 20.5 \times 0.632 = 12.96 \text{ mA}.$$

The current has reached 63.2% of its final value in the interval of one time constant.

$$V_R = 80 \times 0.632 = 50.56 \text{ V}.$$
$$V_L = 80 - 50.56 = 29.44 \text{ V}.$$

The time interval of 12.82 μs is equal to two time constants and therefore

$$I_1 = 20.5\,(1 - e^{-2}) = 20.5 \times 0.865 = 17.72 \text{ mA}.$$
$$V_R = 80 \times 0.865 = 74.8 \text{ V}.$$
$$V_L = 80 - 74.8 = 5.2 \text{ V}.$$

The interval of 19.23 μs is equivalent to three time constants. Then:

$$I_1 = 20.5\,(1 - e^{-3}) = 20.5 \times 0.95 = 19.48 \text{ mA}.$$
$$V_R = 80 \times 0.95 = 76.02 \text{ V}.$$
$$V_L = 80 - 76.02 = 3.98 \text{ V}.$$

The time interval of 25.64 μs is the same as four time constants and therefore,

$$I_1 = 20.5\,(1 - e^{-4}) = 20.5 \times 0.982 = 20.13 \text{ mA}.$$
$$V_R = 80 \times 0.982 = 78.56 \text{ V}.$$
$$V_L = 80 - 78.56 = 1.44 \text{ V}.$$

The time interval of 32.05 μs is equal to five time constants. This virtually completes the transient state in the growth of the current I_1.

$$I_1 = 20.5\,(1 - e^{-5}) = 20.5 \times 0.993 = 20.36 \text{ mA}.$$
$$V_R = 80 \times 0.993 = 79.46 \text{ V}.$$
$$V_L = 80 - 79.46 = 0.54 \text{ V}.$$

The time interval of 64.10 μs is equal to ten time constants. Therefore:

$$I_1 = 20.5\,(1 - e^{-10}) = 20.5 \times 0.99995 = 20.499 \text{ mA}.$$
$$V_R = 80 \times 0.99995 = 79.996 \text{ V}.$$
$$V_L = 80 - 79.996 = 0.004 \text{ V}.$$

Example 43-2

In Example 43-1, it is assumed that the growth transient state has been concluded. If S is now closed in position 2, what are the initial values of I_2, V_R, V_L and their values after time intervals of 6.41 μs and 32.05 μs?

Solution

Initial values of I_2, V_R, V_L are respectively 20.5 mA, 80 V, and ($-$) 80 V.
    After the time interval of 6.41 μs or one time constant.

$$I_2 = 20.5\,e^{-1} = 20.5 \times 0.368 = 7.36 \text{ mA}.$$
$$V_R = 80 \times 0.368 = 29.43 \text{ V}.$$
$$V_L = -29.43 \text{ V}.$$

After the time interval of 32.05 μs or five time constants,

$$I_2 = 20.5\,e^{-5} = 0.00674 \times 20.5 = 0.138 \text{ mA}.$$
$$V_R = 80 \times 0.00674 = 0.539 \text{ V}.$$
$$V_L = -0.539 \text{ V}.$$

The transient decay time is virtually finished.

Chapter 44
Inductors in parallel

IT IS ASSUMED THAT ALL THREE INDUCTORS IN FIG. 44-1 ARE IDEAL, IN THE SENSE that they have no losses and they only possess the property of inductance. Because these components are joined between two common points A and B, the source voltage will be across every inductor. Then each inductor will carry a different rate of growth of current so that its magnetic field will induce the same counter EMF, which exactly balances the source voltage. The sum of the individual rates of current growth will equal the total rate of growth associated with the source voltage.

When the switch, S, is closed, there will be a constant rate of current growth in each inductor and the associated magnetic fields will continue to expand. After a period of time there will be a certain amount of energy stored in each of the three magnetic fields and the sum of these energies will equal the total energy derived from the source voltage. Because the inductors are regarded as ideal components, there is no power *dissipated* in the circuit as heat.

From the mathematical derivations we shall see that the parallel arrangement of inductors reduces the value of the total inductance. The reasons for connecting inductors in parallel are to increase the current capability and to obtain nonstandard values from standard components.

The principles of inductors in series (Chapter 43) and parallel may be extended to series-parallel arrangements (see Example 44-2).

MATHEMATICAL DERIVATIONS

Rate of growth of current in the inductor $L1$,

$$\frac{\Delta I_1}{\Delta T} = \frac{E}{L_1} \text{ amperes per second} \tag{44-1}$$

Rate of growth of current in the inductor $L2$,

$$\frac{\Delta I_2}{\Delta T} = \frac{E}{L_2} \text{ amperes per second} \tag{44-2}$$

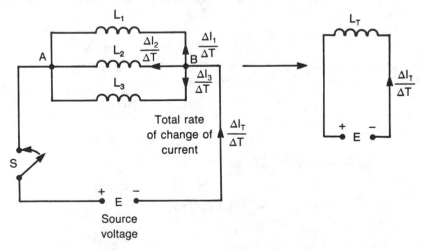

Figure 44-1

Rate of growth of current in the inductor $L3$,

$$\frac{\Delta I_3}{\Delta T} = \frac{E}{L_3} \text{ amperes per second} \tag{44-3}$$

For the meaning of the "Δ" sign, see Chapter 42.

Total rate of growth of the source current,

$$\frac{\Delta I_T}{\Delta T} = \frac{\Delta I_1}{\Delta T} + \frac{\Delta I_2}{\Delta T} + \frac{\Delta I_3}{\Delta T} \tag{44-4}$$

$$= \frac{E}{L_1} + \frac{E}{L_2} + \frac{E}{L_3}$$

$$= E \times \left(\frac{1}{L_1} + \frac{1}{L_2} + \frac{1}{L_3} \right) \text{ A/s}$$

If L_T is the total inductance,

$$\frac{\Delta I_T}{\Delta T} = \frac{E}{L_T} = E \times \frac{1}{L_T} \text{ A/s} \tag{44-5}$$

Comparing Equations 44-4, 44-5:

$$\frac{1}{L_T} = \frac{1}{L_1} + \frac{1}{L_2} + \frac{1}{L_3} \text{ or } L_T = \frac{1}{\dfrac{1}{L_1} + \dfrac{1}{L_2} + \dfrac{1}{L_3}} \text{ henrys} \tag{44-6}$$

For N inductors in parallel:

$$L_T = \frac{1}{\dfrac{1}{L_1} + \dfrac{1}{L_2} + \dfrac{1}{L_3} + \cdots + \dfrac{1}{L_N}} \text{ henrys} \tag{44-7}$$

If all N inductors are of equal value, which is L henrys,

$$L_T = L/N \text{ henrys} \tag{44-8}$$

For *two* inductors in parallel,

$$L_T = \frac{1}{\dfrac{1}{L_1} + \dfrac{1}{L_2}} = \frac{L_1 \times L_2}{L_1 + L_2} \text{ ("product-over-sum" formula)} \tag{44-9}$$

This yields

$$L_1 = \frac{L_2 \times L_T}{L_2 - L_T} \text{ and } L_2 = \frac{L_1 \times L_T}{L_1 - L_T} \tag{44-10}$$

Consequently, the equations for inductors in parallel are comparable with those for resistors in *parallel*.

Energy stored in the magnetic field surrounding the $L1$ inductor,

$$W_1 = \frac{1}{2}L_1 I_1{}^2 \text{ joules} \tag{44-11}$$

Energy stored in the magnetic field surrounding the $L2$ inductor,

$$W_2 = \frac{1}{2}L_2I_2{}^2 \text{ joules} \tag{44-12}$$

Energy stored in the magnetic field surrounding the $L3$ inductors,

$$W_3 = L_3I_3{}^2 \text{ joules}$$

Total energy derived from the source,

$$W_T = W_1 + W_2 + W_3 \tag{44-13}$$

$$= \frac{1}{2}L_TI_T{}^2 \text{ joules}$$

Example 44-1

In Fig. 44-1, $L_1 = 4$ H, $L_2 = 8$ H, $L_3 = 20$ H, and $E = 43$ V.
Calculate the values of

$$\frac{\Delta I_1}{\Delta T}, \frac{\Delta I_2}{\Delta T}, \frac{\Delta I_3}{\Delta T}, \frac{\Delta I_T}{\Delta T}, \text{ and } L_T.$$

If the switch is closed for a period of four seconds, what is the energy stored in the magnetic field surrounding each inductor and what is the total energy derived from the source?

Note: The solution for this example should be compared with the solution to Example 43-1 in which the same inductor and source values are used.

Solution

Rate of the current growth in the $L1$ inductor,

$$\frac{\Delta I_1}{\Delta T} = \frac{43 \text{ V}}{4 \text{ H}} \tag{44-1}$$

$$= 10.75 \text{ A/s}$$

Rate of the current growth in the $L2$ inductor,

$$\frac{\Delta I_2}{\Delta T} = \frac{43 \text{ V}}{8 \text{ H}} \tag{44-2}$$

$$= 5.375 \text{ A/s}$$

Rate of the current growth in the $L3$ inductor,

$$\frac{\Delta I_3}{\Delta T} = \frac{43 \text{ V}}{20 \text{ H}} \tag{44-3}$$

$$= 2.15 \text{ A/s}$$

Total rate of growth of the source current,

$$\frac{\Delta I_T}{\Delta T} = 10.75 + 5.375 + 2.15 \tag{44-4}$$

$$= 18.275 \text{ A/s}$$

Total equivalent inductance,

$$L_T = \frac{E}{\Delta I_T / \Delta T} \qquad (44\text{-}5)$$

$$= \frac{43\ \text{V}}{18.275\ \text{A/s}}$$

$$= 2.35\ \text{H}$$

Total inductance check:

$$L_T = \frac{1}{\dfrac{1}{L_1} + \dfrac{1}{L_2} + \dfrac{1}{L_3}} \qquad (44\text{-}6)$$

$$= \frac{1}{\dfrac{1}{4} + \dfrac{1}{8} + \dfrac{1}{20}}$$

$$= 2.353\ \text{H}$$

After a period of four seconds,

$L1$ inductor current, $I_1 = 10.75$ A/s $\times$ 4 s $= 43$ A.
$L2$ inductor current, $I_2 = 5.375$ A/s $\times$ 4 s $= 21.5$ A.
$L3$ inductor current, $I_3 = 2.15$ A/s $\times$ 4 s $= 8.6$ A.
Source current, $I_T = 43 + 21.5 + 4.6 = 73.1$ A.

Energy stored in the $L1$ inductor,

$$W_1 = \frac{1}{2} L_1 I_1^2 \qquad (44\text{-}11)$$

$$= \frac{1}{2} \times 4\ \text{H} \times (43\ \text{A})^2$$

$$= 3698\ \text{joules.}$$

Energy stored in the $L2$ inductor,

$$W_2 = \frac{1}{2} L_2 I_2^2 \qquad (44\text{-}12)$$

$$= \frac{1}{2} \times 8\ \text{H} \times (21.5\ \text{A})^2$$

$$= 1849\ \text{joules.}$$

Energy stored in the $L3$ inductor,

$$W_3 = \frac{1}{2} L_3 I_3^2 \qquad (44\text{-}13)$$

$$= \frac{1}{2} \times 20\ \text{H} \times (8.6\ \text{A})^2$$

$$= 739.6\ \text{joules.}$$

Total energy stored,

$$W_T = W_1 + W_2 + W_3 \qquad (44\text{-}14)$$
$$= 3698 + 1849 + 739.6$$
$$= 6286.6 \text{ joules.}$$

Energy check:

Total energy derived from the source,

$$= \frac{1}{2} L_T I_T{}^2 \qquad (44\text{-}14)$$

$$= \frac{1}{2} \times 2.353 \text{ H} \times (73.1 \text{ A})^2$$

$$= 6286.8 \text{ joules.}$$

Example 44-2

In Fig. 44-2, calculate the total value of the inductance between the points A and B.

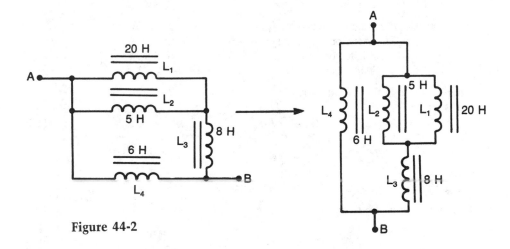

Figure 44-2

Solution

Inductance of

$$5 \text{ H} \, // \, 20 \text{ H} = \frac{5 \times 20}{5 + 20} = 4 \text{ H} \qquad (44\text{-}9)$$

Combined inductance in the right-hand branch between the points A,B

$$4 + 8 = 12 \text{ H} \qquad (44\text{-}6)$$

Inductance in the left-hand branch between the points $A,B = 6$ H.

Total inductance between the points A and B,

$$L_T = \frac{12 \times 6}{12 + 6}$$

$$= \frac{72}{18} = 4.0 \text{ H.}$$

(44-9)

Chapter 45
Electric flux—Coulomb's law

WHEN TWO NEUTRAL INSULATORS SUCH AS A RUBBER ROD AND A PIECE OF FUR are rubbed together, electrons will literally be wiped off from one insulator to the other. This is referred to as the *triboelectric effect* or *friction effect*. Electrons are transferred from the fur to the rod so that the fur has a deficit of electrons and is positively charged while the rod has a surplus of electrons and carries an equal negative charge. If the two charged insulators are positioned near to one another, there is no electron flow so that the charges cannot move. This phenomenon is therefore called *static electricity* or *electrostatics*. If the two insulators are then brought into contact, equalizing currents flow in the form of sparks, which produce a crackling sound and continue until the insulators are again electrically neutral.

In magnetism, like poles repel while unlike poles attract. A similar law applies to electrostatics because charges with the same polarity (for example, two negative charges) repel each other but charges with the opposite polarity (a positive charge and a negative charge) are subjected to a force of attraction. The force of attraction or repulsion is due to the influence that a charged body exerts on its surroundings. This influence is called the electric or electrostatic field, which terminates on a charged body and extends between positive and negative charges. Such a field can exist in a vacuum (free space) or in insulators of air, glass, waxed paper, etc.

As with the magnetic field, an electric field is represented by lines of force that show the direction and strength of the field so that a strong field is shown as a large number of lines close together. The total number of force lines is called the electric flux, whose letter symbol is the Greek letter psi (ψ). In the SI system a single flux line is assumed to emanate from a positive charge of one coulomb and to terminate on an equal negative charge. The number of flux lines is the same as the number of coulombs and therefore can be measured directly in coulombs.

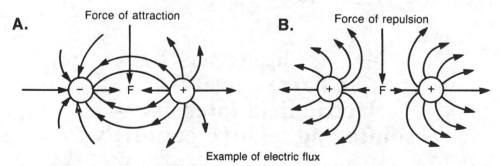

Example of electric flux

Figure 45-1

Figures 45-1A and B show the electric flux patterns for the forces of attraction and repulsion. Coulomb discovered that the magnitude of the force between two charged bodies is directly proportional to the product of the charges but is inversely proportional to the square of their separation.

MATHEMATICAL DERIVATIONS
Coulomb's law
Force of attraction of repulsion,

$$F = \frac{8.99 \times 10^9 \times Q_1 \times Q_2}{d^2} \text{ newtons} \qquad (45\text{-}1)$$

where: Q_1, Q_2 = charges (C)
$\qquad d$ = separation (m).

Example 45-1

In free space a negative charge of 8 μC is positioned 5 cm away from another negative charge of 8 μC. What are the values of the total electrical flux and the force exerted on the charges? Is the force one of attraction or repulsion?

Solution

Total electrical flux, $\psi = 8$ μC

$$\text{Force, } F = \frac{8.99 \times 10^9 \times Q_1 \times Q_2}{d^2} \text{ N} \qquad (45\text{-}1)$$

$$= \frac{8.99 \times 10^9 \times (8 \times 10^{-6})^2}{(5 \times 10^{-2})^2}$$

$$= \frac{8.99 \times 64 \times 10}{25}$$

$$= 230 \text{ N}.$$

Because the charges are of the same polarity, the force will be one of *repulsion*.

Chapter 46
Charge density—
electric field intensity—
absolute and relative permittivity

FIGURE 46-1 ILLUSTRATES TWO IDENTICAL RECTANGULAR PLATES, WHICH ARE made from a conducting material such as silver, copper, aluminum, etc. The area of one side of one plate is A square meters, and the two plates are separated by a distance of d meters in free

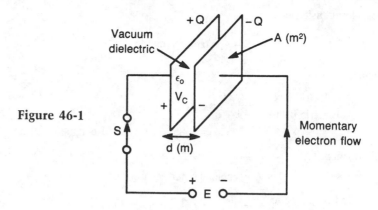

Figure 46-1

space. These plates are charged to a potential difference of V_C volts so that one plate carries a negative charge of Q coulombs and the other plate has an equal positive charge. The total electric flux, ψ, is therefore Q coulombs and consists of Q lines of force, which are spread over an area of A square meters. The flux density, whose letter symbol is D, is found by dividing ψ by A and is measured in coulombs per square meter.

The flux density, D, is the result of the electric field intensity, $\mathcal{E}$, which exists between the two plates. Increasing the value of V_C and/or reducing the separation, d, between the plates, causes a stronger electric field so that the electric field intensity, $\mathcal{E}$, is expressed in terms of the voltage gradient, V_C/d, which is measured in volts per meter. The ratio of $D : \mathcal{E}$ is termed the permittivity of free space whose letter symbol is ϵ_o (also known as the absolute permittivity of a vacuum) and whose value is 8.85×10^{-12} farad per meter in SI units.

If a negative charge (coulombs) is placed in the vacuum, it will be repelled from the negative plate and will be accelerated towards the positive plate. The accelerating force in newtons is directly proportional to the magnitude of the charge and the value of the electric field intensity.

If the free space between the plates is replaced by an insulating material such as mica, waxed paper, etc., the flux density for a given value of electric field intensity, is increased by a factor called the relative permittivity, ϵ_r. This factor has no units and is defined by:

$$\text{Relative permittivity, } \epsilon_r = \frac{\text{electric flux density in the insulating material for a given field intensity}}{\text{electric flux density in free space for the same field intensity}}$$

Values for ϵ_r are 1.0006 for air, 3 to 6 for mica and 3.5 for waxed paper.

It follows that, for an insulating material, the ratio of $D : \mathcal{E}$ is $\epsilon_o \epsilon_r = \epsilon$, which is the insulator's absolute permittivity, measured in farads per meter.

MATHEMATICAL DERIVATIONS

Flux density,

$$D = \frac{\psi}{A} \text{ coulombs per square meters} \tag{46-1}$$

Total electric flux,

$$\psi = D \times A \text{ coulombs} \tag{46-2}$$

151

Electric field intensity,

$$\mathcal{E} = \frac{V}{d} \text{ volts per meter} \tag{46-3}$$

In a vacuum,

$$\frac{\text{Electric flux density, } D}{\text{Electric field intensity, } \mathcal{E}} = \epsilon_o \tag{46-4}$$

$$= 8.85 \times 10^{-12} \text{ farad per meter}$$

Electric flux density,

$$D = \mathcal{E} \, \epsilon_o \text{ coulombs per square meter} \tag{46-5}$$

Force exerted on a charge,

$$F = Q\mathcal{E} \text{ newtons} \tag{46-6}$$

For an insulator other than free space,

$$\text{Relative permittivity, } \epsilon_r = \frac{D \text{ in the insulating material}}{D \text{ in free space}} \tag{45-7}$$

For a given electric field intensity,

$$\frac{\text{Electric flux density, } D}{\text{Electric field intensity, } \mathcal{E}} = \epsilon_o \epsilon_r \tag{46-8}$$

$$= \epsilon \text{ farads per meter}$$

Example 46-1

In Fig. 46-1 the area of one side of each plate is 4 cm² and their separation in free space is 0.8 cm. If the voltage between the plates is 50 V, what are the values of the electric field intensity, the electric flux density and the charge on each plate? If an insulator whose relative permittivity is 3.5 is inserted between the plates, what are the new values of the electric flux density and the charge on each plate?

Solution
Free space
Electric field intensity,

$$\mathcal{E} = \frac{V}{d} \tag{46-3}$$

$$= \frac{50 \text{ V}}{0.8 \times 10^{-2} \text{ m}}$$

$$= 6250 \text{ V/m.}$$

Flux density,

$$D = \mathcal{E} \times \epsilon_o \tag{46-5}$$

$$= 6250 \times 8.85 \times 10^{-12}$$

$$= 5.53 \times 10^{-8} \text{ C/m}^2.$$

Charge on each plate,

$$Q = \text{flux}, \ \psi = D \times A \qquad (46\text{-}2)$$
$$= 5.53 \times 10^{-8} \ \text{C/m}^2 \times 4 \times 10^{-4} \ \text{m}^2$$
$$= 22.1 \ \text{pC}.$$

Insulator

There is no change in the value of the electric field intensity.
Flux density,

$$D = \mathcal{E} \times \epsilon_o \epsilon_r \qquad (46\text{-}8)$$
$$= 5.53 \times 10^{-8} \times 3.5$$
$$= 1.935 \times 10^{-7} \ \text{C/m}^2.$$

Charge on each plate,

$$Q = 1.935 \times 10^{-7} \ \text{C/m}^2 \times 4 \times 10^{-4} \ \text{m}^2$$
$$= 77.4 \ \text{pC}.$$

Chapter 47
Capacitance and the capacitor

CAPACITANCE IS THAT PROPERTY OF AN ELECTRICAL CIRCUIT THAT PREVENTS ANY sudden change in *voltage* and limits the rate of change in the voltage. In other words, the voltage across a capacitance, whose letter symbol is C, cannot change instantaneously.

The property of capacitance is possessed by a capacitor, which is a device for storing charge and consists of two conducting surfaces (copper, silver, aluminum, tin foil, etc.) which are separated by an insulator or dielectric (air, mica, ceramic, etc.).

When a dc voltage, E, is switched across the capacitor, C, in Fig. 47-1, electrons are drawn off the right-hand plate, flow through the voltage source and are then deposited on the left-hand plate. This electron flow is only momentary and ceases when the voltage, V_C, between the plates exactly balances the source voltage, E. The charge then stored by the capacitor directly depends on the value of the applied voltage and also on some constant of the capacitor called the capacitance, which is determined by the component's physical construction. The unit for the capacitance is the farad, F. The capacitance is one farad, if when the voltage applied across the capacitor's plates is one volt, the charge stored is one coulomb. Unfortunately, the farad is far too large a unit for practical purposes so that capacitances are normally either measured in microfarads (μF) or picofarads (pF). Note that a picofarad may alternatively be referred to as a micromicrofarad ($\mu\mu$F).

We already know that a particular electric field intensity (which is inversely proportional to the separation between the plates) produces a certain flux (which is directly proportional to the area of one of the plates). This means that a capacitor's capacitance is directly proportional to the area of the plates, but is inversely proportional to their distance apart. It is true that the capacitance is increased if the conductor plates are brought closer together. However, if the dielectric is made too thin, the voltage between the plates may cause arcing to occur so that the capacitor will

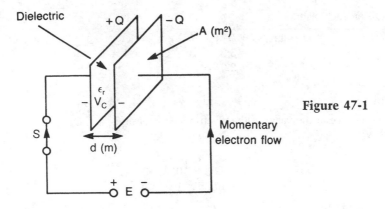

Figure 47-1

be damaged. Each type of insulator has a certain dielectric strength, which is a measure of the insulator's ability to withstand a high electric field intensity. For example, the dielectric strength of mica is typically 50 kV/mm so that in order to avoid breakdown, no more than 50 kV should be applied across a 1 mm thickness of mica. It follows that each capacitor is rated for a particular dc working voltage (WVdc).

MATHEMATICAL DERIVATIONS

For the capacitor in Fig. 47-1,

$$V_C \text{ (volts)} = -\frac{Q \text{ (coulombs)}}{C \text{ (farads)}} \tag{47-1}$$

Using KVL,

$$E + V_C = 0.$$

Therefore,

$$E = \frac{Q}{C}, \; Q = CE, \; C = \frac{Q}{E} \tag{47-2}$$

Flux density,

$$D = \frac{Q}{A} \text{ coulombs per square meter} \tag{47-3}$$

Electric field intensity,

$$\mathcal{E} = \frac{E}{d} \text{ volts per meter} \tag{47-4}$$

For a capacitor with a vacuum dielectric,
Permittivity of free space,

$$\epsilon_o = \frac{D}{\mathcal{E}} \tag{47-5}$$

$$= \frac{Q/A}{E/d}$$

$$= \frac{Q}{E} \times \frac{d}{A}$$

$$= \frac{C \times d}{A}$$

Equation 47-5 yields,
    Capacitance,

$$C = \epsilon_o \times \frac{A}{d} \text{ farads} \qquad (47\text{-}6)$$

If the dielectric has a relatively permittivity of ϵ_r,

$$\text{Capacitance, } C = \frac{\epsilon_o \epsilon_r A}{d} = \frac{\epsilon A}{d} \text{ farads} \qquad (47\text{-}7)$$

where ϵ is the dielectric's absolute permittivity, which is measured in farads per meter.
    Since $Q = CE$,

$$I = \frac{dQ}{dt} = C \frac{dE}{dt}$$

or

$$\frac{dE}{dt} = \frac{I}{C} \text{ volts per second} \qquad (47\text{-}8)$$

Because dE/dt is inversely proportional to C, the capacitance will limit the rate of change in the voltage.

Example 47-1

Two parallel aluminum plates are separated by 0.8 mm in air. If the area of one side of each plate is 50 cm^2, what is the value of the capacitance?

Solution

Capacitance,

$$C = \frac{\epsilon_o \epsilon_r A}{d} \qquad (47\text{-}7)$$

$$= \frac{8.85 \times 10^{-12} \times 1.0006 \times 50 \times 10^{-4}}{0.8 \times 10^{-3}} \text{ F}$$

$$= 55 \text{ pF.}$$

Example 47-2

A paper capacitor consists of two sheets of tin foil, each side of which has an area of 1500 cm^2. The sheets are separated by a 0.15 mm thickness of a waxed paper dielectric with a relative permittivity of 2.7. What is the value of the capacitance? If the capacitor is charged to 120 V, what are the values of the electric field intensity, the electric flux density and the charge stored?

Solution

Capacitance,

$$C = \frac{\epsilon_o \epsilon_r A}{d} \qquad (47\text{-}7)$$

$$= \frac{8.85 \times 10^{-12} \times 2.7 \times 1500 \times 10^{-4}}{0.15 \times 10^{-3}} \text{ F}$$

$$= 0.0239 \ \mu\text{F}.$$

Charge stored,

$$Q = CE$$
$$= 0.0239 \ \mu\text{F} \times 120 \text{ V}$$
$$= 2.87 \ \mu\text{C}.$$

Electric flux density,

$$D = \frac{\psi}{A} = \frac{Q}{A} = \frac{2.87 \ \mu\text{C}}{1500 \times 10^{-4} \text{ m}^2}$$

$$= 19.1 \ \mu\text{C/m}^2.$$

Electric field intensity,

$$\mathcal{E} = \frac{V_C}{d}$$

$$= \frac{120 \text{ V}}{0.15 \times 10^{-3} \text{ m}} = 8 \times 10^5 \text{ V/m}.$$

Chapter 48
Energy stored in the
electric field of a capacitor

CONSIDER THAT A CAPACITOR IS BEING CHARGED FROM A CONSTANT CURRENT source of I amperes over a period of t seconds (Fig. 48-1A). The voltage across the capacitor then increases in a linear manner as shown in the voltage-versus-time graph of Fig. 48-1B. At any time after the switch is closed, one of the plates has acquired a positive charge while the other is negatively charged; the charge stored by the capacitor is given by Q (coulombs) = C (capacitance in farads) $\times V_C$ (volts); see equation 47-2.

When the capacitor is charged, an electric field (ψ), also measured in coulombs, exists between the plates. This represents the stored energy that has been derived from the constant current source. In another sense, the same energy is related to the force of attraction between the positive and negative plates.

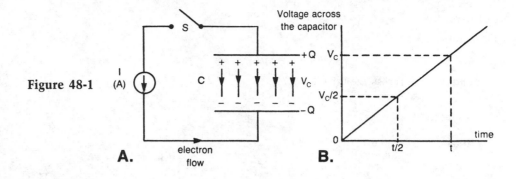

Figure 48-1

A.

electron flow

B.

MATHEMATICAL DERIVATIONS

Final Charge,

$$Q = I \times t \text{ coulombs} \tag{48-1}$$

$$\text{Voltage across the capacitor, } V_C = \frac{Q}{C} \text{ volts} \tag{48-2}$$

$$\text{Average voltage across the capacitor} = \frac{V_C}{2} \text{ volts} \tag{48-3}$$

$$\text{Average power at the source} = I \times \frac{V_C}{2} \text{ watts} \tag{48-4}$$

Total energy supplied from the source during the charging process

$$= \text{average power} \times \text{time} = I \times \frac{V_C}{2} \times t$$

$$= \frac{1}{2} \times V_C Q = \frac{1}{2} \times \frac{Q^2}{C} = \frac{1}{2} \times C V_C{}^2 \text{ joules} \tag{48-5}$$

Alternatively:

The work done in moving a charge through a potential difference, δV_C, is $Q \times \delta V_C$ (joules = coulombs $\times$ volts) $= C V_C \delta V_C$ $\hspace{2cm}$ (48-6)

Integrating over a total period of t seconds,
Total work done or total energy stored

$$W = \int_o^{V_C} C V_C \, dV_C \tag{48-7}$$

$$= \frac{1}{2} C V_C{}^2 \text{ joules}$$

Example 48-1

In Fig. 48-1, the switch, S, is closed and an $8\mu F$ capacitor is charged for a period of 7 seconds from a $5\mu A$ constant current source. At the end of the charging period, what is the amount of charge acquired by the capacitor and what is the corresponding amount of energy stored in the electric field?

157

Solution

Final charge,

$$Q = I \times t = 5\mu A \times 7s = 35 \ \mu C \tag{48-1}$$

Final voltage across the capacitor,

$$V_C = \frac{Q}{C} = \frac{35\mu C}{8\mu F} = 4.75 \ V \tag{48-2}$$

Total energy stored in the capacitor

$$\frac{1}{2} \times V_C \times Q = \frac{1}{2} \times 4.75 \ V \times 35 \ \mu C \tag{48-5}$$

$$= 83.125 \ \mu J$$

Notice that the amount of energy stored in a capacitor is normally very small.

Chapter 49
Capacitors in series

BECAUSE THE CAPACITORS ARE JOINED END-TO-END IN FIG. 49-1, THERE IS ONLY one path for current flow. Consequently, when $S1$ is closed in position 1, the charging current for each capacitor is the same and therefore each capacitor must store the same charge, Q (coulombs). But there are two problems that are worth considering. First, since the electron flow cannot pass through the dielectrics, which theoretically are open circuits, how does the middle capacitor become charged? The explanation lies in electrostatic induction. When the elec-

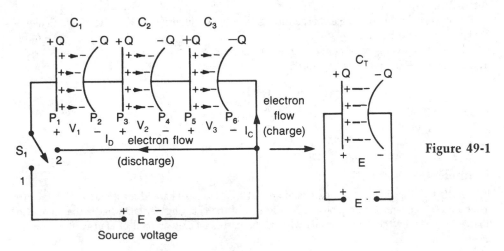

Figure 49-1

tron flow causes the plate, $P1$, to acquire a positive charge, an equal negative charge is induced on the plate, $P2$. This causes another positive charge of the same size to appear on the plate, $P3$, and so on. As the result of this process, the total charge stored between the plates $P1$ and $P6$ is Q (*not* $3 Q$). Secondly, we are told that a current can only flow in a closed circuit, but clearly, the dielectrics represent discontinuities in the circuit of Fig. 49-1. James Clerk Maxwell resolved this difficulty by proposing the so-called *displacement current,* which is not an electron flow but is the result of a rate of change in the electric flux. On the SI system the electric flux in a dielectric consists of lines where one line is assumed to emanate from a positive charge of one coulomb and terminate on an equal negative charge. Consequently, the number of lines is the same as the number of coulombs and the electric flux can therefore be measured in coulombs. When the capacitor is being charged, there is a change in the electric flux and its rate of change will be equivalent to coulombs per second or amperes. This naturally results in the displacement current being measured in the same units as the electron flow.

After the capacitors are charged, there is a voltage across each capacitor and the sum of these voltages must exactly balance the source voltage. Since $Q = VC$ (equation 47-2) and each capacitor carries the same charge, it follows that the *highest* voltage is across the lowest *value* of capacitance and vice-versa.

Each capacitor stores a certain amount of energy in the form of the electric field between its plates. When the switch is moved to position 2, the capacitors discharge and the total energy stored is released in the form of the spark at the contacts of the switch. Incidentally, the spark creates an electromagnetic wave and this was the principle behind the early spark-gap transmitter.

Finally, what is the purpose of connecting capacitors in series? As we shall see in the mathematical derivations, connecting capacitors in series *reduces* the total capacitance so that the total capacitance is less than the value of the smallest capacitance in series. Basically, this is because the series arrangement effectively increases the distance between the end plates, $P1$ and $P6$, connected to the source voltage, and the capacitance is inversely proportional to this distance. Because the capacitor has a dc working voltage (WVdc), the series arrangement may be used to distribute the source voltage between the capacitors so that the voltage across an individual capacitor does not exceed its rating.

MATHEMATICAL DERIVATIONS

$$V_1 = \frac{Q}{C_1}, \ Q = C_1V_1, \ C_1 = \frac{Q}{V_1} \tag{49-1}$$

$$V_2 = \frac{Q}{C_2}, \ Q = C_2V_2, \ C_2 = \frac{Q}{V_2} \tag{49-2}$$

$$V_3 = \frac{Q}{C_3}, \ Q = C_3V_3, \ C_3 = \frac{Q}{V_3} \tag{49-3}$$

Source voltage,

$$E = V_1 + V_2 + V_3 \tag{49-4}$$

$$= \frac{Q}{C_1} + \frac{Q}{C_2} + \frac{Q}{C_3}$$

$$= Q \times \left(\frac{1}{C_1} + \frac{1}{C_2} + \frac{1}{C_3} \right) \text{ volts}$$

If C_T is the total equivalent capacitance,

$$E = \frac{Q}{C_T} = E \times \frac{1}{C_T} \text{ volts} \qquad (49\text{-}5)$$

Comparing Equations 49-4, and 49-5.

$$\frac{1}{C_T} = \frac{1}{C_1} + \frac{1}{C_2} + \frac{1}{C_3} \qquad (49\text{-}6)$$

For N capacitors in series,

$$C_T = \frac{1}{\dfrac{1}{C_1} + \dfrac{1}{C_2} + \dfrac{1}{C_3} + \cdots + \dfrac{1}{C_N}} \text{ farads} \qquad (49\text{-}7)$$

If the N capacitors are all of equal value C,

$$C_T = \frac{C}{N} \text{ farads} \qquad (49\text{-}8)$$

The equations for capacitors in series may therefore be compared with those for resistors in parallel. Consequently connecting capacitors in series *reduces* the total capacitance.

Voltage division

Source voltage,

$$E = Q \times \frac{1}{C_T} = C_1 V_1 \times \frac{1}{C_T} \text{ volts} \qquad (49\text{-}9)$$

Therefore,

$$V_1 = E \times \frac{C_T}{C_1}, \; V_2 = E \times \frac{C_T}{C_2}, \; V_3 = E \times \frac{C_T}{C_3} \text{ volts} \qquad (49\text{-}10)$$

Energy stored in the $C1$ capacitor,

$$W_1 = \frac{1}{2} C_1 V_1{}^2$$

$$= \frac{1}{2} \frac{Q^2}{C_1} = \frac{1}{2} Q V_1 \text{ joules} \qquad (49\text{-}11)$$

Energy stored in the $C2$ capacitor,

$$W_2 = \frac{1}{2} C_2 V_2{}^2 \qquad (49\text{-}12)$$

$$= \frac{1}{2} \frac{Q^2}{C_2} = \frac{1}{2} Q V_2 \text{ joules}$$

Energy stored in the $C3$ capacitor,

$$W_3 = \frac{1}{2} C_3 V_3{}^2 \qquad (49\text{-}13)$$

$$= \frac{1}{2} \frac{Q^2}{C_3} = \frac{1}{2} Q V_3 \text{ joules}$$

Total energy stored,

$$W_T = \frac{1}{2} QE = \frac{1}{2} C_T E^2 \qquad (49\text{-}14)$$

$$= \frac{1}{2} \frac{Q^2}{C_T} \text{ joules}$$

For two capacitors in series,
Total capacitance,

$$C_T = \frac{1}{\dfrac{1}{C_1} + \dfrac{1}{C_2}} = \frac{C_1 \times C_2}{C_1 + C_2} \text{ farads} \qquad (49\text{-}15)$$

This yields

$$C_1 = \frac{C_2 \times C_T}{C_2 - C_T}, \ C_2 = \frac{C_1 \times C_T}{C_1 - C_T} \text{ farads} \qquad (49\text{-}16)$$

$$V_1 = E \times \frac{C_T}{C_1} = E \times \frac{C_2}{C_1 + C_2}, \ V_2 = E \times \frac{C_1}{C_1 + C_2} \text{ volts} \qquad (49\text{-}17)$$

Example 49-1

In Fig. 49-1, $C1 = 20$ μF, $C2 = 5$ μF, $C3 = 4$ μF, and $E = 30$ V. Calculate the values of C_T, V_1, V_2, V_3, and the energy stored in each capacitor.

Solution

Total capacitance,

$$C_T = \frac{1}{\dfrac{1}{C_1} + \dfrac{1}{C_2} + \dfrac{1}{C_3}} \qquad (49\text{-}4)$$

$$= \frac{1}{\dfrac{1}{20} + \dfrac{1}{5} + \dfrac{1}{4}} = 2 \ \mu\text{F}$$

Voltage across the $C1$ capacitor,

$$V_1 = E \times \frac{C_T}{C_1} = 30 \text{ V} \times \frac{2 \ \mu\text{F}}{20 \ \mu\text{F}} \qquad (49\text{-}10)$$

$$= 3 \text{ V}.$$

Voltage across the $C2$ capacitor,

$$V_2 = E \times \frac{C_T}{C_2} \qquad (49\text{-}10)$$

$$= 30 \text{ V} \times \frac{2 \ \mu\text{F}}{5 \ \mu\text{F}}$$

$$= 12 \text{ V}$$

Voltage across the $C3$ capacitor,

$$V_3 = E \times \frac{C_T}{C_3} \tag{49-10}$$

$$= 30 \text{ V} \times \frac{2 \text{ } \mu\text{F}}{4 \text{ } \mu\text{F}}$$

$$= 15 \text{ V}$$

Voltage check:
  Source voltage,

$$E = V_1 + V_2 + V_3 \tag{49-4}$$

$$= 3 + 12 + 15$$

$$= 30 \text{ V}$$

Charge stored by each capacitor,

$$Q = \text{total charge stored} \tag{49-5}$$
$$= E \times C_T$$
$$= 30 \text{ V} \times 2 \text{ } \mu\text{F}$$
$$= 60 \text{ } \mu\text{C}$$

Energy stored by the $C1$ capacitor,

$$W_1 = \frac{1}{2} QV_1 \tag{49-11}$$

$$= \frac{1}{2} \times 60 \text{ } \mu\text{C} \times 3 \text{ V}$$

$$= 90 \text{ } \mu\text{J}$$

Energy stored by the $C2$ capacitor,

$$W_2 = \frac{1}{2} QV_2 \tag{49-12}$$

$$= \frac{1}{2} \times 60 \text{ } \mu\text{C} \times 12 \text{ V}$$

$$= 360 \text{ } \mu\text{J}$$

Energy stored by the $C3$ capacitor,

$$W_3 = \frac{1}{2} QV_3 \tag{49-13}$$

$$= \frac{1}{2} \times 60 \text{ } \mu\text{C} \times 15 \text{ V}$$

$$= 450 \text{ } \mu\text{J}$$

Total energy stored = 90 + 360 + 450 = 900 μJ.
Energy check:

$$\text{Total energy stored} = \frac{1}{2} QE \qquad (49\text{-}14)$$

$$= \frac{1}{2} \times 60 \ \mu\text{C} \times 30 \ \text{V}$$

$$= 900 \ \mu\text{J}$$

Example 49-2

In Fig. 49-2, calculate the values of V_1 and V_2.

Figure 49-2

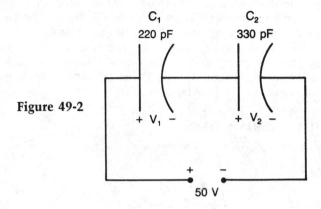

Solution

Voltage,

$$V_1 = E \times \frac{C_2}{C_1 + C_2} \qquad (49\text{-}17)$$

$$= 50 \ \text{V} \times \frac{330 \ \text{pF}}{220 \ \text{pF} + 330 \ \text{pF}}$$

$$= 30 \ \text{V}.$$

$$V_2 = E \times \frac{C_1}{C_1 + C_2}$$

$$= 50 \ \text{V} \times \frac{220 \ \text{pF}}{220 \ \text{pF} + 330 \ \text{pF}}$$

$$= 20 \ \text{V}.$$

Voltage check:

Source Voltage, $E = V_1 + V_2 = 30 + 20 = 50 \ \text{V}.$

Chapter 50
Capacitors in parallel

CAPACITORS IN PARALLEL ARE CONNECTED ACROSS TWO COMMON POINTS (A, B) or lines so that the source voltage is applied across each of the capacitors (Fig. 50-1). Because $Q = CV$, it follows that a different charge must be stored by each capacitor. This is due to the different *momentary* charging currents that exist at the instant the switch, S, is closed. The total charge is then equal to the sum of the individual charges stored by each capacitor.

As we shall see in the mathmatical derivations, the purpose of connecting capacitors in parallel is to increase the total capacitance. In Fig. 50-1 we have connected together on one side the plates $P1$, $P2$, $P3$, and, on the other side, the plates $P4$, $P5$, $P6$. The result is to increase the effective surface area to which the capacitance is directly proportional.

Each of the three capacitors in Fig. 50-1, stores energy in the form of the electric field between its plates. The total energy stored is the sum of the capacitor's individual energies.

The principles of capacitors in series (Chapter 49) and capacitors in parallel may be extended to series-parallel arrangements (see Example 50-3).

MATHEMATICAL DERIVATIONS

Charge $Q_1 = C_1E$, $Q_2 = C_2E$, $Q_3 = C_3E$ coulombs $\hspace{2cm}$ (50-1)

Total charge stored,

$$Q_T = Q_1 + Q_2 + Q_3 \hspace{4cm} \text{(50-2)}$$
$$= C_1E + C_2E + C_3E$$
$$= E \times (C_1 + C_2 + C_3) \text{ coulombs}$$

If C_T is the total equivalent capacitance,

$$\text{Total charge, } Q_T = E \times C_T \text{ coulombs} \hspace{2cm} \text{(50-3)}$$

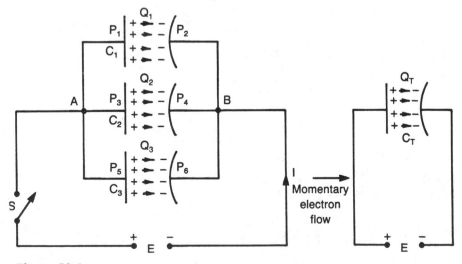

Figure 50-1

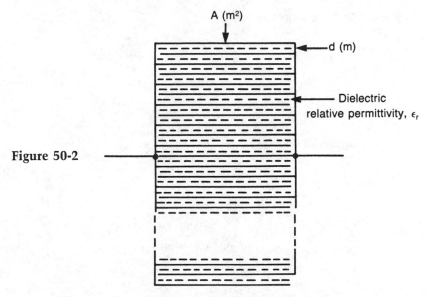

A (m²)

d (m)

Dielectric
relative permittivity, ϵ_r

Figure 50-2

Diagram of a multiplate capacitor.

Comparing Equations 50-2 and 50-3, it follows that

$$\text{Total capacitance, } C_T = C_1 + C_2 + C_3 \text{ farads} \tag{50-4}$$

Consequently, adding capacitors in parallel increases the total capacitance. If a multiplate capacitor is made up of N parallel plates with alternate plates connected as in Fig. 50-2, the total capacitance is

$$C_T = \frac{\epsilon_o \epsilon_r \, (N-1) \, A}{d} \text{ farads} \tag{50-5}$$

where: $\epsilon_o = 8.85 \times 10^{-12} F/m$,
        ϵ_r = relative permittivity
        A = area of *one* side of *one* plate (m²)
        d = distance between adjacent plates (m)

For N capacitors in parallel,

$$C_T = C_1 + C_2 + C_3 + \cdots + C_N \text{ farads} \tag{50-6}$$

If the N capacitors are all of equal value, C,

$$C_T = NC \text{ farads} \tag{50-7}$$

Energy stored by the $C1$ capacitor,

$$W_1 = \frac{1}{2} Q_1 E \tag{50-8}$$

$$= \frac{1}{2} C_1 E^2$$

$$= \frac{1}{2} Q_1^2 / C_1 \text{ joules}$$

Energy stored by the $C2$ capacitor,

$$W_2 = \frac{1}{2}Q_2E \tag{50-9}$$

$$= \frac{1}{2}C_2E^2$$

$$= \frac{1}{2}Q_2{}^2/C_2 \text{ joules}$$

Energy stored by the $C3$ capacitor,

$$W_3 = \frac{1}{2}Q_3E \tag{50-10}$$

$$= \frac{1}{2}C_3E^2$$

$$= \frac{1}{2}Q_3{}^2/C_3 \text{ joules}$$

Total energy stored,

$$W_T = \frac{1}{2}Q_TE = \frac{1}{2}C_TE^2 = \frac{1}{2}Q_T{}^2/C_T \text{ joules} \tag{50-11}$$

Example 50-1

In Fig. 50-1, $C1 = 20\ \mu F$, $C2 = 5\ \mu F$, $C3 = 4\ \mu F$, and $E = 30$ V. Calculate the values of Q_1, Q_2, Q_3, Q_T, C_T, the energy stored by each capacitor and the total energy derived from the source voltage. Note: The solution to this example should be compared with the solution to example 49-1, in which the series circuit contains the same values for the capacitors and the source voltage.

Solution

$$\text{Charge, } Q_1 = C_1E = 20\ \mu F \times 30\ V = 600\ \mu C. \tag{50-1}$$

$$\text{Charge, } Q_2 = C_2E = 5\ \mu F \times 30\ V = 150\ \mu C. \tag{50-1}$$

$$\text{Charge, } Q_3 = C_3E = 4\ \mu F \times 30\ V = 120\ \mu C. \tag{50-1}$$

Total charge stored,

$$Q_T = Q_1 + Q_2 + Q_3 \tag{50-2}$$
$$= 600 + 150 + 120$$
$$= 870\ \mu C.$$

Total capacitance,

$$C_T = C_1 + C_2 + C_3 \tag{50-4}$$
$$= 20 + 5 + 4$$
$$= 29\ \mu F.$$

Charge check:
Total charge,

$$Q_T = C_T \times E = 29 \ \mu\text{F} \times 30 \ \text{V} = 870 \ \mu\text{C} \qquad (50\text{-}3)$$

Energy stored by the $C1$ capacitor,

$$W_1 = \frac{1}{2} Q_1 \times E \qquad (50\text{-}8)$$

$$= \frac{1}{2} \times 600 \ \mu\text{C} \times 30 \ \text{V}$$

$$= 9000 \ \mu\text{J}$$

Energy stored by the $C2$ capacitor,

$$W_2 = \frac{1}{2} Q_2 \times E \qquad (50\text{-}9)$$

$$= \frac{1}{2} \times 150 \ \mu\text{C} \times 30 \ \text{V}$$

$$= 2250 \ \mu\text{J}$$

Energy stored by the $C3$ capacitor,

$$W_3 = \frac{1}{2} Q_3 \times E \qquad (50\text{-}10)$$

$$= \frac{1}{2} \times 120 \ \mu\text{C} \times 30 \ \text{V}$$

$$= 1800 \ \mu\text{J}$$

Total energy stored,

$$W_T = W_1 + W_2 + W_3$$
$$= 9000 + 2250 + 1800$$
$$= 13050 \ \mu\text{J}.$$

Energy check:
Total energy stored and therefore derived from the source,

$$\frac{1}{2} Q_T E = \frac{1}{2} \times 875 \ \mu\text{C} \times 30 \ \text{V} \qquad (50\text{-}11)$$

$$= 13050 \ \mu\text{J}$$

Example 50-2

A 40 μF capacitor is charged from a 150 V source. Another 20 μF capacitor is charged from a 100 V source. These capacitors are correctly paralleled immediately after they are disconnected from the sources. What is (a) the voltage across the combination and (b) the total energy stored before and after the parallel connection?

Solution

(a) Charge stored in the 40 μF ($C1$) capacitor,

$$Q_1 = C_1E_1$$
$$= 40 \ \mu F \times 150 \ V \tag{50-1}$$
$$= 6000 \ \mu C$$

Charge stored in the 20 μF ($C2$) capacitor,

$$Q_2 = C_2E_2$$
$$= 20 \ \mu F \times 100 \ V \tag{50-1}$$
$$= 2000 \ \mu C$$

Total charge,

$$Q_T = Q_1 + Q_2 = 6000 + 2000 = 8000 \ \mu C \tag{50-2}$$

Total capacitance of the parallel combination,

$$C_T = C_1 + C_2 \tag{50-6}$$
$$= 40 + 20$$
$$= 60 \ \mu F$$

Voltage across the parallel combination,

$$E = \frac{Q_T}{C_T} \tag{50-3}$$
$$= \frac{8000 \ \mu C}{60 \ \mu F}$$
$$= 133.3 \ V.$$

(b) Energy stored in the 40 μF capacitor,

$$W_1 = \frac{1}{2}Q_1E_1 \tag{50-8}$$
$$= \frac{1}{2} \times 6000 \ \mu C \times 150 \ V$$
$$= 0.45 \ J$$

Energy stored in the 20 μF capacitor,

$$W_2 = \frac{1}{2}Q_2E_2 \tag{50-8}$$
$$= \frac{1}{2} \times 2000 \ \mu C \times 100 \ V$$
$$= 0.10 \ J$$

Total energy stored *before* the parallel combination,

$$0.45 + 0.10 = 0.550 \ J.$$

Total energy stored *after* the parallel combination,

$$1/2(Q_TE) = 1/2 \times 8000 \ \mu C \times 133.3 \ V$$
$$= 0.533 \ J.$$

Notice that energy has been lost as the result of the parallel connection. Neglecting the resistance of any connecting wires, the lost energy appears in the form of the spark that occurs when the parallel connection is made.

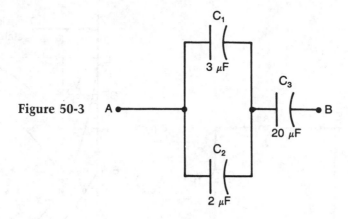

Figure 50-3

Example 50-3

In Fig. 50-3, calculate the value of the total capacitance between the points A and B.

Solution

Capacitance of 2 μF $\parallel$ 3 μF = 2 + 3 = 5 μF $\hspace{5em}$ (50-6)

Total capacitance between the points A and B,

$$C_T = \frac{5 \times 20}{5 + 20} \hspace{5em} \text{(49-13)}$$

$$= 4 \ \mu\text{F}$$

Chapter 51
The RC time constant

FROM CHAPTER 47 WE KNOW THAT THE PROPERTY OF CAPACITANCE PREVENTS ANY sudden change of voltage and also limits the rate at which the voltage can change. Consequently, when the switch, S, is closed in position 1 (Fig. 51-1A), the capacitor must take a certain time before it acquires its full charge. We are therefore going to discuss the factors that determine the duration of the so-called *transient* or *changing* state. This is the interval between the time at which the switch, S, is closed in position 1 and the time when the final or steady-state conditions are reached.

Immediately after the switch is closed in position 1, the capacitor cannot charge instantaneously so that the initial value of V_C must be zero. Consequently, to satisfy KVL, the voltage drop across the resistor, V_R, must balance the source voltage E. The initial current must therefore be at its maximum level of E/R amperes.

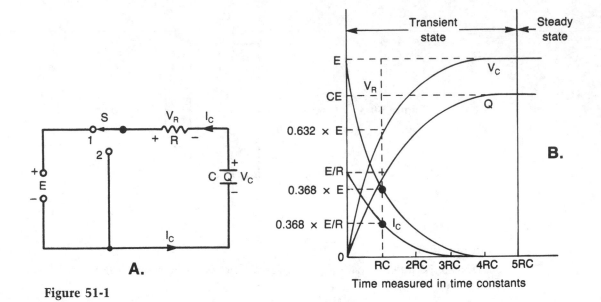

Figure 51-1

As the capacitor charges, its voltage, V_C, rises and therefore the resistor's voltage, V_R, falls because at all times the sum of these voltages exactly balances the source voltage, E. When the voltage drop, V_R, falls, the current is correspondingly less and there is a decrease in the rate at which the capacitor charges. Consequently, we have a situation in which the more the capacitor is charged, the less is its rate of charging. Theoretically, it would take infinite time for the capacitor to charge fully but as we shall see in the mathematical derivations, the capacitor acquires its ultimate charge of CE coulombs to within 1% after a finite time interval, which is determined by the values of the capacitance and the resistance.

Remember that the initial current is E/R amperes (coulombs per second) while the capacitor's final charge is CE coulombs. If the initial current had been maintained, the capacitor would have been fully charged after a time of CE coulombs $/(E/R)$ coulombs per second = RC seconds, which is the *time constant* of the circuit. However, because the rate at which the capacitor charges falls off, the capacitor only acquires 63.2% of its final charge after one time constant and it required *five* time constants before the capacitor is charged to within 1% of CE coulombs (Fig. 51-1B). Only then can we assume that the transient charging period has been concluded.

Note that the conditions at the start of the transient interval can be predicted by regarding the capacitor as a short circuit while the final steady-state values can be obtained by considering the capacitor to be an open circuit.

Assuming that the steady-state conditions have been reached, let us now move the switch, S, to position 2 (Fig. 51-2A). The capacitor cannot discharge instantaneously so that V_C must still be equal to E volts. By the KVL rule the voltage drop across the resistor must abruptly rise to E volts but its polarity is reversed because the discharge current, I_D, is in the opposite direction to the original charging current, I_C.

As the capacitor discharges, V_C and V_R fall together in order to maintain the voltage balance around the closed loop. When V_R decreases, the current is reduced so that the capacitor discharges more slowly. Theoretically, it would take infinite time for the capacitor to discharge fully but after one time constant the capacitor has *lost* 63.2% of its initial charge and its charge has

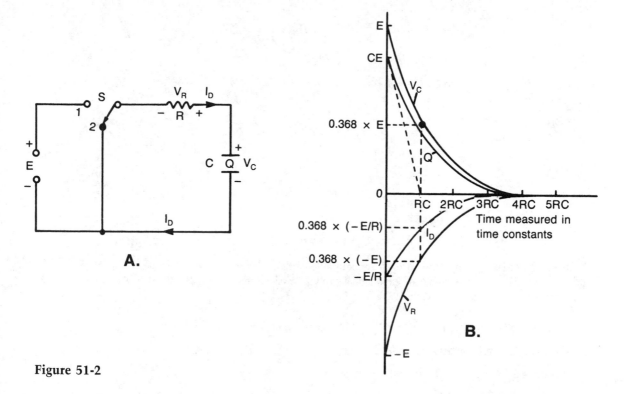

Figure 51-2

therefore dropped to 36.8% of CE coulombs. After five time constants we assume that the transient discharge period has been concluded in a sense that I_D, V_C, V_R have all fallen to less than 1% of their original values (Fig. 51-2B).

MATHEMATICAL DERIVATIONS

In Fig. 51-1A the switch, S, is closed in position 1. The initial conditions are

$$Q = 0 \text{ coulombs, } V_C = 0 \text{ volts, } V_R = E \text{ volts, } I_C = \frac{E}{R} \text{ amperes.}$$

At all times,

$$V_R + V_C = E \tag{51-1}$$

$$I_C \times R + \frac{Q}{C} = E$$

Differentiating the equation throughout with respect to t,

$$R\frac{dI_C}{dt} + \frac{1}{C} \times \frac{dQ}{dt} = 0$$

$$R\frac{dI_C}{dt} + \frac{I_C}{C} = 0$$

$$\int dt = -CR \int \frac{1}{I_C}\, dI_C$$

$$t + k = -CR \ln I_C$$

When $t = 0$, $I_C = \dfrac{E}{R}$ and $k = -RC \ln \dfrac{E}{R}$

Therefore,

$$t = -CR \ln \frac{E}{RI_C}$$

$$I_C = \frac{E}{R} e^{-t/RC} \text{ amperes} \tag{51-2}$$

$$V_R = I_C \times R = E\, e^{-t/RC} \text{ volts} \tag{51-3}$$

$$V_C = E - V_R = E(1 - e^{-t/RC}) \text{ volts} \tag{51-4}$$

where $e = 2.7183 \ldots$ and is the base of the natural logarithms.

Equation 51-4 is an example of exponential growth, while Equations 51-2 and 51-3 are expressions of exponential decay,

When $t = RC$ seconds, $I_C = \dfrac{E}{R} \times e^{-1} = 0.368\, \dfrac{E}{R}$ amperes,

$V_R = 0.368\, E$ volts and $V_C = 0.632\, E$ volts.

After $t = 5\, RC$ seconds, $I_C = 0.007\, \dfrac{E}{R}$ amperes,

$V_R = 0.007\, E$ volts and $V_C = 0.993\, E$ volts.

In Fig. 51-2A, the capacitor is assumed to be fully charged and the switch, S, is then moved to position 2. The initial conditions are:

$$I_D = -\frac{E}{R} \text{ amperes, } V_C = E \text{ volts and } V_R = -E \text{ volts}$$

At all times,

$$V_R + V_C = 0.$$

$$I_D \times R + \frac{Q}{C} = 0.$$

$$R\frac{dI_D}{dt} + \frac{I_D}{C} = 0.$$

This leads to

$$I_D = -\frac{E}{R} e^{-t/RC} \text{ amperes} \tag{51-5}$$

$$V_R = -Ee^{-t/RC} \text{ volts} \tag{51-6}$$

$$V_C = Ee^{-t/RC} \text{ volts} \tag{51-7}$$

When $t = RC$ seconds, $I_D = -0.368\, E/R$, $V_R = -0.368\, E$, and $V_C = 0.368\, E$. After $t = 5\, RC$ seconds, the values of I_D, V_R, V_C have all dropped to 0.7% of their original levels; the transient discharge period is therefore virtually finished.

Example 51-1

In Fig. 51-1A, $C = 100$ pF, $R = 150$ kΩ, and $E = 80$ V. From the instant that S is closed in position 1, what are the values of Q, V_R, V_C, and I_C after time intervals of $t = 0$, 10.5, 15, 30, 45, 60, 75, and 150 μs?

Solution

Time constant, $RC = 150$ k$\Omega \times 100$ pF $= 15$ μs.
When $t = 0$, the initial conditions are:

$$Q = 0 \text{ C}, \ V_R = 80 \text{ V}, \ V_C = 0 \text{ V}, \ I_C = \frac{80 \text{ V}}{150 \text{ k}\Omega} = 0.533 \text{ mA}.$$

When $t = 10.5$ μs $= 10.5/15 = 0.7 \times RC$,

$$Q = CE(1 - e^{-t/RC}) = 100 \text{ pF} \times 80 \text{ V}(1 - e^{-0.7})$$
$$= 8000 \times 0.503 \text{ pC}$$
$$= 4024 \text{ pC}.$$

$$V_C = \frac{Q}{C} = \frac{4024 \text{ pC}}{100 \text{ pF}} = 40.24 \text{ V}.$$

$$V_R = 80 - 40.24 = 39.76 \text{ V}.$$

$$I_C = \frac{V_R}{R} = \frac{39.76 \text{ V}}{150 \text{ k}\Omega} = 0.265 \text{ mA}.$$

When

$$t = 15.0 \ \mu\text{s} = 1 \text{ time constant},$$
$$Q = 8000 \ (1 - e^{-1}) = 8000 \times 0.632 = 5056 \text{ pC}.$$
$$V_C = \frac{5056 \text{ pC}}{100 \text{ pF}} = 50.56 \text{ V}.$$
$$V_R = 80 - 50.56 = 29.44 \text{ V}.$$
$$I_C = \frac{29.44 \text{ V}}{150 \text{ k}\Omega} = 0.196 \text{ mA}.$$

When

$$t = 30 \ \mu\text{s} = 2 \times \text{time constants},$$
$$Q = 8000 \ (1 - e^{-2}) = 8000 \times 0.865 = 6917 \text{ pC}.$$
$$V_C = \frac{6917 \text{ pC}}{100 \text{ pF}} = 69.17 \text{ V}.$$
$$V_R = 80 - 69.17 = 10.83 \text{ V}.$$
$$I_C = \frac{10.83 \text{ V}}{150 \text{ k}\Omega} = 0.072 \text{ mA}.$$

When

$$t = 45 \ \mu\text{s} = 3 \times \text{time constants},$$
$$Q = 8000 \ (1 - e^{-3}) = 8000 \times 0.950 = 7602 \text{ pC}.$$

$$V_C = \frac{7602 \text{ pC}}{100 \text{ pF}} = 76.02 \text{ V.}$$

$$V_R = 80 - 76.02 = 3.98 \text{ V.}$$

$$I_C = \frac{3.98 \text{ V}}{150 \text{ k}\Omega} = 0.0265 \text{ mA.}$$

When

$$t = 60 \text{ }\mu\text{s} = 4 \times \text{time constants,}$$

$$Q = 8000 \text{ } (1 - e^{-4}) = 8000 \times 0.982 = 7853 \text{ pC.}$$

$$V_C = \frac{7853 \text{ pC}}{100 \text{ pF}} = 78.53 \text{ V.}$$

$$V_R = 80 - 78.53 = 1.47 \text{ V.}$$

$$I_C = \frac{1.47 \text{ V}}{150 \text{ k}\Omega} = 0.00977 \text{ mA.}$$

When

$$t = 75 \text{ }\mu\text{s} = 5 \times \text{time constants,}$$

$$Q = 8000 \text{ } (1 - e^{-5}) = 8000 \times 0.993 = 7944 \text{ pC.}$$

$$V_C = \frac{7944 \text{ pC}}{100 \text{ pF}} = 79.44 \text{ V.}$$

$$V_R = 80 - 79.44 \text{ V} = 0.56 \text{ V.}$$

$$I_C = \frac{0.56 \text{ V}}{150 \text{ k}\Omega} = 0.00373 \text{ mA.}$$

The transient charging state is now assumed to be completed.
When

$$t = 150 \text{ }\mu\text{s} = 10 \times \text{time constants,}$$

$$Q = 8000(1 - e^{-10}) = 8000 \times 0.9999546 = 7999.6368 \text{ pC.}$$

$$V_C = \frac{7999.6368 \text{ pC}}{100 \text{ pF}} = 79.9963668 \text{ V.}$$

$$V_R = 80 - 79.996368 \text{ V} = 0.003632 \text{ V.}$$

$$I_C = \frac{0.003632 \text{ V}}{150 \text{ k}\Omega} = 0.0242 \text{ }\mu\text{A.}$$

Example 51-2

In Fig. 51-2A, $C = 100$ pF, $R = 150$ kΩ and $E = 80$ V. It is assumed that the capacitor is fully charged to 80 V and that the switch, S, is then closed in position 2. What are the values of Q, V_C, V_R and I_D after time intervals of $t = 0$, 10.5, 15, 30, 15, 60, 75 and 150 μs?

Solution

From Example 51-1, the time constant is 15 μs. When $t = 0$, the initial conditions are:

$$Q = 8000 \text{ pC, } V_C = 80 \text{ V, } V_R = -80 \text{ V, and } I_D = -\frac{80 \text{ V}}{150 \text{ k}\Omega} = -0.533 \text{ mA}$$

When $t = 10.5$ μs $= 0.7 \times$ time constant,

$$Q = \text{CE } e^{-t/RC} = 8000 \ e^{-0.7} = 3973 \text{ pC}.$$

$$V_C = \frac{3973 \text{ pC}}{100 \text{ pF}} = 39.73 \text{ V}.$$

$$V_R = -39.73 \text{ V}.$$

$$I_D = \frac{-39.73 \text{ V}}{150 \text{ k}\Omega} = -0.265 \text{ mA}.$$

When

$$t = 15 \ \mu\text{s} = 1 \times \text{ time constant},$$

$$Q = 8000 \ e^{-1} = 2943 \text{ pC}.$$

$$V_C = \frac{2943 \text{ pC}}{100 \text{ pF}} = 29.43 \text{ V}.$$

$$V_R = -29.43 \text{ V}.$$

$$I_D = \frac{-29.43 \text{ V}}{150 \text{ k}\Omega} = -0.196 \text{ mA}.$$

When

$$t = 30 \ \mu\text{s} = 2 \times \text{ time constants},$$

$$Q = 8000 \ e^{-2} = 1083 \text{ pC}.$$

$$V_C = \frac{1083 \text{ pC}}{100 \text{ pF}} = 10.83 \text{ V}.$$

$$V_R = -10.83 \text{ V}.$$

$$I_D = \frac{-10.83 \text{ V}}{150 \text{ k}\Omega} = -0.072 \text{ mA}.$$

When

$$t = 45 \ \mu\text{s} = 3 \times \text{ time constants},$$

$$Q = 8000 \ e^{-3} = 398.3 \text{ pC}.$$

$$V_C = \frac{398.3 \text{ pC}}{100 \text{ pF}} = 3.983 \text{ V}.$$

$$V_R = -3.983 \text{ V}.$$

$$I_D = \frac{-3.983 \text{ V}}{150 \text{ k}\Omega} = -0.2655 \text{ mA}.$$

When

$$t = 60 \ \mu\text{s} = 4 \times \text{ time constants},$$

$$Q = 8000 \ e^{-4} = 146.5 \text{ pC}.$$

$$V_C = \frac{146.5 \text{ pC}}{100 \text{ pF}} = 1.465 \text{ V}.$$

$$V_R = -1.465 \text{ V}.$$

$$I_D = \frac{-1.465 \text{ V}}{150 \text{ k}\Omega} = -0.00977 \text{ mA}.$$

When

$$t = 75 \ \mu s = 5 \times \text{time constants},$$
$$Q = 8000 \ e^{-5} = 53.9 \ \text{pC}.$$
$$V_C = \frac{53.9 \ \text{pC}}{100 \ \text{pF}} = 0.539 \ \text{V}.$$
$$V_R = -0.539 \ \text{V}.$$
$$I_D = \frac{-0.539 \ \text{V}}{150 \ \text{k}\Omega} = -3.59 \ \mu \text{A}.$$

The transient decay state is now assumed to have ended.
When

$$t = 150 \ \mu s = 10 \times \text{time constants},$$
$$Q = 8000 \times e^{-10} = 0.363 \ \text{pC}.$$
$$V_C = \frac{0.363 \ \text{pC}}{100 \ \text{pF}} = 0.00363 \ \text{V}.$$
$$V_R = -0.00363 \ \text{V}.$$
$$I_D = \frac{-0.00363 \ \text{V}}{150 \ \text{k}\Omega} = -0.0242 \ \mu \text{A}.$$

Chapter 52
Moving-coil (D'Arsonval) meter movement

THE MOVING COIL METER MOVEMENT IS AN ELECTROMECHANICAL DEVICE whose action depends on the motor effect (Chapter 35). One of its main applications is in the voltmeter—ohmmeter—milliammeter or VOM, which is a multimeter capable of measuring voltage, current, and resistance. Such instrument does not require an external power supply for its operation.

The meter movement itself essentially consists of a rectangular coil that is made from fine insulated copper wire wound on a light aluminum frame (Fig. 52-1A). This frame is carried by a spindle that pivots in jeweled bearings. The current that creates the deflection is lead into and out of the coil by the spiral hair springs, HH', which behave as the controlling device and provide the restoring torque. The whole assembly is mounted between the poles of a permanent magnet so that the amount of current flowing through the coil, creates the deflecting torque. The coil is then free to move in the gaps between the permanent magnetic pole pieces, P, and the soft-iron cylinder, A, which is normally carried by a nonmagnetic bridge attached to P (Fig. 52-1A). The

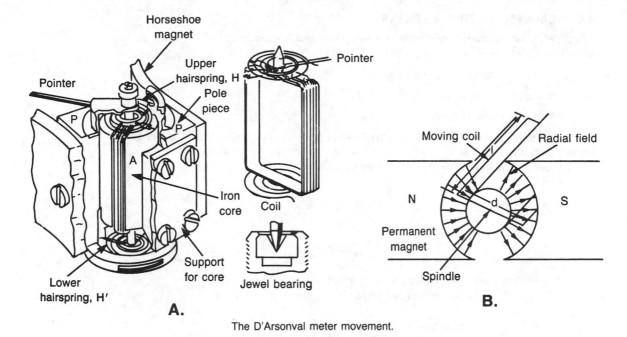

Figure 52-1

purpose of the soft-iron cylinder is to concentrate the magnetic flux and to produce a radial magnetic field with a uniform flux density. For a given current the deflecting torque will then be constant over a wide arc (Fig. 52-1B).

The motor effect (Equation 35-1) shows that the deflecting torque is directly proportional to the current so that the divisions on the scale of the VOM are equally spaced.

When the coil rotates towards the equilibrium position where the deflecting and restoring torques are balanced, there could be an overswing and the coil could oscillate before finally coming to rest. This unwanted oscillation is prevented by eddy current damping of the aluminum frame (Lenz's Law).

The sensitivity of the moving meter movement is inversely related to the amount of the current required to produce full-scale deflection. A high sensitivity will require many turns of fine copper wire to be mounted on a light aluminum frame and then attached to delicate hair springs. In particular, the sensitivity will depend on the strength of the meter's permanent magnet because with a high flux density, even a weak current passing through the coil will be able to produce an appreciable deflection torque.

Any flow of current through the deflecting copper coil will raise its temperature and increase its resistance. A swamping resistor with a negative temperature coefficient is therefore added in series with the coil to provide a suitable total resistance (for example, 50 Ω) which is virtually independent of temperature.

The accuracy of the meter is given as the percentage error related to the full-scale current. For example if the error of a 0.1 mA meter movement is $\pm2\%$, the accuracy for a reading of 1 mA is $\pm2/100 \times 1$ mA $= \pm0.02$ mA, but if the reading is only 0.1 mA, the accuracy is still ±0.02 mA and the percentage error rises to $(\pm0.02/0.1) \times 100 = \pm20\%$.

MATHEMATICAL DERIVATIONS

In Fig. 52-1B, the coil consists of N turns each with a length of l meters and a width of d meters. If the radial magnetic field has a uniform flux of B teslas,

$$\text{Deflection torque, } T = BINld = BINA \text{ newton-meters} \qquad (52\text{-}1)$$

where: I = current flowing through the coil (A)

$\quad\quad$ A = area of the coil (m²).

The restoring torque, T', provided by the hair springs is directly proportional to the deflection angle, $\theta°$, and therefore $T' = k\theta°$, where k is the constant for the hairspring system. In the final rest position of the needle, the two torques exactly balance.

Therefore

$$T = T' \qquad (52\text{-}2)$$

$$BINA = k\theta°$$

$$I = \frac{k}{BAN} \times \theta° = K\theta°$$

where K is the constant of proportionality for the entire meter movement.

If the full-scale deflection current is I amperes.

$$\text{Sensitivity} = \frac{1}{\text{full-scale deflection current}} \qquad (52\text{-}3)$$

$$= \frac{1}{I \text{ amperes or volts-per-ohm}}$$

$$= \frac{1}{I} \text{ ohms-per-volt}$$

$$\text{Full-scale deflection current in amperes} = \frac{1}{\text{Sensitivity in ohms-per-volt}} \qquad (52\text{-}4)$$

Example 52-1

A moving coil meter movement has a sensitivity of 20,000 Ω/V. What is the full-scale deflection current in μA?

Solution

$$\text{Full-scale deflection current} = \frac{1}{\text{Sensitivity}} \qquad (52\text{-}4)$$

$$= \frac{1}{20,000 \text{ Ω/V}}$$

$$= \frac{1,000,000}{20,000} \text{ μA}$$

$$= 50 \text{ μA.}$$

Example 52-2

A moving coil meter movement has a full-scale deflection current of 1 mA. What is the value of its sensitivity?

Solution

$$\text{Sensitivity} = \frac{1}{\text{Full-scale deflection current}}$$ (52-3)

$$= \frac{1}{1 \text{ mA}}$$

$$= \frac{1}{1/1000 \text{ A}}$$

$$= 1000 \ \Omega/\text{V}.$$

❧

Chapter 53
The milliammeter

LET US ASSUME THAT WE HAVE A BASIC MOVING COIL METER MOVEMENT WHOSE total resistance (coil resistance and swamping resistance) is 50 Ω and whose full-scale deflection is 50 μA (sensitivity = 1/50 μA = 20000 Ω/V). Clearly such an instrument as it stands would not be capable of measuring a current of more than 50 μA, whereas we might be required to measure currents of hundreds of milliamperes or even amperes. To solve this problem, a shunt resistor, R_{sh}, is connected across the series combination of the meter movement and the swamp resistor (Figure 53-1A). This shunt resistor is a low value precision type which is usually made from constantan wire with its negligible temperature coefficient. As a result of the shunt's low resistance, most of the current to be measured will be diverted through R_{sh} and only a small part of the current will pass through the meter movement.

Figure 53-1B, shows an arrangement that has a number of shunts for different current ranges. However, in order to take a current reading, the circuit must first be broken. The instrument is then inserted in the break so that it is directly in the path of the current to be measured. When switching from one current range to another, the meter movement could momentarily be placed in the circuit without the protection of the shunt; as a result the meter

Figure 53-1

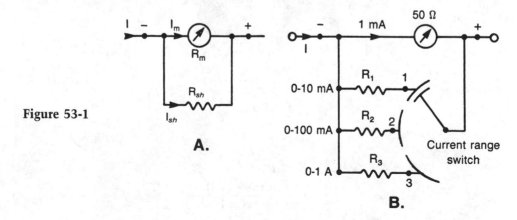

179

movement might be damaged. This could be avoided by using a switch of the make-before-break type.

Because of its very low resistance, a current meter must never be placed directly across a voltage source. After your current measurements have been completed, you should, for safety's sake, switch to a high voltage range. Moreover if you have no idea of the value of the current to be measured, you should start with the highest current range and then, if necessary, move to lower ranges until an appropriate deflection is obtained.

MATHEMATICAL DERIVATIONS

In Fig. 53-1A, using KCL,

$$I = I_m + I_{sh} \text{ milliamperes} \tag{53-1}$$

Because the shunt resistance and the total resistance of the meter movement are in parallel,

$$I_{sh} \times R_{sh} = I_m \times R_m \text{ volts} \tag{53-2}$$

Combining Equations 53-1 and 53-2,

$$R_{sh} = \frac{I_m \times R_m}{I_{sh}} = \frac{I_m \times R_m}{I - I_m} \text{ ohms} \tag{53-3}$$

where: I = current to be measured (mA or A)

I_m = current flow through the meter movement (mA or μA)

R_{sh} = shunt resistance (Ω)

R_m = total resistance of the meter movement (Ω).

Example 53-1

In Fig. 53-1B the sensitivity of the meter movement is 1000 Ω/V and its total resistance is 50 Ω. Calculate the values of the required shunt resistors for the ranges (a) 0 − 10 mA, (b) 0 − 100 mA, and (c) 0 − 1 A.

Solution

(a) Shunt resistor,

$$R_{sh} = \frac{I_m \times R_m}{I - I_m} \tag{53-3}$$

$$= \frac{1 \text{ mA} \times 50 \text{ }\Omega}{10 \text{ mA} - 1 \text{ mA}}$$

$$= \frac{50}{9}$$

$$= 5.555 \text{ }\Omega.$$

(b) Shunt resistance,

$$R_{sh} = \frac{I_m \times R_m}{I - I_m} \tag{53-3}$$

$$= \frac{1 \text{ mA} \times 50 \text{ }\Omega}{100 \text{ mA} - 1 \text{ mA}}$$

$$= \frac{50}{99}$$

$$= 0.505 \ \Omega.$$

(c) Shunt resistance,

$$R_{sh} = \frac{I_m \times R_m}{I - I_m} \tag{53-3}$$

$$= \frac{1 \ \text{mA} \times 50 \ \Omega}{1000 \ \text{mA} - 1 \ \text{mA}}$$

$$= \frac{50}{999}$$

$$= 0.05005 \ \Omega.$$

Example 53-2

If the meter movement of Example 53-1 is used to measure the current in Fig. 53-2, what will be the meter reading?

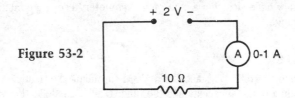

Figure 53-2

Solution

Because the theoretical current is 2 V/10 Ω = 0.2 A, the meter must be switched to the 0–1 A range. The total resistance of the meter movement and the shunt is 1 mA × 50 Ω/1000 mA = 0.05 Ω. The meter reading will therefore be 2 V/(10 Ω + 0.05 Ω) = 0.199 A. This indicates the loading effect of the ammeter on the circuit being monitored. Such a loading effect is negligible with high resistance circuits, which are normally encountered in electronics.

Chapter 54
The voltmeter-loading
effect of a voltmeter

A VOLTMETER IS PLACED IN PARALLEL WITH (SHUNTING ACROSS) THE COMPONENT whose potential difference is to be measured. This means that the voltmeter must possess as high a resistance as possible so that it has minimum loading effect on the circuit being monitored. The moving-coil meter movement by itself is basically a millivoltmeter or microvoltmeter so that, in order to adapt the movement for higher voltages, it is necessary to connect a series

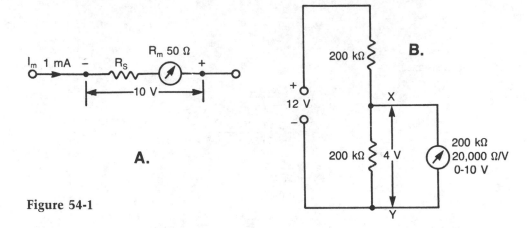

Figure 54-1

"multiplier" resistor, R_S (Fig. 54-1A). Most of the voltage to be measured is then dropped across this high-value series resistor and only a small amount appears across the meter movement to provide the necessary deflection. For a multirange voltmeter, a separate multiplier resistor may be used for each range.

THE VOLTMETER

On a particular range the total resistance of a VOM (including the multiplier resistor) may be found from the product of the sensitivity and the range's full-scale deflection voltage. For example, if the sensitivity is 20,000 Ω/V and the full-scale deflection voltage is 10 V, the voltmeter's resistance is 20,000 Ω/V × 10 V = 200,000 Ω = 200 kΩ. On the 0–100 V range, the voltmeter's resistance would be 20,000 Ω/V × 100 V = 2,000,000 Ω = 2 MΩ.

LOADING EFFECT OF A VOLTMETER

An ideal voltmeter would have infinite resistance so that the instrument does not load the circuit being monitored. However, we have already seen that the resistance of a moving-coil voltmeter is far from infinite and moreover its resistance changes with the particular voltage range which you select. The worst loading effect occurs with a low-voltage, high-resistance circuit such as shown in Fig. 54-1B.

With no voltmeter present the voltage existing between the points A and B is 6 V so that we would switch to the 0–10 V range. Assuming that the voltmeter's sensitivity is 20,000 Ω/V, the voltmeter's resistance is 200 kΩ. To read the voltage between the points A and B the voltmeter is connected across the lower 200 kΩ resistor. However, the parallel combination of this resistor and the voltmeter only has a total resistance of 200 kΩ/2 = 100 kΩ and therefore, by using the VDR, the voltmeter reading will only be 12 V × 100 kΩ/(100 kΩ + 200 kΩ) = 4 V. The loading effect of the voltmeter has therefore reduced the voltage between A and B from 6 V down to 4 V.

Most solid-state voltmeters have very high resistances (several megohms) which are independent of the voltage range chosen; such voltmeters have a minimum loading effect on the circuit being monitored.

MATHEMATICAL DERIVATIONS

In Fig. 54-1A,

$$\text{Total resistance of the voltmeter} = \frac{V}{I_m} \text{ ohms} \qquad (54\text{-}1)$$

$$= V \times (\text{sensitivity in ohms-per-volt}) \qquad (54\text{-}2)$$

$$\text{Resistance of the series "multiplier" resistor, } R_S = \frac{V}{I_m} - R_m \text{ ohms} \qquad (54\text{-}3)$$

Also,

$$\text{Resistance, } R_S = V \times (\text{sensitivity in ohms per volt}) - R_m \text{ ohms} \qquad (54\text{-}4)$$

where: R_S = resistance of series "multiplier" resistor (Ω)

V = full-scale deflection voltage (V)

I_m = total deflection current (A)

R_m = total resistance of the meter movement (Ω).

Example 54-1

In Fig. 54-2 the potential difference between the points A and B is measured by a 1000 Ω/V voltmeter, which is switched to the 0–10 V range. What is the reading of the voltmeter?

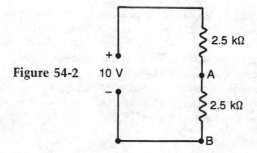

Figure 54-2   10 V

Solution

Total resistance of voltmeter

$$= V \times (\text{sensitivity in } \Omega/V) \qquad (54\text{-}2)$$

$$= 10 \text{ V} \times 1000 \text{ } \Omega/V$$

$$= 10000 \text{ } \Omega$$

$$= 10 \text{ k}\Omega.$$

Total equivalent resistance between the points A and B with the voltmeter included, is (10 × 2.5)/(10 + 2.5) = 2 kΩ.

By the VDR,

Voltmeter reading = 10 V × 2 kΩ/(2 kΩ + 2.5 kΩ) = 4.444 V. This reading compares with the 5 V which exists between the points A and B in the absence of the voltmeter.

Chapter 55
The ohmmeter

BECAUSE ALL RESISTANCE MEASUREMENTS MUST BE MADE WITH POWER REMOVED from the circuit, it is necessary to provide one or more primary cells to create the necessary deflection. These cells are normally included within the instrument itself.

In its simplest form (Fig. 55-1A) an ohmmeter would consist of a 1.5 V primary cell, a rheostat and a 0–1 mA moving-coil meter movement with its 50 Ω total resistance. The unknown resistance, R_x, is then connected between the probes of the ohmmeter.

The steps involved with taking a resistance reading are:

1. Place the probes apart so that an open circuit (infinite resistance) exists and there is zero current. The needle will then be on the far left-hand side of the scale (Fig. 55-1A) so that the reading is zero current but infinite ohms. The symbol for infinity is ∞ which is normally marked on the ohms scale. If the needle is not correctly on the zero current mark, the torque provided by the hairsprings can be mechanically adjusted.
2. Short the probes together (Fig. 55-1B) to create zero resistance and then adjust the rheostat until the full-scale deflection current of 1 mA flows and the needle is on the far right of the scale. The circuit's total resistance is now 1.5 V/1 mA = 1500 Ω and the value of the rheostat, R_{rh}, is then 1500 Ω − 50 Ω = 1450 Ω. As the cell ages, its terminal voltage decreases and the rheostat can be adjusted to bring the current back up to 1 mA.
3. Assume that a 1500 Ω resistor, R_x, is now connected between the probes (Fig. 55-1C). The current would then be 1.5 V/(1450 Ω + 50 Ω + 1500 Ω) = 0.5 mA and the needle would be in the center of the scale. Consequently, we have infinite ohms on the left of the scale, zero ohms on the right and 1500 ohms in the center. The scale is therefore nonlinear because the deflection current is inversely dependent on the value of the unknown resistance. The deflection marks are expanded on the low resistance side but crowded on the high resistance end. Clearly such a scale could not be used for accurate measurements of high resistances (greater than 10 kΩ).

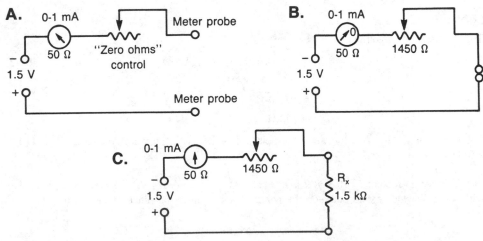

Figure 55-1

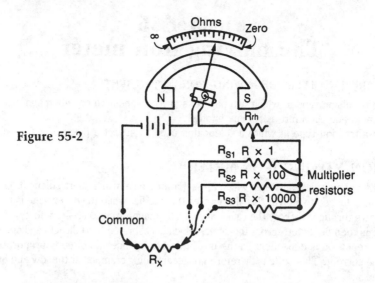

Figure 55-2

With the VOM it is common practice to have three resistance ranges: $R \times 1$, $R \times 100$, $R \times 10000$. This is achieved by including the multiplier resistors R_s, as shown in Fig. 55-2. On the $R \times 1$ range the multiplier resistor, R_{s1}, has a high resistance so that if R_x has a value of 150 kΩ (for example), the needle would be close to the infinite mark and an accurate reading would be impossible. However, if we now switch to the $R \times 100$ range with its lower multiplier resistor, R_{s2}, the needle is deflected to the middle of the scale and the reading would be 1500 Ω $\times$ 100 = 150000 Ω = 150 kΩ.

Note that since the multiplier resistors have different values, you must always readjust the "zero ohms" rheostat when switching from one resistance range to another.

Example 55-1

From Fig. 55-1A what is the deflecting current if the unknown resistance is (a) 500 Ω and (b) 4.5 kΩ?

Solution

(a) Total circuit resistance = 1500 Ω + 500 Ω = 2000 Ω.

$$\text{Deflection current, } I_m = \frac{1.5 \text{ V}}{2000 \text{ }\Omega} = 0.75 \text{ mA}.$$

(b) Total circuit resistance = 1500 Ω + 4.5 kΩ = 6 kΩ.

$$\text{Deflection current, } I_m = \frac{1.5 \text{ V}}{6 \text{ k}\Omega} = 0.25 \text{ mA}.$$

Chapter 56
The moving iron meter

THERE ARE TWO TYPES OF MOVING-IRON INSTRUMENT:

1. The repulsion type in which two parallel soft-iron vanes are equally magnetized inside a solenoid and therefore repel one another.
2. The attraction type in which a solenoid is used to attract a piece of soft iron.

REPULSION MOVING IRON METER

Figure 56-1A shows two iron vanes which are separated axially in a short solenoid. One vane is moveable and attached to a pivot that also carries the needle, while the other vane is fixed. When a current flows through the coil, the vanes are equally magnetized to create a force of repulsion. The pointer is then deflected across the scale, which is calibrated for a direct reading. However, the force of repulsion is dependent on the flux of each iron vane and directly proportional to the *square* of the current. This scale is therefore nonlinear, being cramped at the low end but open at the high end.

The restoring torque of the controlling device is provided either by a spring or by gravity. An air piston is commonly used to behave as a damping device.

Because the amount of the deflection torque depends on the number of ampere-turns for the coil, it is possible to achieve various ranges by winding different numbers of turns on the coil. In addition we can vary the type of wire used so that the meter's resistance can be changed and there is no need for swamping and shunt resistances.

ATTRACTION MOVING-IRON METER

This type of meter is illustrated in Fig. 56-1B. The current to be measured flows through the coil and establishes a flux, which magnetizes the soft-iron disk. As a result the soft-iron disk is drawn into the center of the coil and the pointer moves across the scale. However, the force acting on the iron disk depends on the flux density and the coil's magnetic field intensity, which are both proportional to the current. The deflection is therefore not proportional to the current alone (as in the moving-coil meter movement) but to the square of the current. Consequently, the meter has a nonlinear scale that is cramped at the low end but open at the high end. It is possible to improve the linearity by accurate shaping of the iron disk.

The moving-iron milliammeter may be converted to a voltmeter by adding a suitable noninductive series resistor. When compared with moving-coil instruments, moving-iron meters are

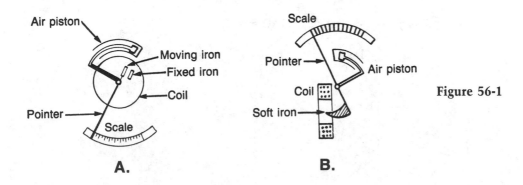

Figure 56-1

A.

B.

relatively inexpensive and robust but have low sensitivity. The accuracy of moving-iron meters is normally ±5% of the full-scale deflection value. Unlike the moving-coil meter, the moving-iron instruments can be used for direct ac measurements.

Example 56-1

A moving-iron meter requires 550 ampere-turns to provide full-scale deflection. How many turns on the coil will be required for the 0–10 A range? For a voltage scale of 0–250 V with a full-scale deflection current of 25 mA, how many turns are required and what is the total resistance of the instrument?

Solution

$$\text{Number of turns required for the 0–10 A range} = \frac{550 \text{ AT}}{10 \text{ A}}$$

$$= 55 \text{ turns.}$$

$$\text{Number of turns for the 0–250 V range} = \frac{550}{0.025}$$

$$= 22000 \text{ turns.}$$

$$\text{Total resistance} = \frac{250 \text{ V}}{25 \text{ mA}} = 10 \text{ k}\Omega.$$

Chapter 57
The wattmeter

BECAUSE THE MOVING COIL METER MOVEMENT WORKS IN CONJUNCTION WITH A permanent magnet, this type of meter can only be used for direct dc measurements (to measure an ac voltage, for example, the alternating voltage must first be rectified and thereby converted to a dc voltage, which is then applied to the moving-coil instrument). However, if the permanent magnet is replaced by an electromagnet, the direction of the flux may be reversed and direct ac measurements are then possible. This electrodynamometer movement of Fig. 57-1A consists of two fixed coils, F and F', which form the electromagnet. The third moving coil, M, is carried by a spindle and the controlling torque is exerted by spiral hairsprings H and H', which also serve to lead the current into and out of the moving coil.

For dc measurements the electrodynamometer movement has no advantage over the D'Arsonval type. It is better to measure the source voltage and the source current separately and then calculate their product to obtain the dc power. Compared with the ac moving-iron instruments of Chapter 56, dynamometer ammeters and voltmeters are less sensitive and more expensive so that they are rarely used. The most important application is the dynamometer wattmeter, which is the common way of measuring power directly in ac circuits.

Figure 57-1B illustrates the manner in which the wattmeter is connected into the circuit. The two fixed coils, F and F', are joined in series with the load and therefore they carry the load

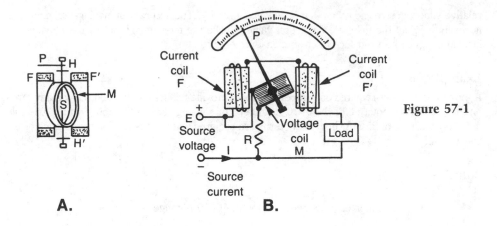

Figure 57-1

A.   B.

current, I. The third moving coil is connected in series with a high value multiplier resistor, R. This combination is connected across the source so that the current through the moving coil is directly proportional to the source voltage, E. Consequently, the torque on the moving coil is directly proportional to the product of the current through the fixed (current) coils and the current through the moving (voltage) coil. It then follows that this torque and its resulting deflection are directly proportional to the power absorbed by the load. Because the controlling torque provided by the hairsprings is also directly proportional to the deflection, electrodynamometer wattmeters have linear scales with the markings evenly spaced.

When used for ac measurements, the moving coil cannot follow the rapid fluctuations in the level of the instantaneous power, so that it assumes a mean or average position that corresponds to the true power in the circuit. In other words, the electrodynamometer movement automatically takes into account the circuit's power factor.

MATHEMATICAL DERIVATIONS
Power measurement: dc

$$\text{Wattmeter Reading} = E \times I \text{ watts} \qquad (57\text{-}1)$$

where: E = source voltage (V)

I = source current (A)

Power measurement: ac

$$\text{Wattmeter Reading} = E \times I \times \text{Power Factor} \qquad (57\text{-}2)$$

where: E = rms source voltage (V)

I = rms source current (A).

Example 57-1

An electrodynamometer wattmeter has a full-scale deflection of 100 W. What value of power will correspond to 65% of the full-scale deflection?

Solution

Because the electrodynamometer wattmeter has a linear scale, the power corresponding to a 65% deflection is 65/100 × 100 W = 65 W.

Chapter 58
The cathode-ray oscilloscope

THE BASIC TRIGGERED SWEEP CATHODE RAY OSCILLOSCOPE (CRO) IS PROBABLY the most flexible, general-purpose measuring instrument in use today and, perhaps, for some time in the future. The key element of the CRO is the electrostatic cathode-ray tube, which operates with a high degree of vacuum and in which a stream of electrons is produced by a heated cathode. This acts together with other metallic elements to form the *electron gun*. The stream of electrons is accelerated by high voltages towards the inside surface of the tube face. Light is produced when the electrons strike the fluorescent screen that coats this inside surface.

Between the beginning of the electron gun and the fluorescent screen are a number of electrodes held at positive potentials with respect to the cathode. Some of these serve the purpose of focusing the electron beam to a point at the screen. A control on the front panel can be used to improve the focus and affect the size and appearance of the spot seen by the operator. Other electrodes cause the electron beam to be deflected. One pair of electrodes, called the *X* plates, causes a horizontal deflection, which is proportional to the impressed voltage. A second pair, the *Y* plates, causes vertical deflection (Fig. 58-1A) whose amount depends on the length of the vertical plates and their separation, the values of the vertical deflection voltage and the horizontal accelerating voltage, as well as the distance between the vertical plates and the screen.

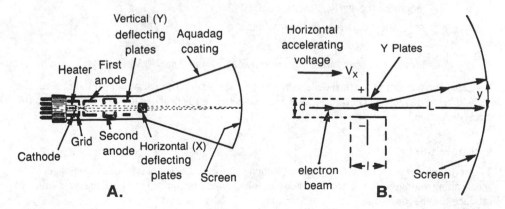

Figure 58-1

An electrostatic CRT commonly has a sensitivity that is about 1 mm of vertical deflection for each volt applied to the Y plates with 1 kV of horizontal accelerating voltage. If $V_Y = 100$ V and $V_X = 10$ kV, the vertical deflection is $1 \times (100/10)$ mm = 1 cm. In other words the sensitivity is 100 V/cm for a horizontal accelerating voltage of 10 kV.

In the most commonly used mode of operation, the horizontal deflection is linear and is produced by a sawtooth voltage waveform applied to the X plates. The beam, and the spot that it produces, move equal distances in equal intervals of time. For slow movements, the spot can be seen moving to the right, leaving a rapidly disappearing trail. Upon reaching the end of its travel on the right, the spot disappears and quickly reappears on the left to repeat its linear journey. With fast movements of the beam, required for most electronic measurements, the beam cannot be seen, but its trace, repeated hundreds, thousands, or even millions of times per second, appears as a solid horizontal line.

The vertical deflection of the beam is usually caused by an amplified signal to be observed and measured. The amount of amplification, and hence the deflection, is controlled by a knob on the CRO panel.

The combined horizontal sweep motion and vertical signal deflection give a graphical picture of the signal waveform. If a dc voltage is applied to the vertical plates, the appearance on the screen is a horizontal line, which is deflected upwards or downwards according to the polarity of the deflecting voltage.

MATHEMATICAL DERIVATIONS
In Fig. 58-1B,
Vertical deflection,

$$y = \frac{Ll\,V_Y}{2d\,V_X} = K \times V_Y \text{ millimeters} \tag{58-1}$$

Sensitivity of the CRT,

$$K = \frac{Ll}{2d\,V_X} \begin{array}{l}\text{millimeters per deflection volt for each 1 kV of} \\ \text{horizontal accelerating voltage,}\end{array} \tag{58-2}$$

where: y = vertical deflection (mm)

L = distance between the vertical plates and the screen (mm)

l = length of the vertical plates (mm)

d = separation between the plates (mm)

V_Y = vertical deflection voltage (V)

V_X = horizontal accelerating voltage provided by the "gun" (kV).

Note that the amount of vertical deflection depends neither on the charge nor on the mass of the particle being deflected. Therefore, electrons and negative residual gas ions are subjected to the same deflection.

Example 58-1

The sensitivity of a CRT is 0.6 mm per deflection volt for each 1 kV of accelerating voltage. If the accelerating voltage is 8 kV, what value of dc voltage must be applied to the Y plates in order to produce a deflection of 2.7 cm?

Solution

Required deflection voltage,

$$V_Y = \frac{y}{K} \qquad\qquad (58\text{-}1)$$

$$= \frac{27 \text{ mm} \times 10 \text{ kV} \times 1 \text{ V}}{0.6 \text{ mm} \times 1 \text{ kV}}$$

$$= 450 \text{ V}.$$

2
PART

Alternating current principles

Chapter 59
Introduction to alternating current (ac)

PREVIOUS CHAPTERS HAVE BEEN CONCERNED WITH DIRECT CURRENT IN WHICH the electron flow has always been in the same direction although the magnitude of the flow has not necessarily been constant.

With alternating current, the current reverses its direction periodically and has an average value of zero. If the average value of the waveform is not zero, it is regarded as a combination of a dc component and an ac component. The alternating current may assume a variety of waveforms some of which are shown in Fig. 59-1A, B, C, and D. In every case a single complete waveform is known as a cycle and the number of cycles occurring in one second is the frequency, which is therefore the rate at which the waveform repeats.

The letter symbol for the frequency is f and its SI unit is the hertz (Hz) which is equivalent to one cycle per second.

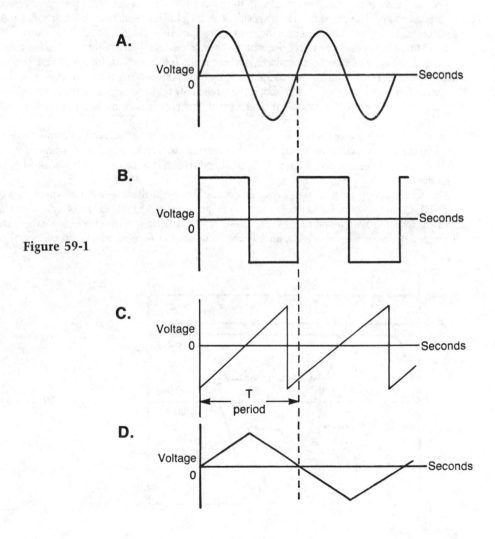

Figure 59-1

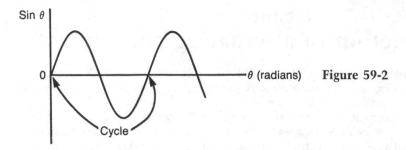

Sin θ

0 ————————————→ θ (radians)   **Figure 59-2**

Cycle

The time interval taken by a complete cycle is the period, which is equal to the reciprocal of the frequency. The letter symbol for the period is T and its basic unit is the second (s). As an example the commercial ac line voltage has a frequency of 60 Hz and therefore its period is 1/60th of a second.

The sine wave is not only associated with the commercial line voltage but is also widely used for communications. The curve itself is the result of plotting the mathematical sine function, sin θ, versus the angle θ, measured in either degrees or radians (Fig. 59-2).

The vertical axis measures only one of a number of electrical parameters such as voltage, current, and power, while the horizontal axis is a time scale. It follows that, for every sine wave, a certain number of radians on the horizontal scale must correspond to a particular time interval. The angular frequency whose letter symbol is the Greek lowercase omega, ω, is found by dividing the number of radians by the equivalent time interval and is therefore measured in radians per second (rad/s).

One cycle of a sine wave consists of a "positive" alternation and a "negative" alternation (Fig. 59-3). The maximum excursion from its average or zero line is called the peak or maximum (E_{max}) value while the distance between the wave's crest and trough is a measure of the peak-to-peak (E_{p-p}) value.

Sine waves are of particular importance because Fourier analysis allows the waveforms of Fig. 59-1B, C, and D, to be broken down into a series of sine waves that consist of a fundamental component together with its harmonics (Fig. 59-4). The harmonic of a sine wave is another sine wave whose frequency is a whole number times the original or fundamental frequency. For

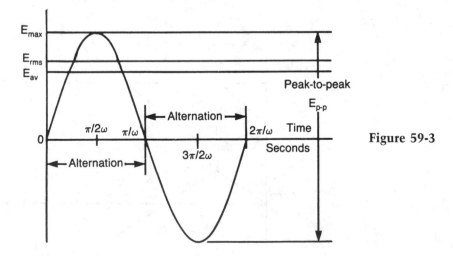

Figure 59-3

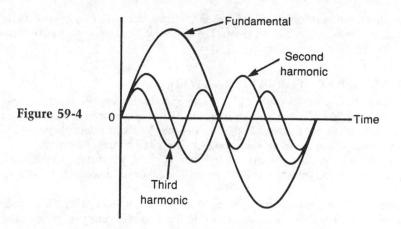

Figure 59-4

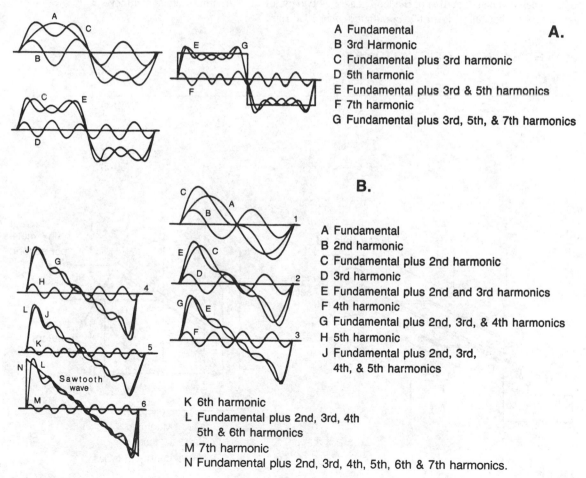

A.

A Fundamental
B 3rd Harmonic
C Fundamental plus 3rd harmonic
D 5th harmonic
E Fundamental plus 3rd & 5th harmonics
F 7th harmonic
G Fundamental plus 3rd, 5th, & 7th harmonics

B.

A Fundamental
B 2nd harmonic
C Fundamental plus 2nd harmonic
D 3rd harmonic
E Fundamental plus 2nd and 3rd harmonics
F 4th harmonic
G Fundamental plus 2nd, 3rd, & 4th harmonics
H 5th harmonic
J Fundamental plus 2nd, 3rd,
    4th, & 5th harmonics

K 6th harmonic
L Fundamental plus 2nd, 3rd, 4th
    5th & 6th harmonics
M 7th harmonic
N Fundamental plus 2nd, 3rd, 4th, 5th, 6th & 7th harmonics.

Figure 59-5

197

example if the fundamental frequency is 1 kHz (f), the second harmonic has a frequency of 2 kHz ($2f$), the third harmonic frequency is 3 kHz ($3f$) etc. The sine wave syntheses of a square wave and a sawtooth wave are illustrated in Fig. 59-5A, B.

THE SIMPLE AC GENERATOR (ALTERNATOR)

The ac generator is a machine that is capable of converting mechanical energy into alternating electrical energy. In its simplest form (Fig. 59-6A), a shaft DD' is driven around by mechanical energy. Attached to the shaft but insulated from it, is a copper loop that rotates between the poles, PP', of a permanent magnet or an electromagnet (energized by direct current).

The ends of the rotating loop are joined to two separate slip rings, SS' that make continuous contact with the stationary carbon brushes, CC'; these brushes are then connected to the electrical load.

The poles' pieces, PP', are especially shaped to provide a constant flux density, B, in which the loop rotates (Fig. 59-6B). One complete rotation of the loop will generate one cycle of alternating voltage, which is applied across the load. If the output frequency is 60 Hz, the loop's speed of rotation is $60 \times 60 = 3600$ revolutions per min. However, if the alternator has four poles, each complete rotation of the loop generates two cycles of alternating voltage and therefore the speed required for a 60 Hz output is only 1800 rpm.

MATHEMATICAL DERIVATIONS

Frequency,

$$f = \frac{1}{T} \text{ Hz} \qquad (59\text{-}1)$$

Period,

$$T = \frac{1}{f} \text{ seconds} \qquad (59\text{-}2)$$

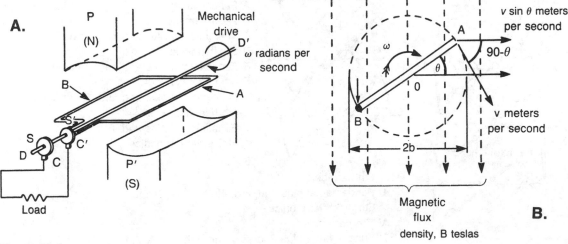

Figure 59-6

One radian is the angle at the center of a circle, which subtends (is opposite to) an arc (l) equal in length to the radius (r).

Angle,

$$\theta = \frac{l}{r} \text{ radians} \tag{59-3}$$

Because the total length of the circle's circumference is $2\pi\, r$ radians, the angle of 360° must be equivalent to $2\pi\, r/r = 2\pi$ radians. Therefore,

$$\theta \text{ (degrees)} = \frac{360}{2\pi} \times \theta \text{ radians} = \frac{180}{\pi} \times \theta \text{ radians} \tag{59-4}$$

$$\theta \text{ (radians)} = \frac{\pi}{180} \times \theta \text{ degrees} \tag{59-5}$$

It follows that

$$1 \text{ radian} = \frac{180}{\pi} \text{ degrees} = 57.296° = 57°17'45'' \tag{59-6}$$

Because the angular velocity or angular frequency, ω, is measured in radians per second,

Angular frequency,

$$\omega = \frac{\theta}{t} = \frac{2\pi}{T} = 2\pi f \text{ radians per second} \tag{59-7}$$

$$\text{Angle, } \theta = 2\pi ft \text{ radians} \tag{59-8}$$

where t is the time in seconds.

Peak-to-peak voltage,

$$E_{p\text{-}p} = 2 \times E_{max} \text{ volts} \tag{59-9}$$

Assume that the loop of N turns in Fig. 59-6B moves with an angular velocity of ω radians per seconds so that the linear velocity of each of the conductors, A, B, is $v = \omega b$ where $2b$ is the width of the loop in meters. The component of the velocity at right angles to the magnetic flux is $v \sin \theta = v \sin \omega t$. From Faraday's law (Equation 36-3), the instantaneous total voltage induced in the loop is:

Instantaneous voltage,

$$e = 2BlNv \sin \theta \tag{59-10}$$

$$= 2BlbN\omega \sin \omega t$$

$$= BAN\omega \sin \omega t$$

$$= E_{max} \sin \omega t = E_{max} \sin 2\pi ft \text{ volts}$$

where: B = uniform flux density (T)

$2b$ = width of load (m)

N = number of turns

l = length of each conductor (m)

A = area of each turn (m^2)

t = time (s)

$E_{max} = BAN\omega$ = peak or maximum voltage value (V)

$$\text{Generated frequency, } f = \frac{np}{60} \text{ hertz} \qquad (59\text{-}11)$$

$$\text{Speed of rotation, } \sim = \frac{60f}{p} \text{ revolutions per minute} \qquad (59\text{-}12)$$

where: p = number of pairs of poles

$\sim$ = speed of rotation in rpm.

Example 59-1

A coil of 2000 turns is rotated at 3600 rpm in a uniform magnetic field with a flux density of 0.05 tesla. The average area of each turn is 25 cm² and the axis of rotation is at right angles to the direction of the flux. Calculate the values of (a) the frequency (b) the period (c) the angular frequency and (d) the peak-to-peak voltage. Write down the trigonometrical expression for the instantaneous voltage and calculate its value when the coil has rotated through nine radians from the position of zero voltage.

Solution

(a) Frequency,

$$f = \frac{np}{60} \qquad (59\text{-}11)$$

$$= \frac{3600 \times 1}{60}$$

$$= 60 \text{ Hz.}$$

(b) Period,

$$T = \frac{1}{f} \qquad (59\text{-}2)$$

$$= 1/60 \text{ s}$$

$$= 16.67 \text{ ms}$$

(c) Angular frequency,

$$\omega = 2\pi f \qquad (59\text{-}7)$$

$$= 377 \text{ rad./s}$$

(d) Peak voltage,

$$E_{max} = BAN\omega \qquad (59\text{-}10)$$

$$= 0.05 \times 25 \times 10^{-4} \times 2000 \times 377$$

$$= 94.25 \text{ V}$$

Peak-to-peak voltage,

$$E_{p\text{-}p} = 2 \times 94.25 \text{ V} = 188.5 \text{ V}$$

Instantaneous output voltage,

$$e = E_{max} \sin \omega t \qquad (59\text{-}10)$$

$$= E_{\max} \sin \theta$$
$$= 94.25 \sin 377t \text{ volts}$$

If $\theta = 9$ radians,
    Instantaneous voltage,

$$e = E_{\max} \sin \theta \qquad\qquad\qquad (59\text{-}10)$$
$$= 94.25 \sin (9 \text{ rad.})$$
$$= 38.84 \text{ V}$$

✍

Chapter 60
The root-mean-square (rms) or effective value of an ac voltage or current

IN CHAPTER 59 WE OBSERVED THAT THE MAGNITUDE OF A SINEWAVE VOLTAGE or current could be measured in terms of its peak or peak-to-peak value. In fact it is common practice to use a cathode-ray oscilloscope (CRO) to obtain the peak-to-peak value of an alternating voltage. Because the peaks of a sine wave only occur twice instantaneously during a cycle, we would normally prefer to use a value that reflects the whole of the cycle. However, we cannot use the average over the complete cycle since this value is always zero. Instead we turn to the root-mean-square (rms) or effective value. Nonmathematically this is defined as that ac value that will provide the same heating or power effect as its equivalent dc value. In other words, a 100 W 100 V bulb that is connected to a 100 V rms 60 Hz supply will appear to have the same brightness as an identical bulb joined to a 100 Vdc source.

The bulb's tungsten filament, which is operated from the dc source, will glow continuously at a constant brightness. However, when the identical bulb is connected to the ac supply, the filament with its low thermal capacity, alternately glows very brightly and is extinguished at the rate of 120 times per second. Due to the persistence of vision the eye cannot directly observe a flicker effect whose rate exceeds 30 times per second; consequently the brightness is "averaged out" and appears to be the same as that of the bulb operated from the dc source. The flicker effect may indirectly be perceived with the aid of a stroboscope.

In Fig. 60-1A an ac sinewave source with a peak value of 10 V is connected across a 1 Ω resistor, while in Fig. 60-1B another 1 Ω resistor is joined to a dc source with a terminal voltage, E. If the average power in the ac circuit over the complete cycle is the same as the constant power of the dc circuit, E must equal the rms value of the ac voltage. When the ac voltage is instantaneously zero, the instantaneous power is also zero but when the voltage is 10 V, the instantaneous power rises to its peak value of $(10 \text{ V})^2/1 \ \Omega = 100$ W (this result is independent of the voltage's polarity, which changes with each alternation). The instantaneous power therefore fluctuates between zero and 100 W and has an average value of 50 W. Notice that the frequency of the power curve is twice that of the ac voltage.

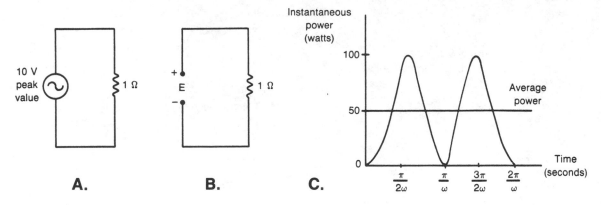

Figure 60-1

If the power in the dc circuit is 50 W, the terminal voltage $E = \sqrt{P \times R} = \sqrt{50 \text{ W} \times 1 \text{ } \Omega} = 7.07$ V. Consequently, an ac voltage with a peak value of 10 V has an effective or rms value of 7.07 V. This value is independent of the numbers chosen, as will be shown in the mathematical derivations. Therefore, for a *sine* wave, the effective value = $0.707 \times$ the peak value or the peak value = $1/0.707 \times$ the effective value = $1.414 \times$ the effective value. These relationships are only true for the sine wave. For example, the effective value of the square wave in Fig. 59-1B must be the same as the peak value while for the triangular wave of Fig. 59-1D the effective value is only $0.577 \times$ the peak value.

Finally, what is the meaning of the term *root mean square,* abbreviated rms? The value of 7.07 V was obtained from the expression

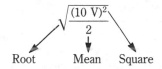

Root   Mean   Square

In other words, we squared the peak value to obtain a value which was proportional to the peak power. We then divided the result by 2 in order to obtain the average or mean power and finally we took the square root to obtain the effective value.

Most alternating voltage and current measurements are in terms of effective values. However, the insulation in ac circuits must normally be able to withstand the peak voltage. For example, the commercial 110 V rms, 60 Hz line voltage has a peak value of 110 V $\times$ 1.414 = 155.5 V and a peak-to-peak value of 2×155.5 V = 311 V.

MATHEMATICAL DERIVATIONS

Let an ac voltage whose instantaneous value is $e = E_{max} \sin \omega t$, be applied across a resistor, R. The instantaneous power dissipated is

Instantaneous power,

$$p = \frac{e^2}{R} = \frac{E^2_{max} \sin^2 \omega t}{R} \qquad (60\text{-}1)$$

$$= \frac{E^2_{max}(1 - \cos 2 \omega t)}{2R} \text{ watts}$$

202

The term "cos 2 ωt" indicates that the frequency of the power curve is twice that of the voltage waveform. The instantaneous power ranges from zero to a peak value of E^2_{max}/R while the average value of "cos 2 ωt" over a complete cycle is zero.
Therefore

Average power,

$$P_{AV} = \frac{E^2_{max}}{2R} \qquad (60\text{-}2)$$

$$= \frac{(E_{max}/\sqrt{2})^2}{R}$$

$$= \frac{(E_{max}/1.414)^2}{R}$$

$$= \frac{(0.707\ E_{max})^2}{R}$$

$$= \frac{E^2_{rms}}{R}\ \text{watts}$$

Effective value,

$$E_{rms} = 0.707 \times E_{max} \qquad (60\text{-}3)$$

$$= \frac{E_{max}}{1.414}$$

$$= \frac{E_{max}}{\sqrt{2}}\ \text{volts}$$

Peak value,

$$E_{max} = \sqrt{2} \times E_{rms} \qquad (60\text{-}4)$$

$$= 1.414 \times E_{rms}\ \text{volts}$$

Similarly,
Effective current,

$$I_{rms} = E_{rms}/R = 0.707 \times I_{max}\ \text{amperes} \qquad (60\text{-}5)$$

Peak current,

$$I_{max} = E_{max}/R = 1.414 \times I_{rms}\ \text{amperes} \qquad (60\text{-}6)$$

Alternatively

$$\text{rms value, } E_{rms} = \sqrt{\frac{\int_0^{2\pi/\omega} (E_{max}\ \sin\ \omega t)^2\ dt}{\int_0^{2\pi/\omega} dt}}$$

$$= \sqrt{\frac{\int_0^{2\pi/\omega} (E^2_{max}/2)(1 - \cos\ 2\omega t)\ dt}{\int_0^{2\pi/\omega} dt}}$$

$$= \sqrt{\frac{(E^2_{max}/2) \times 2\pi/\omega}{2\pi/\omega}}$$

$$= \frac{E_{max}}{\sqrt{2}} = \frac{E_{max}}{1.414} = 0.707 \times E_{max}\ \text{volts}$$

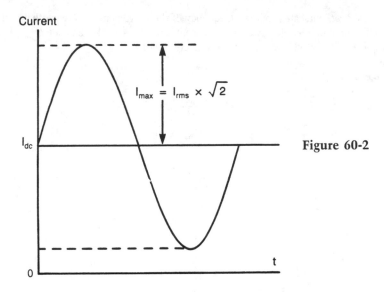

Current

$I_{max} = I_{rms} \times \sqrt{2}$

I_{dc}

t

0

Figure 60-2

If the waveform of the current flowing through a resistor contains both dc and ac components (Fig. 60-2), the effective value of the total current is $\sqrt{I_{dc}^2 + I_{rms}^2}$ where I_{dc} is the dc component and I_{rms} is the effective value of the ac component.

Example 60-1

A sinewave voltage whose peak-to-peak value is 32 V is applied across a 68 Ω resistor. Calculate the values of the peak voltage, the effective voltage, the peak current, the effective current, the peak power and the average power dissipated over the cycle.

Solution

Peak voltage,

$$E_{max} = \frac{E_{p\text{-}p}}{2}$$

$$= \frac{32 \text{ V}}{2}$$

$$= 16 \text{ V}$$

Effective voltage,

$$E_{rms} = 0.707 \times E_{max} \qquad (60\text{-}3)$$

$$= 0.707 \times 16 \text{ V}$$

$$= 11.3 \text{ V}$$

Peak current,

$$I_{max} = \frac{E_{max}}{R} \qquad (60\text{-}6)$$

$$= \frac{16 \text{ V}}{68 \text{ } \Omega}$$

$$= 235 \text{ mA}$$

Effective current,

$$I_{rms} = \frac{E_{rms}}{R} \qquad (60\text{-}5)$$

$$= \frac{11.3 \text{ V}}{68 \text{ } \Omega}$$

$$= 166 \text{ mA}$$

Peak power,

$$P_{max} = \frac{E_{max}^2}{R} \qquad (60\text{-}1)$$

$$= \frac{(16 \text{ V})^2}{68 \text{ } \Omega}$$

$$= 3.76 \text{ W}$$

Average power,

$$P_{AV} = \frac{P_{max}}{2} \qquad (60\text{-}2)$$

$$= \frac{3.76}{2}$$

$$= 1.88 \text{ W}$$

Chapter 61
The average value of a sine wave and its form factor

THE FULL CYCLE OF A SINE WAVE IS COMPOSED OF TWO ALTERNATIONS (FIG. 61-1). One "positive" alternation extends from 0° to 180° while the other "negative" alternation stretches from 180° to 360°. Although the average value of the sine wave over the complete cycle is zero, the average value over an alternation can be found by the use of calculus (see mathematical derivations) and is important in any discussion of half-wave and full-wave rectifier circuits.

Because the "positive" alternation is symmetrical about the 90° mark, it is possible to find the approximate relationship between the average and peak values by computing the mean of the sine values between 0° and 90°. Table 61-1 shows the values for 5° steps and the computation of the average value. In fact the average value = 0.63662 × peak value ≈ 0.637 × peak value.

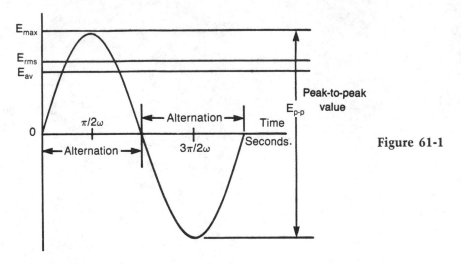

Figure 61-1

The ratio of the effective value to the average value of an alternating current or voltage is called the "form factor" of the waveform. For the sine wave,

$$\text{Form factor} = \frac{\text{rms value}}{\text{average value}} = \frac{0.707 \times \text{peak value}}{0.637 \times \text{peak value}}$$

$$= 1.11$$

Angle $\theta°$	Sin θ
0°	0.000000
5°	0.087156
10°	0.173648
15°	0.258819
20°	0.342020
25°	0.422618
30°	0.500000
35°	0.573576
40°	0.642788
45°	0.707107
50°	0.766044
55°	0.819152
60°	0.866025
65°	0.906308
70°	0.939693
75°	0.965926
80°	0.984808
85°	0.996195
90°	1.000000
	11.951883

Table 61-1

The 19 Values Total 11.951883. Average Value = 11.951883/19 = 0.63 × Peak Value.

The square wave has a form factor of 1.0, while the form factor of a triangular wave is 0.557/0.5 = 1.12.

Compared with other ac waveforms the sine wave is relatively easy to generate and has a high rms value with a good form factor.

MATHEMATICAL DERIVATIONS

Consider a sine wave voltage whose instantaneous value is $e = E_{max} \sin \omega t$. Average value of the sine wave voltage over the "positive" alternation,

$$E_{AV} = \frac{\int_0^{\pi/\omega} E_{max} \sin \omega t \, dt}{\int_0^{\pi/\omega} dt} \qquad (61\text{-}1)$$

$$= \frac{\left[-\dfrac{E_{max}}{\omega} \cos \omega t \right]_0^{\pi/\omega}}{\dfrac{\pi}{\omega}}$$

$$= \frac{\dfrac{2}{\omega} \times E_{max}}{\dfrac{\pi}{\omega}}$$

$$= \frac{2}{\pi} \times E_{max} = 0.63662 \times E_{max} \text{ volts}$$

Peak value,

$$E_{max} = \frac{E_{max}}{0.63662} = 1.5708 \times E_{AV} \text{ volts} \qquad (61\text{-}2)$$

Similarly,

$$I_{AV} = 0.63662 \times I_{max} \text{ amperes} \qquad (61\text{-}3)$$

$$I_{max} = 1.5708 \times I_{AV} \text{ amperes} \qquad (61\text{-}4)$$

Example 61-1

A sinewave alternating current has a peak-to-peak value of 37 mA. What is its average value over an alternation?

Solution

Peak value,

$$I_{max} = \frac{I_{p\text{-}p}}{2} = \frac{37 \text{ mA}}{2} = 18.5 \text{ mA}.$$

Average value,

$$I_{AV} = 0.63662 \times I_{max} \qquad (61\text{-}3)$$

$$= 0.63662 \times 18.5$$

$$= 11.78 \text{ mA}.$$

Chapter 62
Phasor representation
of an ac voltage or current

IF TWO COMPONENTS ARE IN SERIES ACROSS AN AC SOURCE, THERE WILL BE AN alternating voltage across each component. If these two voltages are then added, their resultant exactly balances the source voltage (Kirchhoff's voltage law). In the case of a parallel circuit the individual branch currents must be added to obtain the total current drawn from the source. In previous sections we have expressed a sine wave either as a trigonometrical equation, $e = E_{max} \sin \omega t$ or graphically by a waveform. When we need to add or subtract ac quantities, the trigonometrical equations become difficult to manipulate and the combination of waveforms tends to be tedious.

A third method involves the representation of ac quantities by means of phasors. If the equation of an ac voltage is $e = E_{max} \sin \omega t$, its phasor is a line whose length is a measure of E_{max}. By convention the line phasor is assumed to rotate in the positive or counterclockwise direction with an angular velocity of $\omega = 2\pi f$ radians per second, where f is the voltage's frequency in hertz. The vertical projection, PN, of the phasor on the horizontal reference line is then equal to the instantaneous value, e, of the ac voltage (Fig. 62-1). Therefore as the phasor rotates, the extremity of the line can be said to trace out the voltage's sinusoidal waveform with a frequency that is equal to the phasor's speed of rotation in revolutions per second. One complete rotation of the phasor then traces out one cycle of the sine wave. The phasor diagram therefore contains the same information as the waveform presentation but it is obviously easier to work with lines rather than sine waves.

Because there is a constant relationship between effective and peak values (effective value = 0.707 × peak value), the phasor's length may also be used to indicate the rms value.

Prior to about 1960, the word *vector* was used rather than a *phasor*. A vector is a quantity that possesses both magnitude and direction so that a mechanical force is an example of a vector. By contrast a scalar quantity such as mass possesses magnitude only. The rules for adding or subtracting mechanical vectors are the same as those for adding or subtracting ac phasors. However, the vector rules for multiplication and division are totally different from the corresponding phasor rules. For this reason *vectors* were no longer used to represent ac quantities and instead the word *phasor* was introduced.

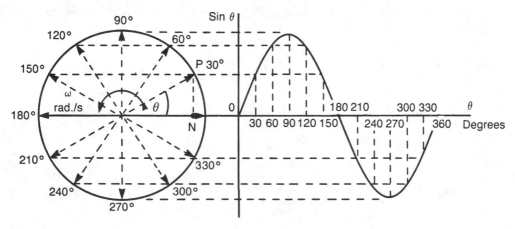

Figure 62-1

MATHEMATICAL DERIVATIONS

In Fig. 62-1

Instantaneous voltage,

$$e = PN = OP \sin \theta$$
$$= E_{max} \sin \theta \quad\quad\quad\quad (62\text{-}1)$$
$$= E_{max} \sin \omega t$$
$$= E_{max} \sin 2\pi \, ft \text{ volts}$$

where: E_{max} = peak voltage (V)

θ = angle measured relative to the horizontal reference line (rad)

ω = angular velocity or frequency (rad/s)

f = frequency (Hz)

t = time (s)

Example 62-1

An ac voltage has an rms value of 6.7 V and a frequency of 400 kHz. What are the values of the peak voltage and the angular velocity? What is the instantaneous voltage after an interval of 2.73 μs as measured from the horizontal reference line?

Solution

Peak voltage,

$$E_{max} = 1.414 \times E_{rms}$$
$$= 1.414 \times 6.7 \text{ V}$$
$$= 9.47 \text{ V}$$

Angular velocity,

$$\omega = 2\pi f$$
$$= 2 \times \pi \times 400 \times 10^3$$
$$= 2.51 \times 10^6 \text{ rad/s}$$

Instantaneous voltage,

$$e = E_{max} \sin 2\pi \, ft \quad\quad\quad\quad (62\text{-}1)$$
$$= 9.47 \sin (2 \times \pi \times 400 \times 10^3 \times 2.73 \times 10^{-6})$$
$$= 5.17 \text{ V}$$

Chapter 63
Phase relationships

WHEN TWO DC VOLTAGES, E_1 AND E_2, ARE IN SERIES, THEY ARE EITHER IN series-aiding or in series-opposing so that only addition or subtraction is necessary in order to obtain their combined voltage (Chapter 12). However, two ac sine wave voltages of the same frequency may not reach similar points in their cycles at the same time. As an example (Chapter 70), when an inductor and a resistor are in series across an ac source, the voltage across the inductor reaches its peak value when the voltage across the resistor is zero. In other words, the two voltages are a quarter of a cycle apart. The amount by which the two sine waves are out of step is referred to as their phase difference, which is either measured in degrees or radians. In our example the shift of one quarter of a cycle is equivalent to a phase difference of 90° or $\pi/2$ radians.

In Fig. 63-1 the e_2 waveform reaches its peak, X, earlier in time than the e_1 waveform with its corresponding peak, Y. The difference between the two waveforms is ϕ radians with e_2 leading e_1 (or e_1 lagging e_2). Phase differences may therefore be either leading or lagging with their angles usually extending up to 180°. In the particular case where ϕ is 180° or π radians, the terms "leading" or "lagging" are not used since 180° leading has the same meaning as 180° lagging. In most cases there is little point in using angles greater than 180° since, for example, a phase difference of 270° leading is equivalent to 90° lagging.

It is impossible to add together two or more series ac voltages (or currents if some form of parallel circuit is involved) without knowing their peak values and their phase relationships. Figure 63-2 shows the results of adding two ac voltages, each of 10-V peak, but with phase differences that are in turn 0°, 90°, 120°, 180°; the resultant (sum) voltages have corresponding peak values of 20 V, 14.4 V, 10 V, and zero. Notice that in each case the frequency of $e_1 + e_2$ is the same as the frequency of e_1 and e_2.

As the mathematical derivations will show, the frequency of $e_1 + e_2$ is unchanged. The peak value of $e_1 + e_2$ is obtained from the peak values of e_1, e_2 and their phase relationship. The same factors determine the phase relationships between $e_1 + e_2$ and e_1, e_2.

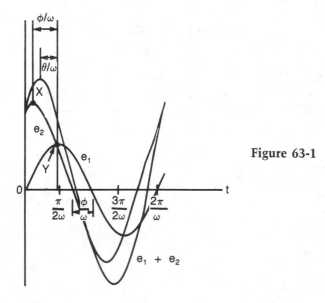

Figure 63-1

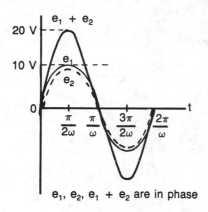

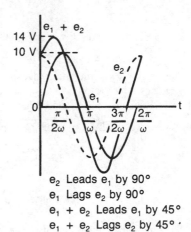

e_2 Leads e_1 by 90°
e_1 Lags e_2 by 90°
$e_1 + e_2$ Leads e_1 by 45°
$e_1 + e_2$ Lags e_2 by 45° ·

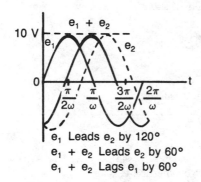

e_1 Leads e_2 by 120°
$e_1 + e_2$ Leads e_2 by 60°
$e_1 + e_2$ Lags e_1 by 60°

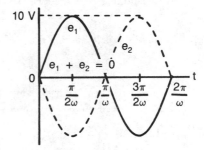

e_1 and e_2 are 180° out of phase

Figure 63-2

MATHEMATICAL DERIVATIONS

In Fig. 63-1, e_2 leads e_1 by an angle of ϕ radians. Therefore,
Instantaneous voltage,

$$e_1 = E_{1max} \sin \omega t \text{ volts} \tag{63-1}$$

Instantaneous voltage,

$$e_2 = E_{2max} \sin (\omega t + \phi) \text{ volts} \tag{63-2}$$

The sum of these two ac voltages is

$$e_1 + e_2 = E_{1max} \sin \omega t + E_{2max} \sin (\omega t + \phi) \tag{63-3}$$
$$= E_{1max} \sin \omega t + E_{2max} \sin \omega t \cos \phi + E_{2max} \cos \omega t \sin \phi$$
$$= (E_{1max} + E_{2max} \cos \phi) \sin \omega t + E_{2max} \sin \phi \cos \omega t \text{ volts}$$

$$\text{Let } E_{1max} + E_{2max} \cos \phi = E_{max} \cos \theta \text{ volts} \tag{63-4}$$

and

$$E_{2max} \sin \phi = E_{max} \sin \theta \text{ volts} \tag{63-5}$$

Then

$$e_1 + e_2 = E_{max} \sin \omega t \cos \theta + E_{max} \cos \omega t \sin \theta \qquad (63\text{-}6)$$
$$= E_{max} \sin (\omega t + \theta) \text{ volts}$$

Therefore E_{max} is the peak value of $e_1 + e_2$ while $e_1 + e_2$ leads e_1 by θ radians but lags e_2 by $\phi - \theta$ radians.

From Equations 63-4, and 63-5,

$$E_{max} = \sqrt{(E_{1max} + E_{2max} \cos \phi)^2 + (E_{2max} \sin \phi)^2} \qquad (63\text{-}7)$$
$$= \sqrt{E_{1max}^2 + E_{2max}^2 + 2\,E_{1max}\,E_{2max} \cos \phi} \text{ volts}$$

and

$$\theta = \text{inv tan} \left[\frac{E_{2max} \sin \phi}{E_{1max} + E_{2max} \cos \phi} \right] \text{ radians} \qquad (63\text{-}8)$$

The sine wave representing $e_1 + e_2$ is shown in Fig. 63-1.

By substituting $-E_2$ for $+E_2$ we can obtain the results of subtracting e_2 from e_1, namely $e_1 - e_2$.

$$E_{max} = \sqrt{E_{1max}^2 + E_{2max}^2 - 2E_{1max}\,E_{2max} \cos \phi} \text{ volts} \qquad (63\text{-}9)$$

and

$$\theta = \text{inv tan} \left[\frac{-E_{2max} \sin \phi}{E_{1max} - E_{2max} \cos \phi} \right] \text{ radians} \qquad (63\text{-}10)$$

Example 63-1

Two alternating voltages are represented by $e_1 = 8 \sin \omega t$ V and $e_2 = 12 \sin (\omega t - \pi/3)$ V. What is the phase relationship between e_1 and e_2? Obtain the trigonometrical results for $e_1 + e_2$.

Solution

e_2 lags e_1 by $\pi/3$ radians or $60°$. Alternatively, e_1 leads e_2 by $60°$.

Peak value of $e_1 + e_2$, $E_{max} = \sqrt{E_{1max}^2 + E_{2max}^2 + 2E_{1max}\,E_{2max} \cos \phi)} \qquad (63\text{-}7)$

$$= \sqrt{8^2 + 12^2 + 2 \times 8 \times 12 \times \cos 60}$$
$$= 17.4 \text{ V}$$

Phase angle of $e_1 + e_2$, $\theta = \text{inv tan} \left[\dfrac{E_{2max} \sin \phi}{E_{1max} + E_{2max} \cos \phi} \right] \qquad (63\text{-}8)$

$$= \text{inv tan} \left(\frac{12 \sin (-60°)}{8 + 12 \cos (-60°)} \right)$$
$$= \text{inv tan} \left(-\frac{10.4}{14} \right)$$
$$= -36.6°$$

Chapter 64
Addition and subtraction of phasors

IN THE MATHEMATICAL DERIVATIONS OF CHAPTER 63, WE ADDED TOGETHER TWO voltage sine waves that were represented by the equations $e_1 = E_{1max} \sin \omega t$ and $e_2 = E_{2max} \sin (\omega t + \phi)$. In this chapter we are going to show how these same voltages are added together in a phasor diagram. Because e_2 leads e_1 by ϕ radians, we can use e_1 as the reference phasor; the e_2 phasor will appear in the first quadrant with its direction inclined at the angle ϕ to the horizontal line (Fig. 64-1).

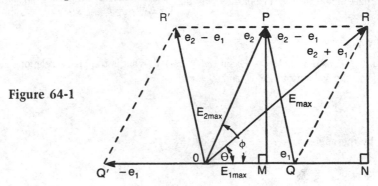

Figure 64-1

The sine waves of e_1 and e_2 are added together to produce the $e_1 + e_2$ sine wave which has a peak value of E_{max} and leads e_1 by the angle θ. The corresponding $e_1 + e_2$ phasor is found by completing the parallelogram whose sides are OP, OQ; the diagonal OR is then the $e_1 + e_2$ phasor while the other diagonal QP is the $e_2 - e_1$ phasor.

MATHEMATICAL DERIVATIONS
In Fig. 64-1

$$RN = E_{2max} \sin \phi \tag{64-1}$$

$$OM = E_{2max} \cos \phi \tag{64-2}$$

$$ON = OM + MN = E_{2max} \cos \phi + E_{1max} \tag{64-3}$$

Using the Phythagorean theorem,

$$OR^2 = RN^2 + ON^2$$
$$E_{max}^2 = (E_{2max} \sin \phi)^2 + (E_{1max} + E_{2max} \cos \phi)^2$$
$$= E_{1max}^2 + E_{2max}^2 \sin^2 \phi + 2E_{1max} E_{2max} \cos \phi + E_{2max}^2 \cos^2 \phi$$
$$= E_{1max}^2 + E_{2max}^2 + 2E_{1max} E_{2max} \cos \phi$$

because

$$\sin^2 \phi + \cos^2 \phi = 1$$

Therefore,

$$E_{max} = \sqrt{E_{1max}^2 + E_{2max}^2 + 2E_{1max} E_{2max} \cos \phi} \text{ volts} \tag{64-4}$$

$$\tan \theta = \frac{RN}{ON} = \frac{E_{2max} \sin \phi}{E_1 + E_{2max} \cos \phi}$$

and

$$\theta = \text{inv tan} \left[\frac{E_{2max} \sin \phi}{E_1 + E_{2max} \cos \phi} \right] \text{ radians} \tag{64-5}$$

These equations match those which appear in the mathematical derivations of Chapter 63. This establishes the validity of the geometrical construction used in the parallelogram of phasors.

A similar analysis leads to the following results:

$$\text{Peak value of } e_2 - e_1 \text{ phasor} = \sqrt{E_{1max}^2 + E_{max}^2 - 2E_{1max} E_{2max} \cos \phi} \text{ volts} \tag{64-6}$$

$$\text{Phase angle of } e_2 - e_1 \text{ phasor} = \text{inv tan} \left[\frac{E_{2max} \sin \phi}{-E_{1max} + E_{2max} \cos \phi} \right] \text{ radians} \tag{64-7}$$

Example 64-1

If $e_1 = 7 \sin(\omega t + \pi/6)$ V and $e_2 = 11 \sin(\omega t + \pi/4)$ V, what is the phase relationship between e_1 and e_2? Find the values of $e_1 + e_2$ and $e_2 - e_1$.

Solution

e_2 leads e_1 by the angle $\phi = \dfrac{\pi}{4} - \dfrac{\pi}{6} = \dfrac{\pi}{12}$ or $15°$

The peak value of $e_1 + e_2$,

$$E = \sqrt{E_{1max}^2 + E_{2max}^2 + 2E_{1max} E_{2max} \cos \phi} \tag{64-4}$$
$$= \sqrt{7^2 + 11^2 + 2 \times 7 \times 11 \times \cos 15°}$$
$$= 17.85 \text{ V}$$

Phase angle of $e_1 + e_2$,

$$\theta = \text{inv tan} \left[\frac{E_{2max} \sin \phi}{E_{1max} + E_{2max} \cos \phi} \right] \tag{64-5}$$

$$= \text{inv tan} \left[\frac{11 \sin 15°}{7 + 11 \cos 15°} \right]$$

$$= 9.17°$$
$$= 0.16 \text{ radians}$$

Therefore,

$$e_1 + e_2 = 17.85 \sin \left[\omega t + \left(\frac{\pi}{6} + 0.16 \right) \text{ radian} \right]$$

$$= 17.85 \sin(\omega t + 0.684 \text{ radian})$$

$$\text{Peak value of } e_2 - e_1 = \sqrt{7^2 + 11^2 - 2 \times 7 \times 11 \times \cos 15°}$$

$$= 4.61 \text{ V}.$$

$$\text{Phase angle of } e_2 - e_1 = \text{inv tan} \left[\frac{+11 \sin 15}{-7 + 11 \cos 15} \right]$$

$$= +38.1°$$
$$= 0.665 \text{ radians}$$

Therefore,

$$e_2 - e_1 = 4.61 \sin \left[\omega t + \left(\frac{\pi}{6} + 0.665 \right) \text{ radian} \right]$$

$$= 4.61 \sin (\omega t + 1.189 \text{ radian})$$

Chapter 65
Resistance in the ac circuit

IF A SINE WAVE IS CONNECTED ACROSS A RESISTOR, R (FIG. 65-1A), OHM'S LAW applies throughout the cycle of the alternating voltage, e. When e is instantaneously zero, there is zero current flowing in the circuit but when the applied voltage reaches one of its peaks, X, the current is also a maximum in one direction. When we reach the peak, Y of the other alternation, the current is again at its maximum value but its direction is reversed. At all times $e/i = R$ and it therefore follows that the sine waves of e and i are in phase (Fig. 65-1B) with a phase difference of zero degrees. This is indicated by the e and i phasors, which lie along the same horizontal line (Fig. 65-1C).

The instantaneous power, p, in the circuit is the product of the instantaneous voltage, e, and the instantaneous current, i. When e and i are simultaneously zero, the instantaneous power is also zero, but when both e and i reach their peaks together, the power reaches its peak value of $E_{max} \times I_{max}$ watts. When the voltage reverses polarity, the current reverses direction but the resistor continues to dissipate (lost) power in the form of heat. The whole of the instantaneous power curve must be drawn above the zero line and its frequency is twice that of the applied voltage. The mean value of the power curve is a measure of the average power dissipated over the voltage cycle.

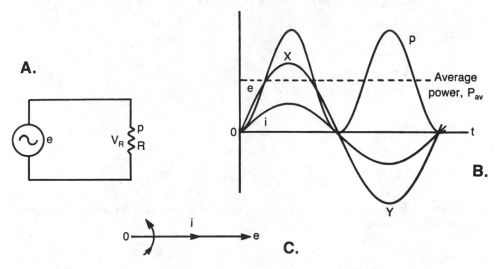

Figure 65-1

MATHEMATICAL DERIVATIONS

Let the equation of the applied ac voltage be $e = E_{max} \sin \omega t$. Since Ohm's law applies throughout the cycle,

Instantaneous current,

$$i = \frac{e}{R} \tag{65-1}$$

$$= \frac{E_{max} \sin \omega t}{R}$$

$$= I_{max} \sin \omega t \text{ amperes}$$

where $I_{max} = E_{max}/R$ and is the peak value of the alternating current. Therefore,

$$\frac{E_{max}}{I_{max}} = \frac{\sqrt{2} \times E_{rms}}{\sqrt{2} \times I_{rms}} \tag{65-2}$$

$$= \frac{E_{rms}}{I_{rms}}$$

$$= R \text{ ohms}$$

The instantaneous power, p, is given by:

Instantaneous power,

$$p = e \times i \tag{65-3}$$

$$= E_{max} \sin \omega t \times I_{max} \sin \omega t$$

$$= E_{max} \times I_{max} \sin^2 \omega t$$

$$= \frac{E_{max} \times I_{max}}{2} (1 - \cos 2\omega t) \text{ watts}$$

The term "$\cos 2\omega t$" indicates that the instantaneous power curve is a second harmonic of the voltage waveform. Since the mean value of $\cos 2\omega t$ over the complete cycle is zero, the average power is:

Average power,

$$P_{AV} = \frac{E_{max} \times I_{max}}{2} \tag{65-4}$$

$$= \frac{E_{max}}{\sqrt{2}} \times \frac{I_{max}}{\sqrt{2}}$$

$$= E_{rms} \times I_{rms}$$

$$= I_{rms}^2 \times R$$

$$= \frac{E_{rms}^2}{R} \text{ watts}$$

By using the effective values, the ac equations for resistors in series and parallel are the same as those for dc.

Example 65-1

In Fig. 65-1, let $e = 18 \sin (\omega t - \pi/3)$ V and $R = 2.7$ kΩ. Calculate the values of the effective current, the peak power and the average power over the cycle. What is the trigonometrical expression for the instantaneous current?

216

Solution

Effective voltage,

$$E_{rms} = \frac{E_{max}}{\sqrt{2}}$$
$$= 0.707 \times E_{max}$$
$$= 0.707 \times 18$$
$$= 12.7 \text{ V}.$$

Effective current,

$$I_{rms} = \frac{E_{rms}}{R} \qquad \text{(65-2)}$$
$$= \frac{12.7 \text{ V}}{2.7 \text{ k}\Omega}$$
$$= 4.71 \text{ mA}.$$

Peak power,

$$P_{max} = E_{max}^2/R$$
$$= (18 \text{ V})^2/2.7 \text{ k}\Omega$$
$$= 120 \text{ mW}.$$

Average power,

$$P_{AV} = E_{rms} \times I_{rms} \qquad \text{(65-4)}$$
$$= 12.7 \text{ V} \times 4.71 \text{ mA}$$
$$= 60 \text{ mW}.$$

Because e and i are in phase, the instantaneous current,

$$i = \frac{18 \text{ V}}{2.7 \text{ k}\Omega} \sin\left(\omega t - \frac{\pi}{3}\right)$$
$$= 6.67 \sin\left(\omega t - \frac{\pi}{3}\right) \text{ mA}$$

Chapter 66
Conductance in the ac circuit

IN CHAPTER 65 WE DECIDED THAT THE SAME RESISTOR FORMULAS APPLIED TO both dc and ac circuits provided the formulas used effective values of voltage and current. This is also true for the property of conductance, which is defined as the reciprocal of the resistance and is the measure of the "ease" with which current flows in an electrical circuit. In the series circuit of Fig. 66-1A the reciprocal formula will determine the total conductance, G_T. This will

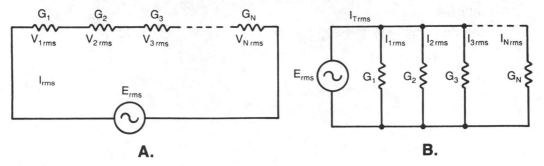

Figure 66-1

then enable us to calculate the effective current, I_{rms} and the effective voltages $V_{1\text{rms}}$, $V_{2\text{rms}}$, $V_{3\text{rms}}$, . . . $V_{N\text{rms}}$.

The main advantage of using conductance (as opposed to resistance) lies with parallel circuits (Fig. 66-1B) because the total conductance is simply found by adding the individual conductances. We can then easily calculate the effective value of the total current, $I_{T\text{rms}}$.

MATHEMATICAL DERIVATIONS
In Fig. 66-1A

Total conductance,

$$G_T = \frac{1}{\dfrac{1}{G_1} + \dfrac{1}{G_2} + \dfrac{1}{G_2} + \cdots + \dfrac{1}{G_N}} \text{ siemens} \tag{66-1}$$

Circuit current,

$$I_{\text{rms}} = E_{\text{rms}} \times G_T \text{ amperes} \tag{66-2}$$

Effective voltage across the R_N resistor,

$$V_{N\text{rms}} = I_{\text{rms}} \times G_N \text{ volts} \tag{66-3}$$

Power dissipated in the R_N resistor,

$$\begin{aligned}
P_N &= V_{N\text{rms}} \times I_{\text{rms}} \\
&= I_{\text{rms}}^2 \times R_N \\
&= I_{\text{rms}}^2 / G_N \\
&= V_{N\text{rms}}^2 / R_N \\
&= V_{N\text{rms}}^2 \times G_N \text{ watts}
\end{aligned} \tag{66-4}$$

In Fig. 66-1B

Total conductance,

$$G_T = G_1 + G_2 + G_3 + \cdots + G_N \text{ siemens} \tag{66-5}$$

Total current,

$$I_{T\text{rms}} = E_{\text{rms}} \times G_T \text{ amperes}$$

Effective current flowing through the R_N resistor,

$$I_{N\text{rms}} = E_{\text{rms}} \times G_N \text{ amperes} \qquad (66\text{-}7)$$

Power dissipated in the R_N resistor,

$$\begin{aligned}
P_N &= E_{\text{rms}}^2 \times G_N \qquad (66\text{-}8)\\
&= E_{\text{rms}} \times I_{N\text{rms}}\\
&= I_{N\text{rms}}^2 / G_N\\
&= E_{N\text{rms}}^2 / R_N\\
&= I_{N\text{rms}}^2 \times R_N \text{ watts}
\end{aligned}$$

Example 66-1

In Fig. 66-2 calculate the value of I_{rms}.

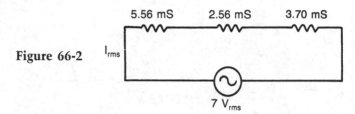

Figure 66-2

Solution

Total conductance,

$$G_T = \cfrac{1}{\cfrac{1}{G_1} + \cfrac{1}{G_2} + \cfrac{1}{G_3}} \qquad (66\text{-}1)$$

$$= \cfrac{1}{\cfrac{1}{5.56 \text{ mS}} + \cfrac{1}{2.56 \text{ mS}} + \cfrac{1}{3.70 \text{ mS}}}$$

$$= 1.19 \text{ mS}$$

Effective current,

$$\begin{aligned}
I_{\text{rms}} &= E_{\text{rms}} \times G_T \qquad (66\text{-}5)\\
&= 7 \text{ V} \times 1.19 \text{ mS}\\
&= 8.33 \text{ mA.}
\end{aligned}$$

Example 66-2

In Fig. 66-3, calculate the value of $I_{T\text{rms}}$.

Solution

Total conductance,

$$\begin{aligned}
G_T &= G_1 + G_2 + G_3 \qquad (66\text{-}5)\\
&= 5.56 + 2.56 + 3.70\\
&= 11.82 \text{ mS}
\end{aligned}$$

219

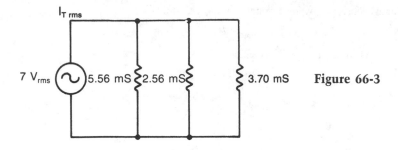

Figure 66-3

Total effective current,

$$I_{Trms} = E_{rms} \times G_T \qquad (66\text{-}6)$$
$$= 7 \text{ V} \times 11.82 \text{ mS}$$
$$= 82.74 \text{ mA}.$$

∽

Chapter 67
Inductive reactance

LET US ASSUME IN FIG. 67-1A THAT THE INDUCTOR ONLY POSSESSES THE PROPERTY of inductance and that we can ignore its resistance. When an alternating current is flowing through the coil, the surrounding magnetic field is continually expanding and collapsing so that there is an induced voltage, v_L, whose value must at all times exactly balance the instantaneous value of the source voltage, e. Therefore, when e is instantaneously at its peak value, the value of v_L is also at its peak and since:

$$v_L = -L \frac{di}{dt} \text{ (equation 39-1)}$$

the *rate of change of the current* is at its highest level. This occurs when the slope of the current sine wave is steepest at the point of zero time and zero current. Consequently, when e is at its peak value, i is instantaneously zero and therefore e, i are 90° out of phase (Figs. 67-1B and C). This may be remembered by the word "*eLi*"; for an inductor, L, the instantaneous voltage, e, leads the instantaneous current, i, by 90°.

We cannot obtain the opposition to the current flow by dividing the instantaneous voltage by the instantaneous current. For example, at the zero degree point, the value of e/i is infinite while at the 90° mark, e/i is zero. However, we can attempt to find the value of E_{rms}/I_{rms} because the effective values are derived from the complete cycles of the voltage and the current.

Assume that the source voltage and the inductance are kept constant but the frequency is raised. The voltage induced in the coil must remain the same but the magnetic field is expanding and collapsing more rapidly so that the required magnetic flux is less. As a result, the current is reduced and is inversely proportional to the frequency. For example, if the frequency is doubled, the value of the effective current is halved.

Now we will keep the source voltage and the frequency constant but the inductor is replaced by another coil with a higher inductance. Because the induced voltage, $v_L = -L \ di/dt$ is the same,

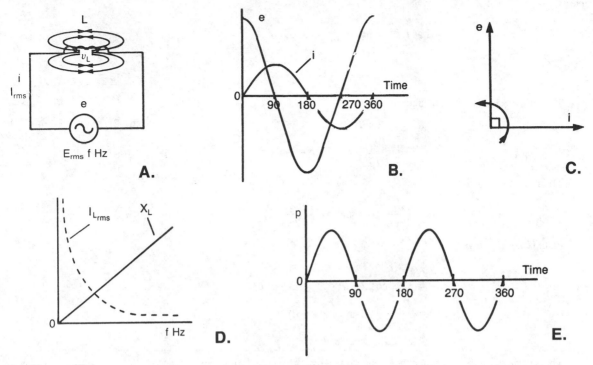

Figure 67-1

the value of di/dt must be less and therefore the effective current is again reduced. Finally by raising the value of the source voltage while the frequency and the inductance are unchanged, the induced voltage must be increased and consequently the effective current is greater.

Summarizing, the effective current is directly proportional to the effective voltage, but it is inversely proportional to the frequency and the inductance. The opposition to the alternating current is therefore determined by the product of the frequency and the inductance. This opposition is called the inductance reactance, $X_L = E_{rms}/I_{rms}$, and is measured in ohms. As shown in the mathematical derivation, $X_L = 2\pi fL$ Ω and because X_L is directly proportional to the frequency, the graph of X_L versus frequency is a straight line (Fig. 67-1D).

The instantaneous power, p, is equal to the product of e and i. The graph of the instantaneous power, p, versus time is a second harmonic sine wave (Fig. 67-1E) whose average value is zero over the source voltage's cycle. During the first quarter cycle the magnetic field is established around the inductor and the energy is drawn from the source. However, during the second quarter cycle the magnetic field collapses and the energy is returned to the source. This action is repeated during the third and fourth quarter cycles so that the average power over the complete cycle is zero.

This highlights the difference between resistance and reactance. Both resistance and reactance limit the value of the alternating current but while resistance dissipates (lost) power in the form of heat, reactance does not lose any power at all.

MATHEMATICAL DERIVATIONS

Let a sinewave current, $i = I_{max} \sin 2\pi$ ft flow through the inductor, L.

Then $v_L = -L$ di/dt and, by Kirchhoff's voltage law, $e + v_L = 0$.

This yields

$$e = L \frac{di}{dt} = 2\pi fL \, I_{max} \cos 2\pi ft \qquad (67\text{-}1)$$

$$= E_{max} \sin \left(2\pi ft + \frac{\pi}{2}\right)$$

Therefore e leads i by $90°$ and

$$E_{max} = 2\pi fL \, I_{max} \qquad (67\text{-}2)$$

$$\frac{E_{max}}{I_{max}} = \frac{E_{rms}}{I_{rms}} = 2\pi fL = X_L \, \Omega$$

Then

$$f = \frac{X_L}{2\pi L} \text{ Hz} \quad \text{and} \quad L = \frac{X_L}{2\pi f} \text{ H} \qquad (67\text{-}3)$$

Instantaneous power,

$$p = e \times i$$

$$= E_{max} \cos 2\pi ft \times I_{max} \sin 2\pi ft$$

$$= \frac{E_{max} I_{max}}{2} \sin (2 \times 2\pi ft)$$

$$= E_{rms} I_{rms} \sin 4\pi ft.$$

The average value of the instantaneous power over the complete cycle is zero.

Example 67-1

In Fig. 67-1A, $L = 150 \ \mu\text{H}$, $f = 820$ kHz and $E_{rms} = 8$ V. Find the value of I_{rms}.

Solution

The inductive reactance is given by

$$X_L = 2 \times \pi \times 820 \times 10^3 \times 150 \times 10^{-6} = 773 \ \Omega \qquad (67\text{-}2)$$

The current is given by

$$I_{rms} = \frac{E_{rms}}{X_L} = \frac{8 \text{ V}}{773 \ \Omega} = 10.4 \text{ mA.} \qquad (67\text{-}2)$$

Example 67-2

In Fig. 67-1A, $f = 15$ kHz, $E_{rms} = 20$ V, $I = 12.4$ mA. Calculate the value of the inductance, L.

Solution

The inductive reactance is given by

$$X_L = \frac{E_{rms}}{I_{rms}} = \frac{20 \text{ V}}{12.4 \text{ mA}} = 1613 \ \Omega \qquad (67\text{-}2)$$

The equation for the inductance is

$$L = \frac{X_L}{2\pi f} = \frac{1613}{2 \times \pi \times 15 \times 10^3} \text{ H} = 17.1 \text{ mH} \qquad (67\text{-}3)$$

Chapter 68
Capacitive reactance

WHEN AN ALTERNATING CURRENT, *I*, IS FLOWING IN THE CIRCUIT OF FIG. 68-1A, the capacitor is continuously charging and discharging so that the voltage, v_C, across the capacitor must at all times exactly balance the instantaneous value of the source voltage, *e*. Therefore, when *e* is momentarily at its peak value, the value of v_C is also at its peak and the capacitor is fully charged. The current is then instantaneously zero and therefore *e* and *i* are 90° out of phase (Fig. 68-1 B, C). This may be remembered by the word *iCe*; for an ideal capacitor, *C*, the instantaneous current, *i*, leads the instantaneous voltage, *e*, by 90°. We then combine "iCe" with the word "eLi" for the inductor and create "eLi, the iCe man!"

Let us now derive the factors that determine the capacitor's opposition to the flow of alternating current. Assume that the source voltage and the capacitance are kept constant but the frequency is raised. This reduces the period so that the capacitor must acquire or lose the same amount of charge in a shorter time. The charging or discharging current must therefore be

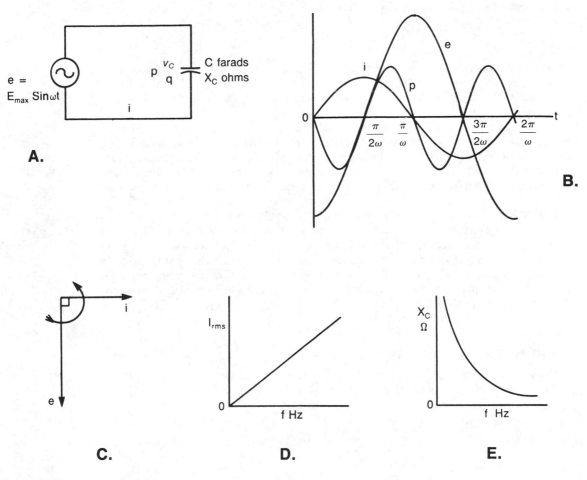

Figure 68-1

223

greater and, in fact, the effective current is directly proportional to the frequency, so that if the frequency is doubled, the effective current is also doubled. By contrast, when the frequency is doubled for an *inductor*, the effective current is halved (Chapter 67).

Now we will keep the source voltage and the frequency constant but we will raise the capacitance. The new capacitor will have to store or lose more charge in the same time so that the effective current is again increased. The values of the capacitance and the effective current are directly proportional so that if the capacitance is halved, the effective current is also halved. Finally by raising the source voltage while the frequency and the capacitance are unchanged, the voltage across the capacitor must be greater, the capacitor must store more charge in the same period of time and consequently the effective current is once more increased.

Summarizing, the value of the effective current is directly proportional to the voltage, the capacitance and the frequency. The opposition to the alternating current flow is therefore *inversely* determined by the product of the frequency and the capacitance. This opposition is called the capacitive reactance,

$$X_C = \frac{E_{\mathrm{rms}}}{I_{\mathrm{rms}}}$$

and is measured in ohms. As shown in the mathematical derivations,

$$X_C = \frac{1}{2\pi fC} \; \Omega$$

and because X_C is inversely proportional to the frequency, the graph of X_C versus frequency is the rectangular hyperbolic curve of Fig. 68-1E. By contrast, the effective current is directly proportional to the frequency so that the graph of I_{rms} versus frequency is a straight line (Fig. 68-1D).

The instantaneous power, p, is equal to the product of e and i. The graph of the instantaneous power, p, versus time is a second harmonic sine wave (Fig. 68-1B) whose average value is zero over the source voltage's cycle. During one quarter cycle the capacitor charges so that energy is drawn from the source and appears in the form of the electric field between the capacitor's plates. However, during the following quarter cycle, the capacitor discharges and the energy is returned to the source.

Like the inductor, the importance of the capacitor lies in the fact that its opposition to alternating current depends on frequency; as the frequency is raised, the effective current increases from zero to infinity (Fig. 67-1D) while the capacitive reactance decreases from infinity towards zero (Fig. 68-1E). Consequently, for given values of L and C there must always be a (resonant) frequency for which $X_L = X_C$; this fact is used in the L, C tuning circuit which, for example, is capable of selecting a particular station through its ability to distinguish between one frequency and another (Chapters 80 and 82).

MATHEMATICAL DERIVATIONS

Let a sine-wave voltage, $e = E_{\max} \sin 2\pi ft$, be applied to the capacitor, C.

Then $V_c = -q/C$ and, by Kirchhoff's voltage law, $e + V_C = 0$.

This yields

$$e = \frac{q}{C} \text{ or } q = Ce.$$

Then

$$\frac{dq}{dt} = i = C\frac{de}{dt} = 2\pi fCE_{max}\cos 2\pi ft \qquad (68\text{-}1)$$

$$= 2\pi fCE_{max}\sin\left(2\pi ft + \frac{\pi}{2}\right)$$

$$= I_{max}\sin\left(2\pi ft + \frac{\pi}{2}\right)\text{ amperes}$$

Therefore i leads e by 90° and

$$I_{max} = 2\pi fCE_{max} \qquad (68\text{-}2)$$

or

$$\frac{E_{max}}{I_{max}} = \frac{E_{rms}}{I_{rms}} = \frac{1}{2\pi fC} = X_C\ \Omega$$

Then

$$f = \frac{1}{2\pi C X_C} = \frac{0.159}{CX_C}\text{ and }C = \frac{1}{2\pi f X_C} = \frac{0.159}{fX_C}\text{ F} \qquad (68\text{-}3)$$

Instantaneous power,

$$p = e \times i \qquad (68\text{-}4)$$

$$= E_{max}\sin 2\pi ft \times I_{max}\cos 2\pi ft$$

$$= \frac{E_{max} \times I_{max}}{2}\sin 4\pi ft$$

$$= E_{rms}I_{rms}\sin 4\pi ft$$

The average value of the instantaneous power over the complete cycle is zero.

Example 68-1

In Fig. 68-1A, $C = 250$ pF, $f = 822$ kHz, and $E_{rms} = 8$ V. Find the value of I_{rms}.

Solution

Capacitive reactance,

$$X_C = \frac{1}{2 \times \pi \times 822 \times 10^3 \times 250 \times 10^{-12}} \qquad (68\text{-}2)$$

$$= 774\ \Omega$$

Current,

$$I = \frac{8\text{ V}}{774\ \Omega} \qquad (68\text{-}2)$$

$$= 10.3\text{ mA}$$

These results should be compared with the results of Example 67-1.

Example 68-2

In Fig. 68-1A, $f = 15$ kHz, $E_{rms} = 20$ V, $I_{rms} = 12.4$ mA. Calculate the value of the capacitance.

Solution

Capacitive reactance,

$$X_C = \frac{E_{rms}}{I_{rms}} = \frac{20 \text{ V}}{12.4 \text{ mA}} = 1613 \ \Omega \qquad (68\text{-}2)$$

Capacitance,

$$C = \frac{1}{2 \times \pi \times f \times X_C} \qquad (68\text{-}3)$$

$$= \frac{1}{2 \times \pi \times 15 \times 10^3 \times 1613} \text{ F}$$

$$= 0.00658 \ \mu\text{F}$$

Chapter 69
The general ac circuit—
impedance—power factor

PROVIDED THERE IS ONLY A SINGLE SINEWAVE SOURCE, ALL AC CIRCUITS, however complex, may ultimately be analyzed into a resistance in series with a reactance, which may be either inductive or capacitive. The total opposition to the flow of alternating current is therefore a resistance/reactance combination that is called the impedance, z. See Fig. 69-1.

The impedance, z, is a phasor that is defined as the ratio of the source voltage phasor to the source current phasor. The magnitude, Z, of the impedance phasor is measured in ohms and is equal to the ratio of $E_{rms} : I_{rms}$ where E_{rms} and I_{rms} are the effective values of the source voltage and the source current.

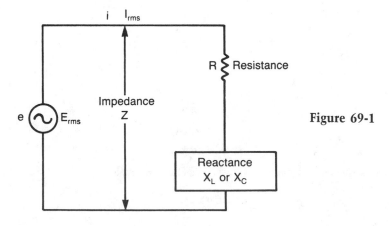

Figure 69-1

The phase angle, ϕ, of the impedance is the angle between the direction of z and the horizontal reference line. The value of this angle normally ranges from $-90°$ (capacitive reactance only) through $0°$ (resistance only) to $+90°$ (inductive reactance only).

In the general ac circuit the resistive component dissipates (lost) power in the form of heat while the reactive component stores power in the form of either an electric field (capacitive reactance) or a magnetic field (inductive reactance). Therefore, the product of $E_{rms} \times I_{rms}$ does not represent the dissipated power and is instead referred to as the *apparent* power as measured at the source in volt-amperes (VA). By contrast the dissipated power is called the *true* power and the ratio of true power to apparent power is the power factor.

If an ac circuit is *inductive,* the phase angle is positive and the power factor is *lagging* (the source current lags the source voltage). With a *capacitive* circuit the phase angle is negative and the power factor is *leading* (the source current leads the source voltage). The value of the power factor extends from zero (reactance only) to unity (resistance only).

MATHEMATICAL DERIVATIONS

Let the instantaneous current be represented by

$$i = I_{max} \sin \omega t \text{ amperes.}$$

Then the instantaneous source voltage,

$$e = E_{max} \sin (\omega t \pm \phi) \text{ volts.}$$

Magnitude of the impedance,

$$Z = \frac{E_{max}}{I_{max}} = \frac{E_{rms}}{I_{rms}} \text{ ohms} \tag{69-1}$$

This yields

$$E_{rms} = I_{rms} \times Z, \; I_{rms} = E/Z \tag{69-2}$$

Instantaneous power,

$$p = e \times i \tag{69-3}$$

$$= E_{max} I_{max} \sin \omega t \sin (\omega t \pm \phi)$$

$$= \frac{E_{max} I_{max}}{2} [\cos \phi - \cos (2\omega t \pm \phi)]$$

The mean value of $\cos (2\omega t \pm \phi)$ is zero over the complete cycle of the source voltage. Therefore the average power over the complete cycle is:

True power,

$$P_{AV} = \frac{E_{max} \times I_{max} \cos \phi}{2} \tag{69-4}$$

$$= \frac{E_{max}}{\sqrt{2}} \times \frac{I_{max}}{\sqrt{2}} \times \cos \phi \tag{69-4}$$

$$= E_{rms} \times I_{rms} \times \cos \phi$$

$$= \text{Apparent power} \times \cos \phi \text{ watts}$$

$$\text{The power factor} = \frac{\text{True power (watts)}}{\text{Apparent power (volt-amperes)}} = \cos \phi \tag{69-5}$$

Phase angle,

$$\phi = \text{inverse cos (power factor)} \tag{69-6}$$

The power factor is therefore the cosine of the phase angle.

Example 69-1

An alternating source has an instantaneous voltage of $e = 19 \sin(\omega t + \pi/3)$ V. If the instantaneous source current is $i = 3.6 \sin(\omega t - \pi/12)$ mA, calculate the values of the impedance, the phase angle and the power factor. Is the circuit inductive or capacitive? What are the values of the true power and the apparent power?

Solution

The voltage, e, leads the current, i, by an angle of $\pi/4 - (-\pi/12) = +\pi/3$ radians. The circuit is therefore inductive.

Magnitude of the impedance,

$$Z = \frac{E_{\text{rms}}}{I_{\text{rms}}} = \frac{E_{\text{max}}}{I_{\text{max}}} = \frac{19 \text{ V}}{3.6 \text{ mA}} = 5.28 \text{ k}\Omega$$

The power factor is $\cos(+\pi/3) = 0.5$, lagging.

Apparent power $= 19.4 \text{ V} \times 3.6 \text{ mA} = 69.84 \text{ mVA}$.

$$\text{True power} = \text{Apparent power} \times \text{power factor} \tag{69-5}$$
$$= 69.84 \times 0.5$$
$$= 34.92 \text{ mW}.$$

Chapter 70
Sine-wave input voltage
to R and L in series

LET US ASSUME THAT THE VALUE OF $X_L(= 2\pi f L \ \Omega)$ IS GREATER THAN THE VALUE of R. Since R and L are in series, the same alternating current must flow through each component (Fig. 70-1A). The instantaneous voltage drop (v_R) across the resistor and the instantaneous current (i) through the resistor are in phase (Fig. 70-1B) so that their phasors lie along the same horizontal line (Fig. 70-1C). By contrast the instantaneous voltage (v_L) across the inductor leads the instantaneous current (i) by 90° and consequently their phasors are perpendicular. The phasor sum of V_R and V_L is the supply voltage, e; the current, I, then lags the source voltage, e, by the phase angle, ϕ. This inductive circuit is then considered to have a lagging power factor (resistance, R/impedance, Z) and the phase angle is *positive*.

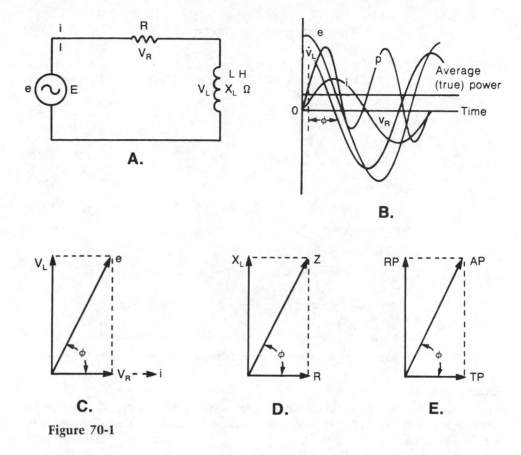

Figure 70-1

The total opposition to the alternating current flow is measured by the impedance phasor, z, which is defined as the ratio of the e phasor to the i phasor and is equal to the phasor sum of R and X_L (Fig. 70-1D).

The true power (TP in watts) is the power dissipated or lost as heat in the resistor; it is also the average value of the instantaneous power curve (Fig. 70-1B). The reactive, idle or wattless power (RP) in volt-amperes reactive (VAr) is the power stored by the inductor as a magnetic field during one quarter of the ac cycle; this power is subsequently returned to the source during the next quarter cycle. The apparent power (AP) in volt-amperes (VA) is the product of the source voltage and the source current, and is the phasor sum of the true power and the reactive power (Fig. 70-1E). The power factor is the ratio of the true power to the apparent power and is equal to the cosine of the phase angle.

MATHEMATICAL DERIVATIONS

Impedance phasor,

$$z = Z \angle\phi = R + jX_L = R + j \times 2\pi fL \text{ ohms} \tag{70-1}$$

In terms of the effective values (capital letters):

$$\frac{E}{I} = Z, \ E = I \times Z, \ I = \frac{E}{Z} \tag{70-2}$$

229

$$\phi = \text{inv} \tan \frac{X_L}{R} \text{ and is a positive angle} \tag{70-3}$$

(when $R = X_L$, $\phi = +45°$).

$$Z = \sqrt{R^2 + X_L^2}, \ X_L = \sqrt{Z^2 - R^2}, \ R = \sqrt{Z^2 - X_L^2} \text{ ohms} \tag{70-4}$$

$$V_R = I \times R, \ R = \frac{V_R}{I}, \ I = \frac{V_R}{R} \tag{70-5}$$

$$V_L = I \times X_L, \ X_L = \frac{V_L}{I}, \ I = \frac{V_L}{X_L} \tag{70-6}$$

$$E = \sqrt{V_R^2 + V_L^2}, \ V_L = \sqrt{E^2 - V_R^2}, \ V_R = \sqrt{E^2 - V_L^2} \text{ volts} \tag{70-7}$$

$$\text{True power (TP)} = I \times V_R = I^2R = V_R^2/R \text{ watts} \tag{70-8}$$

$$\text{Reactive power (RP)} = I \times V_L = I^2X_L = V_L^2/X_L \text{ VArs} \tag{70-9}$$

$$\text{Apparent power (AP)} = I \times E = I^2Z = E^2/Z \text{ VA} \tag{70-10}$$

$$\text{AP} = \sqrt{TP^2 + RP^2}, \ \text{RP} = \sqrt{AP^2 - TP^2}, \ \text{TP} = \sqrt{AP^2 - RP^2} \tag{70-11}$$

$$\text{Power factor} = \frac{TP}{AP} = \frac{I^2R}{I^2Z} = \frac{R}{Z} = \cos \phi, \text{ and is lagging} \tag{70-12}$$

$$\text{Phase angle, } \phi = \text{inv} \cos \text{ (power factor) and is a positive angle} \tag{70-13}$$

$$\text{True power} = \text{apparent power} \times \text{power factor} = EI \cos \phi \text{ watts} \tag{70-14}$$

Example 70-1

In Fig. 70-2, $R = 680 \ \Omega$, $L = 150 \ \mu H$ and the supply voltage is 2 V at a frequency of 800 kHz. Calculate the values of Z, I, V_R, V_L, true power, reactive power, apparent power, power factor and the phase angle, ϕ.

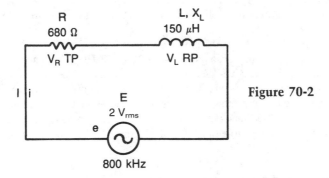

Figure 70-2

Solution

Inductive reactance,

$$X_L = 2\pi fL = 2 \times \pi \times 800 \times 10^3 \times 150 \times 10^{-6} = 754 \ \Omega \tag{70-1}$$

Impedance,

$$Z = \sqrt{R^2 + X_L{}^2} = \sqrt{680^2 + 754^2} = 1015 \ \Omega \qquad (70\text{-}4)$$

Current,

$$I = \frac{E}{Z} = \frac{2 \ \text{V}}{1015 \ \Omega} = 1.97 \ \text{mA} \qquad (70\text{-}2)$$

Voltage across the resistor,

$$V_R = I \times R = 1.97 \ \text{mA} \times 680 \ \Omega = 1.34 \ \text{V}. \qquad (70\text{-}5)$$

Voltage across the inductor,

$$V_L = I \times X_L = 1.97 \ \text{mA} \times 754 \ \Omega = 1.49 \ \text{V}. \qquad (70\text{-}6)$$

Voltage check:

$$E = \sqrt{V_R{}^2 + V_L{}^2} = \sqrt{1.34^2 + 1.49^2} = 2.0 \ \text{V} \qquad (70\text{-}7)$$

$$\text{True Power} = I \times V_R = 1.97 \ \text{mA} \times 1.34 \ \text{V} = 2.64 \ \text{mW} \qquad (70\text{-}8)$$

$$\text{Reactive Power} = I \times V_L = 1.97 \ \text{mA} \times 1.49 \ \text{V} = 2.93 \ \text{mVAr} \qquad (70\text{-}9)$$

$$\text{Apparent Power} = I \times E = 1.97 \ \text{mA} \times 2 \ \text{V} = 3.94 \ \text{mVA} \qquad (70\text{-}10)$$

Power check:

$$\text{AP} = \sqrt{\text{TP}^2 + \text{RP}^2} = \sqrt{2.64^2 + 2.93^2} = 3.94 \ \text{mVA} \qquad (70\text{-}6)$$

$$\text{Power factor} = \frac{\text{TP}}{\text{AP}} = \frac{2.64 \ \text{mW}}{3.94 \ \text{mVA}} = 0.67, \text{ lagging} \qquad (70\text{-}12)$$

$$\text{Phase angle, } \phi = \text{inv cos (power factor)} = \text{inv cos } 0.67 = +47.9° \qquad (70\text{-}13)$$

Chapter 71
Sine-wave input voltage
to *R* and *C* in series

LET US ASSUME THAT THE VALUE OF $X_C = 1/2\pi fC \ \Omega$ IS GREATER THAN THE VALUE OF *R*. Since *R* and *C* are in series, the same alternating current must flow through each component (Fig. 71-1A). The instantaneous voltage drop (v_R) across the resistor and the instantaneous current (i) through the resistor are in phase (Fig. 71-1B) so that their phasors lie along the same horizontal reference line (Fig. 71-1C). By contrast the instantaneous voltage (v_C) across the capacitor lags the instantaneous current (i) by 90° and consequently their phasors are perpendicular. The phasor sum of v_R and v_C is the supply voltage, *e*; the current, *i*, then leads the source voltage, *e*, by the phase angle, ϕ. This capacitive circuit is considered to have a *leading* power factor (resistance, *R*/impedance, *Z*) and the phase angle is *negative*.

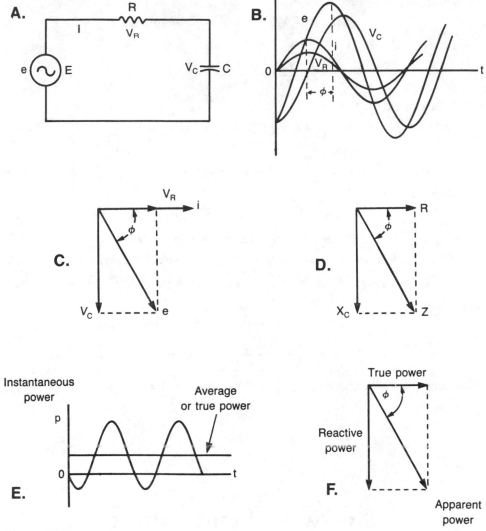

Figure 71-1

The total opposition to the alternating current flow is measured by the impedance phasor, z, which is defined as the ratio of the e phasor to the i phasor, and is equal to the phasor sum of R and X_C (Fig. 71-1D).

The true power (TP in watts) is the power dissipated or lost as heat in the resistor; it is also the average value of the instantaneous power curve (Fig. 71-1E). The reactive, idle or wattless power (RP) in volt-amperes reactive (VAr) is the power stored by the capacitor as an electric field during one quarter of the ac cycle; this power is subsequently returned to the source during the next quarter cycle. The apparent power (AP) in volt-amperes (VA) is the product of the source voltage and the source current, and is the phasor sum of the true power and the reactive power (Fig. 71-1F). The power factor is the ratio of the true power to the apparent power and is equal to the cosine of the phase angle.

MATHEMATICAL DERIVATIONS

Impedance phasor,

$$z = Z \angle -\phi = R - jX_C = R - \frac{j \times 1}{2\pi f C} \text{ ohms} \tag{71-1}$$

In terms of the effective values (capital letters):

$$\frac{E}{I} = Z, \ E = I \times Z, \ I = \frac{E}{Z} \tag{71-2}$$

$$\phi = \text{inv tan} \left(\frac{-X_C}{R} \right) \text{ and is a negative angle} \tag{71-3}$$

(when $R = X_C$, $\phi = -45°$)

$$Z = \sqrt{R^2 + X_C^2}, \ X_C = \sqrt{Z^2 - R^2}, \ R = \sqrt{Z^2 - X_C^2} \text{ ohms} \tag{71-4}$$

$$V_R = I \times R, \ R = \frac{V_R}{I}, \ I = \frac{V_R}{R} \tag{71-5}$$

$$V_C = I \times X_C, \ X_C = \frac{V_C}{I}, \ I = \frac{V_C}{X_C} \tag{71-6}$$

$$E = \sqrt{V_R^2 + V_C^2}, \ V_C = \sqrt{E^2 - V_R^2}, \ V_R = \sqrt{E^2 - V_C^2} \text{ volts} \tag{71-7}$$

$$\text{True power (TP)} = I \times V_R = I^2 R = V_R^2/R \text{ watts} \tag{71-8}$$

$$\text{Reactive power (RP)} = I \times V_C = I^2 X_C = V_C^2/X_C \text{ VArs} \tag{71-9}$$

$$\text{Apparent power (AP)} = I \times E = I^2 Z = E^2/Z \text{ VA} \tag{71-10}$$

$$\text{AP} = \sqrt{\text{TP}^2 + \text{RP}^2}, \ \text{RP} = \sqrt{\text{AP}^2 - \text{TP}^2}, \ \text{TP} = \sqrt{\text{AP}^2 - \text{RP}^2} \tag{71-11}$$

$$\text{Power factor} = \frac{\text{TP}}{\text{AP}} = \frac{I^2 R}{I^2 Z} = \frac{R}{Z} = \cos \phi \text{ and is leading} \tag{71-12}$$

$$\text{Phase angle, } \phi = \text{inv cos (power factor) and is a negative angle} \tag{71-13}$$

$$\text{True power} = \text{apparent power} \times \text{power factor} = EI \cos \phi \text{ watts} \tag{71-14}$$

Example 71-1

In Fig. 71-2, $R = 820 \ \Omega$, $C = 0.068 \ \mu\text{F}$ and the supply voltage is 30 V at the frequency of 4.5 kHz. Calculate the values of Z, I, V_R, V_C, true power, reactive power, apparent power, power factor and the phase angle ϕ.

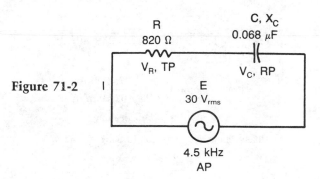

Figure 71-2

Capacitive reactance,

$$X_C = \frac{1}{2\pi f C} \tag{71-1}$$

$$= \frac{1}{2 \times \pi \times 4.5 \times 10^3 \times 0.068 \times 10^{-6}}$$

$$= 520 \ \Omega$$

Impedance,

$$Z = \sqrt{R^2 + X_C^2} = \sqrt{820^2 + 520^2} \tag{71-4}$$

$$= 971 \ \Omega$$

Current,

$$I = \frac{E}{Z} = \frac{30 \text{ V}}{971 \ \Omega} \tag{71-2}$$

$$= 30.9 \text{ mA}$$

Voltage across the resistor,

$$V_R = I \times R = 30.9 \text{ mA} \times 820 \ \Omega \tag{71-5}$$

$$= 25.34 \text{ V}$$

Voltage across the capacitor,

$$V_C = I \times X_C = 30.9 \text{ mA} \times 520 \ \Omega \tag{71-6}$$

$$= 16.09 \text{ V}$$

Voltage check:
Source voltage,

$$E = \sqrt{V_R^2 + V_C^2} = \sqrt{25.34^2 + 16.09^2} \tag{71-7}$$

$$= 30.0 \text{ V}$$

$$\text{True Power (TP)} = I \times V_R = 30.9 \text{ mA} \times 25.34 \text{ V} \tag{71-8}$$

$$= 783.0 \text{ mW}$$

$$\text{Reactive Power (RP)} = I \times V_C = 30.9 \text{ mA} \times 16.09 \text{ V} \tag{71-9}$$

$$= 497.2 \text{ mVAr}$$

$$\text{Apparent Power (AP)} = I \times E = 30.9 \text{ mA} \times 30 \text{ V} \tag{71-10}$$

$$= 927.0 \text{ mVA}$$

Power check:

$$\text{AP} = \sqrt{\text{TP}^2 + \text{RP}^2} = \sqrt{783.0^2 + 497.2^2} \tag{71-11}$$

$$= 927.5 \text{ mVA}$$

$$\text{Power factor} = \frac{\text{TP}}{\text{AP}} = \frac{783.0 \text{ mW}}{927.0 \text{ mVA}} \tag{71-12}$$

$$= 0.845, \text{ leading}$$

$$\text{Phase angle, } \phi = \text{inv cos } (0.845) = -32.4° \tag{71-13}$$

Chapter 72
Sine-wave input voltage
to *L* and *C* in series

L ET US ASSUME THAT THE VALUE OF $X_L (=2\pi f L\ \Omega)$ IS GREATER THAN THE VALUE of $X_C = 1/(2\pi f C)\ \Omega$. Because L and C are in series, the same alternating current must be associated with both the inductor and the capacitor (Fig. 72-1A). The instantaneous voltage (v_L) across the inductor leads the instantaneous current (i) by 90° while the instantaneous voltage across the capacitor (v_C) lags i by 90° (Fig. 72-1B). Consequently, v_L and v_C are 180° out of phase and their phasors are pointing in opposite directions (Fig. 72-1C). Since we have assumed that $X_L > X_C$, $v_L > v_C$ so that the supply voltage, e, (which is the phasor sum of v_L and v_C) is in phase with v_L but has a magnitude of $V_L - V_C$. Because i lags e by 90°, the circuit behaves inductively and has a lagging power factor with a phase angle of +90°. The magnitude of the impedance, Z, is found by combining X_L and X_C so that $Z = X_L - X_C\ \Omega$ (Fig. 72-1D).

Because the inductor and the capacitor are considered to be ideal components, there is no resistance present in the circuit and no true power is dissipated. The inductive reactive power (IRP) is greater than the capacitive reactive power (CRP) and the apparent power (AP) is entirely reactive with a value equal to IRP − CRP (Fig. 72-1E).

If, by contrast, we lowered the frequency to the point where X_C became greater than X_L, E would be equal to $V_C - V_L$ and Z would be $X_C - X_L$. The current, i, would then lead e by 90°, the

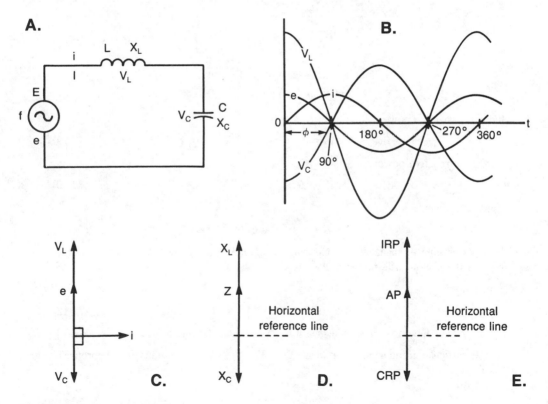

Figure 72-1

235

circuit would behave capacitively with a leading power factor and a phase angle of $-90°$. The apparent power would equal CRP $-$ IRP.

In the particular case where the frequency is chosen so that X_L is equal to X_C, V_L is equal to V_C and the impedance, Z, is zero. The current would then be theoretically infinite.

MATHEMATICAL DERIVATIONS

Impedance phasor,

$$z = j\,(X_L - X_C) = j\left(2\pi fL - \frac{1}{2\pi fC}\right) \text{ ohms} \tag{72-1}$$

In terms of the effective values (capital letters)

$$\frac{E}{I} = Z,\ E = I \times Z,\ I = \frac{E}{Z} \tag{72-2}$$

$$\phi = +90° \text{ if } X_L > X_C \text{ and } \phi = -90° \text{ if } X_C > X_L \tag{72-3}$$

$$\text{Impedance, } Z = X_L \sim X_C \text{ ohms} \tag{72-4}$$

The sign "$\sim$" means the "difference"; you are required to subtract the smaller quantity from the larger quantity so that the result is always positive. This takes into account the two cases, $X_L > X_C$ and $X_C > X_L$.

$$V_L = I \times X_L,\ X_L = \frac{V_L}{I},\ I = \frac{V_L}{X_L} \tag{72-5}$$

$$V_C = I \times X_C,\ X_C = \frac{V_C}{I},\ I = \frac{V_C}{X_C} \tag{72-6}$$

$$\text{Source voltage, } E = V_L \sim V_C \text{ volts} \tag{72-7}$$

$$\text{True power (TP)} = 0 \tag{72-8}$$

Inductive reactive power,

$$\text{IRP} = I \times V_L = I^2 \times X_L = V_L{}^2/X_L \text{ VArs} \tag{72-9}$$

Capacitive reactive power,

$$\text{CRP} = I \times V_C = I^2 \times X_C = V_C{}^2/X_C \text{ VArs} \tag{72-10}$$

Apparent power,

$$\text{AP} = I \times E = I^2 \times Z = E^2/Z \text{ VA(r)s} \tag{72-11}$$

It is permissible to measure the apparent power in VArs because the power is entirely reactive and there is no true power in the circuit.

Apparent power,

$$\text{AP} = \text{IRP} \sim \text{CRP VA(r)s} \tag{72-12}$$

$$\text{Power factor} = 0 \text{ and is lagging if } X_L > X_C \text{ but is leading if } X_C > X_L. \tag{72-13}$$

At the particular frequency, f, for which $X_L = X_C$, $V_L = V_C$, $Z = 0$ and I is theoretically infinite. Then

$$2\pi fL = \frac{1}{2\pi fC}$$

This yields

$$f = \frac{1}{2\pi\sqrt{LC}} \text{ Hz}, \ L = \frac{1}{4\pi^2 f^2 C} \text{ H}, \ C = \frac{1}{4\pi^2 f^2 L} \text{ F} \tag{72-14}$$

Example 72-1

In Fig. 72-2, $L = 125 \ \mu\text{H}$, $C = 275$ pF, and the source voltage is 50 mV at 750 kHz. Calculate the values of Z, I, V_L, TP, IRP, CRP, AP, power factor and the phase angle, ϕ. At which frequency will the inductive and capacitive reactances be equal?

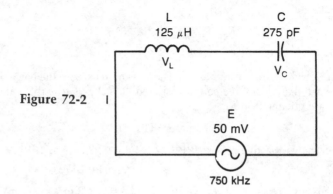

Figure 72-2

Solution

Inductive reactance,

$$X_L = 2\pi f L = 2 \times \pi \times 750 \times 10^3 \times 125 \times 10^{-6} \tag{72-1}$$
$$= 589 \ \Omega$$

Capacitive reactance,

$$X_C = \frac{1}{2\pi f C} \tag{72-1}$$

$$= \frac{1}{2 \times \pi \times 750 \times 10^3 \times 275 \times 10^{-12}}$$
$$= 772 \ \Omega$$

Impedance,

$$Z = X_C - X_L = 772 - 589 \tag{72-4}$$
$$= 183 \ \Omega \text{ and is capacitive}$$

Current,

$$I = \frac{E}{Z} = \frac{50 \text{ mV}}{183 \ \Omega} \tag{72-2}$$
$$= 0.273 \text{ mA}$$

Voltage across the inductor,

$$V_L = I \times X_L = 0.273 \text{ mA} \times 589 \ \Omega \qquad (72\text{-}5)$$
$$= 161 \text{ mV}$$

Voltage across the capacitor,

$$V_C = I \times X_C = 0.273 \text{ mA} \times 772 \ \Omega \qquad (72\text{-}6)$$
$$= 211 \text{ mV}$$

Voltage check:

Source voltage,

$$E = V_C - V_L = 211 - 161 \qquad (72\text{-}7)$$
$$= 50 \text{ mV}$$

Notice that both V_L and V_C are greater than E. This is allowed because the power in the circuit is entirely reactive and there is no true power dissipated. It is always true that either V_L or V_C or both V_L and V_C are greater than E.

$$\text{True power (TP)} = 0 \qquad (72\text{-}8)$$

$$\text{Inductive reactive power (IRP)} = I \times V_L = 0.273 \text{ mA} \times 161 \text{ mV} \qquad (72\text{-}9)$$
$$= 44.0 \ \mu\text{VAr}$$

$$\text{Capacitive reactive power (CRP)} = I \times V_C = 0.273 \text{ mA} \times 211 \text{ mV} \qquad (72\text{-}10)$$
$$= 57.6 \ \mu\text{VAr}$$

$$\text{Apparent power (AP)} = I \times E = 0.273 \times 50 \text{ mV} \qquad (72\text{-}11)$$
$$= 13.6 \ \mu\text{VAr}$$

Power check:

$$\text{AP} = \text{CRP} - \text{IRP} = 57.6 - 44.0 \qquad (72\text{-}12)$$
$$= 13.6 \ \mu\text{VAr}$$

$$\text{Power factor} = 0, \text{ leading} \qquad (72\text{-}13)$$

$$\text{Phase angle, } \phi = -90° \qquad (72\text{-}14)$$

Frequency for which the reactances are equal,

$$f = \frac{1}{2\pi\sqrt{LC}} = \frac{1}{2\pi\sqrt{125 \times 10^{-6} \times 275 \times 10^{-12}}} \text{ Hz} \qquad (72\text{-}15)$$
$$= 858 \text{ kHz}$$

Notice that the Pythagorean theorem is never used in the solution of this type of problem.

Chapter 73
Sine-wave input voltage
to *R*, *L*, and *C* in series

THIS CIRCUIT MUST CONTAIN ALL OF THE INFORMATION CONTAINED IN THE previous three chapters (70, 71, and 72). For example, if the capacitor were eliminated, the results would then be the same as for R and L in series.

Let us assume that X_L is greater than X_C and that R is less than $X_L - X_C$. Since all three components are in series, the instantaneous current, i, is the same throughout the circuit. In terms of phase relationships, v_R is in phase with i, v_L leads i by 90° while v_C lags i by 90° (Fig. 73-1B and C). The combined voltage across the inductor and the capacitor is represented by the phasor v_X, which is in phase with v_L and leads i by 90°. The current i lags the source voltage e which is the phasor sum of v_X and v_R. The circuit is therefore overall inductive, the power factor is lagging and the phase angle, ϕ, is positive.

The net reactance, X, is equal to $X_L - X_C$ which is then combined with R to produce the impedance, Z (Fig. 73-1D). The total reactive power is the phasor sum of the inductive reactive power and the capacitive reactive power and when the total reactive power is combined with the

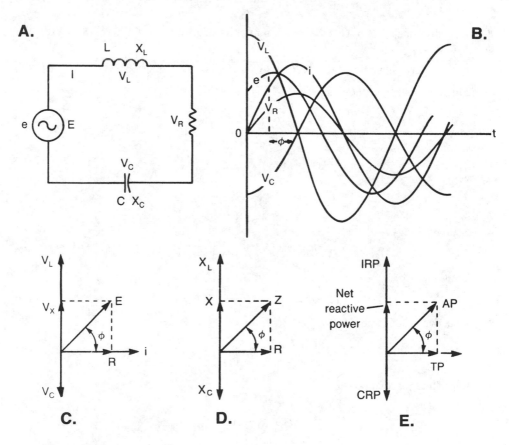

Figure 73-1

true power, the result is the apparent power (Fig. 73-1E).

If X_C were greater than X_L, v_X would be in phase with v_C, i would lead e by the phase angle, ϕ. The power factor would then be leading and ϕ would be a negative angle.

At the particular frequency, $f = 1/(2\pi\sqrt{LC})$ Hz, the reactances are equal and therefore cancel each other. The impedance is then equal to the resistance and the phase angle is zero. As a result the values of the true power and the apparent power are the same.

MATHEMATICAL DERIVATIONS

Impedance phasor,

$$z = R + jX = R + j(X_L - X_C) \tag{73-1}$$

$$= R + j\left(2\pi fL - \frac{1}{2\pi fC}\right) \text{ ohms}$$

In terms of the effective values (capital letters):

$$\frac{E}{I} = Z, \; E = I \times Z, \; I = \frac{E}{Z} \tag{73-2}$$

$$\phi = \text{inv tan}\left(\frac{X_L - X_C}{R}\right) \text{ degrees} \tag{73-3}$$

ϕ is a positive angle when $X_L > X_C$, but is a negative angle if $X_C > X_L$. When $R = X_L - X_C$, $\phi = +45°$.

$$\text{Impedance, } Z = \sqrt{R^2 + (X_L \sim X_C)^2} = \sqrt{R^2 + X^2} \text{ ohms} \tag{73-4}$$

For the meaning of the difference "~" sign, see Chapter 72.

$$V_R = I \times R, \; R = \frac{V_R}{I}, \; I = \frac{V_R}{R} \tag{73-5}$$

$$V_L = I \times X_L, \; X_L = \frac{V_L}{I}, \; I = \frac{V_L}{X_L} \tag{73-6}$$

$$V_C = I \times X_C, \; X_C = \frac{V_C}{I}, \; I = \frac{V_C}{X_C} \tag{73-7}$$

Source voltage,

$$E = \sqrt{V_R^2 + (V_L \sim V_C)^2} = \sqrt{V_R^2 + V_X^2} \text{ volts} \tag{73-8}$$

True power,

$$\text{TP} = I \times V_R = I^2R = V_R^2/R \text{ watts} \tag{73-9}$$

Inductive reactive power,

$$\text{IRP} = I \times V_L \tag{73-10}$$

$$= I^2 \times X_L = V_L^2/X_L \text{ VArs}$$

Capacitive reactive power,

$$\text{CRP} = I \times V_C \tag{73-11}$$

$$= I^2 \times X_C = V_C^2/X_C \text{ VArs}$$

Apparent power

$$AP = I \times E = I^2Z = E^2/Z \text{ VA} \qquad (73\text{-}12)$$

Apparent power,

$$AP = \sqrt{TP^2 + (IRP \sim CRP)^2} \qquad (73\text{-}13)$$

Power factor,

$$PF = \frac{TP}{AP} = \frac{I^2R}{I^2Z} = \frac{R}{Z} = \cos \phi \text{ and is lagging} \qquad (73\text{-}14)$$

Phase angle,

$$\phi = \text{inv cos (PF) and is a } positive \text{ angle} \qquad (73\text{-}15)$$

True power

$$TP = E \times I \times \cos \phi \text{ watts} \qquad (73\text{-}16)$$

Example 73-1

In Fig. 73-2, $R = 350$ Ω, $L = 2.5$ H, $C = 5$ μF, and the source voltage is 55 V, 60 Hz. Calculate the values of Z, I, V_R, V_L, V_C, true power, inductive reactive power, capacitive reactive power, apparent power, power factor and the phase angle, ϕ.

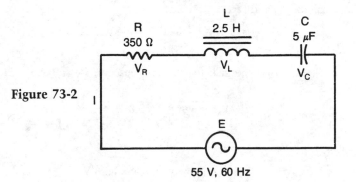

Figure 73-2

Solution

Inductive reactance,

$$X_L = 2\pi fL = 2 \times \pi \times 60 \times 2.5 \qquad (73\text{-}1)$$
$$= 942.5 \ \Omega$$

Capacitive reactance,

$$X_C = \frac{1}{2\pi fC} \qquad (73\text{-}1)$$

$$= \frac{1}{2 \times \pi \times 60 \times 5 \times 10^{-6}}$$

$$= 530.5 \ \Omega$$

Total reactance,

$$X = X_L - X_C = 942.5 - 530.5 \tag{73-1}$$
$$= 412.0 \ \Omega$$

Impedance,

$$Z = \sqrt{R^2 + X^2} = \sqrt{350^2 + 412^2} \tag{73-4}$$
$$= 514 \ \Omega$$

Current,

$$I = \frac{E}{Z} = \frac{55 \ V}{541 \ \Omega}$$
$$= 0.1017 \ A.$$

Voltage across the resistor,

$$V_R = I \times R = 0.1017 \ A \times 350 \ \Omega \tag{73-5}$$
$$= 35.6 \ V$$

Voltage across the inductor,

$$V_L = I \times X_L = 0.1017 \ A \times 942.5 \ \Omega \tag{73-6}$$
$$= 95.85 \ V$$

Voltage across the capacitor,

$$V_C = I \times X_C = 0.1017 \ A \times 530.5 \ \Omega \tag{73-7}$$
$$= 54.0 \ V$$

Combined voltage across the inductor and the capacitor, $V_X = 95.85 - 54.0 = 41.85$ V. Voltage check:

$$E = \sqrt{V_R^2 + V_X^2} = \sqrt{35.6^2 + 41.85^2} \tag{73-8}$$
$$= 54.9 \ V$$

True power,

$$TP = I \times V_R = 0.1017 \ A \times 35.6 \ V \tag{73-9}$$
$$= 3.62 \ W$$

Inductive reactive power,

$$IRP = I \times V_L = 0.1017 \ A \times 95.85 \ V \tag{73-10}$$
$$= 9.75 \ VAr$$

Capacitive reactive power,

$$CRP = I \times V_C = 0.1017 \ A \times 54.0 \ V \tag{73-11}$$
$$= 5.49 \ VAr$$

Apparent power,

$$AP = I \times E = 0.1017 \ A \times 55 \ V \tag{73-12}$$
$$= 5.59 \ VA$$

Power check:

$$AP = \sqrt{TP^2 + (IRP - CRP)^2} \tag{73-13}$$
$$= \sqrt{3.62^2 + (9.75 - 5.49)^2}$$
$$= 5.59 \text{ VA}$$

$$\text{Power factor} = \frac{TP}{AP} = \frac{3.62 \text{ W}}{5.59 \text{ VA}} = 0.647, \text{ lagging} \tag{73-14}$$

$$\text{Phase angle, } \phi = \text{inv cos } 0.647 = +49.6° \tag{73-15}$$

Chapter 74
Sine-wave input voltage
to R and L in parallel

BECAUSE R AND L ARE IN PARALLEL, THE SOURCE VOLTAGE, e, IS ACROSS EACH of the components and the source or supply current, i_S, will be the phasor sum of the two branch currents, i_R and i_L (Fig. 74-1A and B). If we assume that the value of R is greater than the value of X_L, the resistor current ($I_R = E/R$) will be less than the inductor current ($I_L = E/X_L$). Because i_R is in phase with e while i_L lags e by 90°, the source current will lag the source voltage by the phase angle, ϕ (Fig. 74-1B and C). The power factor will be lagging and the phase angle is positive (note that in both the series and parallel combinations of R and L, the source current lags the source voltage so that in each case the phase angle is positive).

The "ease" with which the source current flows, is measured by the admittance, $Y_T (= I_S/E)$ which is the reciprocal of the total impedance, Z_T, and is measured in siemens. Since $I_R = E \times G$, $I_L = E \times B_L$ and $I_S = E \times Y_T$, the admittance, Y_T, is the phasor sum of the conductance, G, and the inductive susceptance, B_L (Fig. 74-1D).

The true power, TP, is equal to E^2/R watts and is independent of the frequency. The reactive power, RP, is associated with the inductor while the apparent power (AP) at the source is the phasor sum of the true power and the reactive power (Fig. 74-1E).

MATHEMATICAL DERIVATIONS

In terms of effective values:

$$I_R = \frac{E}{R}, \ E = I_R \times R, \ R = \frac{E}{I_R} \tag{74-1}$$

$$I_L = \frac{E}{X_L}, \ E = I_L \times X_L, \ X_L = \frac{E}{I_L} \tag{74-2}$$

$$I_S = \frac{E}{Z_T}, \ E = I_S \times Z_T, \ Z_T = \frac{E}{I_S} \tag{74-3}$$

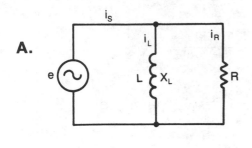

A.

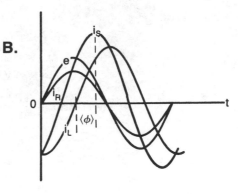

B.

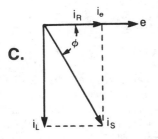

C.

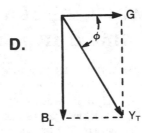

D.

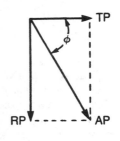

E.

Figure 74-1

Source current,

$$I_S = \sqrt{I_R^2 + I_L^2}$$ (74-4)
$$I_R = \sqrt{I_S^2 - I_L^2}$$
$$I_L = \sqrt{I_S^2 - I_R^2} \text{ amperes}$$

Because $I_S^2 = I_R^2 + I_L^2$,

$$\left(\frac{E}{Z_T}\right)^2 = \left(\frac{E}{R}\right)^2 + \left(\frac{E}{X_L}\right)^2$$

This yields
Total impedance,

$$Z_T = \frac{R \times X_L}{\sqrt{R^2 + X_L^2}} \text{ ohms}$$ (74-5)

Equation 74-5 represents the "product-over-sum" formula when the "Pythagorean sum" is involved.

Because the conductance $G = 1/R$, the inductive susceptance $B_L = 1/X_L$ and the admittance $Y_T = 1/Z_T$.

Admittance, $Y_T = \sqrt{G^2 + B_L^2}$, $G = \sqrt{Y_T^2 - B_L^2}$, $B_L = \sqrt{Y_T^2 - G^2}$ siemens (74-6)

True power,

$$TP = E \times I_R = I_R^2 \times R = E^2/R = E^2G \text{ watts} \qquad (74\text{-}7)$$

Reactive power,

$$RP = E \times I_L = I_L^2 \times X_L = E^2B_L \text{ volt-amperes reactive} \qquad (74\text{-}8)$$

Apparent power,

$$AP = E \times I_S = I_S^2 \times Z_T = E^2/Z_T = E^2Y_T \text{ volt-amperes} \qquad (74\text{-}9)$$

$$AP = \sqrt{TP^2 + RP^2}, \ TP = \sqrt{AP^2 - RP^2}, \ RP = \sqrt{AP^2 - TP^2} \qquad (74\text{-}10)$$

Power factor,

$$PF = \frac{TP}{AP} = \cos \phi = \frac{E^2/R}{E^2/Z_T} = \frac{Z_T}{R} \left(not \ \frac{R}{Z_T} \right), \text{ lagging} \qquad (74\text{-}11)$$

Phase angle,

$$\phi = \text{inv cos (PF) and is a positive angle} \qquad (74\text{-}12)$$

Example 74-1

In Fig. 74-2, calculate the values of I_R, I_L, I_S, Z_T, Y_T, true power, reactive power, apparent power, power factor and the phase angle, ϕ.

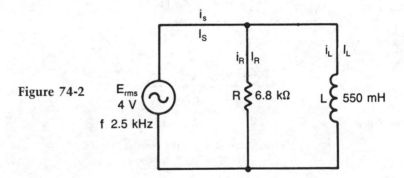

Figure 74-2

Solution

Inductive reactance,

$$X_L = 2\pi fL = 2 \times \pi \times 2.5 \times 10^3 \times 550 \times 10^{-3} \ \Omega = 8.64 \text{ k}\Omega$$

Resistor current,

$$I_R = \frac{E}{R} = \frac{4}{6.8 \text{ k}\Omega} \qquad (74\text{-}1)$$

$$= 0.588 \text{ mA}$$

Inductor current,

$$I_L = \frac{E}{X_L} = \frac{4 \text{ V}}{8.64 \text{ k}\Omega} \qquad (74\text{-}2)$$

$$= 0.463 \text{ mA}$$

Source current,

$$I_S = \sqrt{I_R^2 + I_L^2} = \sqrt{0.588^2 + 0.463^2} \qquad (74\text{-}4)$$
$$= 0.7484 \text{ mA}$$

Total impedance,

$$Z_T = \frac{E}{I_S} = \frac{4 \text{ V}}{0.7484 \text{ mA}} \qquad (74\text{-}3)$$
$$= 5.34 \text{ k}\Omega$$

Impedance check #1:
    Total impedance,

$$Z_T = \frac{R \times X_L}{\sqrt{R^2 + X_L^2}} \qquad (74\text{-}5)$$
$$= \frac{6.8 \times 8.64}{\sqrt{6.8^2 + 8.64^2}}$$
$$= 5.34 \text{ k}\Omega$$

Impedance check #2:
    Conductance,

$$G = \frac{1}{R} = \frac{1}{6.8 \text{ k}\Omega}$$
$$= 0.147 \text{ mS}$$

Inductive susceptance,

$$B_L = \frac{1}{X_L} = \frac{1}{8.64 \text{ k}\Omega}$$
$$= 0.116 \text{ mS}$$

Admittance,

$$Y_T = \sqrt{G^2 + B_L^2} = \sqrt{0.147^2 + 0.116^2} \qquad (74\text{-}6)$$
$$= 0.187 \text{ mS}$$

Total impedance,

$$Z_T = \frac{1}{Y_T} = \frac{1}{0.187 \text{ mS}}$$
$$= 5.34 \text{ k}\Omega$$

True power,

$$\text{TP} = E \times I_R = 4 \text{ V} \times 0.588 \text{ mA} \qquad (74\text{-}7)$$
$$= 2.35 \text{ mW}$$

Reactive power,

$$\text{RP} = E \times I_L = 4 \text{ V} \times 0.463 \text{ mA} \qquad (74\text{-}8)$$
$$= 1.85 \text{ mVAr}$$

Apparent power,

$$AP = E \times I_S = 4 \text{ V} \times 0.7484 \text{ mA} \qquad (74\text{-}9)$$
$$= 2.99 \text{ mVA}$$

Power check:

$$AP = \sqrt{TP^2 + RP^2} = \sqrt{2.35^2 + 1.85^2} \qquad (74\text{-}10)$$
$$= 2.99 \text{ mVA}$$

Power factor,

$$PF = \frac{TP}{AP} = \frac{2.35 \text{ W}}{2.99 \text{ VA}} \qquad (74\text{-}11)$$
$$= 0.786, \text{ lagging}$$

Phase angle,

$$\phi = \text{inv cos } 0.786 = +38.2° \qquad (74\text{-}12)$$

Chapter 75
Sine-wave input voltage
to R and C in parallel

BECAUSE R AND C ARE IN PARALLEL, THE SOURCE VOLTAGE, E, IS ACROSS EACH of the components and the source or supply current, i_S, will be the phasor sum of the two branch currents, i_R and i_C (Fig. 75-1A and B). If we assume that the value of R is greater than the value of X_C, the resistor current ($I_R = E/R$) will be less than the capacitor current ($I_C = E/X_C$). Because i_R is in phase with e while i_C leads e by 90°, the source current will *lead* the source voltage by the phase angle, ϕ, (Figs. 75-1B and C). The power factor will then be *leading* and the phase angle is negative.

The "ease" with which the source current flows is measured by the admittance $Y_T(= I_S/E)$, which is the reciprocal of the total impedance, Z_T, and is measured in siemens. Since $I_R = E \times G$, $I_C = E \times B_C$ and $I_S = E \times Y_T$, the admittance, Y_T, is the phasor sum of the conductance, G, and the capacitive susceptance, B_C (Fig. 75-1D).

The true power, TP, is equal to E^2/R watts and is independent of the frequency. The reactive power, RP, is associated with the capacitor while the apparent power, AP, at the source is the phasor sum of the true power and the reactive power (Fig. 75-1E).

MATHEMATICAL DERIVATIONS

In terms of effective values:

$$I_R = \frac{E}{R}, \ E = I_R \times R, \ R = \frac{E}{I_R} \qquad (75\text{-}1)$$

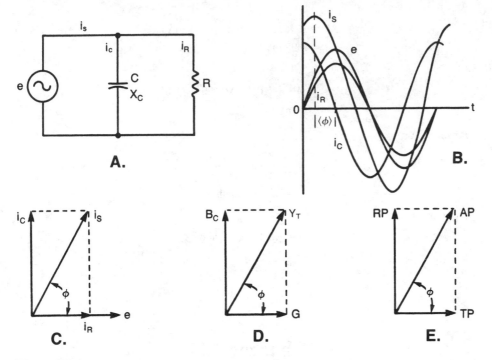

Figure 75-1

$$I_C = \frac{E}{X_C}, \; E = I_C \times X_C, \; X_C = \frac{E}{I_C} \tag{75-2}$$

$$I_S = \frac{E}{Z_T}, \; E = I_S \times Z_T, \; Z_T = \frac{E}{I_S} \tag{75-3}$$

Source current,

$$I_S = \sqrt{I_R{}^2 + I_C{}^2}, \; I_R = \sqrt{I_S{}^2 - I_C{}^2}, \; I_C = \sqrt{I_S{}^2 - I_R{}^2} \text{ amperes} \tag{75-4}$$

Since $I_S{}^2 = I_R{}^2 + I_C{}^2$,

$$\left(\frac{E}{Z_T}\right)^2 = \left(\frac{E}{R}\right)^2 + \left(\frac{E}{X_C}\right)^2$$

This yields

Total impedance,

$$Z_T = \frac{R \times X_C}{\sqrt{R^2 + X_C{}^2}} \text{ ohms} \tag{75-5}$$

Equation 75-5 represents the "product-over-sum" formula when the "Pythagorean sum" is involved.

Because the conductance, $G = 1/R$, the capacitive susceptance, $B_C = 1/X_C$, and the admittance $Y_T = 1/Z_T$,

$$Y_T = \sqrt{G^2 + B_C{}^2}, \; G = \sqrt{Y_T{}^2 - B_C{}^2}, \; B_C = \sqrt{Y_T{}^2 - G^2} \text{ siemens} \tag{75-6}$$

248

True Power,

$$TP = E \times I_R = I_R{}^2 \times R = E^2/R = E^2G \text{ watts} \tag{75-7}$$

Reactive Power,

$$RP = E \times I_C = I_C{}^2 \times X_C = E^2/X_C = E^2B_C \text{ volt-amperes reactive} \tag{75-8}$$

Apparent Power,

$$AP = E \times I_S = I_S{}^2 \times Z_T = E^2/Z_T = E^2Y_T \text{ volt-amperes} \tag{75-9}$$

$$AP = \sqrt{TP^2 + RP^2}, \; TP = \sqrt{AP^2 - RP^2}, \; RP = \sqrt{AP^2 - TP^2} \tag{75-10}$$

Power factor,

$$PF = \frac{TP}{AP} = \cos \phi = \frac{E^2/R}{E^2/Z_T} = \frac{Z_T}{R} \left(not \; \frac{R}{Z_T} \right), \text{ leading} \tag{75-11}$$

Phase angle,

$$\phi = \text{inv cos (PF) and is a negative angle} \tag{75-12}$$

Note that if $R = X_C$, $\phi = -45°$ and the power factor $= 0.707$.

Example 75-1

In Fig. 75-2, calculate the values of I_R, I_C, I_S, Z_T, Y_T, true power, reactive power, apparent power, power factor and the phase angle ϕ.

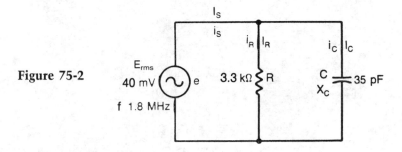

Figure 75-2

Solution

Capacitive reactance,

$$X_C = \frac{1}{2\pi fC}$$

$$= \frac{1}{2 \times \pi \times 1.8 \times 10^6 \times 35 \times 10^{-12}} \; \Omega$$

$$= 2.53 \text{ k}\Omega.$$

Resistor current,

$$I_R = \frac{E}{R} = \frac{40 \text{ mV}}{3.3 \text{ k}\Omega} \tag{75-1}$$

$$= 12.1 \; \mu\text{A}$$

Capacitor current,

$$I_C = \frac{E}{X_C} = \frac{40 \text{ mV}}{2.53 \text{ k}\Omega} \tag{75-2}$$
$$= 1.58 \ \mu\text{A}$$

Source current,

$$I_S = \sqrt{I_R^2 + I_C^2} = \sqrt{12.1^2 + 15.8^2} \tag{75-4}$$
$$= 19.9 \ \mu\text{A}$$

Total impedance,

$$Z_T = \frac{E}{I_S} = \frac{40 \text{ mV}}{19.9 \ \mu\text{A}} \tag{75-3}$$
$$= 2.01 \text{ k}\Omega$$

Impedance check #1:
  Total impedance,

$$Z_T = \frac{R \times X_C}{\sqrt{R^2 + X_C^2}} \tag{75-5}$$
$$= \frac{3.3 \times 2.53}{\sqrt{3.3^2 + 2.53^2}}$$
$$= 2.01 \text{ k}\Omega$$

Impedance check #2:
  Conductance,

$$G = \frac{1}{R} = \frac{1}{3.3 \text{ k}\Omega}$$
$$= 0.303 \text{ mS.}$$

  Capacitive susceptance,

$$B_C = \frac{1}{X_C} = \frac{1}{2.53 \text{ k}\Omega}$$
$$= 0.395 \text{ mS.}$$

Admittance,

$$Y_T = \sqrt{G^2 + B_C^2} = \sqrt{0.303^2 + 0.395^2} \tag{75-6}$$
$$= 0.498 \text{ mS}$$

Total impedance,

$$Z_T = \frac{1}{Y_T} = \frac{1}{0.498 \text{ mS}}$$
$$= 2.01 \text{ k}\Omega.$$

True Power,

$$TP = E \times I_R = 40 \text{ mV} \times 12.1 \ \mu\text{A} \tag{75-7}$$
$$= 0.484 \ \mu\text{W}$$

Reactive Power,

$$RP = E \times I_C = 40 \text{ mV} \times 15.8 \text{ }\mu\text{A} \qquad (75\text{-}8)$$
$$= 0.632 \text{ }\mu\text{VAr}$$

Apparent Power,

$$AP = E \times I_S = 40 \text{ mV} \times 19.9 \text{ }\mu\text{A} \qquad (75\text{-}9)$$
$$= 0.796 \text{ }\mu\text{VA}$$

Power check:

$$AP = \sqrt{TP^2 + RP^2} = \sqrt{0.484^2 + 0.632^2} \qquad (75\text{-}10)$$
$$= 0.796 \text{ }\mu\text{VA}$$

$$\text{Power factor} = \frac{TP}{AP} = \frac{0.484 \text{ }\mu\text{W}}{0.796 \text{ }\mu\text{VA}} \qquad (75\text{-}11)$$
$$= 0.61, \text{ leading}$$

Phase angle,

$$\phi = \text{inv cos } 0.61 = -52.5° \qquad (75\text{-}12)$$

Chapter 76
Sine-wave input voltage
to *L* and *C* in parallel

BECAUSE *L* AND *C* ARE IN PARALLEL, THE SOURCE VOLTAGE, *E*, IS ACROSS EACH of the components and the source or supply current, i_S, will be the phasor sum of the two branch currents, i_L and i_C (Figs. 76-1A and B). If we assume that the value of X_C is *greater* than the value of X_L, the capacitor current ($I_C = E/X_C$) will be *less* than the inductor current ($I_L = E/X_L$). Because i_C *leads* the source voltage by 90° while i_L *lags* the same source voltage by 90°, i_C and i_L are 180° out-of-phase (Fig. 76-1B) and their phasors are pointing in opposite directions (Fig. 76-1C). The phasor sum of i_L and i_C is the source current, i_S, which will be in phase with i_L and will *lag* the source voltage by 90°. The current therefore behaves inductively, the power factor is *lagging* and the phase angle is *positive*. Notice that this opposite to the result obtained from the series *L*, *C* circuit of Chapter 72 where if X_C were greater than X_L, the power factor was leading. This is because, in the parallel circuit, we are concerned with the branch currents, which are *inversely* proportional to the reactances, while in the series circuit, we considered the component voltages, which were *directly* proportional to the reactances.

If X_L is greater than X_C, the capacitor current is greater than the inductor current; i_S is in phase with i_C and leads the source voltage by 90°. The power factor is then leading and the phase angle is negative.

The admittance, $Y_T(= I_S/E)$ is the reciprocal of the total impedance, Z_T, and is measured in siemens. Because $I_L = E \times B_L$, $I_C = E \times B_C$ and $I_S = E \times Y_T$, the admittance, Y_T, is the phasor sum of the inductive susceptance, B_L, and the capacitive susceptance, B_C (Fig. 76-1D).

251

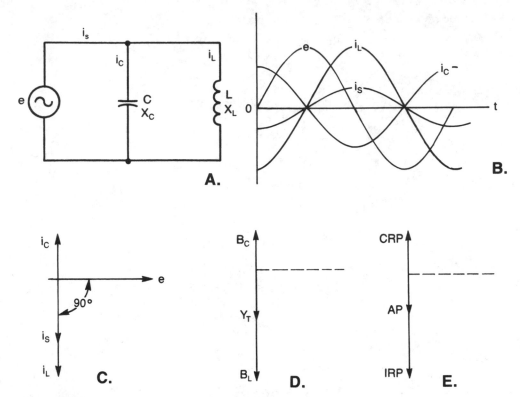

Figure 76-1

Because the components are assumed to be ideal, there are no resistance losses associated with the circuit and the true power, TP, is zero. The apparent power (AP) is the phasor sum of the inductive reactive power (IRP) and the capacitive reactive power (CRP) (Fig. 76-1E).

Notice at the particular frequency, f, for which the reactances are equal, the branch currents are also equal and therefore the supply current is zero. The total impedance is then infinite and the parallel combination behaves as an *open* circuit. This is theoretically possible because the power in the circuit is entirely reactive and there is no true power dissipated as heat.

MATHEMATICAL DERIVATIONS

In terms of effective values:

$$I_L = \frac{E}{X_L}, \; E = I_L \times X_L, \; X_L = \frac{E}{I_L} \tag{76-1}$$

$$I_C = \frac{E}{X_C}, \; E = I_C \times X_C, \; X_C = \frac{E}{I_C} \tag{76-2}$$

$$I_S = \frac{E}{Z_T}, \; E = I_S \times Z_T, \; Z_T = \frac{E}{I_S} \tag{76-3}$$

Source current,

$$I_S = I_L \sim I_C \text{ amperes} \tag{76-4}$$

For the meaning of the "~" difference sign, see Chapter 72.

Because $I_S = I_L \sim I_C$,

$$\frac{E}{Z_T} = \frac{E}{X_L} \sim \frac{E}{X_C}$$

This yields:

Total impedance,

$$Z_T = \frac{X_L \times X_C}{X_L \sim X_C} \text{ ohms} \tag{76-5}$$

Equation 76-5 represents the "product-over-sum" formula when the "difference" sign is involved. From this formula it follows that Z_T must be greater in value than either X_L or X_C or both X_L and X_C.

Because the inductive susceptance, $B_L = 1/X_L$, the capacitive susceptance, $B_C = 1/X_C$, and the admittance $Y_T = 1/Z_T$.

Admittance,

$$Y_T = B_L \sim B_C \text{ siemens} \tag{76-6}$$

True Power,

$$TP = 0 \tag{76-7}$$

Inductive reactive power,

$$IRP = E \times I_L = I_L{}^2 \times X_L \tag{76-8}$$
$$= E^2/X_L$$
$$= E^2 B_L \text{ volt-amperes reactive}$$

Capacitive reactive power,

$$CRP = E \times I_C = I_C{}^2 \times X_C \tag{76-9}$$
$$= E^2/X_C$$
$$= E^2 B_C \text{ volt-amperes reactive}$$

Apparent power,

$$AP = E \times I_S = I_s{}^2 Z_T = E^2/Z_T \tag{76-10}$$
$$= E^2 Y_T \text{ volt-amperes}$$

Apparent power,

$$AP = IRP \sim CRP \text{ volt-amperes} \tag{76-11}$$

$$\text{Power factor} = \frac{TP}{AP} = 0, \text{ lagging} \tag{76-12}$$

Notice that it is theoretically possible for a power factor to be either zero, lagging, or zero, leading. This is illustrated in Fig. 76-2, which shows the scale of a power factor meter.

Phase angle,

$$\phi = \text{inv cos } 0 = +90° \tag{76-13}$$

At the particular frequency, f, for which $X_L = X_C$, $I_L = I_C$, $I_S = 0$ and $Z_T = \infty$ Ω. Then from Chapter 72,

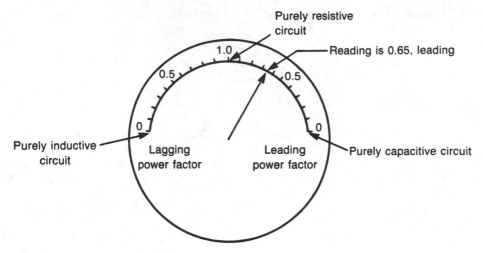

Figure 76-2

$$f = \frac{1}{2\pi\sqrt{LC}} \text{ Hz, } L = \frac{1}{4\pi^2f^2C} \text{ H, } C = \frac{1}{4\pi^2f^2L} \text{ F} \qquad (76\text{-}14)$$

Example 76-1

In Fig. 76-3, calculate the values of I_L, I_C, I_S, Z_T, Y_T, TP, IRP, CRP, AP, power factor and the phase angle, ϕ. What is the particular frequency for which $I_S = 0$?

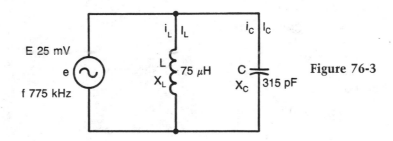

Figure 76-3

Solution

Inductive reactance,

$$X_L = 2\pi f L = 2 \times \pi \times 775 \times 10^3 \times 75 \times 10^{-6}$$
$$= 365 \ \Omega.$$

Capacitive reactance,

$$X_C = \frac{1}{2\pi f C}$$

$$= \frac{1}{2 \times \pi \times 775 \times 10^3 \times 315 \times 10^{-12}}$$

$$= 652 \ \Omega.$$

Inductor current,

$$I_L = \frac{E}{X_L} = \frac{25 \text{ mV}}{365 \text{ }\Omega} \tag{76-1}$$
$$= 68.50 \text{ }\mu\text{A}$$

Capacitor current,

$$I_C = \frac{E}{X_C} = \frac{25 \text{ mV}}{652 \text{ }\Omega} \tag{76-2}$$
$$= 38.34 \text{ }\mu\text{A}$$

Source current,

$$I_S = I_L - I_C = 68.50 - 38.34 \tag{76-4}$$
$$= 30.16 \text{ }\mu\text{A}$$

Total impedance,

$$Z_T = \frac{E}{I_S} = \frac{25 \text{ mV}}{30.16 \text{ }\mu\text{A}} \tag{76-3}$$
$$= 829 \text{ }\Omega$$

To find the total impedance by the branch current method, you may assume any convenient value for the source voltage if none is given.

Impedance check #1:
　　Total impedance,

$$Z_T = \frac{X_L \times X_C}{X_L - X_C} = \frac{652 \times 365}{652 - 365} \tag{76-5}$$
$$= 829 \text{ }\Omega$$

Impedance check #2:
　　Inductive susceptance,

$$B_L = \frac{1}{X_L} = \frac{1}{365 \text{ }\Omega}$$
$$= 2.740 \text{ mS}$$

Capacitive susceptance,

$$B_C = \frac{1}{X_C} = \frac{1}{652 \text{ }\Omega}$$
$$= 1.534 \text{ mS}$$

Admittance,

$$Y_T = B_L - B_C = 2.740 - 1.534 \tag{76-6}$$
$$= 1.206 \text{ mS}$$

Total impedance,

$$Z_T = \frac{1}{Y_T} = \frac{1}{1.206 \text{ mS}}$$
$$= 829 \text{ }\Omega$$

True power,

$$TP = 0 \qquad (76\text{-}7)$$

Inductive reactive power,

$$IRP = E \times I_L \qquad (76\text{-}8)$$
$$= 25 \text{ mV} \times 68.50 \ \mu\text{A}$$
$$= 1.7125 \ \mu\text{VAr}$$

Capacitive reactive power,

$$CRP = E \times I_C \qquad (76\text{-}9)$$
$$= 25 \text{ mV} \times 38.34 \ \mu\text{A}$$
$$= 0.9585 \ \mu\text{VAr}$$

Apparent power,

$$AP = E \times I_S = 25 \text{ mV} \times 30.16 \ \mu\text{A} \qquad (76\text{-}10)$$
$$= 0.7540 \ \mu\text{VAr}$$

Power check:

$$AP = IRP - CRP = 1.7125 - 0.9585 \qquad (76\text{-}11)$$
$$= 0.7540 \ \mu\text{VAr}$$

$$\text{Power factor} = \frac{TP}{AP} = 0, \text{ lagging} \qquad (76\text{-}12)$$

$$\text{Phase angle, } \phi = \text{inv cos } 0 = +90° \qquad (76\text{-}13)$$

The source current, I_S, is zero when the reactances are equal. This occurs at the frequency,

$$f = \frac{1}{2\pi\sqrt{LC}} \qquad (76\text{-}14)$$
$$= \frac{1}{2 \times \pi \times \sqrt{75 \times 10^{-6} \times 315 \times 10^{-12}}} \text{ Hz}$$
$$= 1.03 \text{ MHz}$$

Note that the Pythagorean theorem was not used in the solution of this example.

Chapter 77
Sine-wave input voltage
to R, L, and C in parallel

THIS CIRCUIT MUST CONTAIN ALL THE INFORMATION OF CHAPTERS 74, 75, AND 76 because if, for example, the capacitor were eliminated, we would be left with R and L in parallel (Chapter 74). Because all three components are in parallel across the source voltage, the

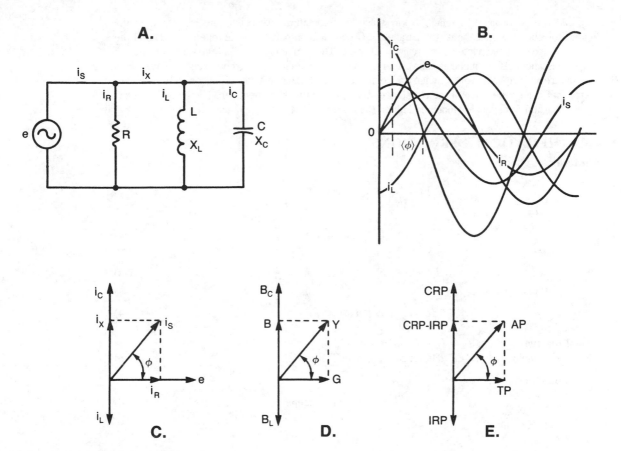

Figure 77-1

source current, i_S, will be the phasor sum of the three branch currents, i_R, i_L and i_C (Figs. 77-1A and B). If we assume that the inductive reactance is greater than the capacitive reactance and that the value of the resistance is relatively large, the capacitor current ($I_C = E/X_C$) will be greater than the inductor current ($I_L = E/X_L$) and the resistor current ($I_R = E/R$) will be small. Since i_C *leads* the source voltage by 90° while i_L *lags* the same source voltage by 90°, i_C and i_L are 180° out-of-phase (Fig. 77-1B) and their phasors are pointing in opposite directions (Fig. 77-1C). The phasor sum of i_L and i_C is the total reactive current, i_X, which will be in phase with i_C. The resistor current, i_R, is then combined with i_X to produce the source current, i_S, which *leads* the source voltage so that the power factor is *leading* and the phase angle, ϕ, is negative.

If X_C is greater than X_L, the inductor current is greater than the capacitor current; i_X is in phase with i_L and i_S *lags* the source voltage by the *positive* phase angle, ϕ. The power factor is then *lagging*.

The admittance $Y_T(= I_S/E)$ is the reciprocal of the total impedance, Z_T, and is measured in siemens. Because $I_R = E \times G$, $I_L = E \times B_L$, $I_C = E \times B_C$ and $I_S = E \times Y_T$, the admittance, Y_T, is the phasor sum of the conductance, G, the inductive susceptance, B_L, and the capacitive susceptance, B_C (Fig. 77-1D). In a similar way the apparent power (AP) is the phasor sum of the true power (TP), the inductive reactive power (IRP) and the capacitive reactive power (CRP) (Fig. 77-1E).

257

At the particular frequency, f, for which the reactances are equal, the inductor and the capacitor currents cancel out (the parallel L, C combination behaves as an open circuit) and the source current is the same as the resistor current. The total impedance is then equal in value to the resistance. At frequencies above and below f, the parallel L, C combination behaves as a certain reactance value which, when placed in parallel with the resistance, produces a total impedance that is less the value of the resistance. When these two statements are combined, it means that the total impedance cannot be higher than the value of the resistance.

MATHEMATICAL DERIVATIONS

In terms of effective values:

$$I_R = \frac{E}{R}, \ E = I_R \times R, \ R = \frac{E}{I_R} \tag{77-1}$$

$$I_L = \frac{E}{X_L}, \ E = I_L \times X_L, \ X_L = \frac{E}{I_L} \tag{77-2}$$

$$I_C = \frac{E}{X_C}, \ E = I_C \times X_C, \ X_C = \frac{E}{I_C} \tag{77-3}$$

$$I_S = \frac{E}{Z_T}, \ E = I_S \times Z_T, \ Z_T = \frac{E}{I_S} \tag{77-4}$$

Total reactive current, $I_X = I_L \sim I_C$ amperes
For the meaning of the "$\sim$" difference sign, see Chapter 72. $\qquad$ (77-5)

Source current,

$$I_S = \sqrt{I_R^2 + (I_L \sim I_C)^2} \text{ amperes} \tag{77-6}$$

Total reactance,

$$X_T = \frac{X_L \times X_C}{X_L \sim X_C} \text{ ohms} \tag{76-5, 77-7}$$

Total impedance,

$$Z_T = \frac{R \times X_T}{\sqrt{R^2 + X_T^2}} \text{ ohms} \tag{74-5, 75-5, 77-8}$$

Because $I_R = EG$, $I_L = EB_L$, $I_C = EB_C$ and $I_S = EY_T$,
Admittance,

$$Y_T = \sqrt{G^2 + (B_L \sim B_C)^2} \text{ siemens} \tag{77-9}$$

True power,

$$\text{TP} = E \times I_R \tag{77-10}$$
$$= I_R^2 \times R$$
$$= E^2/R$$
$$= E^2 G \text{ watts}$$

Inductive reactive power,

$$\text{IRP} = E \times I_L \tag{77-11}$$
$$= I_L^2 \times X_L$$

$$= E^2/X_L$$
$$= E^2B_L \text{ volt-amperes reactive}$$

Capacitive reactive power,

$$CRP = E \times I_C \tag{77-12}$$
$$= I_C{}^2 \times X_C$$
$$= E^2/X_C$$
$$= E^2B_C \text{ volt-amperes reactive}$$

Apparent power,

$$AP = E \times I_S \tag{77-13}$$
$$= I_S{}^2 \times Z_T$$
$$= E^2/Z_T$$
$$= E^2Y_T \text{ volt-amperes}$$

Apparent power,

$$AP = \sqrt{TP^2 + (IRP \sim CRP)^2} \text{ volt-amperes} \tag{77-14}$$

Power factor,

$$PF = \frac{TP}{AP} = \cos \phi = \frac{Z_T}{R}, \text{ leading} \tag{77-15}$$

Phase angle,

$$\phi = \text{inv cos (power factor) and is a negative angle} \tag{77-16}$$

Example 77-1

If Fig. 77-2, calculate the values of I_R, I_L, I_C, I_S, Z_T, Y_T, TP, IRP, CRP, AP, power factor, and the phase angle.

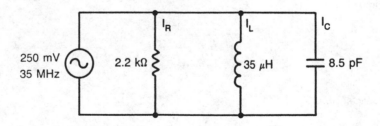

Figure 77-2

Solution

Inductive reactance,

$$X_L = 2\pi fL = 2 \times \pi \times 35 \times 10^6 \times 3.5 \times 10^{-6}$$
$$= 770 \ \Omega.$$

Capacitive reactance,

$$X_C = \frac{1}{2\pi f C}$$

$$= \frac{1}{2 \times \pi \times 35 \times 10^6 \times 8.5 \times 10^{-12}}$$

$$= 535 \ \Omega.$$

Resistor current,

$$I_R = \frac{E}{R} = \frac{250 \text{ mV}}{2.2 \text{ k}\Omega} \qquad (77\text{-}1)$$

$$= 113.6 \ \mu A$$

Inductor current,

$$I_L = \frac{E}{X_L} = \frac{250 \text{ mV}}{770 \ \Omega} \qquad (77\text{-}2)$$

$$= 324.7 \ \mu A$$

Capacitor current,

$$I_C = \frac{E}{X_C} = \frac{250 \text{ mV}}{535 \ \Omega} \qquad (77\text{-}3)$$

$$= 467.3 \ \mu A$$

Total reactive current,

$$I_X = 467.3 - 324.7 \qquad (77\text{-}5)$$

$$= 142.6 \ \mu A$$

Source current,

$$I_S = \sqrt{I_R^2 + I_X^2} = \sqrt{113.6^2 + 142.6^2} \qquad (77\text{-}6)$$

$$= 182.3 \ \mu A$$

Total impedance,

$$Z_T = \frac{E}{I_S} = \frac{250 \text{ mV}}{182.3 \ \mu A} \qquad (77\text{-}4)$$

$$= 1.37 \text{ k}\Omega$$

Impedance check #1:
   Total reactance,

$$X_T = \frac{X_L \times X_C}{X_L - X_C} \qquad (77\text{-}7)$$

$$= \frac{770 \times 535}{770 - 535} \ \Omega$$

$$= 1.753 \text{ k}\Omega$$

Total impedance,

$$Z_T = \frac{R \times X_T}{\sqrt{R^2 + X_T^2}} \qquad (77\text{-}8)$$

$$= \frac{2.2 \times 1.753}{\sqrt{2.2^2 + 1.753^2}}$$
$$= 1.37 \text{ k}\Omega$$

Impedance check #2:
Conductance,

$$G = \frac{1}{2.2 \text{ k}\Omega} = 0.4545 \text{ mS}.$$

Inductive susceptance,

$$B_L = \frac{1}{770 \ \Omega} = 1.299 \text{ mS}.$$

Capacitive susceptance,

$$B_C = \frac{1}{535 \ \Omega} = 1.869 \text{ mS}.$$

Admittance,

$$Y_T = \sqrt{G^2 + (B_C - B_L)^2} = \sqrt{0.4545^2 + (1.869 - 1.299)^2} \qquad (77\text{-}9)$$
$$= 0.729 \text{ mS}.$$

Impedance,

$$Z_T = \frac{1}{Y_T} = \frac{1}{0.729 \text{ mS}}$$
$$= 1.37 \text{ k}\Omega.$$

True power,

$$\text{TP} = E \times I_R = 250 \text{ mV} \times 113.6 \ \mu\text{A} \qquad (77\text{-}10)$$
$$= 28.4 \ \mu\text{W}.$$

Inductive reactive power,

$$\text{IRP} = E \times I_L = 250 \text{ mV} \times 324.7 \ \mu\text{A} \qquad (77\text{-}11)$$
$$= 81.2 \ \mu\text{VAr}.$$

Capacitive reactive power,

$$\text{CRP} = E \times I_C = 250 \text{ mV} \times 467.3 \ \mu\text{A} \qquad (77\text{-}12)$$
$$= 116.8 \ \mu\text{VAr}.$$

Apparent power,

$$\text{AP} = E \times I_S = 250 \text{ mV} \times 182.3 \ \mu\text{A} \qquad (77\text{-}13)$$
$$= 45.58 \ \mu\text{VA}.$$

Power check:

$$\text{AP} = \sqrt{\text{TP}^2 + (\text{CRP} - \text{IRP})^2} = \sqrt{28.4^2 + (116.8 - 81.2)^2} \qquad (77\text{-}14)$$
$$= 45.54 \ \mu\text{VA}.$$

Power factor,

$$PF = \frac{TP}{AP} = \frac{28.4}{45.58} \qquad (77\text{-}15)$$

$$= 0.62, \text{ leading.}$$

Phase angle,

$$\phi = \text{inv cos } 0.62 = -51.5°. \qquad (77\text{-}16)$$

Chapter 78
Maximum power transfer
to the load (ac case)

I N CHAPTER 23 WE LEARNED THAT IN A DC CIRCUIT, THERE IS MAXIMUM POWER transfer to the load when its resistance is matched (made equal to) to the sum of all resistances not associated with the load. In the ac circuit there is the added complication of reactance, which limits the flow of current but does not dissipate any true power.

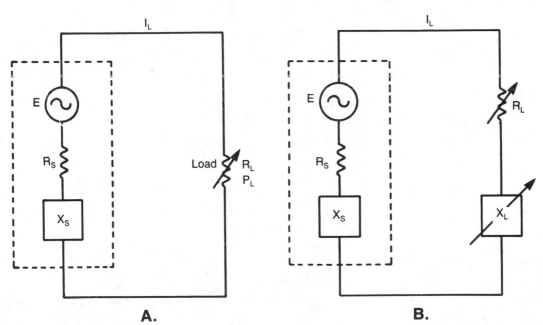

Figure 78-1

There are two separate ac cases to consider. In Fig. 78-1A the ac voltage source has an internal impedance with a resistive component and a reactive component that may be either inductive or capacitive. The mathematical derivations then show that the load power is at maximum when the *load resistance* is varied to match the *magnitude* of the source's internal impedance.

The second case is illustrated in Fig. 78-1B. Both the internal impedance and the load impedance have resistive and reactive components that are either inductive or capacitive. Maximum power to the load is then achieved by a two-stage procedure.

1. The load reactance is varied until its *magnitude* is equal to the *magnitude* of the source's internal reactance. However, if the source reactance is inductive, the load reactance must be capacitive and vice versa. As a result the net reactance of the complete circuit is reduced to zero.
2. Once the load reactance cancels the internal reactance, the remaining load resistance is matched (made equal) to the source resistance.

MATHEMATICAL DERIVATIONS
In Fig. 78-1A

In the following equations, all the voltages and currents are effective values,
Magnitude of the total impedance of the circuit,

$$Z_T = \sqrt{(R_S + R_L)^2 + X_S^2} \text{ ohms} \tag{78-1}$$

Load current,

$$I_L = \frac{E}{Z_T} = \frac{E}{\sqrt{(R_S + R_L)^2 + X_S^2}} \text{ amperes} \tag{78-2}$$

Load power,

$$P_L = I_L^2 \times R_L = \frac{E^2 R_L}{(R_S + R_L)^2 + X_S^2} \text{ volts} \tag{78-3}$$

For the load power to reach its maximum value,

$$\frac{dP_L}{dR_L} = E^2 \left[\frac{(R_S + R_L)^2 + X_S^2 - R_L[2(R_S + R_L)]}{[(R_S + R_L)^2 + X_S^2]^2} \right] = 0$$

Therefore

$$(R_S + R_L)^2 + X_S^2 = 2R_L R_S + 2R_L^2 \tag{78-4}$$
$$R_L = \sqrt{R_S^2 + X_S^2} \text{ ohms}$$

Therefore the load resistance is matched to the magnitude of the internal impedance.
From Equations 78-3, 78-4,
Value of the maximum power,

$$P_{Lmax} = \frac{E^2 R_L}{(R_S + R_L)^2 + X_S^2} \tag{78-5}$$

$$= \frac{E^2 R_L}{(R_S + R_L)^2 + (R_L^2 - R_S^2)}$$

$$= \frac{E^2 R_L}{2(R_L{}^2 + R_L R_S)}$$

$$= \frac{E^2}{2(R_L + R_S)} \text{ watts}$$

In Fig. 78-1B

Magnitude of the total circuit impedance,

$$Z_T = \sqrt{(R_L + R_S)^2 + (X_L + X_S)^2} \text{ ohms} \tag{78-6}$$

where X_L and X_S may be either inductive or capacitive. The load current is

$$I_L = \frac{E}{Z_T} = \frac{E}{\sqrt{(R_L + R_S)^2 + (X_L + X_S)^2}} \text{ amperes} \tag{78-7}$$

The power in the load is

$$P_L = I_L{}^2 R_L = \frac{E^2 R_L}{(R_L + R_S)^2 + (X_L + X_S)^2} \text{ watts} \tag{78-8}$$

If X_L is varied, P_L will reach its maximum value when $X_L + X_S = 0$ or $X_L = -X_S$. The load reactance X_L will then have the same magnitude as X_S, but the two reactances are required to cancel so that, if, for example, X_S is inductive, X_L is made capacitive and vice-versa.

After the reactance in the circuit has been reduced to zero,

$$P_L = \frac{E^2 R_L}{(R_L + R_S)^2} \text{ watts} \tag{78-9}$$

The load power, P_L will then have its maximum value when R_L is matched to R_S:

$$P_{L\text{max}} = \frac{E^2}{4R_S} = \frac{E^2}{4R_L} \text{ watts} \tag{78-10}$$

and the corresponding load current is:

$$I_L = \frac{E}{2R_S} \text{ amperes} \tag{78-11}$$

When $R_L = R_S$ and $X_L = -X_S$, the load and internal impedances are said to be conjugates (Chapter 89).

Example 78-1

In Fig. 78-1A, $R_S = 2$ kΩ, $X_S = 3$ kΩ and $E = 10$ V rms. Determine the load power when R_L is (a) 2 kΩ and (b) 3 kΩ. What is the value of R_S for which the load power is a maximum? Calculate the amount of the maximum load power.

Solution

(a) Total impedance,

$$Z_T = \sqrt{(R_S + R_L)^2 + X_S{}^2} = \sqrt{(2 + 2)^2 + 3^2}$$

$$= 5 \text{ k}\Omega.$$

Load current,

$$I_L = \frac{E}{Z_T} = \frac{10 \text{ V}}{5 \text{ k}\Omega}$$

$$= 2 \text{ mA.}$$

Load power,

$$P_L = I_L{}^2 \times R_L = (2 \text{ mA})^2 \times 2 \text{ k}\Omega$$

$$= 8 \text{ mW.}$$

(b) Total impedance,

$$Z_T = \sqrt{(R_L + R_S)^2 + X_S{}^2} = \sqrt{(2 + 3)^2 + 3^2}$$

$$= 5.83 \text{ k}\Omega.$$

Load current,

$$I_L = \frac{10 \text{ V}}{5.83 \text{ k}\Omega}$$

$$= 1.71 \text{ mA.}$$

Load power,

$$P_L = (I_L)^2 \times R_L = (1.71 \text{ mA})^2 \times 3 \text{ k}\Omega$$

$$= 8.8 \text{ mW.}$$

The load power is a maximum when:

$$R_L = \sqrt{R_S{}^2 + X_S{}^2} = \sqrt{2^2 + 3^2} \qquad (78\text{-}4)$$

$$= 3.61 \text{ k}\Omega.$$

Total load impedance,

$$Z_T = \sqrt{(R_L + R_S)^2 + X_S{}^2} = \sqrt{(2 + 3.61)^2 + 3^2}$$

$$= 6.36 \text{ k}\Omega.$$

Load current,

$$I_L = \frac{10 \text{ V}}{6.36 \text{ k}\Omega}$$

$$= 1.57 \text{ mA.}$$

Maximum load power

$$P_{Lmax} = (I_L^2) \times R_L$$

$$= (1.57 \text{ mA})^2 \times 3.61 \text{ k}\Omega$$

$$= 8.89 \text{ mW.}$$

Example 78-2

In Fig. 78-1B $E = 18$ V rms and $R_S = 4$ kΩ. The source reactance, X_S, is capacitive and equal to 5 kΩ. Determine the values of R_L and X_L which will allow maximum power transfer to the load and calculate the amount of the maximum load power.

Solution

For maximum power transfer to the load, the load impedance is the conjugate of the source impedance. Therefore $R_L = 4$ kΩ and X_L is an inductive reactance of 5 kΩ.

Maximum load power

$$P_{L\text{max}} = \frac{E^2}{4R_L} \text{ watts} \tag{78-10}$$

$$= \frac{(18 \text{ V})^2}{4 \times 4 \text{ k}\Omega}$$

$$= 20.25 \text{ mW}.$$

Chapter 79
Resonance in a series LCR circuit

THE CONDITION OF ELECTRICAL RESONANCE IN *ALL* CIRCUITS IS DEFINED AS follows: *any two*-terminal (single source) network containing resistance and reactance is said to be in resonance when the source voltage and the current drawn from the source are in phase.

It follows from this definition that a resonant circuit has a phase angle of zero and a power factor of unity.

If the series LCR circuit of Fig. 79-1A is at resonance, the values of the inductive reactance and the capacitive reactance must be equal. Therefore the phasor sum of v_L and v_C is zero (Fig. 79-1B and C) so that $e = v_R$ and the circuit is purely resistive. The total impedance of the circuit is equal to the resistance and is at its minimum value (Fig. 79-1D); for this reason the series resonant LCR combination is sometimes referred to as an acceptor circuit. It follows that, at resonance, the circuit current is at its maximum value which is equal to E/R amperes.

Because the values of the inductive reactance and the capacitive reactance are both dependent on the frequency, there must be a particular resonant frequency for which the two reactances are equal. The manner in which the behavior of the series LCR circuit varies with frequency, is illustrated by means of response curves (Fig. 79-2A and B). These are the graphs of certain variables (such as impedance, current, voltages across inductor, capacitor, etc.) versus frequency. Such response curves are important because they show the circuit's ability to distinguish between one frequency and another.

Tuning a series LCR circuit means adjusting the value of the inductor or the capacitor until the resonant frequency is the same as the desired signal frequency. Let us take the amplitude modulation (AM) broadcast band as an example. Each station is assigned a particular operating frequency while the frequencies of the nearest stations either side are 10 kHz away. Induced in the antenna (Fig. 79-3A and B) of an AM receiver are literally hundreds of signals from all the radio waves in the vicinity. The purpose of the tuned circuit is to provide maximum response at the frequency of the wanted signal but much smaller responses at the other unwanted signals. This is achieved by adjusting the capacitor until the resonant frequency is equal to the assigned frequency of the desired station.

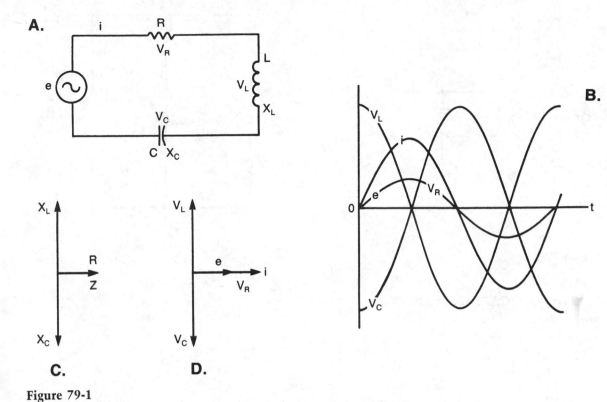

Figure 79-1

Note that the circuit of Fig. 79-3B is a series arrangement because the wanted signal induced in the antenna drives a current through the coil $L1$. The alternating magnetic flux surrounding $L1$ cuts the other coil so that an rf (radio frequency) voltage is induced in $L2$. This voltage is within the loop containing the coil $L2$ and the capacitor C so that these components and their source voltage are in series.

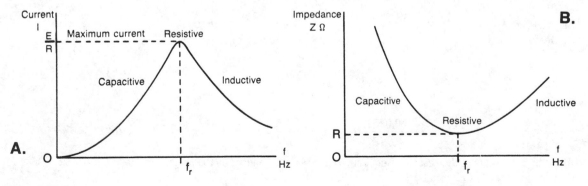

Figure 79-2

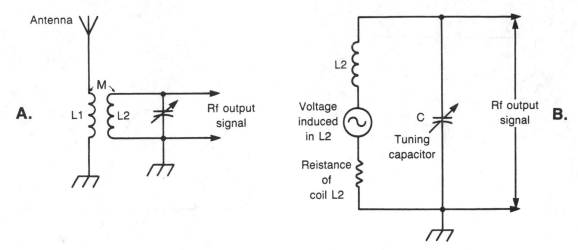

Figure 79-3

MATHEMATICAL DERIVATIONS

At resonance,

Phase angle, ϕ, is zero.

Phasor $e = V_R$.

Phasor sum of v_L and v_C is zero.

Inductive reactance, X_L = capacitive reactance, X_C.

Inductor voltage, V_L = capacitor voltage, V_C.

Impedance, Z, is equal in value to the resistance, R, and is at its *minimum* level.

Current, I, is equal to E/R and is at its *maximum* level.

Because

$$X_L = X_C,$$

$$2\pi f_r L = \frac{1}{2\pi f_r C}$$

This yields

Resonant frequency,

$$f_r = \frac{1}{2\pi\sqrt{LC}} = \frac{0.159}{\sqrt{LC}} \text{ Hz} \qquad (79\text{-}1)$$

Then,

$$L = \frac{0.0253}{f_r^2 C} \text{ H and } C = \frac{0.0253}{f_r^2 L} \text{ F} \qquad (79\text{-}2)$$

Notice that the value of the resonant frequency is inversely dependent on the product of L and C but is independent of R.

At frequencies below the value of the resonant frequency, X_C is greater than X_L, i leads e

and the circuit behaves capacitively. At frequencies above the resonant frequency, X_L is greater than X_C, i lags e and the circuit is overall inductive. Only at resonance does the circuit behave resistively.

$$\text{True Power} = I^2R = E \times I = E^2/R \text{ watts} \qquad (79\text{-}3)$$

Because the phase angle is zero and the power factor is unity, the values of the true power and the apparent power are equal.

Example 79-1

In Fig. 79-1, $R = 10\ \Omega$, $L = 150\ \mu\text{H}$, $C = 250\ \text{pF}$, and $E = 2\ \text{V}$. Determine the value of the resonant frequency and calculate the resonant values of Z, I, V_R, V_L, and V_C.

Solution

Resonant frequency,

$$f_r = \frac{1}{2\pi\sqrt{LC}} \text{ Hz} \qquad (79\text{-}1)$$

$$= \frac{1}{2 \times \pi \times \sqrt{150 \times 10^{-6} \times 250 \times 10^{-12}}} \text{ Hz}$$

$$= 822 \text{ kHz}.$$

At the resonant frequency,

$$X_L = X_C = 2 \times \pi \times 822 \times 10^3 \times 150 \times 10^{-6}$$

$$= 775\ \Omega.$$

When the circuit is at resonance,

$$\text{Impedance, } Z = R = 10\ \Omega.$$

$$\text{Current, } I = \frac{2\text{ V}}{10\ \Omega} = 200 \text{ mA}.$$

$$\text{Resistor voltage, } V_R = E$$

$$= 2 \text{ V}.$$

$$\text{Inductor voltage, } V_L = I \times X_L = 200 \text{ mA} \times 775\ \Omega$$

$$= 155 \text{ V}.$$

$$\text{Capacitor voltage, } V_C = V_L$$

$$= 155 \text{ V}.$$

$$\text{True Power} = I^2R = (200 \text{ mA})^2 \times 10\ \Omega$$

$$= 0.4 \text{ W}.$$

$$\text{Phase angle is zero, power factor is unity.}$$

Note that the equal values of V_L and V_C are many times greater than the source voltage. This is possible because these voltages are 180° out of phase and their phasor sum is zero. In addition both the inductor and the capacitor are reactive components that only store power and do not dissipate power.

Chapter 80
Q—selectivity—bandwidth

WHEN A SERIES LCR CIRCUIT IS AT RESONANCE, THE CURRENT HAS ITS maximum possible value for a given source voltage. Across the inductor and the capacitor are then developed equal but 180° out-of-phase voltages. These voltages may each be many times greater than the source voltage (refer to Example 79-1). The number of times greater is called the voltage magnification factor, which is referred to as Q. The Q factor is just a number and has no units. However, it is a measure of the inductor's merit in the sense that a "good" coil will have a high value of inductive reactance compared with its resistance.

The main importance of Q is its indication of a series tuned circuit's selectivity; this is defined as its ability to distinguish between the signal frequency to which it is resonant and other unwanted signals on nearby frequencies. It therefore follows that the higher the selectivity, the greater is the freedom from adjacent channel interference. The degree of the selectivity is related to the sharpness of the current response curve (the sharper the curve, the greater the selectivity) and may be measured by the frequency separation between two specific points on the curve (Fig. 80-1). The points arbitrarily chosen are those for which the true power in the circuit is one half of the maximum true power, which occurs when the circuit is resonant. These positions on the response curve are also referred to as the 3 decibel (dB) points because a loss of 3 dB is equivalent to a power ratio of one half. The frequency separation between these points is called the bandwidth (or bandpass) of the tuned circuit.

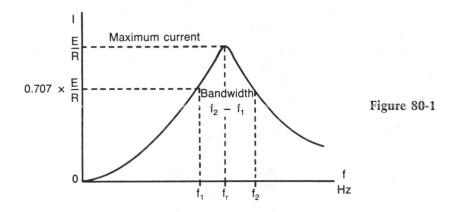

Figure 80-1

At the 3 dB points, the rms circuit current will be $1/\sqrt{2}$ or 0.707 times the maximum rms value of the current at resonance (do not confuse this result with the relationship between the rms and the peak values of a sinewave alternating current). In addition, the overall reactance at the 3 dB points is equal to the circuit's resistance so that the phase angle is 45° and the power factor is cos 45° = 0.707.

The narrower the bandwidth and the higher the resonant frequency, the sharper is the response curve and the greater is the selectivity. The mathematical derivations will show that the Q value is a direct measure of the selectivity.

MATHEMATICAL DERIVATIONS
Voltage magnification factor

At resonance,

$$Q = \frac{V_L}{E} = \frac{V_C}{E} \qquad (80\text{-}1)$$

$$Q = \frac{I \times X_L}{I \times R} = \frac{X_L}{R} = \frac{2\pi f_r L}{R} = \frac{\text{inductor's reactance}}{\text{inductor's resistance}} \qquad (80\text{-}2)$$

Because the resonant frequency,

$$f_r = \frac{1}{2\pi\sqrt{LC}} \text{ Hz} \qquad (80\text{-}3)$$

$$Q = \frac{2\pi L}{2\pi\sqrt{LC} \times R} = \frac{1}{R} \times \sqrt{\frac{L}{C}}$$

Because the resonant frequency depends on the product of L and C, it follows that, for the given resonant frequency, there are an infinite number of possible pairs of values for L and C. However, because the Q value is determined by the ratio of L and C, there are, for particular values of f_r and Q, only a limited range of values for L and C (assuming that the value of R does not alter appreciably). For audio frequency (af) circuits involving a few kHz, Q is of the order of 10 to 20 but with radio frequency (rf) circuits, Q may exceed 100.

Inductor's power factor and figure of merit

$$\text{Inductor's power factor} = \frac{R}{\sqrt{R^2 + X_L{}^2}} \qquad (80\text{-}4)$$

$$= \frac{R}{X_L \times \sqrt{\left(1 + \frac{R^2}{X_L{}^2}\right)}}$$

$$= \frac{1}{Q \times \sqrt{1 + \frac{1}{Q^2}}}$$

The inductor's power factor equals $1/Q$ to within 1%, provided Q is greater than 10. Therefore the values of Q and the power factor for radio frequency inductors are reciprocals. The higher the Q, the lower is the power factor and a low power factor is the feature of a "good" coil. Therefore Q directly measures the inductor's merit.

Bandwidth

At any point on the current response curve,

$$\text{True power} = EI \cos \phi \qquad (80\text{-}5)$$

$$= E \times \frac{E}{Z} \times \frac{R}{Z} = \frac{E^2 R}{Z^2} \text{ watts}$$

At the peak of the response curve,

$$\text{Maximum power} = \frac{E^2}{R} \text{ watts}$$

At the half-power points,

$$\frac{E^2R}{Z^2} = \frac{1}{2} \times \frac{E^2}{R} \qquad (80\text{-}6)$$

$$2R^2 = Z^2 = R^2 + (X_L - X_C)^2$$

$$R^2 = (X_L - X_C)^2$$

$$X_L - X_C = \pm R \text{ ohms}$$

At the frequency f_1 (Fig. 80-1). X_C is greater than X_L. Therefore,

$$X_C - X_L = R$$

$$\frac{1}{2\pi f_1 C} - 2\pi f_1 L = R$$

$$4\pi^2 f_1^2 LC + 2\pi f_1 CR - 1 = 0.$$

This yields

$$f_1 = -\frac{R}{4\pi L} + \frac{1}{2\pi} \sqrt{\frac{1}{LC} + \frac{R^2}{4L^2}} \text{ Hz} \qquad (80\text{-}7)$$

At the frequency f_2, X_L is greater than X_C. Therefore,

$$X_C - X_L = -R$$

This yields

$$f_2 = +\frac{R}{4\pi L} + \frac{1}{2\pi} \sqrt{\frac{1}{LC} + \frac{R^2}{4L^2}} \qquad (80\text{-}8)$$

$$\text{Bandwidth, } f_2 - f_1 = \frac{R}{4\pi L} - \left(\frac{-R}{4\pi L}\right) = \frac{R}{2\pi L} \text{ Hz} \qquad (80\text{-}9)$$

Therefore,

$$\frac{\text{Bandwidth}}{\text{Resonant Frequency}} = \frac{R}{2\pi f_r L} = \frac{R}{X_L} = \frac{1}{Q} \qquad (80\text{-}10)$$

$$\text{Bandwidth} = \frac{\text{resonant frequency}}{Q} \qquad (80\text{-}11)$$

$$Q = \frac{\text{resonant frequency}}{\text{bandwidth}} \qquad (80\text{-}12)$$

Q is a direct measure of the degree of selectivity. The higher the value of Q, the sharper is the current response curve, the greater is the degree of selectivity and the narrower is the bandwidth.

Sharpness of the impedance response curve

Impedance

$$Z = \sqrt{R^2 + (X_L - X_C)^2} \text{ ohms}$$

Except near the trough of the response curve, the value of R can normally be neglected in comparison with the overall reactance.

Therefore,

$$Z \approx X_L - X_C = 2\pi fL - \frac{1}{2\pi fC} \qquad (80\text{-}13)$$

$$\frac{dZ}{df} = 2\pi L + \frac{1}{2\pi Cf^2}$$

$$= \sqrt{\frac{L}{C}} \left(\frac{1}{f_r} + \frac{f_r}{f^2} \right)$$

The higher the value of R the flatter is the impedance response curve and the lower is the selectivity. Therefore the slope of the impedance response is inversely proportional to R but directly proportional to $\sqrt{L/C}$. It follows that the sharpness of the impedance response is directly proportional to Q. These results are illustrated in Fig. 80-2A and B.

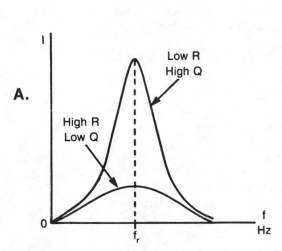
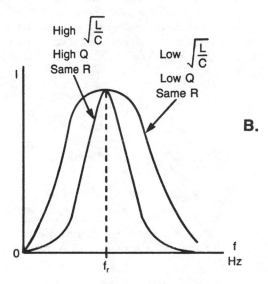

Figure 80-2

Energy relationships

We will now show that the value of Q is equal to 2π times the ratio of the maximum energy stored during the cycle to the mean energy dissipated over the period.

The maximum energy stored during the cycle is in the form of the magnetic field surrounding the inductor; it is also in the form of the electric field associated with the capacitor.

$$\text{Maximum energy stored in the inductor} = \frac{1}{2} L(I_{\text{peak}})^2 \qquad (80\text{-}14)$$

$$= \frac{1}{2} L(\sqrt{2} \times I_{\text{rms}})^2$$

$$= LI_{\text{rms}}^2 \text{ joules}$$

where I_{peak} and I_{rms} are respectively the peak and the effective values of the current at resonance.

The mean energy dissipated over the period, T seconds, is equal to the product of the power and the time. Therefore,

$$\text{Mean energy dissipated} = I_{\text{rms}}^2 \times R \times T \tag{80-15}$$

$$= \frac{I_{\text{rms}}^2 \times R}{f_r} \text{ joules}$$

Then,

$$2\pi \times \frac{\text{maximum energy stored during the cycle}}{\text{mean energy dissipated over the period}} = 2\pi \frac{LI_{\text{rms}}^2}{I_{\text{rms}}^2 \times R/f_r} \tag{80-16}$$

$$= \frac{2\pi f_r L}{R} = Q$$

Note that if two identical series LCR circuits are joined end-to-end with zero mutual coupling between the coils, the total impedance of the series combination at resonance is doubled ($Z = 2R$), but the Q and the resonant frequency are unchanged. However, if a single additional damping resistor, R_d, is connected to the series LCR circuit, the new total impedance at resonance is greater and equal to $R + R_d$. Therefore the new Q is reduced and is given by:

$$Q_{new} = \frac{1}{R + R_d} \times \sqrt{\frac{L}{C}} = \frac{R}{R + R_d} \times Q_{old} \tag{80-17}$$

$$\text{New bandwidth} = \frac{R + R_d}{R} \times \text{old bandwidth} \tag{80-18}$$

The new bandwidth is greater but the resonant frequency remains the same.

Example 80-1

In Fig. 80-1A, $R = 8\ \Omega$, $L = 150\ \mu\text{H}$, $C = 250\ \text{pF}$, and $E = 2\ \text{V}$. What are the values of Q and the circuit's bandwidth? If an additional $10\ \Omega$ resistor is inserted in series, calculate the new values of the resonant frequency, Q and the bandwidth.

Solution

$$Q = \frac{1}{R} \times \sqrt{\frac{L}{C}} \tag{80-3}$$

$$= \frac{1}{8} \times \sqrt{\frac{150 \times 10^{-6}}{250 \times 10^{-12}}}$$

$$= 97$$

Bandwidth,

$$BW = \frac{R}{2\pi L} \tag{80-9}$$

$$= \frac{8}{2 \times \pi \times 150 \times 10^{-6}} \text{ Hz}$$

$$= 8.5 \text{ kHz}$$

Resonant frequency,

$$f_r = \frac{1}{2\pi\sqrt{LC}}$$

$$= \frac{1}{2 \times \pi \times \sqrt{150 \times 10^{-6} \times 250 \times 10^{-12}}} \text{ Hz}$$

$$= 822 \text{ kHz}$$

Check

$$Q = \frac{\text{resonant frequency}}{\text{bandwidth}} = \frac{822 \text{ kHz}}{8.5 \text{ kHz}}$$

$$= 97$$

When an additional 10 Ω resistor is connected in series, the resonant frequency is unchanged at 822 kHz.

The new Q is given by

$$Q = \frac{R}{R + R_d} \times \text{old } Q \qquad\qquad (80\text{-}17)$$

$$= \frac{8 \text{ }\Omega}{8 \text{ }\Omega + 10 \text{ }\Omega} \times 97$$

$$= 43$$

$$\text{New Bandwidth} = \frac{\text{resonant frequency}}{\text{new } Q} \qquad\qquad (80\text{-}11)$$

$$= \frac{822 \text{ kHz}}{43}$$

$$= 19.1 \text{ kHz}.$$

Chapter 81
Parallel resonant LCR circuit

FROM THE DEFINITION OF RESONANCE, THE SUPPLY OR SOURCE CURRENT, i_S, must be in phase with the source voltage, e (Fig. 81-1A). This will occur if the values of the inductor and capacitor currents are equal; only then will the phasor sum, i_x, of their currents be zero (remember that i_L and i_C are 180° out of phase). Then the resistor current (which is independent of the frequency) will be the same as the supply current whose value will be at its minimum level of E/R amperes. It follows that the total impedance, Z_T, of the complete circuit will be entirely resistive and have its maximum value of R ohms.

At resonance the inductive and capacitive reactances are equal and the parallel LC combination behaves theoretically as an open circuit; the total impedance of the circuit is then equal to the resistance. At frequencies other than the resonant frequency, the LC combination behaves as a

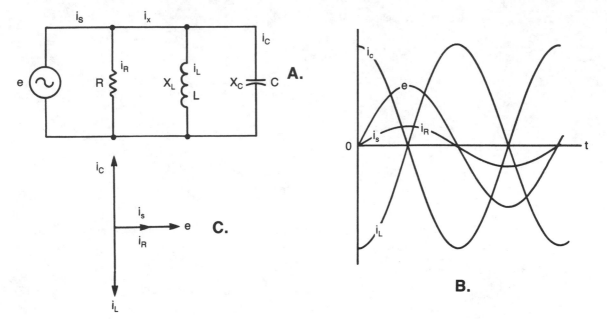

Figure 81-1

certain value of reactance, which when placed in parallel with the resistance, produces a total impedance that is less than the value of the resistance. Combining these two statements, it means that the total impedance can never exceed the value of the resistance.

At frequencies below the resonant frequency, the capacitive reactance is greater than the inductive reactance, the capacitor current is less than the inductor current and the circuit behaves inductively. At frequencies above the resonant frequency, the inductive reactance is greater than the capacitive reactance, the inductor current is less than the capacitor current and the circuit behaves capacitively. These results are the reverse of those for the series LCR circuit, which behaved capacitively for frequencies below f_r and inductively for frequencies above f_r. The current and the impedance response curves are illustrated in Fig. 81-2A and B.

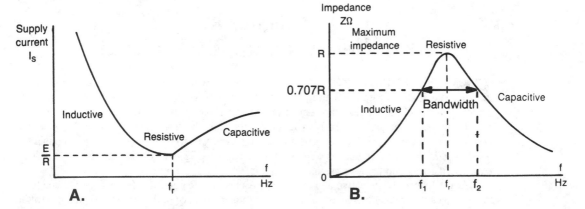

Figure 81-2

In the resonant phasor diagram of Fig. 80-1C, the currents I_L and I_C are equal in magnitude and each may be many times greater than the supply current. The number of times greater is equal to the circuit's Q factor which is a direct measure of the selectivity (Chapter 80).

MATHEMATICAL DERIVATIONS

At resonance,

Phase angle of the entire circuit, $\phi = 0°$.

Inductor current, $I_L (= E/X_L)$ is equal in value to the capacitor current, $I_C (= E/X_C)$ amperes.

Supply current, I_S, is equal to the resistor current, $I_R = E/R$ amperes, which is its minimum value.

The impedance, Z_T, of the circuit is at its maximum level and equal to the value of the resistance, R ohms.

Inductance reactance, X_L = capacitance reactance, X_C.

Therefore,

$$2\pi f_r L = \frac{1}{2\pi f_r C}$$

This yields,

Resonant frequency,

$$f_r = \frac{1}{2\pi\sqrt{LC}} \text{ Hz} \qquad (81\text{-}1)$$

Then,

$$L = \frac{1}{4\pi^2 f_r^2 C} = \frac{0.0253}{f_r^2 C} \text{ H} \qquad (81\text{-}2)$$

and

$$C = \frac{1}{4\pi^2 f_r^2 L} = \frac{0.0253}{f_r^2 C} \text{ F} \qquad (81\text{-}3)$$

These last three equations are the same as the corresponding expressions for series resonance.

Current magnification

At resonance, the currents I_L and I_C are equal in magnitude and each is Q times the supply current, I_S.

Then

$$I_L = \frac{E}{X_L}, \ I_C = \frac{E}{X_C}, \ I_S = I_R = \frac{E}{R} \text{ amperes.}$$

and

$$Q = \frac{I_L}{I_S} = \frac{E/X_L}{E/R} = \frac{R}{X_L} = \frac{R}{X_C} \left(not \ \frac{X_L}{R} \text{ or } \frac{X_C}{R} \right) \qquad (81\text{-}4)$$

Because

$$Q = \frac{R}{X_L} = \frac{R}{2\pi f_r L} \qquad (81\text{-}5)$$

and

$$f_r = \frac{1}{2\pi\sqrt{LC}} \text{ Hz,}$$

$$Q = \frac{R \times 2\pi\sqrt{LC}}{2\pi L} = R \times \sqrt{\frac{C}{L}}$$

Impedance magnification

From Equation 81-4,

$$R = QX_L = QX_C \qquad (81\text{-}6)$$

The impedance at resonance is Q times the reactance of either the inductor or the capacitor.

Selectivity

As in the case of the series LCR circuit, Q determines the sharpness of the response curve and is therefore the direct measure of the selectivity.

$$Q = \frac{\text{resonant frequency}}{\text{bandwidth}} = \frac{f_r}{f_2 - f_1} \qquad (81\text{-}7)$$

The bandwidth is defined from the impedance response curve as the frequency separation between the points where the total impedance of the circuit is 0.707 times the maximum impedance, R. From this definition,

$$\text{Bandwidth} = f_2 - f_1 = \frac{1}{2\pi RC} \text{ Hz} \qquad (81\text{-}8)$$

where C and R are respectively measured in farads and ohms. The concept of the half-power points cannot be used in the case of the parallel LCR circuit because the true power is always E^2/R watts and is independent of the frequency.

Energy relationships

As in the case of the series LCR circuit, Q is equal to

$$2\pi \times \frac{\text{maximum energy stored during the cycle}}{\text{mean energy dissipated during the period, } T.}$$

The maximum energy stored in the capacitor is

$$\frac{1}{2}C(E_{\text{peak}})^2 = \frac{1}{2}C(\sqrt{2}E_{\text{rms}})^2 = CE_{\text{rms}}^2 \text{ joules} \qquad (81\text{-}9)$$

The mean energy dissipated during the period is

$$E_{\text{rms}}^2 \times T/R \text{ joules}$$

Then,

$$Q = 2\pi \times \frac{CE_{\text{rms}}^2}{E_{\text{rms}}^2 \times T/R} = 2\pi f_r C \times R = R/X_C \tag{81-10}$$

Note that if two identical parallel resonant LCR circuits are shunted across each other with zero mutual coupling between the coils, the equivalent resistance is halved. Therefore the impedance at resonance is halved but the Q, the resonant frequency, and the bandwidth remain the same.

If a single damping resistor, R_d, is connected across the parallel LCR circuit, the new impedance at resonance is $R_d \| R = R \times R_d/(R + R_d)$ ohms and is reduced in value.

Then the new value of Q is given by:

$$Q_{\text{new}} = \frac{R \times R_d}{R + R_d} \times \sqrt{\frac{C}{L}} = \frac{R_d}{R + R_d} \times Q_{\text{old}} \tag{81-11}$$

The value of Q has therefore been decreased.

$$\text{New bandwidth} = \frac{R + R_d}{R_d} \times \text{old bandwidth} \tag{81-12}$$

The bandwidth is consequently increased although the resonant frequency remains the same.

Example 81-1

In Fig. 81-1A, $R = 8.2$ kΩ, $L = 3.3$ μH, $C = 6.5$ pF, and $E = 15$ mV. Calculate the resonant frequency and determine the resonant values of I_S, I_R, I_L, I_C and the true power. What are the values of the Q factor and the circuit's bandwidth? If an additional 10 kΩ resistor is added in parallel, what are the new values of the resonant frequency, Q, and bandwidth?

Solution

Resonant frequency,

$$f_r = \frac{1}{2\pi\sqrt{LC}} \tag{81-1}$$

$$= \frac{1}{2 \times \pi \times \sqrt{3.3 \times 10^{-6} \times 6.5 \times 10^{-12}}} \text{ Hz}$$

$$= 34.4 \text{ MHz.}$$

The inductive reactance at resonance,

$$X_L = 2 \times \pi \times f \times L$$

$$= 2 \times \pi \times 34.4 \times 10^6 \times 3.3 \times 10^{-6}$$

$$= 712.5 \ \Omega$$

The capacitive reactance at resonance,

$$X_C = 712.5 \ \Omega.$$

Supply current,

$$I_S = I_R = \frac{E}{R} = \frac{15 \text{ mV}}{8.2 \text{ k}\Omega}$$

$$= 1.83 \ \mu\text{A.}$$

Inductor current,

$$I_L = \frac{E}{X_L} = \frac{15 \text{ mV}}{712.5 \text{ } \Omega}$$
$$= 21.05 \text{ } \mu\text{A}.$$

Capacitor current,

$$I_C = 21.05 \text{ } \mu\text{A}.$$

$$\text{True power} = E^2/R = (15 \text{ mV})^2/8.2 \text{ k}\Omega$$
$$= 27.4 \text{ nW}.$$

$$Q = R \times \sqrt{\frac{C}{L}} \tag{81-5}$$

$$= 8.2 \times 10^3 \times \sqrt{\frac{6.5 \times 10^{-12}}{3.3 \times 10^{-6}}}$$

$$= 11.5$$

Check:

$$Q = \frac{I_C}{I_S} = \frac{21.05 \text{ } \mu\text{A}}{1.83 \text{ } \mu\text{A}}$$
$$= 11.5$$

$$\text{Bandwidth} = \frac{1}{2\pi RC} \tag{81-8}$$

$$= \frac{1}{2 \times \pi \times 8.2 \times 10^3 \times 6.5 \times 10^{-12}} \text{ Hz}$$
$$= 2.99 \text{ MHz}.$$

Check:

$$\text{Bandwidth} = \frac{f_r}{Q} = \frac{34.4 \text{ MHz}}{11.5} \tag{81-7}$$
$$= 2.99 \text{ MHz}.$$

When the additional 10 kΩ resistor is added in parallel, the new total equivalent resistance at resonance is 10 kΩ‖8.2 kΩ = 10 × 8.2/(10 + 8.2) = 4.5 kΩ. The resonant frequency is unchanged but the new value of Q is 11.5 × 4.5 kΩ/8.2 kΩ = 6.3 and the new bandwidth is 2.99 MHz × 8.2 kΩ/4.5 kΩ = 5.45 MHz.

Chapter 82
The parallel resonant "tank" circuit

THE PARALLEL RESONANT TANK CIRCUIT CONSISTS OF A PRACTICAL INDUCTOR with its series resistance, in parallel with a capacitor whose losses are assumed to be negligible (Fig. 82-1A). Such a circuit is commonly used as the collector load of certain transistor rf (radio frequency) amplifiers; in such stages a high value of load impedance is required at the frequency to which the circuit is tuned.

Because the inductor branch contains both inductive reactance and resistance, the current, i_L, lags the source voltage, e, by the phase angle, ϕ. If the frequency is raised, the impedance of the inductor branch increases and the angle, ϕ, moves closer to 90°.

The capacitor current, i_C, leads the source voltage, e, by 90° while the supply current, i_S, is the phasor sum of i_C and i_L. Figure 82-1B shows the changes that occur in the circuit's behavior as the frequency is varied. At all frequencies below the resonant frequency, the supply current lags the source voltage by the *total* phase angle, θ, and the circuit behaves inductively. When the frequency exceeds its resonant value, the supply current leads the source voltage and the circuit is capacitive. These results are similar to those obtained for the parallel resonant LCR circuit of Chapter 81 but are opposite to those for the series LCR circuit of Chapter 79.

At resonance, the supply voltage and the supply current are in phase so that the angle, θ, is zero. As suggested by the phasor diagrams of Fig. 82-1B, the supply current at resonance is at its minimum level and consequently the total impedance at resonance is at its maximum value. This last statement is not mathematically exact but may be regarded as true provided the value of the Q factor is sufficiently high.

The supply current and impedance response curves are illustrated in Fig. 82-2A and B.

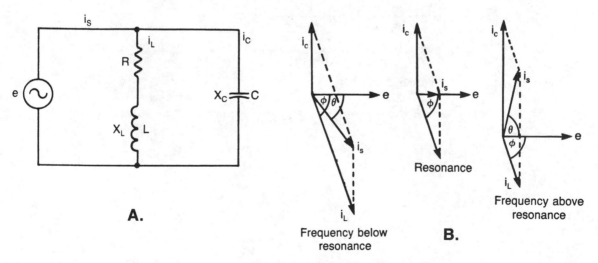

A.

Frequency below resonance

Resonance

B.

Frequency above resonance

Figure 82-1

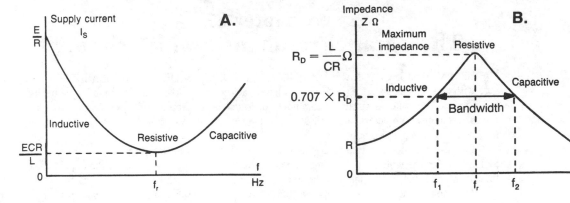

Figure 82-2

MATHEMATICAL DERIVATIONS

In the resonant phasor diagram of Fig. 81-1B,

$$I_C = I_L \sin \phi \text{ amperes} \tag{82-1}$$

Capacitor current,

$$I_C = \frac{E}{X_C} = 2\pi f_r CE \text{ amperes}$$

Inductor current,

$$I_L = \frac{E}{\sqrt{R^2 + X_L^2}} = \frac{E}{\sqrt{R^2 + 4\pi^2 f_r^2 L^2}} \text{ amperes}$$

$$\text{Sin } \phi = \frac{X_L}{\sqrt{R^2 + X_L^2}} = \frac{2\pi f_r L}{\sqrt{R^2 + 4\pi^2 f_r^2 L^2}}$$

Substituting these results in Equation 82-1,

$$2\pi f_r CE = \frac{E}{\sqrt{R^2 + 4\pi^2 f_r^2 L^2}} \times \frac{2\pi f_r L}{\sqrt{R^2 + 4\pi^2 f_r^2 L^2}} \tag{82-2}$$

$$R^2 + 4\pi^2 f_r^2 L^2 = \frac{L}{C}$$

This yields
Resonant frequency,

$$f_r = \frac{1}{2\pi} \sqrt{\frac{1}{LC} - \frac{R^2}{L^2}} \tag{82-3}$$

$$= \frac{1}{2\pi\sqrt{LC}} \sqrt{\frac{1 - R^2 C}{L}} \tag{82-4}$$

$$= \frac{1}{2\pi\sqrt{LC}} \sqrt{1 - \frac{1}{Q^2}} \text{ Hz}$$

where Q is the merit factor of the coil and is equal to X_L/R or $(\sqrt{L/C})/R$.

Notice that the formula for the resonant frequency of a tank circuit differs slightly from the comparable expression for the series and parallel LCR circuits. However, provided the value of Q is sufficiently high, the formula for the tank circuit reduces to $f_r = 1/(2\pi\sqrt{LC})$.

The total impedance at resonance is given by

Total impedance,

$$Z_T = \frac{E}{I_S} = \frac{E}{I_L \cos \phi}$$

Because

$$I_L = \frac{E}{\sqrt{R^2 + X_L^2}}$$

and

$$\cos \phi = \frac{R}{\sqrt{R^2 + X_L^2}},$$

$$Z_T = \frac{E \times (R^2 + X_L^2)}{E \times R} = \frac{R^2 + X_L^2}{R}$$

But from Equation 82-2,

$$R^2 + X_L^2 = \frac{L}{C}$$

Therefore,

$$Z_T = R_D = \frac{L}{CR} \text{ ohms} \tag{82-5}$$

The quantity L/CR Ω is called the *dynamic* resistance, R_D, because it only appears under operating conditions. The value of the dynamic resistance is virtually equal to the maximum level of the impedance at resonance.

The true power in the tank circuit may either be expressed as $I_S^2 \times L/CR$ watts or $I_L^2 \times R$ watts. Therefore

$$\frac{I_L^2}{I_S^2} = \frac{L/CR}{R} = \frac{L}{R^2C} = Q^2$$

or

$$I_L (\approx I_C) = Q \times I_S \text{ amperes} \tag{82-6}$$

The capacitor current is nearly the same as the inductor current. Consequently, at resonance there is a large circulating or "flywheel" current associated with the inductor and the capacitor, and this current is Q times the supply or line current. It follows that the value of the dynamic resistance is approximately equal to Q times the inductive reactance or Q times the capacitive reactance. Moreover because

$$Z_T \approx Q \times X_L$$

and

$$X_L = Q \times R,$$
$$Z_T \approx Q^2 \times R \text{ ohms}.$$

Selectivity

The Q factor is a measure of the tank circuit's selectivity and equals resonant frequency/bandwidth. From the impedance response curve the bandwidth is defined as the frequency separation between those points where the circuit's total impedance is equal to 0.707 times the dynamic resistance (the total impedance at resonance).

Energy relationship

$$Q \text{ factor} = 2\pi \times \frac{\text{maximum energy stored during the cycle}}{\text{mean energy dissipated over the period}}$$

Note that if two identical tank circuits are paralleled with zero mutual coupling between the coils, the total dynamic resistance at resonance is halved, but the Q and the resonant frequency remain unchanged.

If a single damping resistor, R_d, is connected across the tank circuit, the new total impedance at resonance is

$$\text{New total impedance, } Z_{\text{new}} = \frac{R_d \times L/CR}{R_d + L/CR} \text{ ohms} \tag{82-7}$$

and

$$Q_{\text{new}} = \frac{R_d}{R_d + L/CR} \times Q_{\text{old}} \tag{82-8}$$

Therefore, the values of Q and the total impedance are both reduced. The bandwidth is greater but the resonant frequency remains the same (assuming that the value of Q is greater than 10).

Example 82-1

In Fig. 82-1, $R = 12 \ \Omega$, $L = 25 \ \mu\text{H}$, $C = 15 \ \text{pF}$, and $E = 80 \ \text{mV}$. Find the value of the resonant frequency and calculate the resonant values of I_S, I_L, I_C. What are the values of Q and the tank circuit's bandwidth?

Solution

The exact formula for the resonant frequency is:

$$f_r = \frac{1}{2\pi} \times \sqrt{\frac{1}{LC} - \frac{R^2}{L^2}} \text{ Hz} \tag{82-3}$$

However,

$$\frac{1}{LC} = \frac{1}{25 \times 10^{-6} \times 15 \times 10^{-12}}$$
$$= 2.667 \times 10^{15}$$

and

$$\frac{R^2}{L^2} = \frac{144}{(25 \times 10^{-6})^2}$$
$$= 2.304 \times 10^{11}$$

Clearly R^2/L^2 is negligible when compared with $1/LC$

The formula therefore reduces to
Resonant frequency,

$$f_r = \frac{1}{2\pi\sqrt{LC}}$$

$$= \frac{1}{2 \times \pi \times \sqrt{25 \times 10^{-6} \times 15 \times 10^{-12}}} \text{ Hz}$$

$$= 8.22 \text{ MHz.}$$

At the resonant frequency,

$$X_C = \frac{1}{2\pi f_r C}$$

$$= \frac{1}{2 \times \pi \times 8.22 \times 10^6 \times 15 \times 10^{-12}}$$

$$= 1291 \ \Omega.$$

The impedance of the inductor branch is

$$Z_L = \sqrt{R^2 + X_L{}^2} = \sqrt{18^2 + 1291^2}$$

$$\approx 1291 \ \Omega.$$

Circulating current,

$$I_L = I_C = \frac{80 \text{ mV}}{1291 \ \Omega}$$

$$= 62.00 \ \mu\text{A.}$$

Dynamic resistance

$$R_D = \frac{L}{CR} \ \Omega \qquad\qquad (82\text{-}5)$$

$$= \frac{25 \times 10^{-6}}{15 \times 10^{-12} \times 12} \ \Omega$$

$$= 138.9 \text{ k}\Omega.$$

Supply, line or "make-up" current,

$$I_S = \frac{E}{R_D} = \frac{80 \text{ mV}}{138.9 \text{ k}\Omega}$$

$$= 0.576 \ \mu\text{A.}$$

$$Q = \frac{X_L}{R} = \frac{1291 \ \Omega}{12 \ \Omega} = 107.6$$

Check:

$$Q = \frac{I_C}{I_S} = \frac{62.00 \ \mu\text{A}}{0.576 \ \mu\text{A}} \qquad\qquad (82\text{-}6)$$

$$= 107.6$$

Bandwidth,

$$BW = \frac{f_r}{Q} = \frac{8.22 \text{ MHz}}{107.6}$$

$$= 76.4 \text{ kHz.}$$

Chapter 83
Free oscillation in an LC circuit

THE RESONANT CIRCUITS THAT WE DISCUSSED IN PREVIOUS CHAPTERS WERE connected to an ac source that generated the resonant frequency; the resultant oscillation was therefore "forced." By contrast it is possible to create a "free" or "natural" oscillation by removing the source and connecting a coil across a charged capacitor (Figs. 83-1A, B, C, and D).

Figure 83-1A shows the tank circuit immediately after the inductor has been connected. The capacitor then produces a current (electron flow) in the direction shown. The inductor provides a counter EMF that balances the voltage across the capacitor. As the capacitor discharges through the inductor, the inductor is storing energy in the form of a magnetic field. At the time when the capacitor has completed its discharge, there is zero voltage across the inductor and all of the circuit's energy is stored in the inductor's magnetic field.

Figure 83-1B shows the second quarter cycle of the circuit's operation. The magnetic field of the inductor is collapsing in order to maintain the inductor current. The collapsing magnetic field produces a current in the same direction as that previously established. This current now charges the capacitor with the polarity opposite to that which it originally had. When the magnetic field of the inductor is exhausted, the capacitor is fully charged and all of the circuit's energy is stored in the form of an electrostatic field across the capacitor.

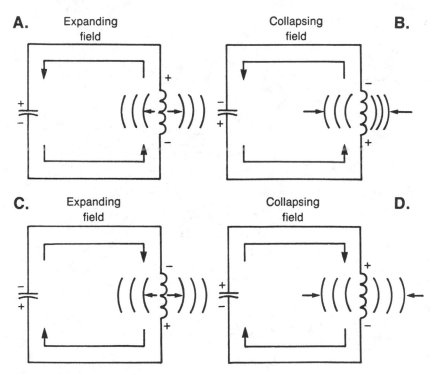

Figure 83-1

286

Figure 83-1C shows the third quarter cycle of operation. The capacitor is now discharging and once again the inductor is building up its magnetic field. The direction of the electron flow is in the direction shown.

In Fig. 83-1D, the capacitor is seen to be fully discharged so that the voltage across the capacitor is again zero. The collapsing magnetic field of the inductor will continue the circuit current and recharge the capacitor back to the original condition shown in Fig. 83-1A. Further cycles of operation will be identical to the one just described.

In this electrical circuit there exists some resistance that is mainly associated with the coil. This resistance will constitute a power loss that will eventually kill the oscillations in the tank circuit. The output of a tank circuit in which the amplitude of oscillations decreases to zero is called a *damped wave*. Damping can be overcome only by resupplying energy to the tuned circuit at a rate comparable to that at which it is being used. This is usually accomplished through the use of a dc source and a transistor, which acts as an amplifying device.

The inductance and capacitance values in the LC circuit determine the natural frequency of oscillation:

$$f_0 = \frac{1}{2\pi} \sqrt{\frac{1}{LC} - \frac{R^2}{4L^2}} \text{ Hz}$$

where: L = coil's inductance (H)

C = capacitance (F)

R = coil's resistance (Ω).

Note that this frequency differs slightly from the "forced" resonant frequencies of the series LCR circuit, the parallel LCR circuit and the tank circuit. However, for high Q circuits the differences are negligible.

MATHEMATICAL DERIVATIONS

In Fig. 83-2, using Kirchhoff's voltage law

$$V_L + V_R + V_C = 0.$$

$$L \frac{di}{dt} + iR + \frac{Q}{C} = 0.$$

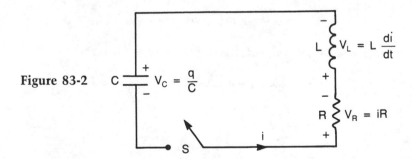

Figure 83-2

Differentiating with respect to t

$$L \frac{d^2i}{dt^2} + R \frac{di}{dt} + \frac{1}{C} \frac{dq}{dt} = 0. \tag{83-1}$$

$$\frac{d^2i}{dt^2} + \frac{R}{L} \times \frac{di}{dt} + \frac{i}{LC} = 0$$

The solution to this differential equation is

$$i = Ae^{m_1 t} + Be^{m_2 t}$$

where m_1 and m_2 are the roots of a quadratic equation.

The roots are:

$$m = \pm \sqrt{\frac{R^2}{4L^2} - \frac{1}{LC}} - \frac{R}{2L}$$

In the practical case, $1/LC \gg R^2/4L^2$

Therefore

$$i = e^{-\alpha t} (F \cos \omega_0 t + G \sin \omega_0 t)$$

where: $\alpha = \dfrac{R}{2L}$

$$\omega_0 = \sqrt{\frac{1}{LC} - \frac{R^2}{4L^2}}$$

When

$$t = 0, \; i = 0, \; \frac{di}{dt} = \frac{E}{L} = \text{amperes per second}$$

Therefore,

$$F = 0, \; G = \frac{E}{\omega_0 L} \text{ amperes.}$$

This yields
"Free" oscillation current,

$$i = \frac{E}{2\pi f_0 L} e^{-Rt/L} \sin 2\pi f_0 t \text{ amperes} \tag{83-2}$$

This is the equation of the damped oscillation (Fig. 84-3) where:

$$\text{Initial peak current value} = \frac{E}{2\pi f_0 L} \text{ amperes} \tag{83-3}$$

Decay factor,

$$\alpha = \frac{R}{L} \text{ sec}^{-1} \tag{83-4}$$

Natural frequency of oscillation,

$$f_0 = \frac{1}{2\pi} \sqrt{\frac{1}{LC} - \frac{R^2}{4L^2}} \text{ Hz}$$

The damped oscillation principle is used in the "ringing" coil which is capable of generating a short duration pulse.

Example 83-1

A 150 μH coil has a resistance of 8 Ω and is connected across a charged 500 pF capacitor. Calculate the values of the decay factor and the natural frequency of oscillation. See Fig. 83-3.

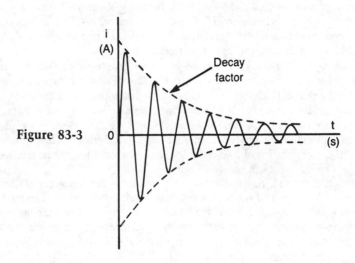

Figure 83-3

Solution
Natural frequency of oscillation,

$$f_0 = \frac{1}{2\pi} \sqrt{\frac{1}{LC} - \frac{R^2}{4L^2}} \qquad (83\text{-}3)$$

The value of $\dfrac{1}{LC}$ is $\dfrac{1}{150 \times 10^{-6} \times 500 \times 10^{-12}} = 1.33 \times 10^{13}$

The value of $\dfrac{R^2}{4L^2}$ is $\dfrac{8^2}{4 \times (50 \times 10^{-6})^2} = 7.11 \times 10^8$

Therefore $1/LC \gg R^2/4L^2$ and the expression for the natural frequency reduces to:

$$f_0 = \frac{\sqrt{1.33 \times 10^{13}}}{2 \times \pi} \text{ Hz} = 580 \text{ kHz}.$$

Decay factor,

$$\alpha = \frac{R}{L} = \frac{8}{150 \times 10^{-6}} \qquad (83\text{-}4)$$

$$= 5.33 \times 10^4 \text{ s}^{-1}$$

Chapter 84
Mutually coupled coils

$\mathbf{I}$N PREVIOUS CHAPTERS WE HAVE DISCUSSED THE INDUCTORS THAT HAVE A magnetic flux linked with their *own* turns and therefore possess the property of *self*-induct-ance. We are now going to consider the situation where the magnetic flux associated with one coil links with the turns of another coil. Such inductors are said to be magnetically, inductively, mutually, or transformer coupled and possess the property of mutual inductance, whose letter symbol is M and whose unit is the henry (H). In most of the circuits of the previous chapters there were a pair of input terminals (2-port network). We now have a pair of input and another pair of output terminals; this is referred to as a 4-port network.

The alternating, i_1, in the coil #1 (Fig. 84-1) creates an alternating flux, only part of which links with coil #1 and the remainder links with coil #2. This causes an induced voltage, v_2, whose size depends on the rate of change of the current, i_1, and on the mutual inductance, M, between the coils. The mutual inductance is 1 H if, when the current, i_1, is instantaneously changing at the rate of one ampere per second, the induced voltage, v_2, is 1 volt. The factors which determine the mutual inductance include the number of turns, N_1 and N_2, the cross-sectional area of the coils, their separation, the orientation of their axes and the nature of their cores. Consequently, the induced voltage $v_2 = M \times$ rate of change of i_1.

If the alternating current, i_1, has a sinusoidal waveform with a frequency of f hertz, $V_{2\text{rms}} = 2\pi f M I_{\text{rms}}$ (compare $V_{L\text{rms}} = 2\pi f L I_{\text{rms}}$ for the property of self-inductance). When the two coils are wound in the same sense, v_2 lags i_1 by 90° but if the two coils are wound in the opposite sense, v_2 leads i_1 by 90°. Note that the property of the mutual inductance is reversible in the sense that if the same rate of change in the current, i_1, is flowing in coil #2, then the voltage induced in the coil #1 is $v_1 = M \times$ rate of change in the current, i_1.

The mathematical derivations will show that, in the extreme case where the coils are tightly wound, one on top of the other, with a common soft iron core, the leakage flux is extremely small and can therefore be neglected. Assuming perfect flux linkage between the coils (corresponding to zero flux leakage), the mutual inductance is given by

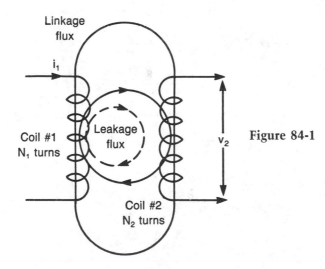

Figure 84-1

Mutual inductance,

$$M = \frac{\mu_0 \mu_r N_1 N_2 A}{l} \text{ H}$$

where: A = cross-sectional area for each coil (m^2).

   μ_r = relative permeability of the soft iron.

   l = the coil's length (m).

   μ_0 = permeability of free space ($4\pi \times 10^{-7}$ H/m).

The self-inductance of coil #1 is:

$$\text{Self-inductance, } L_1 = \frac{\mu_0 \mu_r N_1^2 A}{l} \text{ H}$$

and the self-inductance of coil #2 is:

$$\text{Self-inductance, } L_2 = \frac{\mu_0 \mu_r N_2^2 A}{l} \text{ H}$$

This yields

$$M^2 = L_1 L_2$$

and

$$\text{Mutual inductance, } M = \sqrt{L_1 L_2} \text{ H}.$$

If the leakage flux is not negligible, then only a fraction, k, of the total flux links with the two coils. This fraction k, whose value cannot exceed unity, is called the coefficient of coupling or coupling factor. Its value can be close to unity if a common soft iron core is used for the two coils, but may be very small (less than 0.01) with a nonmagnetic core and the two coils widely separated (loose coupling).

MATHEMATICAL DERIVATIONS

In Fig. 84-1,

   Induced voltage,

$$v_2 = -M \frac{\Delta i_1}{\Delta t} \text{ volts} \tag{84-1}$$

where: M = mutual inductance (H)

   $\dfrac{\Delta i_1}{\Delta t}$ = rate of change in the current, i_1 (A/s)

If the two coils are *tightly* wound,

$$v_2 = -N_2 \frac{\Delta \phi}{\Delta t} \tag{84-2}$$

where: ϕ = linkage flux (Wb).

The minus sign is the direct result of Lenz's law.

But

$$\phi = BA \tag{84-3}$$
$$= \mu_0\mu_r HA$$
$$= \frac{\mu_0\mu_r i_1 N_1 A}{l}$$

where: B = flux density in the common iron core (T).

N_1 = number of turns in coil #1.

A = cross-sectional area of each coil (m$_2$).

l = length of each coil (m).

μ_r = relative permeability of iron core.

μ_0 = permeability of free space ($4\pi \times 10^{-7}$ H/m).

Combining Equations 84-2 and 84-3,

$$v_2 = -\frac{\mu_0\mu_r N_1 N_2 A}{l} \times \frac{\Delta i_1}{\Delta t} \tag{84-4}$$

Combining equations 84-1 and 84-4,
Mutual inductance,

$$M = \frac{\mu_0\mu_r N_1 N_2 A}{l} H \tag{84-5}$$

Because

$$L_1 = \frac{\mu_0\mu_r N_1{}^2 A}{L} H \text{ and } L_2 = \frac{\mu_0\mu_r N_2{}^2 A}{l} H,$$

it follows that:

$$M^2 = L_1 L_2 \text{ and } M = \sqrt{L_1 L_2} \text{ H} \tag{84-6}$$

The result of Equation 84-6 would be equivalent to a coupling factor, k, of unity. However, in the general case,

Coupling factor,

$$k = \frac{M}{\sqrt{L_1 L_2}} \text{ or } M = k \times \sqrt{L_1 L_2} \text{ H} \tag{84-7}$$

Notice that k is just a number and has no units.

Equation 84-7 yields

$$L_1 = \frac{M^2}{k^2 L_2} \text{ H and } L_2 = \frac{M^2}{k^2 L_1} \text{ H} \tag{84-8}$$

and

$$M = k L_1 \times \frac{N_2}{N_1} = k L_2 \times \frac{N_1}{N_2} H \tag{84-9}$$

If $L1$ and $L2$ are each equal to L,

$$k = \frac{M}{L} \text{ and } M = kL \tag{84-10}$$

Note if a steady direct current is flowing through coil #1, the linkage flux will be constant in magnitude and direction so that the voltage induced in coil #2 will be zero.

Example 84-1

Two inductors whose coupling factor is unity, have self-inductances of 1.7 H and 2.5 H. What is the value of their mutual inductance? If the rate of change in the current flowing through the 1.7 H inductor is 8 A/s, what is the voltage induced in the 2.5 H inductor?

Solution

Mutual inductance,

$$M = \sqrt{L_1 \times L_2} \text{ H} \tag{84-6}$$
$$= 1.7 \times 2.5$$
$$= 2.06 \text{ H.}$$

Voltage induced in the 2.5 H inductor,

$$v_2 = (-)M\frac{\Delta i_1}{\Delta t} \tag{84-1}$$
$$= 2.06 \text{ H} \times 8 \text{ A/s}$$
$$= 16.5 \text{ V.}$$

Example 84-2

Two coils with self-inductances of 85 μH and 115 μH have a mutual inductance of 14 μH. What is the value of their coupling factor?

Solution

Coupling factor,

$$k = \frac{M}{\sqrt{L_1 L_2}} \tag{84-7}$$
$$= \frac{14}{\sqrt{84 \times 115}}$$
$$= 0.14.$$

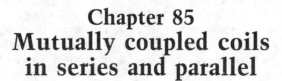

Chapter 85
Mutually coupled coils
in series and parallel

IF TWO MUTUALLY COUPLED COILS HAVE A COMMON AXIS AND ARE CONNECTED in series, their individual fluxes may aid or oppose, depending on the sense in which the coils

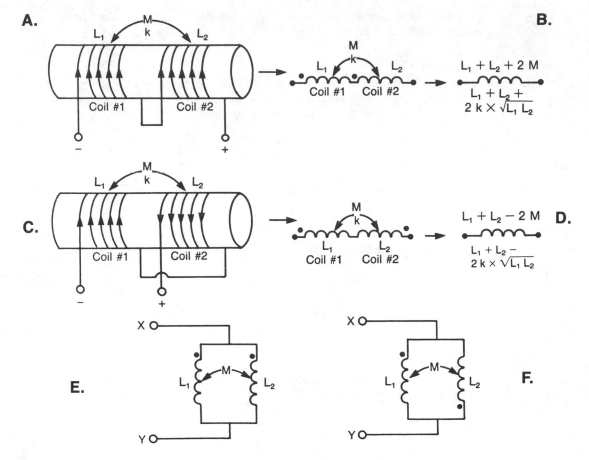

Figure 85-1

are wound. In Fig. 85-1A the directions of the arrows are indicative of the electron current flow and the individual fluxes surrounding $L1$ and $L2$ will be aiding.

MUTUALLY COUPLED COILS IN SERIES

In the series-aiding connection the rate of change in the current, $\Delta I/\Delta T$, is the same throughout the circuit (Fig. 85-1B). The dots at the ends of the coil symbols are called polarity marks; when the current enters the dotted end of $L1$ and leaves, it must enter $L2$ at its dotted end in order for the two fluxes to add. The voltage induced in the coil #1 will be a combination of $L_1(\Delta I/\Delta T)$ volts due to self-inductance and $M(\Delta I/\Delta T)$ volts due to the flux which surrounds coil #2 and links partially with coil #1. In a similar way the total voltage induced in coil #2 will be the addition of $L_2(\Delta I/\Delta T)$ volts due to self-inductance and $M(\Delta I/\Delta T)$ volts as the result of mutual inductance. The effect of this mutual coupling is to increase the total equivalent inductance.

In Fig. 85-1C, the fluxes surrounding, $L1$, $L2$ will subtract; consequently, the coils are connected in series-opposing as indicated by the positions of the polarity marks. The connections to $L2$ are reversed so that the current enters its undotted end (Fig. 85-1D). The mutually induced voltages then oppose the voltages created by self-induction and the total equivalent inductance is reduced.

MUTUALLY COUPLED COILS IN PARALLEL

Figure 85-1E shows two mutually coupled coils in parallel. The connection is parallel-aiding as shown by the positions of the polarity dots. Compared with zero mutual coupling, the effect of the parallel-aiding connection is to increase the total equivalent inductance.

In Fig. 85-1F the connection is parallel-opposing and the result is to reduce the total equivalent inductance.

MATHEMATICAL DERIVATIONS
Series-aiding connection

The total voltage across the series-aiding circuit of Fig. 85-1B is

$$L_1 \frac{\Delta I}{\Delta T} + M \frac{\Delta I}{\Delta T} + L_2 \frac{\Delta I}{\Delta T} + M \frac{\Delta I}{\Delta T} = \left(L_1 + L_2 + 2M \right) \frac{\Delta I}{\Delta T} \text{ volts} \qquad (85\text{-}1)$$

Therefore the total equivalent inductance is

$$L_T{}^+ = L_1 + L_2 + 2M \text{ henrys} \qquad (85\text{-}2)$$

$$L_T{}^+ = L_1 + L_2 + 2k\sqrt{L_1 L_2} \text{ henrys} \qquad (85\text{-}3)$$

The positive sign associated with L_T means "aiding."

Series-opposing connection

If the fluxes are opposing, the sign of M is reversed. The total voltage across the series-opposing circuit of Fig. 85-1C is

$$L_1 \frac{\Delta I}{\Delta T} - M \frac{\Delta I}{\Delta T} + L_2 \frac{\Delta I}{\Delta T} - M \frac{\Delta I}{\Delta T} = \left(L_1 + L_2 - 2M \right) \frac{\Delta I}{\Delta T} \text{ volts} \qquad (85\text{-}4)$$

Therefore the total equivalent inductance is

$$L_T{}^- = L_1 + L_2 - 2M \text{ henrys} \qquad (85\text{-}5)$$

$$L_T{}^- = L_1 + L_2 - 2k\sqrt{L_1 L_2} \text{ henrys} \qquad (85\text{-}6)$$

The negative sign associated with L_T means "opposing."
From equations 85-2, 85-5,

$$L_1 + L_2 = \frac{L_T{}^+ + L_T{}^-}{2} \text{ henrys} \qquad (85\text{-}7)$$

and

$$M = \frac{L_T{}^+ - L_T{}^-}{4} \text{ henrys} \qquad (85\text{-}8)$$

If $L1$ and $L2$ are each equal to L, Equations 85-2 and 85-5 become

$$L_T{}^+ = 2(L + M) = 2L(1 + k) \text{ henrys} \qquad (85\text{-}9)$$

and

$$L_T{}^- = 2(L - M) = 2L(1 - k) \text{ henrys} \qquad (85\text{-}10)$$

Then

$$L = \frac{L_T{}^+ + L_T{}^-}{4} \text{ henrys} \qquad (85\text{-}11)$$

$$M = \frac{L_T{}^+ - L_T{}^-}{4} \text{ henrys} \qquad (85\text{-}12)$$

and

$$k = \frac{L_T{}^+ - L_T{}^-}{L_T{}^+ + L_T{}^-} \qquad (85\text{-}13)$$

Equations 85-12 and 85-13 are used in an experimental determination of the M and k values.

Parallel-aiding and parallel-opposing connections

In Fig. 85-1E, the total equivalent inductance $L_T{}^+$ between the points X and Y is given by:

$$L_T{}^+ = \frac{L_1 L_2 - M^2}{L_1 + L_2 - 2M} \text{ henrys} \qquad (85\text{-}14)$$

In Fig. 85-1F, the total equivalent inductance $L_T{}^-$ between the points X and Y is

$$L_T{}^- = \frac{L_1 L_2 - M^2}{L_1 + L_2 + 2M} \text{ henrys} \qquad (85\text{-}15)$$

Note: The above expressions for $L_T{}^+$ and $L_T{}^-$ are not the same as those derived from

$$\frac{1}{L_T \pm} = \frac{1}{L_1 \pm M} + \frac{1}{L_2 \pm M}$$

which is erroneously quoted in some textbooks.

If there is zero mutual coupling ($M = 0$) between the coils, Equations 85-14 and 85-15 both reduce to

$$L_T = \frac{L_1 L_2}{L_1 + L_2} \text{ henrys}$$

which is the familiar "product-over-sum" formula for two self-inductances in parallel.

If L_1 and L_2 are equal to L, Equations 85-14 and 85-15 become

$$L_T{}^+ = \frac{L^2 - M^2}{2(L - M)} = \frac{L + M}{2} \text{ henrys} \qquad (85\text{-}16)$$

and

$$L_T{}^- = \frac{L - M}{2} \text{ henrys} \qquad (85\text{-}17)$$

Example 85-1

Two mutually coupled coils are joined in a series-aiding arrangement. If the self-inductances are 0.65 H and 0.45 H, and the coupling factor is 0.85, what is the value of the total equivalent inductance? If one of the coils is reversed without changing the magnitude of the coupling factor, what is the new value of the total equivalent inductance?

296

Solution

Mutual inductance,

$$M = k \times \sqrt{L_1 L_2}$$
$$= 0.85 \times \sqrt{0.65 \times 0.45}$$
$$= 0.46 \text{ H.}$$

Total equivalent inductance,

$$L_T{}^+ = L_1 + L_2 + 2M \qquad (85\text{-}2)$$
$$= 0.65 + 0.45 + 2 \times 0.46$$
$$= 2.02 \text{ H.}$$

If one of the coils is reversed, the new connection is series-opposing,
Total equivalent inductance,

$$L_T{}^- = L_1 + L_2 - 2M \qquad (85\text{-}5)$$
$$= 0.65 + 0.45 - 2 \times 0.46$$
$$= 0.18 \text{ H.}$$

Example 85-2

Two coils whose self-inductances are 75 mH and 55 mH are connected in parallel-aiding. If the coupling factor is 0.35, what is the total equivalent inductance of the parallel combination? If one of the coils is now reversed without changing the value of the coupling factor, what is the new value of the total equivalent inductance?

Solution

Mutual inductance,

$$M = k \times \sqrt{L_1 L_2}$$
$$= 0.35 \times \sqrt{75 \times 55}$$
$$= 22.5 \text{ mH.}$$

Total equivalent inductance,

$$L_T{}^+ = \frac{L_1 L_2 - M^2}{L_1 + L_2 - 2M} \qquad (85\text{-}14)$$
$$= \frac{75 \times 55 - 22.5^2}{75 + 55 - 2 \times 22.5}$$
$$= 42.6 \text{ mH.}$$

Total equivalent inductance,

$$L_T{}^- = \frac{L_1 L_2 - M^2}{L_1 + L_2 + 2M} \qquad (85\text{-}15)$$
$$= \frac{75 \times 55 - 22.5^2}{75 + 55 + 2 \times 22.5}$$
$$= 20.7 \text{ mH.}$$

Chapter 86
The power transformer—
transformer efficiency

THE PURPOSE OF A POWER TRANSFORMER IS TO INCREASE OR DECREASE THE value of the ac line or supply voltage without altering the frequency. This operation is achieved with a high level of efficiency.

A power transformer consists of a primary coil and a secondary coil whose number of turns are respectively N_p and N_s (Fig. 86-1A and B). The alternating source voltage, E_P, is applied across the primary coil while the load is connected across the secondary coil. These two coils may each consist of thousands of turns, which are wound on a common soft-iron core so that the leakage flux is reduced to a low value.

In the case of the ideal transformer, the leakage flux is zero so the mutual inductance, $M = \sqrt{L_p L_s}$, and the coupling factor, k, is unity. When an alternating current, I_p, flows in the primary coil, it creates a magnetic flux which links with the secondary coil and induces the secondary voltage, E_s. With zero flux leakage, there are the *same* volts per turn associated with both the primary and secondary coils. It follows that if the number of secondary turns exceeds the number of primary turns, the secondary voltage is greater than the primary voltage and we have a so-called "step-up" transformer. Similarly, if the number of secondary turns is less than the number of primary turns, the secondary voltage is smaller than the primary voltage and the transformer is of the "step-down" type. The terms "step-up" and "step-down" normally refer to

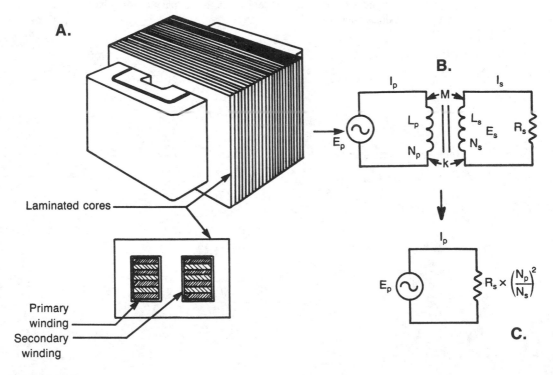

Figure 86-1

the voltage and not to the current. From the phase point of view, the primary and secondary voltages are 180° out of phase provided the two coils are wound in the same sense.

An ideal power transformer has zero power losses and therefore the power input to the primary circuit equals the power output from the secondary circuit. It follows that a step-up of the voltage level from the primary circuit to the secondary circuit must be accompanied by a corresponding reduction in the current levels.

PRACTICAL POWER TRANSFORMER

The practical power transformer suffers from the following losses:

1. The copper loss which is the power dissipated in the resistances of the primary and secondary windings.
2. The iron loss, which is dissipated in the core. This lost energy is subdivided into:
   (a) the eddy-current loss. This is caused by the moving magnetic flux, which cuts the core and induces circulating currents within the metal. The loss is reduced by laminating the core into thin slices with each slice insulated from its neighbor.
   (b) the hysteresis loss, which is caused by the rapid magnetizing, demagnetizing, and remagnetizing of the core during the cycle of the primary current.

For practical power transformers, the total losses are less than 10% of the primary power.

When a load resistance is connected across the secondary coil, the secondary current creates an alternating magnetic flux which, by Lenz's Law, opposes and partially cancels the primary flux. As a result the primary current increases and an effective resistance of value E_p/I_p ohms is presented to the primary source; this ohmic value is called the resistance *reflected* from the secondary circuit into the primary circuit owing to the presence of the secondary load.

Note that if a steady direct current flows in the primary coil, the leakage flux will be constant in magnitude and direction so that the voltage induced in the secondary coil is zero. However, if the steady dc voltage applied to the primary coil is "chopped" to produce a square wave, the transformer responds to this type of input and some form of alternating voltage (not however, a simple sine wave) will be induced in the secondary coil; the secondary voltage is then rectified to provide a final output dc voltage which is larger than the dc input voltage.

MATHEMATICAL DERIVATIONS
Ideal transformer

$$\text{Turns ratio, } \frac{N_p}{N_s} = \frac{E_p}{E_s} \tag{86-1}$$

Because an ideal transformer has no losses,

$$\text{Primary power, } P_p = \text{secondary power, } P_s \tag{86-2}$$
$$E_p \times I_p = E_s \times I_s \text{ watts}$$

Combining Equations 86-1 and 86-2,

$$\frac{E_p}{E_s} = \frac{I_s}{I_p} = \frac{N_p}{N_s} \tag{86-3}$$

Primary voltage,

$$E_p = \frac{E_s I_s}{I_p} = \frac{E_s N_p}{N_s} \text{ volts} \tag{86-4}$$

Primary current,

$$I_p = \frac{E_s I_s}{E_p} = \frac{I_s N_s}{N_p} \text{ amperes} \qquad (86\text{-}5)$$

Secondary voltage,

$$E_s = \frac{E_p I_p}{I_s} = \frac{E_p N_s}{N_p} \text{ volts} \qquad (86\text{-}6)$$

Secondary current,

$$I_s = \frac{E_p I_p}{E_s} = \frac{I_p N_p}{N_s} \text{ amperes} \qquad (86\text{-}7)$$

Primary turns,

$$N_p = \frac{E_p N_s}{E_s} = \frac{I_s N_s}{I_p} \qquad (86\text{-}8)$$

Secondary turns,

$$N_s = \frac{E_s N_p}{E_p} = \frac{I_p N_p}{I_s} \qquad (86\text{-}9)$$

Practical transformer

Transformer efficiency,

$$\eta = \frac{\text{power output from the secondary circuit}}{\text{power input to the primary circuit}} \times 100\% \qquad (86\text{-}10)$$

$$= \frac{E_s I_s}{E_p I_p} \times 100\%$$

Secondary power output,

$$P_s = E_s I_s \qquad (86\text{-}11)$$

$$= P_p \times \frac{\eta}{100}$$

$$= E_p I_p \times \frac{\eta}{100} \text{ watts}$$

Then
Secondary voltage,

$$E_s = \frac{E_p I_p \eta}{I_s \times 100} \text{ volts} \qquad (86\text{-}12)$$

Secondary current,

$$I_s = \frac{E_p I_p \eta}{E_s \times 100} \text{ amperes} \qquad (86\text{-}13)$$

Primary voltage,

$$E_p = \frac{E_s I_s \times 100}{\eta I_p} \text{ volts} \tag{86-14}$$

Primary current,

$$I_p = \frac{E_s I_s \times 100}{\eta E_p} \text{ amperes} \tag{86-15}$$

The total power loss in the transformer is given by:

$$P_{\text{loss}} = \text{primary power} - \text{secondary power} \tag{86-16}$$
$$= E_p \times I_p - E_s \times I_s$$
$$= \left(\frac{100 - \eta}{100}\right) \times P_p$$
$$= \left(\frac{100 - \eta}{\eta}\right) \times P_s \text{ watts}$$

Reflected resistance

If the secondary is loaded with a resistance, R_s

$$R_s = \frac{E_s}{I_s} \text{ ohms} \tag{86-17}$$

Because

$$E_p = \frac{E_s N_p}{N_s} \text{ (Equation 86-4) and } I_p = \frac{I_s N_s}{N_p} \text{ (Equation 86-5)}$$

$$R_p = \frac{E_p}{I_p} \tag{86-18}$$
$$= \frac{E_s N_p / N_s}{I_s N_s / N_p}$$
$$= \frac{E_s}{I_s} \times \left(\frac{N_p}{N_s}\right)^2$$
$$= R_s \times \left(\frac{N_p}{N_s}\right)^2 \text{ ohms}$$

where R_p = effective resistance presented to the primary source.

The equation

$$\frac{E_p}{I_p} = R_s \times \left(\frac{N_p}{N_s}\right)^2$$

is represented by the equivalent circuit of Fig. 86-1C. The expression $R_s \times (N_p/N_s)^2$ is the effective resistive load presented to the primary source and is referred to as the value of the resistance reflected from the secondary circuit into the primary circuit owing to the introduction of the secondary load, R_s. If we choose the turns ratio so that the value of the reflected resistance, $R_s \times (N_p/N_s)^2$, is equal to the internal resistance of the primary source, the secondary load is then matched to the primary source for maximum power transfer to the secondary load.

If R_p is the resistance associated with the primary source, the matched condition is

$$R_p = R_s \times \left(\frac{N_p}{N_s}\right)^2 \text{ and } R_s = R_p \times \left(\frac{N_s}{N_p}\right)^2 \text{ ohms} \qquad (86\text{-}19)$$

or

$$\left(\frac{N_p}{N_s}\right)^2 = \frac{R_p}{R_s} \text{ and turns ratio, } \frac{N_p}{N_s} = \sqrt{\frac{R_p}{R_s}} \qquad (86\text{-}20)$$

Example 86-1

In Fig. 86-1B, $E_p = 110$ V, 60 Hz, $N_p = 1200$ turns, $N_s = 4800$ turns, $R_s = 220\ \Omega$. Assuming that the transformer is 100% efficient and has a coupling factor of unity, what are the values of E_s, I_s, I_p, primary power, secondary power and the reflected resistance?

Solution

Turns ratio, $\dfrac{N_p}{N_s} = \dfrac{1200}{4800} = 1{:}4$

The transformer is of the "step-up" variety.
Secondary voltage,

$$E_s = E_p \times \frac{N_s}{N_p} \qquad (86\text{-}6)$$
$$= 110 \text{ V} \times 4$$
$$= 440 \text{ V}.$$

Secondary current,

$$I_s = \frac{E_s}{R_s}$$
$$= \frac{440 \text{ V}}{220\ \Omega}$$
$$= 2 \text{ A}.$$

Primary current,

$$I_p = I_s \times \frac{N_s}{N_p} \qquad (86\text{-}5)$$
$$= 2 \text{ A} \times 4$$
$$= 8 \text{ A}.$$

Primary power,

$$P_p = \text{secondary power, } P_s \qquad (86\text{-}2)$$
$$= E_p \times I_p$$
$$= 110 \text{ V} \times 8 \text{ A}$$
$$= 880 \text{ W}.$$

Check:

Secondary power,

$$P_s = E_s \times I_s$$
$$= 440 \text{ V} \times 2 \text{ A}$$
$$= 880 \text{ W.}$$

Reflected resistance

$$\frac{E_p}{I_p} = \frac{110 \text{ V}}{8 \text{ A}} = 13.75 \ \Omega.$$

Check:

Reflected resistance

$$R_p = R_s \times \left(\frac{N_p}{N_s}\right)^2 \tag{86-18}$$
$$= 220 \ \Omega \times \left(\frac{1}{4}\right)^2$$
$$= 13.75 \ \Omega.$$

Example 86-2

In Fig. 86-1B, $E_p = 220$ V, 60 Hz, $I_p = 2.3$ A, $E_s = 55$ V, $R_s = 6.5$ Ω. What are the values of the primary power, the secondary power, and the transformer efficiency?

Solution

The transformer of the "step-down" variety with a turns ratio of 220 V : 55 V = 4 : 1.

Primary power,

$$P_p = E_p \times I_p \tag{86-2}$$
$$= 220 \text{ V} \times 2.3 \text{ A}$$
$$= 506 \text{ W.}$$

Secondary power,

$$P_s = \frac{E_s^{\ 2}}{R_s} = \frac{(55 \text{ V})^2}{6.5 \ \Omega}$$
$$= 465 \text{ W.}$$

Transformer efficiency,

$$\eta = \frac{P_s}{P_p} \times 100 \tag{86-10}$$
$$= \frac{465 \text{ W}}{506 \text{ W}} \times 100$$
$$= 92\%.$$

Chapter 87
Complex algebra—
operator j—
rectangular/polar conversions

IN PREVIOUS CHAPTERS AN ALTERNATING VOLTAGE OR CURRENT HAS BEEN represented either by a sine wave or a phasor or a trigonometrical expression. While these representations are adequate for simple series and parallel arrangements, they are too cumbersome for the analysis of more complicated circuits such as series-parallel combinations and those circuits that require the use of the network theorems. What is clearly needed is a form of algebra that can be applied directly to the solution of general ac circuits. Such an algebra must be capable of taking into account the circuit's phase relationships by distinguishing between the resistive and reactive elements. This is achieved in complex algebra by the introduction of the operator j. See Fig. 87-1.

The definition of the operator j is as follows: A phasor when multiplied by the operator j is rotated through 90° or $\pi/2$ radians in the positive or counterclockwise direction but the magnitude of the phasor is unchanged. In a similar way, a phasor when multiplied by the operator $-j$, is rotated through 90° in the clockwise direction.

Consider a simple case in which a resistance of 3 Ω is connected in series with an *inductive* reactance of 4 Ω. The corresponding phasor diagram is shown in Fig. 87-1A. Because the inductive reactance phasor is pointing vertically upwards, it can be said to lie along the "$+j$" axis (assuming the reference line to be horizontal). The phasor equation for the series combination is therefore:

$$\text{Total impedance phasor, } z = 3 + j\,4 \ \Omega$$

This expression for the impedance phasor is known as rectangular notation because the 3 Ω and the $+j4$ Ω phasors are perpendicular. Alternatively, we can say that the magnitude of the impedance phasor is $\sqrt{3^2 + 4^2} = 5$ Ω while its phase angle is inv cos (3/5) = +53.1°. These results may be combined by stating that the impedance phasor is $5 \angle +53.1°$ (a magnitude of 5 Ω with a phase angle of +53.1°): this expression for the impedance is referred to as polar notation.

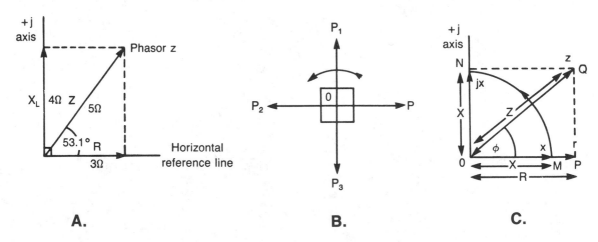

A.   **B.**   **C.**

Figure 87-1

If the resistance of 3 Ω were connected in series with a *capacitance* reactance of 4 Ω, the phasor of the total impedance would be $3 - j4$ (rectangular notation) or 5 Ω $\angle -53.1°$ (polar notation).

We have shown that an impedance phasor of $3 + j4$ Ω (rectangular notation) can be alternatively expressed as 5 Ω $\angle +53.1°$ (polar notation). We are now going to examine the means by which we can convert between rectangular and polar notations (R $\rightarrow$ P key and P $\rightarrow$ R key on a scientific calculator). But, first of all, why do we require both notations?

In the analysis of ac circuits we need to add, subtract, multiply and divide phasors. These operations are involved with the rules of complex algebra (Chapters 88 and 89) which require both rectangular and polar notations. For example if two impedances, z_1 and z_2, are in parallel, the total impedance phasor may be found from the product-over-sum formula, $z_T = (z_1 \times z_2)/(z_1 + z_2)$. Clearly this formula involves the addition of the phasors z_1, z_2, their multiplication and finally the division of the numerator phasor by the denominator phasor. If z_2 and z_T are given, the formula is then rearranged as $z_1 = (z_2 \times z_T)/(z_2 - z_T)$ so that subtraction of phasors is required in the calculation of z_1.

MATHEMATICAL DERIVATIONS

In Fig. 87-1B,

$$\text{Phasor } OP_1 = j \times \text{phasor } OP$$
$$\text{Phasor } OP_2 = j \times \text{phasor } OP_1$$
$$= j^2 \times \text{phasor } OP.$$

Because phasors OP and OP_2 are 180° apart,

$$\text{Phasor } OP_2 = -\text{phasor } OP.$$

Multiplication by -1 rotates a phasor through 180°; of course, the direction of the rotation is immaterial.

But phasor $OP_2 = j^2 \times$ phasor OP.

$$\text{Therefore } j^2 = -1 \tag{87-1}$$

It follows that $+j$ and $-j$ are reciprocals.

Because the square of a "real" positive or negative number is always positive, j cannot be evaluated in terms of "real" numbers and is therefore known as an "imaginary" quantity. The term "imaginary" is used as the opposite of "real" and is not thought of as something that does not exist. We should also point out that a complex number such as $3 + j4$ is composed of a real number and an imaginary number. The word *complex* indicates a combination of real and imaginary quantities and must not be confused with "complicated."

Figure 87-1C represents a phasor diagram in which phasor OP = phasor r, phasor OM = phasor x and phasor OQ = phasor z. Then phasor ON = phasor jx and since phasor OQ = phasor ON + phasor OP.

$$\text{Phasor } z = r + jx \tag{87-2}$$

This is known as the rectangular notation since the phasors r and jx are separated by 90°. Because the phasor z is specified in terms of two phasors whose order is important (phasor $2 + j3$ is *not* the same as phasor $3 + j2$), the representation of the phasor z is referred to as complex algebra.

The polar method of denoting a phasor is in terms of its magnitude and its direction. The magnitude of z is indicated by the length of the line OQ, and its direction is measured by the angle, ϕ, between OQ and the horizontal reference line. In polar notation,

$$z = Z \angle \phi \tag{87-3}$$

where Z is the magnitude of the phasor z and the value of ϕ lies between $0°$ and $\pm180°$.

When the phasor lies in the lower two quadrants, the angle, ϕ, is negative so that $z = Z \angle -\phi$ (sometimes shown as $z = Z \angle \phi$). The phasor $Z \angle \phi$ should not be confused with $-Z \angle \phi$ which equals $Z \angle \phi \pm \pi$.

If R and X are the magnitudes of the phasors r and x, the rectangular $\rightarrow$ polar conversion is achieved by the equations:

$$Z = +\sqrt{R^2 + X^2} \text{ ohms} \qquad (87\text{-}4)$$

and

$$\phi = \text{inv tan} \frac{X}{R} \qquad (87\text{-}5)$$

Note that Z is always considered to be positive.

In the polar $\rightarrow$ rectangular conversion,

$$R = Z \cos \phi \text{ ohms} \qquad (87\text{-}6)$$

and

$$X = Z \sin \phi \text{ ohms} \qquad (87\text{-}7)$$

Many electronic calculators have the capability of rapid conversions between rectangular and polar notations. You should familiarize yourself with the methods used in your own calculator.

Example 87-1

A series circuit contains a 6 kΩ resistor, an inductor whose reactance is 7 kΩ, and a capacitor with a reactance of 4 kΩ. Write down the expression for the total impedance phasor of the series combination.

Solution

Total impedance,

$$z = 6 + j7 - j4$$
$$= 6 + j3 \text{ k}\Omega \text{ (rectangular notation)}$$

The result that $j7 - j4 = j3$, agrees with the rules of complex algebra, as discussed in Chapter 88.

The total impedance has a magnitude of $\sqrt{6^2 + 3^2} = 6.71$ kΩ and a phase angle of inverse cos $(6/6.71) = +26.6°$. Therefore, in polar notation, the impedance phasor is 6.71 k$\Omega \angle +26.6°$.

Example 87-2

Convert (a) $z = 2 - j3$, $z = 5 + j2$, $z = 4 + j7$, $z = -3 - j$ into polar notation and (b) $z = 7 \angle 32°$, $z = 4 \angle 83°$, $z = 5 \angle -17°$, $z = 3 \angle -127°$ into rectangular notation.

Solution

(a) From Equations 87-4 and 87-5,

$$z = 2 - j3 = +\sqrt{2^2 + 3^2} \angle \text{inv tan} (-3/2)$$
$$= +3.61 \angle -56.3°$$
$$z = 5 + j2 = +\sqrt{5^2 + 2^2} \angle \text{inv tan} (2/5)$$

$$= +5.39 \angle 21.8°$$
$$z = 4 + j7 = +\sqrt{4^2 + 7^2} \angle \text{inv tan } (7/4)$$
$$= +8.06 \angle 60.3°$$
$$z = -3 - j = -3 - j1 = +\sqrt{3^2 + 1^2} \angle \text{inv tan } (-1/-3)$$
$$= +3.16 \angle -161.6°$$

Caution should be exercised in the determination of the angle. It is sometimes advisable to use a rough sketch in order to establish the quadrant in which the phasor lies. It should also be remembered that the angle derived mathematically from the calculator, is always measured with respect to the horizontal line and is an acute angle.

Note the following special cases of rectangular to polar conversions:

$$2 + j0 = 2 = 2 \angle 0°$$
$$0 + j5 = +j5 = 5 \angle 90°$$
$$0 - j3 = -j3 = 3 \angle -90°$$
$$1 + j = +\sqrt{1^2 + 1^2} \angle \text{inv tan } (1/1) = \sqrt{2} \angle 45° = 1.414 \angle 45°$$

(b) From Equations 87-6 and 87-7,

$$z = 7 \angle 32° = 7 \cos 32° + j7 \sin 32°$$
$$= 5.94 + j\, 3.71.$$
$$z = 4 \angle 83° = 4 \cos 83° + j4 \sin 83°$$
$$= 0.49 + j\, 3.97$$
$$z = 5 \angle -17° = 5 \cos (-17°) + j\, 5 \sin (-17°)$$
$$= 4.78 - j\, 1.46$$
$$z = 3 \angle -127° = 3 \cos (-127°) + j\, 3 \sin (-127°)$$
$$= -1.81 - j\, 2.40$$

Chapter 88
Equating real and imaginary parts—
addition and subtraction of phasors

IN ELECTRONICS ANALYSIS WE ARE OFTEN FACED WITH AN EQUATION IN WHICH both the left-hand side and the right-hand side of the equation contain "real" terms and "imaginary" terms. Because "real" numbers and "imaginary" numbers lie on axes which are mutually perpendicular, it follows that the "real" terms on the left-hand side must equal the "real" terms on the right-hand side. Similarly, the "imaginary" terms on the left-hand side can be equated with the "imaginary" terms on the right-hand side. The process therefore derives two equations from the original single equation and is known as "equating real and imaginary parts."

So far we have added ac quantities together by means of sine waves and phasor diagrams. When we turn to complex algebra as a means of adding or subtracting phasors, the rectangular notation is much to be preferred. As examples, $(4 - j2) + (3 + j5) = (7 + j3)$ and $(4 - j2) - (3 + j5) = 1 - j7$; however $3 \angle 20° + 2 \angle 35°$ is *not* equal to $5 \angle 55°$. Consequently, if you are asked to add together two phasors which are expressed in polar notation, the best procedure is to convert the two phasors into rectangular notation and then carry out the addition; the answer can afterwards be converted back into polar notation (if necessary).

MATHEMATICAL DERIVATIONS
Rules of complex algebra equating "real" and "imaginary" parts

Let phasor, $z_1 = R_1 + jX_1$ and phasor $z_2 = R_2 + jX_2$.

If phasor $z_1 =$ phasor z_2, then

$$R_1 = R_2 \text{ and } X_1 = X_2 \tag{88-1}$$

This is the process of equating the "real" and the "imaginary" parts of an equation.

Addition and subtraction of phasors

If $z_1 = R_1 + jX_1$ and $z_2 = R_2 + jX_2$, then the sum of z_1 and z_2 is

$$\text{Phasor } z_1 + z_2 = (R_1 + jX_1) + (R_2 + jX_2) \tag{88-2}$$
$$= (R_1 + R_2) + j(X_1 + X_2)$$

When subtracting phasor z_2 from phasor z_1, the result is

$$\text{Phasor } z_1 - z_2 = (R_1 + jX_1) - (R_2 + jX_2) \tag{88-3}$$
$$= (R_1 - R_2) + j(X_1 - X_2)$$

Example 88-1

If phasor $z_1 = 4 - j2$ and phasor $z_2 = 3 + j5$, find the values of the phasors $z_1 + z_2$ and $z_1 - z_2$. Illustrate the results on a phasor diagram.

Solution

$$\text{Phasor } z_1 = 4 - j2 = \sqrt{4^2 + 2^2} \text{ inv tan } (-2/4)$$
$$= 4.47 \angle -26.6°$$
$$\text{Phasor } z_2 = 3 + j5 = \sqrt{3^2 + 5^2} \text{ inv tan } (5/3)$$
$$= 5.83 \angle 59.0°$$
$$\text{Phasor } z_1 + z_2 = (4 - j2) + (3 + j5) \tag{88-2}$$
$$= 7 + j3 = \sqrt{7^2 + 3^2} \angle \text{inv tan } (3/7) \tag{88-3}$$
$$= 7.62 \angle 23.2°$$
$$\text{Phasor } z_1 - z_2 = (4 - j2) - (3 + j5)$$
$$= 1 - j7 = \sqrt{1^2 + 7^2} \angle \text{inv tan } (-7/1)$$
$$= 7.07 \angle -81.9°$$

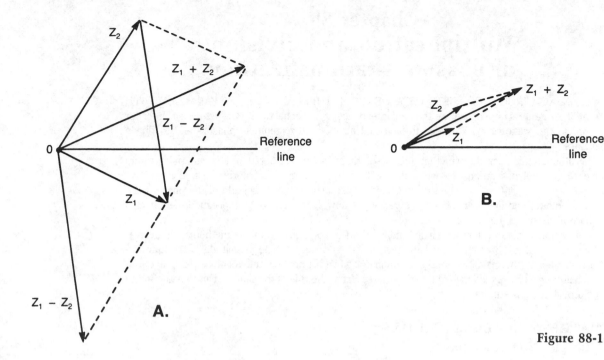

Figure 88-1

These results are illustrated to scale in Fig. 88-1A. This phasor diagram confirms the parallelogram method of obtaining the sum and difference phasors.

Example 88-2
If phasor $z_1 = 3 \angle 20°$ and phasor $z_2 = 2 \angle 35°$, find the value of the phasor $z_1 + z_2$.

Solution

$$\text{Phasor } z_1 = 3 \angle 20° = 3 \cos 20° + j3 \sin 20° \tag{88-2}$$
$$= 2.82 + j1.03$$
$$\text{Phasor } z_2 = 2 \angle 35° = 2 \cos 35° + j2 \sin 35°$$
$$= 1.64 + j1.15$$
$$\text{Phasor } z_1 + z_2 = (2.82 + j1.03) + (1.64 + j1.15)$$
$$= 4.46 + j2.18$$
$$= \sqrt{4.46^2 + 2.18^2} \angle \text{inv tan } (2.18/4.46)$$
$$= 4.96 \angle 26.04°$$

This result is illustrated in Fig. 88-1B.

Chapter 89
Multiplication and division
of phasors—rationalization

IF AN AC CIRCUIT CONSISTS OF A GENERAL IMPEDANCE, z_1, IN PARALLEL WITH another general impedance, z_2, the total impedance, z_T, is equal to $(z_1 \times z_2)/(z_1 + z_2)$. In order to obtain the value of z_T, we need to have the ability both to multiply phasors and to divide phasors.

Although it is possible to multiply phasors in their rectangular notation, it is generally preferable to use polar notation. As we shall see from the mathematical derivations, $2 \angle -35° \times 3 \angle 40° = 2 \times 3 \angle [(-35°) + (+40°)] = 6 \angle 5°$; consequently the magnitude of the phasors are multiplied but their angles are added algebraically. Knowing the rule for multiplication, we can also square a phasor or take its square root.

Division is also preferably carried out with polar notation. From the mathematical derivations, $3 \angle 40°/2 \angle -35° = 1.5 \angle [(+40°) - (-35°)] = 1.5 \angle 75°$. The magnitude of the numerator is divided by the magnitude of the denominator while the angle of the denominator is algebraically subtracted from the angle of the numerator. From the rule of division we can also obtain the reciprocal of a phasor.

MATHEMATICAL DERIVATIONS
Multiplication of phasors

If phasor, $z_1 = R_1 + jX_1$ and phasor, $z_2 = R_2 + jX_2$, then by rectangular notation,
Product phasor,

$$z_1 z_2 = (R_1 + j X_1)(R_2 + j X_2) \tag{89-1}$$
$$= R_1 R_2 + j R_1 X_2 + j X_1 R_2 + j^2 X_1 X_2$$
$$= (R_1 R_2 - X_1 X_2) + j(R_1 X_2 + X_1 R_2)$$

because $j^2 = -1$.

Using polar notation, $z_1 = Z_1 \angle \phi_1$ and $z_2 = Z_2 \angle \phi_2$.
Product phasor,

$$z_1 z_2 = Z_1 (\cos \phi_1 + j \sin \phi_1) \times Z_2 (\cos \phi_2 + j \sin \phi_2) \tag{89-2}$$
$$= Z_1 Z_2 [(\cos \phi_1 \cos \phi_2 - \sin \phi_1 \sin \phi_2) + j(\sin \phi_1 \cos \phi_2 + \cos \phi_1 \sin \phi_2)]$$
$$= Z_1 Z_2 [\cos (\phi_1 + \phi_2) + j \sin (\phi_1 + \phi_2)]$$
$$= Z_1 Z_2 \angle (\phi_1 + \phi_2)$$

When multiplying phasors, the magnitudes are multiplied but the angles are added algebraically.

If a phasor, $z = Z \angle \phi$,

The square of the phasor,

$$z^2 = Z \angle \phi \times Z \angle \phi = Z^2 \angle 2\phi \tag{89-3}$$

Therefore, when squaring a phasor, the magnitude is squared but the angle is doubled. It follows that by reversing the process to obtain the square root of a phasor, you must take the square root of the magnitude and halve the angle.

The square root of the phasor,

$$\sqrt{z} = \sqrt{Z \angle \phi} = \sqrt{Z} \angle (\phi/2) = Z^{\frac{1}{2}} \angle (\phi/2) \tag{89-4}$$

310

Division of phasors

If phasor $z_1 = R_1 + jX_1$ and phasor $z_2 = R_2 + jX_2$,

$$\text{Phasor } \frac{z_1}{z_2} = \frac{R_1 + jX_1}{R_2 + jX_2}$$

In order to separate out the "real" and "imaginary" parts of z_1/z_2, it is necessary to eliminate j from the denominator. This is achieved by the process of *rationalization,* which means multiplying both the numerator and the denominator by the conjugate of the denominator. The conjugate of a phasor is another phasor that has the same magnitude but whose angle, although equal in size, has the opposite sign. Therefore the conjugate of $Z\angle\phi$ is $Z\angle-\phi$ and the conjugate of $R + jX$ is $R - jX$. It is important to recognize that the product of a phasor and its conjugate is always a totally "real" quantity with no j component.

Rationalizing z_1/z_2 yields

$$\text{Phasor } \frac{z_1}{z_2} = \frac{(R_1 + jX_1)(R_2 - jX_2)}{(R_2 + jX_2)(R_2 - jX_2)} \tag{89-5}$$

$$= \frac{R_1R_2 + X_1X_2}{R_2{}^2 + X_2{}^2} + j\frac{(X_1R_2 - X_2R_1)}{R_2{}^2 + X_2{}^2}$$

In polar form, let $z_1 = Z_1 \angle\phi_1$ and $z_2 = Z_2 \angle\phi_2$. Then

$$\frac{z_1}{z_2} = \frac{Z_1(\cos \phi_1 + j \sin \phi_1)}{Z_2(\cos \phi_2 + j \sin \phi_2)}$$

$$= \frac{Z_1}{Z_2} \times \left[\frac{(\cos \phi_1 + j \sin \phi_1)(\cos \phi_2 - j \sin \phi_2)}{(\cos \phi_2 + j \sin \phi_2)(\cos \phi_2 - j \sin \phi_2)} \right]$$

$$= \frac{Z_1}{Z_2} \times \left[\frac{(\cos \phi_1 \cos \phi_2 + \sin \phi_1 \sin \phi_2) + j(\sin \phi_1 \cos \phi_2 - \cos \phi_1 \sin \phi_2)}{\cos^2 \phi_2 + \sin^2 \phi_2} \right]$$

$$= \frac{Z_1}{Z_2} \times \left[\frac{\cos (\phi_1 - \phi_2) + j \sin (\phi_1 - \phi_2)}{1} \right]$$

because $\cos^2 \phi_2 + \sin^2 \phi_1 = 1$.

Therefore

$$\frac{z_1}{z_2} = \frac{Z_1}{Z_2} \angle(\phi_1 - \phi_2) \tag{89-6}$$

The magnitude of the numerator is divided by the magnitude of the denominator, but the angle of the denominator is algebraically subtracted from the angle of the numerator.

In the special case that involves the reciprocal of the phasor, $z = Z \angle\phi$,

$$\frac{1}{z} = \frac{1 \angle0°}{Z \angle\phi} = \frac{1}{Z} \angle-\phi \tag{89-7}$$

Example 89-1

If phasor $z_1 = 1 + j2$ and $z_2 = 3 - j2$, find the values of the phasors z_1z_2, z_1/z_2, $z_1{}^2$, $\sqrt{z_2}$, $1/z_1$, and the conjugate of z_2. Illustrate your results on phasor diagrams.

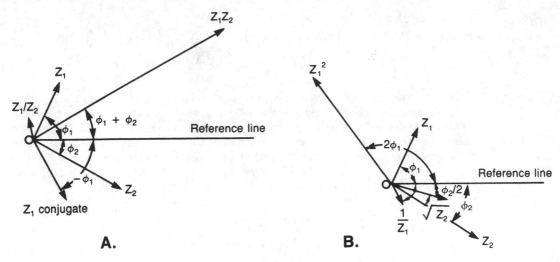

A. **B.**

Figure 89-1

Solution

$$\text{Phasor } z_1 = 1 + j2 = \sqrt{1^2 + 2^2} \; \angle \text{inv tan } (2/1)$$
$$= 2.24 \angle 63.4°$$
$$\text{Phasor } z_2 = 3 - j2 = \sqrt{3^2 + 2^2} \; \angle \text{inv tan } (2/3)$$
$$= 3.61 \angle -33.7°$$
$$\text{Phasor } z_1 z_2 = 2.24 \; \angle 63.4° \times 3.61 \; \angle -33.7°$$
$$= 8.09 \; \angle 29.7°$$
$$\text{Phasor } \frac{z_1}{z_2} = \frac{2.24 \; \angle 63.4°}{3.61 \; \angle -33.7°} = 0.62 \; \angle [(+63.4°) - (-33.7°)]$$
$$= 0.62 \; \angle 97.1°$$
$$\text{Phasor } z_1^2 = (2.24 \; \angle 63.4°)^2$$
$$= 5.02 \; \angle 126.8°$$
$$\text{Phasor } \sqrt{z_2} = \sqrt{3.61 \; \angle -33.7°}$$
$$= 1.90 \; \angle -16.85°$$

$$\text{Phasor } \frac{1}{z_1} = \frac{1}{2.24 \; \angle 63.4°}$$
$$= 0.45 \; \angle -63.4°$$

Phasor conjugate of $z_1 = 2.24 \; \angle -63.4°$

These results are illustrated in the phasor diagrams of Fig. 89-1A and B.

Chapter 90
Analysis of a series-parallel circuit with the aid of the j operator

IN TERMS OF IMPEDANCE PHASORS THE FORMULA FOR THE TOTAL IMPEDANCE IS the same as we earlier derived for comparable resistor networks. In the circuit of Fig. 90-1B, we would start by combining the parallel z_1, z_2 phasors by using the "product-over-sum" formula. To the $z_1\|z_2$ phasor we would then add the series phasor z_3 and the result is the total impedance phasor, z_T. These operations are performed with the rules of complex algebra as outlined in Chapters 88 and 89.

After obtaining the total impedance phasor, the source voltage, e, is divided by z_T to obtain the total current, i_T. Subsequently, we can calculate the individual voltages across the components, the branch currents and the powers associated with each current.

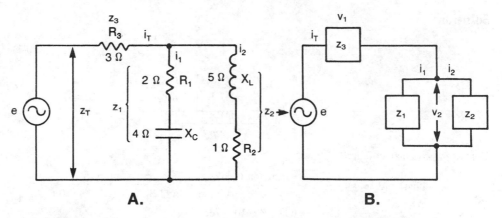

Figure 90-1

MATHEMATICAL DERIVATIONS

In Fig. 90-1 A and B,

$$\text{Phasor } z_1\|z_2 = \frac{z_1 \times z_2}{z_1 + z_2} \tag{90-1}$$

Total impedance phasor,

$$z_T = z_3 + z_1\|z_2 \tag{90-2}$$
$$= z_3 + \frac{z_1 \times z_2}{z_1 + z_2}$$

Total current phasor,

$$i_T = \frac{e}{z_T} \tag{90-3}$$

Voltage phasor,

$$v_1 = i_T \times z_3 \tag{90-4}$$

Voltage phasor,

$$v_2 = e - v_1 \tag{90-5}$$

Current phasor,

$$i_1 = \frac{v_2}{z_1} \tag{90-6}$$

Current phasor,

$$i_2 = \frac{v_2}{z_2} \tag{90-7}$$

Example 90-1

In Fig. 90-1A, $z_1 = 2 - j4$ Ω, $z_2 = 1 + j5$ Ω, $z_3 = 3 + j0$ Ω, $e = 7 \angle 60°$ V. Calculate the values of z_T and i_T.

Solution

$$\text{Phasor } z_1 \| z_2 = \frac{z_1 \times z_2}{z_1 + z_2} \tag{90-1}$$

$$= \frac{(2 - j4) \times (1 + j5)}{(2 - j4) + (1 + j5)}$$

$$= \frac{(2 - j4) \times (1 + j5)}{3 + j1}$$

Converting to polar notations,

$$\text{Phasor, } z_1 \| z_2 = \frac{\sqrt{2^2 + 4^2} \angle \text{inv tan } (-4/2) \times \sqrt{1^2 + 5^2} \angle \text{inv tan } (5/1)}{\sqrt{3^2 + 1^2} \angle \text{inv tan } (1/3)}$$

$$= \frac{4.47 \angle -63.4° \times 5.10 \angle 78.7°}{3.16 \angle 18.4°}$$

$$= 7.21 \angle -3.1°$$

$$= 7.19 - j\,0.39 \ \Omega$$

Phasor,

$$z_T = z_3 + z_1 \| z_2 \tag{90-2}$$

$$= 3 + 7.19 - j\,0.39$$

$$= 10.19 - j\,0.39 \Omega \text{ (rectangular notation)}$$

$$= 10.20 \ \Omega \ \angle -2.19° \text{ (polar notation)}$$

Phasor,

$$i_T = \frac{e}{z_T} \tag{90-3}$$

$$= \frac{7 \angle 60° \text{ V}}{10.20 \ \Omega \ \angle -2.19°}$$

$$= 0.69 \text{ A } \angle[(+60°) - (-2.19°)]$$

$$= 0.69 \text{ A } \angle 62.19°$$

Chapter 91
Analysis of a parallel branch circuit with the aid of the j operator

IN TERMS OF THE IMPEDANCE PHASORS THE FORMULA FOR THE TOTAL IMPEDANCE is the same as we earlier derived for comparable resistor networks. In the circuit of Fig. 91-1B we would use the reciprocal formula to obtain the total impedance phasor, z_T. The calculation of z_T is obtained by using the rules of complex algebra as outlined in Chapters 88 and 89.

The branch currents i_1, i_2 etc., are calculated by dividing the source voltage, e, by each of the branch impedances. The products $e \times i_1$, $e \times i_2$ etc. will then provide the individual powers associated with the branches.

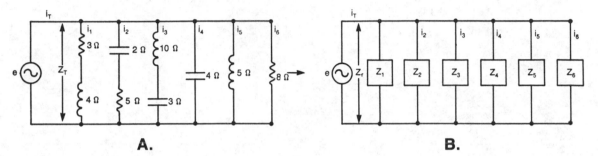

A.   **B.**

Figure 91-1

MATHEMATICAL DERIVATIONS

In Fig. 91-1A and B,

$$\frac{1}{z_T} = \frac{1}{z_1} + \frac{1}{z_2} + \frac{1}{z_3} + \frac{1}{z_4} + \frac{1}{z_5} + \frac{1}{z_6} \tag{91-1}$$

Total impedance phasor,

$$z_T = \frac{1}{\dfrac{1}{z_1} + \dfrac{1}{z_2} + \dfrac{1}{z_3} + \dfrac{1}{z_4} + \dfrac{1}{z_5} + \dfrac{1}{z_6}} \tag{91-2}$$

Total current phasor,

$$i_T = \frac{e}{z_T} \tag{91-3}$$

Branch current phasors,

$$i_1 = \frac{e}{z_1}, \ i_2 = \frac{e}{z_2}, \ \text{etc.} \tag{91-4}$$

Example 91-1

In Fig. 91-1A, $z_1 = 3 + j4$, $z_2 = 5 - j2$, $z_3 = j10 - j3$, $z_4 = 0 - j4$, $z_5 = 0 + j5$, $z_6 = 8 + j0$, $e = 8 \angle 25°$ V. Find the values of the phasors z_T and i_T.

Solution

The total impedance phasor, z_T is given by

$$\frac{1}{z_T} = \frac{1}{z_1} + \frac{1}{z_2} + \frac{1}{z_3} + \frac{1}{z_4} + \frac{1}{z_5} + \frac{1}{z_6} \qquad (91\text{-}1)$$

$$\frac{1}{z_T} = \frac{1}{3 + j4} + \frac{1}{5 - j2} + \frac{1}{j10 - j3} + \frac{1}{0 - j4} + \frac{1}{0 + j5} + \frac{1}{8 + j0}$$

$$= \frac{3 - j4}{3^2 + 4^2} + \frac{5 + j2}{5^2 + 2^2} + \frac{1}{j7} + \frac{1}{-j4} + \frac{1}{j5} + \frac{1}{8}$$

$$= \frac{3 - j4}{25} + \frac{5 + j2}{29} - j0.142 + j0.25 - j0.2 + 0.125$$

$$= 0.12 - j0.16 + 0.172 + j0.069 - j0.142 + j0.25 - j0.2 + 0.125$$

$$= 0.417 - j0.183$$

$$= 0.455 \angle{-23.7°}$$

Total impedance phasor,

$$z_T = \frac{1}{0.455 \angle{-23.7°}}$$

$$= 2.20 \ \Omega \ \angle 23.7°$$

Total current,

$$i_T = \frac{e}{z_T} \qquad (91\text{-}3)$$

$$= \frac{8 \angle 25° \ \text{V}}{2.20 \ \angle{+23.7°}}$$

$$= 3.64 \ \angle[(+25°) - (+23.7°)] \ \text{A}$$

$$= 3.64 \ \angle 1.3° \ \text{A}.$$

Chapter 92
Solution of an ac circuit
by the use of Thévenin's theorem

THE STEPS FOR APPLYING THÉVENIN'S THEOREM TO AC CIRCUITS ARE THE SAME as those we outlined for dc circuits in Chapter 31. The only difference is the use of complex algebra in obtaining the results of each step.

Example 92-1

In the circuit of Fig. 92-1A, calculate the value of the load current, i_L.

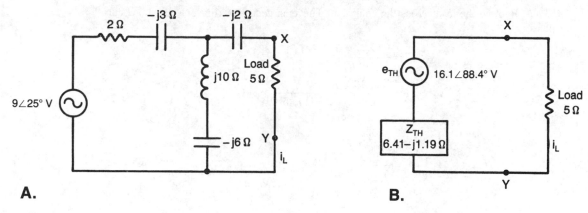

A.

B.

Figure 92-1

Solution

Step 1. Remove the 5 Ω resistive load and obtain the open-circuit voltage between the points X and Y. Using the voltage division rule,

Open-circuit voltage,

$$e_{TH} = 9 \angle 25° \times \frac{(j10 - j6)}{(2 - j3) + (j10 - j6)}$$

$$= 9 \angle 25° \times \frac{j4}{2 + j1}$$

$$= 9 \angle 25° \times \frac{4 \angle 90°}{2.24 \angle 26.6°}$$

$$= 16.1 \angle 88.4° \text{ V}.$$

Step 2. Place a short circuit across the voltage source and calculate the value of the impedance phasor between the terminals X and Y.

Impedance phasor,

$$z_{TH} = -j2 + \frac{(2 - j3) \times (j10 - j6)}{(2 - j3) + (j10 - j6)}$$

$$= -j2 + \frac{(2 - j3)(j4)}{2 + j1}$$

$$= -j2 + \frac{12 + j8}{2 + j1}$$

$$= \frac{-j2 (2 + j1) + 12 + j8}{2 + j1}$$

$$= \frac{14 + j4}{2 + j1}$$

$$= \frac{14.6 \angle 15.9°}{2.24 \angle 26.6°}$$

$$= 6.52 \angle -10.7°$$

$$= 6.41 - j1.19 \text{ Ω}.$$

Step 3. Replace the load across the equivalent Thévenin generator (Fig. 92-1B). Load current phasor,

$$i_L = \frac{e_{TH}}{z_{TH} + R_L}$$

$$= \frac{16.1 \angle 88.4° \text{ V}}{(6.41 - j1.19) \ \Omega + 5 \ \Omega}$$

$$= \frac{16.1 \angle 88.4°}{11.41 - j1.9}$$

$$= \frac{16.1 \angle 88.4°}{11.56 \angle -9.45°}$$

$$= 1.39 \angle 97.85° \text{ A.}$$

Chapter 93
The practical single-phase alternator

IN CHAPTER 59 THE BASIC ALTERNATOR CONTAINED A REVOLVING ARMATURE whose conductors cut the flux of a stationary magnetic field. The alternating voltage generated was then taken through slip-rings to the load. The contacts between these slip-rings and the carbon brushes were subject to friction wear and sparking; moreover they were liable to arc over at high voltages. These problems are largely overcome in the revolving field alternator of Fig. 93-1A.

In the revolving field alternator, a dc source drives a direct current through the slip-rings, brushes and the windings on a rotor, which is driven around by mechanical means. This creates a rotating magnetic field that cuts the conductors embedded in the surrounding stator. An alternating voltage then appears between the ends, S (start) and F (finish) of the stator winding. Since S and F are fixed terminals, there are no sliding contacts and the whole of the stator winding may be continuously insulated.

The alternator of Fig. 93-1A has two poles so that one complete rotation of the rotor generates one cycle of ac in the stator. Therefore, in order to generate 60 Hz, the rotor must turn at 60 revolutions per second or $60 \times 60 = 3600$ rpm. However, in the four pole machine of Fig. 93-1B, one revolution produces two cycles of the ac voltage and therefore it would only require a speed of 1800 rpm to generate a 60 Hz output.

The power delivered to a resistive load by a single phase alternator is fluctuating at twice the line frequency (Chapter 65). This presents a problem because the load on the mechanical source of energy and therefore the necessary torque will not be constant.

MATHEMATICAL DERIVATIONS

For the revolving field alternator,
Generated frequency,

$$f = \frac{N \times p}{60} \text{ Hertz} \tag{93-1}$$

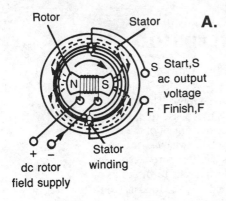

A.

Rotor

Stator

Start,S
ac output
voltage
Finish,F

+ −
dc rotor
field supply

Stator
winding

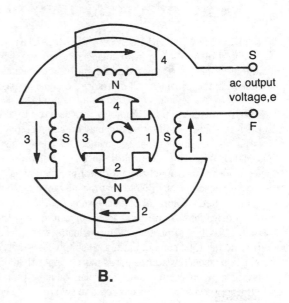

S

ac output
voltage,e

F

B.

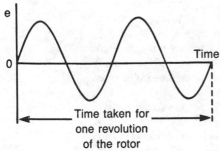

e

0

Time

Time taken for
one revolution
of the rotor

Figure 93-1

where: N = rotor speed (rpm)

p = number of pole pairs.

Example 93-1

In a revolving field alternator a six pole rotor is rotating at 1000 rpm. What is the value of the generated frequency?

Solution

Generated frequency,

$$f = \frac{N \times p}{60} \qquad (93\text{-}1)$$

$$= \frac{1000 \times (6/2)}{60}$$

$$= 50 \text{ Hz}.$$

Chapter 94
The practical two-phase alternator

IN THIS TYPE OF REVOLVING FIELD ALTERNATOR, TWO IDENTICAL COILS ARE mounted on the stator with their axes separated by 90°. Figure 94-1A shows a simplified arrangement of the windings that are cut by the magnetic flux of a two pole rotor. The EMFs e_1, e_2 induced in the windings will therefore be equal in magnitude but 90° out of phase (Fig. 94-1B).

The fact that e_2 leads e_1 by 90° is a result of the assumed clockwise movement of the rotor. This phase relationship is commonly represented by drawing the two stator windings 90° apart (Fig. 94-1C). Each of the windings may be connected to separate loads but it is more convenient to use a neutral line so that the number of lines required is reduced from four to three, as in Fig. 94-1D. Assuming identical (balanced) resistive loads across the windings, the total instantaneous power of the two-phase alternator is constant as opposed to the fluctuating power output of the single phase machine. Note that this advantage is only achieved when the loads are balanced.

If the effective current of each load is I_{rms}, the neutral line current is $1.414\ I_{rms}$ (see mathematical derivations). The total load current in the three lines is therefore $(2 + \sqrt{2})\ I_{rms} = 3.414\ I_{rms}$. If the same two balanced loads were connected in parallel across a single phase alternator, the total current in the two supply lines is $4\ I_{rms}$. Assuming that a neutral line is used, the single phase alternator requires more copper than the comparable two-phase generator when supplying a given voltage to a particular value of total load resistance.

If the two voltages are connected to two pairs of coils whose axes are perpendicular, the two fluxes associated with the coils will combine to produce a rotating magnetic field whose angular velocity is equal to that of the alternating voltage. This principle is used in ac induction and synchronous motors. Compared with a single-phase alternator, a high power two-phase generator requires a smaller stator, is more efficient and is less subject to vibration.

MATHEMATICAL DERIVATIONS
In Fig. 94-1B,

Phase voltage,

$$e_1 = E_{max}\sin \omega t \text{ volts} \tag{94-1}$$

Phase voltage,

$$e_2 = E_{max}\left(\sin \omega t + \frac{\pi}{2}\right) \tag{94-2}$$

$$= E_{max}\cos \omega t \text{ volts}$$

where: ω, angular frequency $= 2\pi f(\text{rad/s})$

E_{max} = peak value of the phase voltage (V)

e_1, e_2 = instantaneous phase voltages (V)

Assuming balanced resistive loads,
Instantaneous power,

$$p_1 = \frac{e_1^{\,2}}{R_L} = \frac{E_{max}^2 \sin^2 \omega t}{R_L} \text{ watts} \tag{94-3}$$

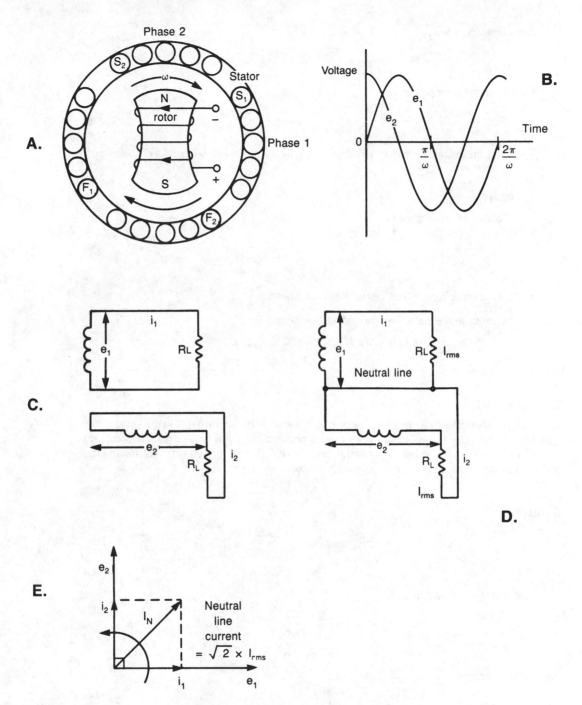

Figure 94-1

Instantaneous power,

$$p_2 = \frac{e_2^2}{R_L} = \frac{E_{max}^2 \cos^2 \omega t}{R_L} \text{ watts} \qquad (94\text{-}4)$$

Total instantaneous power,

$$p_1 + p_2 = \frac{E_{max}^2}{R_L} \text{ watts} \qquad (94\text{-}5)$$

The total instantaneous power is independent of the time and is therefore a constant.

In Fig. 94-1E,

Assuming balanced resistive loads,
    Effective value of the neutral line current,

$$I_N = \sqrt{I_{rms}^2 + I_{rms}^2} \qquad (94\text{-}6)$$
$$= \sqrt{2}\, I_{rms} \text{ ampere}$$

where: I_{rms} is the effective load current for each phase.

Total current in the three lines (including the neutral line):

$$2 I_{rms} + \sqrt{2}\, I_{rms} = (2 + \sqrt{2})\, I_{rms}$$
$$= 3.414\, I_{rms} \text{ amperes}$$

Example 94-1

A two-phase alternator has a four-pole rotor whose speed is 1800 rpm. What is the value of the generated frequency? The effective voltage of each phase is 240 V and the two resistive loads are balanced so that each has an effective resistance of 80 Ω. Calculate the current in the neutral line and the instantaneous power output of the alternator.

Solution

$$\text{Generated frequency} = \frac{1800 \times 2}{60}$$
$$= 60 \text{ Hz.}$$

Load current for each phase,

$$I_{rms} = \frac{240 \text{ V}}{80 \text{ }\Omega}$$
$$= 3 \text{ A rms.}$$

Neutral line current

$$I_N = \sqrt{2}\, I_{rms} \qquad (94\text{-}6)$$
$$= \sqrt{2} \times 3$$
$$= 4.242 \text{ A rms.}$$

Total instantaneous power

$$p_1 + p_2 = \frac{E_{\max}^2}{R_L}$$

(94-5)

$$= \frac{(\sqrt{2} \times 240 \text{ V})^2}{80 \ \Omega}$$

$$= 1440 \text{ W}.$$

The average power delivered over the cycle is also 1440 W.

Note: For the corresponding single phase alternator the total load resistance is 80 Ω/2 = 40 Ω. The maximum instantaneous power is $(\sqrt{2} \times 240 \text{ V})^2/40 \ \Omega = 2880$ W so that the average power over the cycle is 2880 W/2 = 1440 W. The total of the currents in the two supply lines is 2 $\times$ 240 V/40 Ω = 12 A rms; this compares with the two-phase alternator where the total of the currents in the two supply lines and the neutral line is $(2 \times 3) + 4.242 = 10.242$ A rms.

Chapter 95
The three-phase alternator

THE THREE-PHASE ALTERNATOR SHOWN IN FIG. 95-1A HAS THREE SINGLE-PHASE windings which are equally spaced on the stator so that the voltage induced in each winding is 120°(2 π/3 radians) out of phase with the voltage induced in the other two windings (Fig. 95-1B). These windings are independent of each other and could be connected to three separate loads. Such connections would require a six-line system (Fig. 95-1C).

Assume that the three loads are balanced in the sense that they will all have the same ohms value and the same phase angle. It follows that they would all have identical power factors, which in our example are assumed to be lagging. The three load currents will also be equal in magnitude, but 120° out of phase so that their phasor sum is zero (Fig. 95-2A). It is then possible to join the corresponding ends of the three windings to a common point and to connect the other ends to separate terminals. A single neutral line is attached to the common point to produce a four-wire wye (Y) system (Fig. 95-2B).

Apart from the advantage of reducing the number of wires from six to four, the theoretical current carried by the neutral line is zero. The thin neutral line only requires a low current capacity because the current it carries, only exists as a result of some imbalance between the loads. Alternatively the neutral line may be replaced by a ground return. The saving in the copper required by the supply lines is therefore greater for a balanced three-phase system than for a balanced two-phase system.

The mathematical derivations show that the total instantaneous power for the balanced three phase system is constant and therefore independent of the time.

MATHEMATICAL DERIVATIONS

In Fig. 95-1B, the phase voltages v_1, v_2, v_3 are shown as three phasors that are equally separated

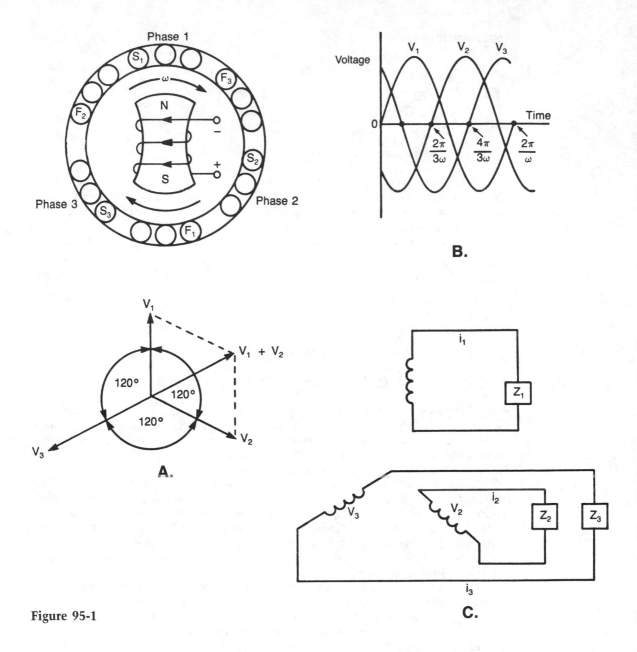

Figure 95-1

by $2\pi/3$ radians. If v_1 is used as the reference phasor and the rotor is assumed to be revolving in the clockwise direction,

Phase #1 voltage,

$$v_1 = E_{max} \sin \omega t \text{ volts} \tag{95-1}$$

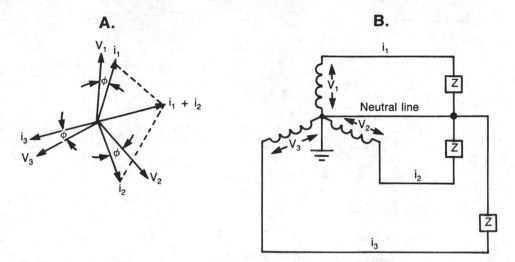

Figure 95-2

Phase #2 voltage,

$$v_2 = E_{max} \sin\left(\omega t - \frac{2\pi}{3}\right) \tag{95-2}$$

$$= -\frac{E_{max} \sin \omega t}{2} - \frac{\sqrt{3}}{2} E_{max} \cos \omega t \text{ volts}$$

Phase #3 voltage,

$$v_3 = E_{max} \sin\left(\omega t - \frac{4\pi}{3}\right) \tag{95-3}$$

$$= -\frac{E_{max} \sin \omega t}{2} + \frac{\sqrt{3}}{2} E_{max} \cos \omega t \text{ volts}$$

where: E_{max} = peak value of the phase voltage (V)

ω = angular frequency (rad/s)

By adding Equations 95-1, 95-2, and 95-3,

$$v_1 + v_2 + v_3 = 0 \tag{95-4}$$

For three balanced loads with the same lagging power factor, each load current will lag its corresponding phase voltage by the same phase angle, ϕ (Fig. 95-2A).

It follows that:

$$i_1 + i_2 + i_3 = 0 \tag{95-5}$$

If each of the balanced loads is equal to a resistance R_L, the instantaneous powers delivered by the three phases are:

Instantaneous power,

$$p_1 = \frac{E_{max}^2}{R_L} \sin^2 \omega t \text{ watts} \tag{95-6}$$

Instantaneous power,

$$p_2 = \frac{E_{max}^2}{R_L} \sin^2\left(\omega t - \frac{2\pi}{3}\right) \tag{95-7}$$

$$= \frac{E_{max}^2}{R_L}\left(\frac{-\sin \omega t + \sqrt{3} \cos \omega t}{2}\right)^2 \text{ watts}$$

Instantaneous power,

$$p_3 = \frac{E_{max}^2}{R_L} \sin^2\left(\omega t + \frac{2\pi}{3}\right) \tag{95-8}$$

$$= \frac{E_{max}^2}{R_L}\left(\frac{\sin \omega t + \sqrt{3} \cos \omega t}{2}\right)^2 \text{ watts}$$

Total instantaneous power,

$$p_T = p_1 + p_2 + p_3$$

$$= \frac{E_{max}^2}{R_L}\left[\sin^2 \omega t + \frac{\sin^2 \omega t}{4} + \frac{\sin^2 \omega t}{4} + \frac{3}{4}\cos^2 \omega t + \frac{3}{4}\cos^2 \omega t\right] \tag{95-9}$$

$$= \frac{3E_{max}^2}{2R_L} = \frac{3E_{rms}^2}{R_L} \text{ watts}$$

The total instantaneous power, p_T, for a balanced three-phase system is therefore constant and is independent of the time.

Example 95-1

A three-phase alternator has a four-pole rotor whose speed is 1500 rpm. What is the value of the generated frequency? The effective voltage of each phase is 120 V and the resistive loads are balanced so that each has an effective resistance of 60 Ω. Calculate the total of the line and neutral currents and determine the total instantaneous power output of the alternator.

Solution

Generated frequency,

$$f = \frac{1500 \times 2}{60} = 50 \text{ Hz.}$$

Load current for each phase,

$$I_L = \frac{120 \text{ V}}{60 \text{ } \Omega} = 2 \text{ A.}$$

Neutral line current is *zero* for a balanced load system.
Total of the load and neutral currents

$$I_T = 3 \times 2 \text{ A} = 6 \text{ A.}$$

Total instantaneous power,

$$p_T = \text{average power over the cycle} \tag{95-9}$$

$$= \frac{3E_{rms}^2}{R}$$

$$= 3 \times \frac{(120 \text{ V})^2}{60 \text{ }\Omega}$$

$$= 720 \text{ W}.$$

Note: For the equivalent two-phase alternator the load on each phase would be $2 \times 60 \text{ }\Omega/3 = 40 \text{ }\Omega$. The current for each phase is 120 V/40 Ω = 3 A and the neutral line current is $3 \times \sqrt{2} = 4.242$ A. The total of the line and neutral currents is $(2 \times 3) + 4.242 = 10.242$ A. Total instantaneous power

$$p_T = \text{average power over the cycle}$$

$$= 2 \times \frac{(120 \text{ V})^2}{40 \text{ }\Omega}$$

$$= 720 \text{ W}.$$

With the comparable single-phase alternator the total load resistance would be 60 Ω/3 = 20 Ω. The total current in the supply lines is 2×120 V/20 Ω = 12 A and the average power over the cycle is $(120 \text{ V})^2$/20 Ω = 720 W.

Chapter 96
The line voltage of the three-phase wye connection

THE ALTERNATOR EMF APPLIED ACROSS ONE OF THE THREE BALANCED LOADS in Chapter 95, was the voltage which existed between ground (or the neutral line) and the line connected to the load. This EMF is normally called the phase voltage, V_P. However, with three-phase systems, it is normal practice to make use of the line voltage, V_L, as well as the phase voltage. As illustrated in Fig. 96-1A, the line voltage is the alternating EMF which exists between points X and Y, and is related to the voltages generated in the first and second phases of the stator windings.

To determine the relationship between V_L and V_P, it is important to realize that the correct stator terminals must be connected to the common point if the phase voltages are to be represented as 120° ($2\pi/3$ radians) apart. For example, if the two terminals of one phase winding are reversed, its phase voltage is shifted by 180°. The phase voltages v_x, v_y, v_z are the alternating (phasor) voltages monitored at the points X, Y, Z, with respect to the common point 0. Consequently the (phasor) line voltage, v_{xy}, is the alternating voltage at the point X with respect to the point Y and is therefore the voltage difference between the two points (Fig. 96-1B). The phasor line voltages are

$$v_{xy} = v_x - v_y$$
$$v_{yz} = v_y - v_z$$
$$v_{zx} = v_z - v_x$$

By adding together these three equations, it follows that the phasor sum of the line voltages is zero. The mathematical derivations will also prove that the phasors of the three line voltages

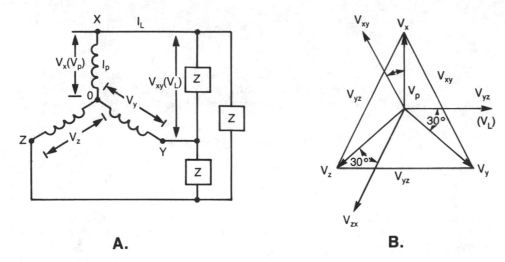

Figure 96-1

are 120° apart and are shifted by 30° from the phasors of the phase voltages.

When three balanced loads are connected between the lines, the phasors of the three line currents are also equally spaced by 120°. Because each line is directly connected to a terminal of a phase winding, the line current, i_L, must be equal to the phase current, i_P.

MATHEMATICAL DERIVATIONS

In Fig. 96-1B, let v_x be the reference phase voltage so that:

Phase voltage,

$$v_x = E \angle 0 \text{ volts} \tag{96-1}$$

Therefore,
Phase voltage,

$$v_y = E \angle -2\pi/3 \text{ volts} \tag{96-2}$$

Phase voltage,

$$v_z = E \angle +2\pi/3 \text{ volts} \tag{96-3}$$

where E is the effective value of the phase voltage.

Then,
Line voltage,

$$
\begin{aligned}
v_{xy} &= v_x - v_y \\
&= E \angle 0 - E \angle -2\pi/3 \\
&= E - E[(-1/2) + j(-\sqrt{3}/2)] \\
&= E (3/2 + j \sqrt{3}/2) \\
&= \sqrt{3}E \angle \pi/6 = \sqrt{3}E \angle 30° \text{ volts}
\end{aligned}
\tag{96-4}
$$

This means that the line voltage, v_{xy}, has a magnitude of $\sqrt{3}E$ and leads v_x by 30°.
Line voltage,

$$v_{yz} = v_y - v_z \tag{96-5}$$
$$= E \angle -2\pi/3 - E \angle +2\pi/3$$
$$= E[(-1/2) + j(-\sqrt{3}/2)] - E[(-1/2) + j(+\sqrt{3}/2)]$$
$$= \sqrt{3}E \angle -\pi/2 \text{ volts}$$

The line voltage, v_{yz}, has a magnitude of $\sqrt{3}E$ volts and leads v_y by 30°.
Line voltage,

$$v_{zx} = v_z - v_x \tag{96-6}$$
$$= E \angle +2\pi/3 - E \angle 0$$
$$= E[(-1/2) + j(+\sqrt{3}/2)] - E$$
$$= E[(-3/2) + j(+\sqrt{3}/2)]$$
$$= \sqrt{3}E \angle 5\pi/6 \text{ volts}$$

The line voltage, v_{zx}, has a magnitude of $\sqrt{3}E$ volts and leads v_z by 30°. Equations 96-4, 96-5, 96-6 all show that the line voltage is $\sqrt{3} \times$ the phase voltage or $1.732 \times$ the phase voltage. In addition the line phasors are 120° apart and each line voltage phasor is shifted by 30° from one of the phase voltage phasors.

Example 96-1

In a three-phase system the reference phase voltage is $120 \angle 50°$ volts, 60 Hz. Balanced loads, each equal to $15 \angle 20° \ \Omega$, are connected between the supply lines. Express the line voltages and the phase currents in their polar forms.

Solution

The phase voltages are $120 \angle 50°$ volts, $120 \angle (50° + 120°) = 120 \angle 170°$ volts and $120 \angle (50° - 120°) = 120 \angle -70°$ volts.

The line voltages are

$$v_{xy} = \sqrt{3}E \angle (50° + 30°) \tag{96-4}$$
$$= 1.732 \times 120 \angle 80°$$
$$= 209 \angle 80° \text{ volts.}$$

Line voltage,

$$v_{yz} = \sqrt{3}E \angle (170° + 30°) \tag{96-5}$$
$$= 1.732 \times 120 \angle 200°$$
$$= 209 \angle -160° \text{ volts.}$$

Line voltage,

$$v_{zx} = \sqrt{3}E \angle (-70° + 30°) \tag{96-6}$$
$$= 1.732 \times 120 \angle -40°$$
$$= 209 \angle -40° \text{ volts.}$$

The phase currents are equal to the line currents, which are
Phase current,

$$i_x = \frac{209 \angle 80° \text{ V}}{15 \angle 20° \text{ }\Omega}$$
$$= 13.9 \angle 60° \text{ A.}$$

Phase current,

$$i_y = \frac{209 \angle -160° \text{ V}}{15 \angle 20° \text{ }\Omega}$$
$$= 13.9 \angle 180° \text{ A.}$$

Phase current,

$$i_z = \frac{209 \angle -40° \text{ V}}{15 \angle 20° \text{ }\Omega}$$
$$= 13.9 \angle -60° \text{ A}$$

The phasor of the three phase currents are also equally separated by 120°.

Chapter 97
The three-phase delta connection

BECAUSE THE SUM OF THE THREE INSTANTANEOUS PHASE VOLTAGES IS AT ALL times zero, it is possible to connect the phase windings in a delta (Δ) formation. This is illustrated in Fig. 97-1A and provided the correct connections are made, there will be zero circulating current in the delta loop. Three supply lines may then be connected to the corners of the delta formation and these are joined to three loads also arranged in delta. It is clear that the voltage applied across any one of the loads must be the same as the voltage generated in the corresponding phase winding across which that particular load is connected. Therefore, in the three-phase delta system, the line voltage, V_L, is equal to the phase voltage, V_P.

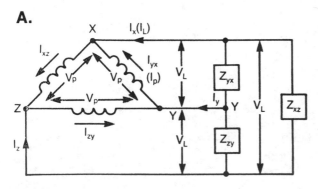
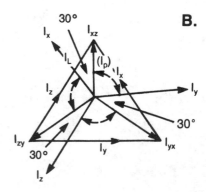

Figure 97-1

By contrast with the voltage relationship, a particular line current phasor, i_x, is associated with two phase currents, i_{yx} and i_{xz}. Each of the load currents will then be equal to its corresponding phase current. However, each line current will be equal to the phasor difference between two of the load currents so that the related equations are:

$$i_x = i_{xz} - i_{yx}$$
$$i_y = i_{yx} - i_{zy}$$
$$i_z = i_{zy} - i_{xz}$$

The sum of the three line currents is therefore zero; this is true irrespective of whether the loads are balanced or not. However, if the loads are balanced, the currents in the three windings are separated from each other by a phase difference of 120°. The relationships between the line and phase currents of the delta system are then comparable with the equations relating the line and phase voltages in the wye arrangement (Chapter 96). Therefore, in the delta system, the line current, I_L, is equal to $\sqrt{3} \times$ the phase current, I_P (or $1.732 \times I_P$). In addition each line current is shifted by 30° from one of the phase currents (Fig. 97-1B).

The total power output from the alternator can be measured in terms of the line voltage and current values together with the power factor for each of the balanced loads. As the mathematical derivations will show, the expression for the total power is the same for both wye and delta connections.

MATHEMATICAL DERIVATIONS

In a balanced wye or delta system the power in each load is the same and therefore the total power is:

$$\text{Total power, } P_T = 3 \times E_{\text{rms}} \text{ (load)} \times I_{\text{rms}} \text{ (load)} \times \cos \phi \text{ watts} \qquad (97\text{-}1)$$

where $\cos \phi$ is the power factor for each of the balanced loads.

For the wye connection:

$$E_{\text{rms}} \text{ (line)} = \sqrt{3} \times E_{\text{rms}} \text{ (load) and } I_{\text{rms}} \text{ (line)} = I_{\text{rms}} \text{ (load)} \qquad (97\text{-}2)$$

Combining Equations 97-1 and 97-2,

Total power,

$$P_T = 3 \times \frac{E_{\text{rms}} \text{ (line)}}{\sqrt{3}} \times I_{\text{rms}} \text{ (line)} \times \cos \phi \qquad (97\text{-}3)$$

$$= \sqrt{3} \times E_{\text{rms}} \text{ (line)} \times I_{\text{rms}} \text{ (line)} \times \cos \phi \text{ watts}$$

For the delta connection:

$$E_{\text{rms}} \text{ (line)} = E_{\text{rms}} \text{ (load) and } I_{\text{rms}} \text{ (line)} = \sqrt{3} \, I_{\text{rms}} \text{ (load)}$$

Combining Equations 97-1 and 97-3,

Total power,

$$P_T = 3 \times E_{\text{rms}} \text{ (line)} \times \frac{I_{\text{rms}} \text{ (line)}}{\sqrt{3}} \times \cos \phi \qquad (97\text{-}4)$$

$$= \sqrt{3} \times E_{\text{rms}} \text{ (line)} \times I_{\text{rms}} \text{ (line)} \times \cos \phi \text{ watts}$$

The expressions for the total power are the same for both the wye and delta connections.

Example 97-1

In a three-phase delta-connected alternator the phase voltages are 110 $\angle 0°$ V, 110 $\angle 120°$ V and 110 $\angle -120°$ V. Each of the three balanced loads has an impedance of 5.5 $\angle 25°$ Ω. Calculate the values of the line currents and the total power delivered from the three phase alternator.

Solution

The phase currents are:

$$I_{xz} = \frac{110 \angle 0° \text{ V}}{5.5 \angle 25° \text{ } \Omega} = 20 \angle -25° \text{ A}$$

$$I_{zy} = \frac{110 \angle 120° \text{ V}}{5.5 \angle 25° \text{ } \Omega} = 20 \angle 95° \text{ A}$$

$$I_{yx} = \frac{110 \angle -120° \text{ V}}{5.5 \angle 25° \text{ } \Omega} = 20 \angle -145° \text{ A}$$

The line currents are:

$$i_x = i_{xz} - i_{yx}$$
$$= 20 \angle -25° - 20 \angle -145°$$
$$= 18.1 - j8.45 + 16.4 + j11.47$$
$$= 34.5 + j3.02$$
$$= 34.6 \angle 5° \text{ A}.$$

The same result may be derived from

$$i_x = 20 \times \sqrt{3} \angle [(-25°) + (+30°)] = 34.6 \angle 5° \text{ A}$$

Similarly

$$i_y = 34.6 \angle [(+95°) + (+ 30°)]$$
$$= 34.6 \angle 125° \text{ A}$$
$$= -19.84 + j28.35 \text{ A}$$

and

$$i_z = 34.6 \angle [(-145°) + (+30°)]$$
$$= 34.6 \angle -115° \text{ A}$$
$$= -14.66 - j31.37 \text{ A}.$$

The sum of the line currents is $(+34.5 + j3.02) + (-19.84 + j28.35) + (-14.66 - j31.37) = 0$ A. Each of the line currents has a magnitude of $20\sqrt{3} = 20 \times 1.732 = 34.6$ A and each line current is shifted by 30° from one of the phase currents.

The total power delivered from the alternator is:
Total power,

$$P_T = \sqrt{3} \times E_{\text{rms}} \text{ (line)} \times I_{\text{rms}} \text{ (line)} \times \cos \phi \qquad (97\text{-}3)$$
$$= \sqrt{3} \times 110 \text{ V} \times 34.6 \text{ A} \times \cos 20°$$
$$= 6194 \text{ W}$$
$$= 6.194 \text{ kW}.$$

Example 97-2

In a three-phase delta connected alternator the phase voltages are 160 $\angle 80°$ V, 160 $\angle -160°$ V and 160 $\angle -40°$ V. The corresponding three unbalanced loads are 20 $\angle 20°$ Ω, 40 $\angle -30°$ Ω and 80 $\angle 130°$ Ω.

Solution

The phase currents are:

$$\frac{160 \angle 80° \text{ V}}{20 \angle 20° \ \Omega} = 8 \angle 60° \text{ A.}$$

$$\frac{160 \angle -160° \text{ V}}{40 \angle -30° \ \Omega} = 4 \angle -130° \text{ A}$$

$$\frac{160 \angle -40° \text{ V}}{80 \angle 130° \ \Omega} = 2 \angle -170° \text{ A}$$

The corresponding line currents are:

$$8 \angle 60° - 2 \angle -170° = (4 + j6.928) - (-1.97 - j0.347)$$
$$= 5.97 + j7.27$$
$$= 9.41 \angle 50.6° \text{ A}$$

$$2 \angle -170° - 4 \angle -130° = (-1.97 - j0.347) - (-2.57 - j3.064)$$
$$= 0.60 + j2.72$$
$$= 2.79 \angle 77.6° \text{ A}$$

$$4 \angle -130° - 8 \angle 60° = (-2.57 - j3.064) - (4 + j6.928)$$
$$= -6.57 - j9.99$$
$$= 11.96 \angle -123.3° \text{ A}$$

Although the loads are unbalanced, the total of the three line currents is

$$(5.97 + j7.27) + (0.60 + j2.72) + (-6.57 - j9.99) = 0$$

Chapter 98
The decibel

ON MANY OCCASIONS ELECTRONICS AND COMMUNICATIONS ARE CONCERNED with the transmission of alternating power from one position to another. The various lines and items of equipment (which constitute the transmission system) introduce both power gains and losses.

Consider a network that joins an alternating source (for example, an rf generator) to a load. Let the input power be P_i and the output power be P_o. The ratio of the output power to the input power is the power ratio P_o/P_i (Fig. 98-1A). A network such as an attenuator introduces a loss, in which case P_o/P_i is less than unity. By contrast an amplifier provides a gain so that P_o/P_i is greater

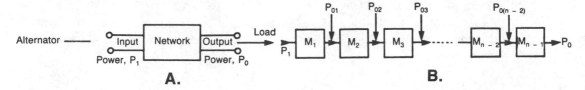

Figure 98-1

than unity. If a number of these networks are connected in cascade (Fig. 98-1B), and the individual power ratios are known, the overall power ratio is obtained by multiplying together the individual power ratios.

In a complex system containing a variety of networks each contributing a gain or loss, the calculation of the overall power ratio may become extremely laborious. To simplify the calculation, the individual power ratios are expressed by a logarithmic unit, enabling the algebraic sum to be employed in place of multiplication. The logarithmic unit employed is the decibel (abbreviated to dB) and the power gain or loss of a network is then expressed as 10 log (P_o/P_i) dB where "log" is used as an abbreviation for *common logarithm to the base 10*.

The common logarithm of a number is the power or exponent to which the base (10) must be raised in order that the result is equal to the value of the number. For example, log 100000 = 5 because 10^5 = 100000 whereas log 0.01 = 1/100 = -2 since 10^{-2} = $1/10^2$ = 1/100. Therefore, a power gain of 100000 is equivalent to 10 log 100000 = 10 × 5 = 50 dB, while a power loss ratio of 0.01 corresponds to 10 log 0.01 = 10 × (-2) = -20 dB. If an attenuator whose power loss is 0.01 is followed by an amplifier with a power gain ratio of 100000, the overall power ratio is 100000 × 0.01 = 1000 or +30 dB. The same result is obtained by taking the algebraic sum of +50 dB and -20 dB [(+50 dB) + (-20 dB) = +30 dB]. Corresponding values of power ratios and decibels are shown in Table 98-1.

Voltage E_o/E_i ratio, or current I_o/I_i ratio,	Power ratio, P_o/P_i	N decibels (dB)	
1000	1000000	60	
316	100000	50	
100	10000	40	Power
31.6	1000	30	Gain
10	100	20	
3.16	10	10	
1	1	0	
0.316	0.1	-10	
0.1	0.01	-20	
0.0316	0.001	-30	Power
0.01	0.0001	-40	Loss
0.00316	0.00001	-50	
0.001	0.000001	-60	

Table 98-1

A larger logarithmic unit is the bel which is equivalent to 10 decibels and named after Alexander Graham Bell. The decibel was originally related to acoustics and was regarded as the smallest change in sound intensity that could be detected by the human ear.

The decibel is essentially a unit of power ratio and not of absolute power; but if some standard reference level of power is assumed to represent 0 dB, then any value of absolute power can be expressed as so many dB above or below this reference standard. While various other standards are occasionally encountered, the standard most commonly adopted is 1 mW (0.001 W) which is delivered along a standard 600 Ω program transmission line. This means that one milliwatt of single tone audio power is 0 dBm (dB with respect to 1 mW). Then one watt, for example, is 10 log (1 W/1 mW) = 10 log 1000 = 30 dBm, while one microwatt is 10 log (1 μW/1 mW) = 10 log (1/1000) = −30 dBm.

Zero VU (volume unit) has the same reference level of 1 mW and it therefore appears that the dBm and the VU are identical. However, the use of the dBm is normally confined to single frequencies (tones) while the VU is reserved for complex audio signals such as speech or music. When it is desired to compare the powers developed in equal input and output resistances, it is sufficient to use their associated voltages and currents. The power gain or loss in decibels is then equal to twenty times the common logarithm of the voltage or current ratio.

MATHEMATICAL DERIVATIONS
In Fig. 98-1A,

The power gain or loss of a network is expressed as

$$N = 10 \log. \ (P_0/P_i) \ \text{dB} \tag{98-1}$$

where: N = power gain or loss (dB)

P_0 = output power (W)

P_i = input power (W)

Positive and negative values of N respectively represent power gains and losses.

Power ratio,

$$\frac{P_0}{P_i} = \text{inv} \log \frac{N}{10} = 10^{N/10} \tag{98-2}$$

In Fig. 98-1B,

Overall power ratio,

$$\frac{P_0}{P_i} = \frac{P_{01}}{P_i} \times \frac{P_{02}}{P_{01}} \times \frac{P_{03}}{P_{02}} \times \cdots \times \frac{P_0}{P_{0(n-1)}} \tag{98-3}$$

Total dB gain,

$$N_T = N_1 + N_2 + N_3 \cdots + N_N \tag{98-4}$$

where N_T is the algebraic sum of the individual gains and losses.

Absolute power

Absolute power,

$$P = 10 \log \frac{P}{1 \ \text{mW}} \ \text{dBm} \tag{98-5}$$

Voltage and current ratios

Consider equal input and output resistances for a particular stage such as an amplifier or an attenuator. If the input and output voltages have rms values of E_i and E_0 and are associated with input and output rms currents of I_i and I_0,

Input power,

$$P_i = E_i \times I_i = I_i^2 \times R = E_i^2/R \text{ watts} \tag{98-6}$$

Output power,

$$P_0 = E_0 \times I_0 = I_0^2 \times R = E_0^2/R \text{ watts} \tag{98-7}$$

Therefore the power gain in decibels is given by

$$N = 10 \log \left(\frac{P_0}{P_i}\right) = 10 \log \left(\frac{I_0^2 R}{I_i^2 R}\right) = 10 \log \left(\frac{E_0^2/R}{E_i^2/R}\right) \tag{98-8}$$

$$= 20 \log \left(\frac{I_0}{I_i}\right) = 20 \log \left(\frac{E_0}{E_i}\right) \text{dB}$$

The power gain in decibels is equal to 20 log (current ratio) or 20 log (voltage ratio) provided that the input and output resistances are the same.

If the input and output powers are associated with input and output impedances of $Z_i = R_i + jX_i$ and $Z_0 = R_0 + jX_0$ respectively, the formula $N = 20 \log (I_0/I_i)$ and $20 \log (V_0/V_i)$ may still be used provided $R_i = R_0$. If the input and output resistances are not equal, the formulas become

$$N = 10 \log \left(\frac{E_0^2/R_0}{E_i^2/R_i}\right) \tag{98-9}$$

$$= 20 \log \left(\frac{E_0}{E_i}\right) - 10 \log \left(\frac{R_0}{R_i}\right) \text{dB}$$

and

$$N = 10 \log \left(\frac{I_0^2 R_0}{I_i^2 R_i}\right) \tag{98-10}$$

$$= 20 \log \left(\frac{I_0}{I_i}\right) + 10 \log \left(\frac{R_0}{R_i}\right) \text{dB}$$

Example 98-1

1. Convert power ratios of (a) 273 and (b) 0.0469 into decibels.
2. Convert (a) +37 dB and (b) −14.6 dB into their corresponding power ratios.

Solution

1. (a) Number of decibels,

$$N = 10 \log \left(\frac{P_0}{P_i}\right) \text{dB} \tag{98-1}$$

$$= 10 \log 273$$

$$= 10 \times 2.436$$

$$= +24.36 \text{ dB}$$

(b) Number of decibels,

$$N = 10 \log \left(\frac{P_0}{P_i} \right) \text{ dB} \qquad (98\text{-}1)$$

$$= 10 \log 0.0469$$

$$= -13.3 \text{ dB}$$

2. (a) Power ratio,

$$\frac{P_0}{P_i} = \text{inv log} \left(\frac{N}{10} \right) \qquad (98\text{-}2)$$

$$= \text{inv log} \left(\frac{37}{10} \right)$$

$$= \text{inv log } 3.7 = 5012$$

(b) Power ratio,

$$\frac{P_0}{P_i} = \text{inv log } \frac{N}{10} \qquad (98\text{-}2)$$

$$= \text{inv log} \left(\frac{-14.6}{10} \right)$$

$$= 0.0347$$

Example 98-2

In Fig. 98-2, calculate the value of the overall power ratio. Express each of the individual power ratios as well as the overall power ratio in terms of dB.

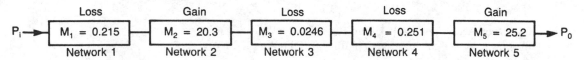

Figure 98-2

Solution

Overall power ratio

$$N_T = \frac{P_{01}}{P_i} \times \frac{P_{02}}{P_{01}} \times \frac{P_{03}}{P_{02}} \times \frac{P_{04}}{P_{03}} \times \frac{P_0}{P_{04}} \qquad (98\text{-}3)$$

$$= M_1 \times M_2 \times M_3 \times M_4 \times M_5$$

$$= 0.215 \times 20.3 \times 0.0246 \times 0.251 \times 25.2$$

$$= 0.679$$

The overall power ratio is less than unity and therefore represents a loss of $10 \log 0.679 = -1.69$ dB.

In terms of decibels

$$N_1 = 10 \log M_1 = 10 \log 0.215 = -6.68 \text{ dB.}$$

$$N_2 = 10 \log M_2 = 10 \log 20.3 = +13.07 \text{ dB.}$$

$$N_3 = 10 \log M_3 = 10 \log 0.0246 = -16.09 \text{ dB.}$$
$$N_4 = 10 \log M_4 = 10 \log 0.251 = -6.00 \text{ dB.}$$
$$N_5 = 10 \log M_5 = 10 \log 25.2 = +14.01 \text{ dB.}$$
$$N_T = 10 \log M_T = 10 \log 0.679 = -1.68 \text{ dB.}$$

Then

$$
\begin{aligned}
N_T &= N_1 + N_2 + N_3 + N_4 + N_5 \\
&= (-6.68) + (+13.07) + (-16.09) + (-6.00) + (+14.01) \\
&= -1.69 \text{ dB.}
\end{aligned}
$$

The overall gain or loss in dB for a number of cascaded stages is the algebraic sum of the decibel gains and losses in the individual stages.

Example 98-3

1. Express power levels of (a) 6.38 W and (b) 26.7 μW in terms of dBm.
2. What power levels are represented by (a) +17 dBm and (b) −8.6 VU?

Solution

1. (a) A power level of 6.38 W is equivalent to 6380 mW. Therefore,

$$6.38 \text{ W} = 10 \log(6380 \text{ mW/1 mW})$$
$$= 38.05 \text{ dBm.}$$

(b) A power level of 26.7 μW is equivalent to 0.0267 mW. Therefore,

$$26.7 \ \mu\text{W} = 10 \log(0.0267 \text{ mW/1 mW})$$
$$= -15.73 \text{ dBm.}$$

2. (a) Power level = inv log(+17 dBm/10) = 50.12 mW.
   (b) Power level = inv log(−8.6 VU/10) = 0.138 mW = 138 μW.

Example 98-4

1. Assuming that the input and output resistances are equal, convert (a) +43.7 dB and (b) −27.6 dB in their corresponding voltage (or current) ratios.
2. Assuming equal input and output resistances, convert (a) a voltage ratio of 56.3 and (b) a current ratio of 0.00353 into decibels.

Solution

1. (a) Voltage (or current) ratio,

$$\frac{V_0}{V_i} = \text{inv log}\left(\frac{43.7}{20}\right)$$
$$= 153.$$

(b) Voltage (or current) ratio,

$$\frac{V_0}{V_i} = \text{inv log}\left(\frac{-27.6}{20}\right)$$
$$= 0.0417.$$

2. (a) Number of decibels,

$$N = 20 \log\left(\frac{E_0}{E_i}\right) \qquad (98\text{-}8)$$

$$= 20 \log 56.3$$

$$= 35.01 \text{ dB.}$$

(b) Number of decibels,

$$N = 20 \log\left(\frac{I_0}{I_i}\right) \qquad (98\text{-}8)$$

$$= 20 \log 0.00353$$

$$= -49.04 \text{ dB.}$$

Example 98-5

The input resistance of an amplifier is 1.3 kΩ while its output resistance is 17 kΩ. If the amplifier voltage gain is 35, calculate its power gain in dB.

Solution

Power gain,

$$\frac{P_0}{P_i} = 20 \log\left(\frac{E_0}{E_i}\right) - 10 \log\left(\frac{R_0}{R_i}\right) \qquad (98\text{-}9)$$

$$= 20 \log 35 - 10 \log\left(\frac{17}{1.3}\right)$$

$$= 30.88 - 8.59$$

$$= 22.33 \text{ dB}$$

Chapter 99
The neper

THE DECIBEL IS FUNDAMENTALLY A UNIT OF POWER RATIO, BUT IT CAN BE USED to express current ratios when the resistive components of the impedances through which the currents flow are equal, and the voltage ratios when the conductive components of those impedances are equal. The neper is fundamentally a unit of current ratio, but it can be used to express power ratios when the resistive components of the impedances are equal.

The loss of power in an electrical network is known as attenuation. Attenuation may be measured using either the decibel or the neper notation.

If the power entering a network is P_i and the power leaving is P_o, then the attenuation in decibels is defined as $10 \log (P_o/P_i)$. If the current entering a network is I_i and the current leaving

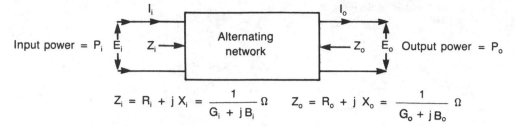

Figure 99-1

is I_o, then the attenuation in nepers is defined as $\ln (I_o/I_i)$ where ln is the natural or Napierian logarithm which is based on the exponential quantity, $e = 2.7183 \ldots$ (Fig. 99-1).

Because of its derivation from the exponential e, the neper is the most convenient unit for expressing attenuation in theoretical work. By contrast the decibel is defined in terms of common logarithms and is a more convenient unit in practical calculations which use the decimal system.

MATHEMATICAL DERIVATIONS
Attenuation in dB,

$$N = 10 \log \left(\frac{P_o}{P_i}\right) \tag{99-1}$$

$$= 20 \log \left(\frac{I_o}{I_i}\right), \text{ provided that } R_i = R_o \tag{99-2}$$

$$= 20 \log \left(\frac{E_o}{E_i}\right), \text{ provided that } G_i = G_o \tag{99-3}$$

Attenuation in nepers,

$$N' = \ln \left(\frac{I_o}{I_i}\right) = \ln \left(\frac{E_o}{E_i}\right), \text{ provided that } Z_i = Z_o \tag{99-4}$$

$$= \frac{1}{2} \ln \left(\frac{P_o}{P_i}\right), \text{ provided that } R_i = R_o \tag{99-5}$$

If the resistive components of the impedances at the input and output of the network are equal, the amount of attenuation may be readily converted from one unit to another.
Attenuation in dB,

$$N = 20 \log \left(\frac{I_o}{I_i}\right) \tag{99-6}$$

$$= 20 \ln \left(\frac{I_o}{I_i}\right) \times \log e$$

$$= 8.686 \times \ln \left(\frac{I_o}{I_i}\right)$$

$$= 8.686 \times \text{(attenuation in nepers)}$$

Therefore the attenuation in dB $= 8.686 \times$ (attenuation in nepers), provided that $R_i = R_o$, or the attenuation in nepers $= 0.1151 \times$ (attenuation in dB), provided that $R_i = R_o$.

340

Example 99-1

The input to an electrical network is 27 mA and the output current is 75 μA. Assuming that the input and output resistances are equal, calculate the attenuation of the network in (a) decibels and (b) nepers.

Solution

(a)
$$\text{Current ratio} = \frac{I_o}{I_i} = \frac{75 \ \mu\text{A}}{27 \ \text{mA}} \tag{99-2}$$
$$= 2.77 \times 10^{-3}$$

Attenuation in dB,

$$N = 20 \log (2.77 \times 10^{-3})$$
$$= -51.15 \text{ dB}$$

(b) Attenuation in nepers,

$$N' = \ln (2.77 \times 10^{-3}) \tag{99-4}$$
$$= -5.889 \text{ nepers.}$$

Check:

$$\frac{\text{Attenuation in dB}}{\text{Attenuation in nepers}} = \frac{51.15}{5.889} = 8.686, \text{ rounded off.}$$

Chapter 100
Waveform analysis

ANY SINGLE-VALUED, FINITE AND CONTINUOUS FUNCTION, *F(T)* HAVING A PERIOD of $2\pi/\omega$ seconds, may be expressed in the following form:

FOURIER'S THEOREM

$$f(t) = a_0 + A_1 \sin (\omega t + \theta_1) + A_2 \sin (2\omega t + \theta_2) \tag{100-1}$$
$$+ A_3 \sin (3\omega t + \theta_3) + \cdots$$

where: $\omega = 2\pi f$ radians per second

t = time in seconds

Because

$$A \sin (\omega t + \theta) = A \sin \omega t \cos \theta + A \cos \omega t \sin \theta$$
$$= a \cos \omega t + b \sin \omega t$$

where: $a = A \sin \theta$

$b = A \cos \theta,$

the Fourier expansion may be expressed as:

$$f(t) = a_0 + a_1 \cos \omega t + a_2 \cos 2\omega t + a_3 \cos 3\omega t + \cdots \qquad (100\text{-}2)$$
$$+ \, b_1 \sin \omega t + b_2 \sin 2\omega t + b_3 \sin 3\omega t + \cdots$$

This means that the waveform may be regarded as composed of a mean level (a_0) together with fundamental sine and cosine waves as well as their harmonics.

In order to use the expansion to analyze a complex waveform, it is necessary to determine the values of the coefficients a_0, a_1, $a_2 \ldots$, b_1, $b_2 \ldots$. This is done by using integral calculus which is outlined in the mathematical derivations. However, the following is the quoted result for

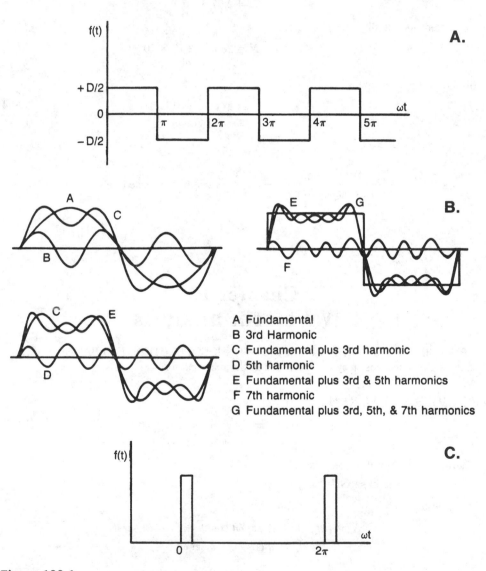

Figure 100-1

342

the symmetrical square wave (Fig. 100-1A) which is frequently encountered in communications. For such a square wave whose mean level is zero, the expansion is:

$$f(t) = \frac{2D}{\pi} \left[\sin \omega t + \frac{1}{3} \sin 3\omega t + \frac{1}{5} \sin 5\omega t + \cdots + \frac{1}{n} \sin n\omega t + \cdots \right] \qquad (100\text{-}3)$$

Due to the symmetry of the square wave the expansion contains neither cosine terms nor even harmonic sine terms.

Notice that the amplitude of the nth harmonic is $1/n$ so that the amplitudes of the higher harmonics only decrease slowly. It follows that the symmetrical square wave contains a large number of strong harmonics.

The expansion of equation 100-3 may be verified by adding the fundamental sinewave component and its odd harmonics. This is called waveform synthesis which is illustrated in Fig. 100-1B. However, waveform "G" is far from being a symmetrical square and many more odd harmonics would be necessary before achieving a good approximation to the required waveform.

If the square waveform is made very asymmetrical, its appearance is that of a repeating pulse of short duration (Fig. 100-1C); as an example this would be the modulation waveform in a pulsed radar set. Both odd and even harmonics then appear in the expansion but their amplitudes decrease very slowly. This waveform is therefore extremely rich in harmonics and to achieve a good approximation in its synthesis, we might well have to include harmonics higher than the thousandth.

MATHEMATICAL DERIVATIONS

Figure 100-2 shows a square waveform which is a single-valued repeating function of time with a period of $2\pi/\omega$ seconds; consequently, it may be analyzed by Fourier's Theorem.

From $\omega t = 0$ to $\omega t = 2\pi$, the equation of the function is $f(t) = D$. From $\omega t = \pi$ to $\omega t = 2\pi$, the equation of the function is $f(t) = 0$.

Figure 100-2

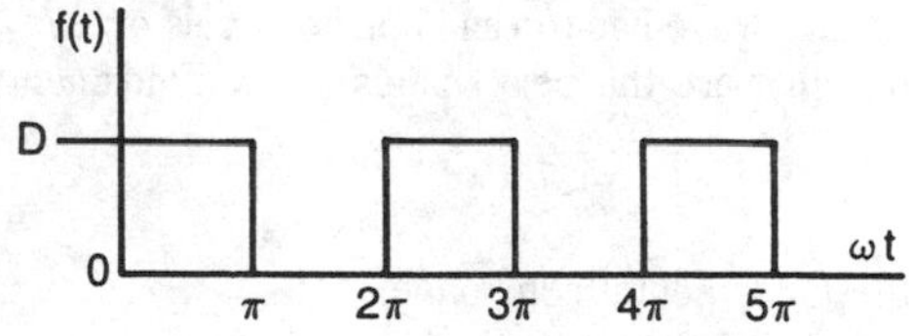

Then

$$a_0 = \frac{1}{2\pi} \int_0^{2\pi} f(t) \, d(\omega t) \qquad (100\text{-}4)$$

$$= \frac{1}{2\pi} \int_0^{\pi} f(t) \, d(\omega t) + \frac{1}{2\pi} \int_{\pi}^{2\pi} f(t) \, d(\omega t)$$

$$= \frac{1}{2\pi} \times D \times \pi + 0 = \frac{D}{2}$$

$$a_n = \frac{1}{\pi} \int_0^{2\pi} f(t) \cos n\omega t \, d(\omega t) \qquad (100\text{-}5)$$

$$= \frac{1}{\pi} \left[\int_0^{\pi} D \cos n\omega t \, d(\omega t) + \int_{\pi}^{2\pi} 0 \times \cos n\omega t \, d(\omega t) \right]$$

$$= \frac{1}{\pi} \left[\frac{D \sin n\omega t}{n} \right]_0^\pi$$

$$= 0$$

All cosine terms are therefore zero.

$$b_n = \frac{1}{\pi} \int_0^{2\pi} f(t) \sin n\omega t \, d(\omega t) \qquad (100\text{-}6)$$

$$= \frac{1}{\pi} \left[\int_0^\pi D \sin n\omega t \, d(\omega t) + \int_\pi^{2\pi} 0 \times \sin n\omega t \, d(\omega t) \right]$$

$$= \frac{1}{\pi} \left[\frac{-D \cos n\omega t}{n} \right]_0^\pi$$

$$= \frac{D}{n\pi} (1 - \cos n\pi)$$

When n is an odd number, $(1 - \cos n\pi) = 2$. When n is an even number, $(1 - \cos n\pi) = 0$.
The required equation for the square wave of Fig. 100-2 is

$$f(t) = \frac{D}{2} + \frac{2D}{\pi} \left(\sin \omega t + \frac{1}{3} \sin 3\omega t + \frac{1}{5} \sin 5\omega t + \cdots \frac{1}{n} \sin n\omega t + \cdots \right) \qquad (100\text{-}7)$$

For the square wave of Fig. 100-1A, the average value is zero and its Fourier equation is

$$f(t) = \frac{2D}{\pi} \left(\sin \omega t + \frac{1}{3} \sin 3\omega t + \frac{1}{5} \sin 5\omega t + \cdots + \frac{1}{n} \sin n\omega t + \cdots \right) \qquad (100\text{-}8)$$

which matches with Equation 100-3.

Example 100-1

A symmetrical square wave has top and bottom levels of $+7$ V and -7 V. In the Fourier analysis of the waveform, what are the peak values of the fundamental and fifth harmonic components?

Solution

Peak value of the fundamental component,

$$= \frac{2D}{\pi} \qquad (100\text{-}7)$$

$$= \frac{2 \times (2 \times 7) \text{ V}}{\pi}$$

$$= 8.91 \text{ V}.$$

Peak value of the fifth harmonic component,

$$= \frac{8.91 \text{ V}}{5} \qquad (100\text{-}7)$$

$$= 1.78 \text{ V}.$$

3
PART

Solid-state devices and their associated circuits

Chapter 101
The pn junction diode

THE MODERN POWER RECTIFIER DEVICE IS THE SILICON DIODE, WHICH basically uses a pn junction to provide the "one-way" action. The p-type material is produced by doping pure silicon with an acceptor or a trivalent impurity such as indium, gallium, or boron. The n-type silicon is created by doping with a donor or pentavalent impurity of arsenic, antimony, or phosphorus. Although both n-type and p-type materials are electrically neutral, the majority charge carriers in the n-type and p-type semiconductors are respectively, negative electrons and positive holes.

When the silicon pn junction is formed, there is a movement of the majority carriers across the junction. At room temperature this creates an internal potential barrier of about 0.7 V across a depletion region, which is in the immediate vicinity of the junction and is devoid of majority charge carriers. In order to forward-bias the junction, it is necessary to connect the positive terminal of a voltage source to the p region and the negative terminal to the n region (Fig. 101-1A). However, the applied bias must exceed the 0.7 V internal potential before appreciable forward current will flow (Fig. 101-1B).

For a general-purpose silicon diode, with the p region as the anode and the n region as the cathode, a forward voltage, V_F, of about 1 V will correspond to a forward current, I_F, of several hundred milliamperes. The small voltage drop across the conducting silicon diode is its real advantage when compared with other rectifier devices because even with high forward currents, the power dissipation will be low.

To create a reverse bias (Fig. 101-1C), the positive terminal of a voltage source is connected to the n region (cathode) and the negative terminal to the p region (anode). If only the majority carriers were involved, there would be zero reverse current; however thermal energy creates minority carriers (electrons in the p region and holes in the n region) so that there is a small reverse current, I_{co}, whose value rises rapidly and saturates at about 1 μA. However, the saturated value of I_{co} depends on the diode's temperature and will approximately double for every 10° Celsius rise. Typically a reverse bias, V_R, of about 250 V to 500 V may be applied with only a small reverse current, I_R. However, with excessive reverse voltage there is a breakdown point at which there is a rapidly increasing flow of reverse current; this action is the principle behind the zener diode which is used for voltage regulation purposes.

MATHEMATICAL DERIVATIONS

In the roughest or first approximation, the forward-biased diode is regarded as a short circuit. To a second approximation we take into account the 0.7 V drop across the diode and in the third approximation we include the bulk resistance of the semiconductor material from which the diode is made. The equation for the third approximation is

$$I_F = \frac{V_F - 0.7\text{ V}}{R_B} \text{ amperes} \qquad (101\text{-}1)$$

where: I_F = forward current (A)

V_F = forward voltage (V)

R_B = bulk resistance (Ω)

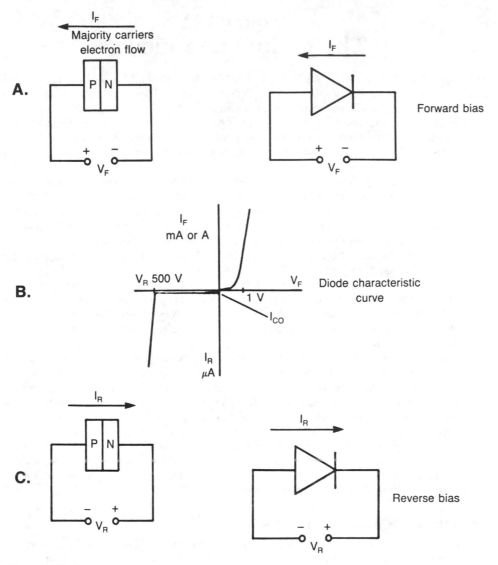

Figure 101-1

$$\text{Power dissipated in the diode, } P_D = V_F \times I_F \text{ watts} \qquad (101\text{-}2)$$

Example 101-1

For a particular silicon diode, V_F is 0.9 V and the bulk resistance is 0.5Ω. What is the corresponding value of I_F?

Solution

Forward current,

348

$$I_F = \frac{V_F - 0.7\text{ V}}{R_B}$$

$$= \frac{0.2\text{ V}}{0.5\ \Omega}$$

$$= 400\text{ mA}$$

(101-1)

Chapter 102
The zener diode

THE BASIC ZENER DIODE VOLTAGE REGULATOR CIRCUIT IS SHOWN IN FIG. 102-1A. Its purpose is to stabilize the value of the load voltage, V_L, against changes in the source voltage, E, and/or changes in the load current, I_L. The diode operates with reverse bias in the breakdown region (Fig. 102-1B) where to a first approximation, the diode's reverse current, I_Z, is independent of the diode voltage, V_Z, which is equal to the load voltage, V_L. The value of the resistor, R, is such that it allows the diode to operate well within the breakdown region and at the same time regulates the load voltage by dropping the difference between V_Z and the unregulated source voltage, E.

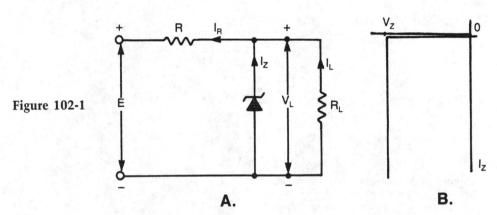

Figure 102-1

A.   B.

MATHEMATICAL DERIVATIONS

Using Kirchhoff's current law (KCL),
    Resistor current,

$$I_R = I_L + I_Z \text{ milliamperes}$$

(102-1)

In addition,
    Load current,

$$I_L = \frac{V_L}{R_L} = \frac{V_Z}{R_L} \text{ milliamperes}$$

(102-2)

349

Consequently,

Resistor current,

$$I_R = \frac{E - V_Z}{R} = \frac{E - V_L}{R} \text{ milliamperes} \qquad (102\text{-}3)$$

The above equations refer to the roughest or first approximation. However, because the breakdown curve is not completely vertical, a second approximation will involve the zener diode's resistance, R_Z, which is derived from the slope of its characteristic. Therefore:

Load voltage,

$$V_L = V_Z + I_Z R_Z \text{ volts} \qquad (102\text{-}4)$$

Finally, the power dissipation in the zener diode is given by:

Zener diode power dissipated,

$$P_Z = I_Z V_Z \text{ watts} \qquad (102\text{-}5)$$

The value of P_Z must not exceed the diode's power rating.

Example 102-1

In Fig. 102-1A, the zener diode's breakdown voltage is 9 V. If $E = 50$ V, $R = 4$ kΩ, $R_L = 2.5$ kΩ, calculate the values of I_R, I_Z, I_L, and P_Z. If E changes to 60 V, recalculate the values of I_R, I_Z, I_L, and P_Z. If R_L changes to 5 kΩ, recalculate the values of I_R, I_Z, I_L, and P_Z.

Note: Use the first or roughest approximation throughout and neglect any effect of R_Z.

Solution

The dropping resistor current is given by:

$$I_R = \frac{E - V_Z}{R} = \frac{50 \text{ V} - 9 \text{ V}}{4 \text{ k}\Omega} \qquad (102\text{-}3)$$

$$= 10.25 \text{ mA}$$

The load current,

$$I_L = \frac{V_Z}{R_L} = \frac{9 \text{ V}}{2.5 \text{ k}\Omega} \qquad (102\text{-}2)$$

$$= 3.6 \text{ mA}$$

Zener diode current,

$$I_Z = I_R - I_L = 10.25 - 3.6 \qquad (102\text{-}1)$$

$$= 6.65 \text{ mA}$$

Power dissipation,

$$P_Z = I_Z V_Z = 6.65 \text{ mA} \times 9 \text{ V} \qquad (102\text{-}5)$$

$$= 60 \text{ mW, rounded off}$$

If $E = 60$ V,

$$I_R = \frac{60 \text{ V} - 9 \text{ V}}{4 \text{ k}\Omega}$$

$$= 12.75 \text{ mA}$$

$$I_L = 3.6 \text{ mA}$$

$$I_Z = 12.75 - 3.6 = 9.15 \text{ mA}$$

$$P_Z = 9.15 \text{ mA} \times 9 \text{ V} = 82 \text{ mW}$$

The diode current has risen by $9.15 - 6.65 = 2.5$ mA. The increase in the voltage drop across is $2.5 \text{ mA} \times 4 \text{ k}\Omega = 10$ V which absorbs the change in the source voltage from 50 V to 60 V.

If $R_L = 5 \text{ k}\Omega$,

$$I_R = \frac{50 \text{ V} - 9 \text{ V}}{4 \text{ k}\Omega}$$

$$= 10.25 \text{ mA}$$

$$I_L = \frac{9 \text{ V}}{5 \text{ k}\Omega}$$

$$= 1.8 \text{ mA}$$

$$I_Z = 10.25 - 1.8$$

$$= 8.45 \text{ mA}$$

$$P_Z = 8.45 \text{ mA} \times 9 \text{ V}$$

$$= 76 \text{ mW, rounded off}$$

The decrease of $3.6 - 1.8 = 1.8$ mA in the load current is balanced by an equal increase of $8.45 - 6.65 = 1.8$ mA in the diode current. In his example a zener diode with a power rating of 1/2 W would be adequate.

Chapter 103
The bipolar transistor

THE ACTION OF THE BIPOLAR TRANSISTORS INVOLVES TWO SETS OF CHARGE carriers; these are the positive charge carriers (holes) and the negative charge carriers (electrons). Bipolar transistors are also of two types, pnp and npn, whose symbols are shown in Fig. 103-1A. Each of these types possesses three sections which are called the emitter, the base, and the collector; the arrow in each symbol indicates the direction of conventional flow (the opposite direction to that of the electron flow).

The bipolar transistor may be roughly compared to a triode with the emitter, base, and collector corresponding to the cathode, control grid, and plate. For normal conditions, the voltages are applied to the transistor so as to forward-bias the emitter/base junction but to reverse bias the collector/base junction. For the pnp transistor (Fig. 103-1B), this requires that the emitter is positive with respect to the base, which is in turn positive with respect to the collector. These polarities are reversed for the npn transistor. Some of the methods of producing the correct dc bias voltage for the emitter/base and collector/base junctions are discussed in Chapters 105, 106, and 107. Under either static or signal conditions, emitter current (I_E), base current

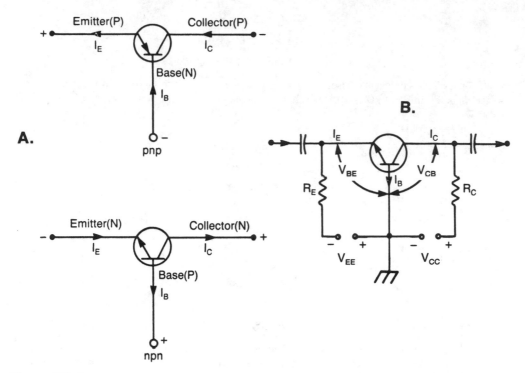

Figure 103-1

(I_B) and collector current (I_C) flow in a bipolar transistor circuit (Fig. 103-1A). Applying Kirchhoff's current law:

$$I_E = I_C + I_B, \ I_C = I_E - I_B, \ I_B = I_E - I_C$$

Under normal operating conditions, I_B is typically less than 5% of either I_E or I_C, and I_C is only slightly less than I_E.

MATHEMATICAL DERIVATIONS

In the npn circuit of Fig. 103-1B, the emitter/base junction is forward-biased by V_{EE}, while the collector/base junction is reversed-biased by V_{CC}. The actual forward bias applied between the emitter and base is only a few tenths of a volt, so that:

Emitter current,

$$I_E = \frac{V_{EE} - V_{BE}}{R_E} \text{ milliamperes} \qquad (103\text{-}1)$$

or to a first approximation,

Emitter current,

$$I_E \approx \frac{V_{EE}}{R_E} \text{ milliamperes} \qquad (103\text{-}2)$$

The voltage between the collector and base is:

$$V_{CB} = V_{CC} - \text{(the voltage drop across } R_C)\qquad(103\text{-}3)$$
$$= V_{CC} - I_C R_C \text{ volts}$$

Under quiescent conditions, the static current gain, α dc is defined as the ratio of the (output) collector current, I_C, to the (input) emitter current, I_E.

Then

Current gain,

$$\alpha_{dc} = \frac{I_C}{I_E} = \frac{I_E - I_B}{I_E}\qquad(103\text{-}4)$$

$$= 1 - \frac{I_B}{I_E}$$

and

Collector/base voltage,

$$V_{CB} = V_{CC} - I_C R_C\qquad(103\text{-}5)$$
$$= V_{CC} - \alpha_{dc} I_E R_C \text{ volts}$$

Because I_C, is less than I_E, the value of α dc is less than 1 and is typically 0.95 to 0.99.

In other circuit arrangements, the static current gain, β_{dc}, is defined as the ratio of the collector current to the base current.

Therefore,

Static emitter gain,

$$\beta_{dc} = \frac{I_C}{I_B} = \frac{I_C}{I_E - I_C} = \frac{I_C/I_E}{1 - I_C/I_E}\qquad(103\text{-}6)$$

$$= \frac{\alpha_{dc}}{1 - \alpha_{dc}}$$

This yields

$$\alpha_{dc} = \frac{\beta_{dc}}{1 + \beta_{dc}}\qquad(103\text{-}7)$$

Typical values of β_{dc} vary between 50 and 200. β_{dc} is, however, an unstable quantity and changes both with temperature and the choice of the operating conditions.

Example 103-1

Assume that the circuit of Fig. 103-1B, is being operated under static conditions with $V_{EE} = 12$ V, $R_E = 22$ kΩ, $V_{BE} = 0.7$ V, $V_{CC} = 15$ V, $\alpha_{dc} = 0.97$. Calculate the values of I_E, I_C and V_{CB}.

Solution

Emitter current,

$$I_E = \frac{V_{EE} - V_{BE}}{R_E}\qquad(103\text{-}1)$$

$$= \frac{12 \text{ V} - 0.7 \text{ V}}{22 \text{ k}\Omega}$$

$$= 0.514 \text{ mA}$$

Collector current,

$$I_C = \alpha_{dc} I_E \qquad (103\text{-}4)$$

$$= 0.97 \times 0.514$$

$$= 0.4986 \text{ mA}$$

Base current, $I_B = I_E - I_C = 0.514 - 0.4986$

$$= 0.0154 \text{ mA}$$

Collector-base voltage,

$$V_{CB} = V_{CC} - I_C R_C \qquad (103\text{-}3)$$

$$= 15 \text{ V} - 0.4986 \text{ mA} \times 15 \text{ k}\Omega$$

$$= 7.5 \text{ V}$$

Chapter 104
Base bias

FOR THE NORMAL OPERATION OF A BIPOLAR TRANSISTOR THE EMITTER/BASE junction must be forward biased while the collector/base junction is reverse biased. The circuit shown in Fig. 104-1A contains an npn transistor whose emitter/base junction is forward biased by the source, V_{BB}. This voltage is connected in series with R_B between *base* and ground and, for this reason, the arrangement is referred to as base bias. In order that the amplifier should behave in a linear manner, it is necessary that the dc or quiescent (no signal) conditions should remain as stable as possible. Unfortunately, the mathematical derivations show that base bias is sensitive to the value of β_{dc} which in turn is affected by temperature variations.

MATHEMATICAL DERIVATIONS

The KVL equation for the loop GBEG is

$$-V_{BE} + I_E R_B + V_{BB} + I_E R_E = 0 \qquad (104\text{-}1)$$

However, the base current,

$$I_B = \frac{I_C}{\beta_{dc}} = \frac{I_E}{\beta_{dc}}$$

Therefore, the emitter current,

$$I_E = \frac{V_{BB} - V_{BE}}{R_E + \dfrac{R_B}{\beta_{dc}}} \text{ milliamperes} \qquad (104\text{-}2)$$

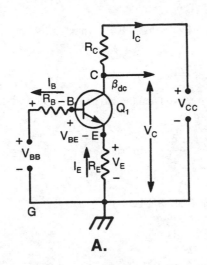

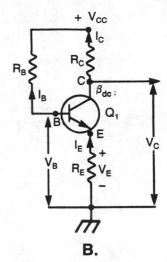

Figure 104-1

A.

B.

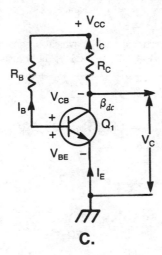

C.

If $Q1$ is a silicon transistor,

$$V_{BE} \approx 0.7 \text{ V.}$$

This yields

$$I_E = \frac{V_{BB} - 0.7}{R_E + \dfrac{R_B}{\beta_{dc}}} \text{ milliamperes} \tag{104-3}$$

Emitter voltage,

$$V_E = I_E R_E \text{ volts} \tag{104-4}$$

Collector voltage,

$$V_C = V_{CC} - I_C R_C \text{ volts} \tag{104-5}$$

Base voltage,

$$V_B = V_E + V_{BE} = V_E + 0.7 \text{ V volts} \qquad (104\text{-}6)$$

Collector/base voltage,

$$V_{CB} = V_C - V_B \text{ volts} \qquad (104\text{-}7)$$

In Fig. 104-1B, the source V_{BB} is removed and the circuit is operated with a single source, V_{CC}. Equation 104-2 then becomes

$$I_E = \frac{V_{CC} - 0.7 \text{ V}}{R_E + \dfrac{R_B}{\beta_{dc}}} \text{ milliamperes} \qquad (104\text{-}8)$$

If R_E is replaced by a short circuit (Fig. 104-1C).

$$I_E = \frac{V_{CC} - 0.7 \text{ V}}{R_B/\beta_{dc}} \qquad (104\text{-}9)$$

$$= \beta_{dc} \times \left(\frac{V_{CC} - 0.7 \text{ V}}{R_B}\right) \text{ milliamperes}$$

The emitter base current, I_E, is proportional to β_{dc}, which is sensitive to temperature variations. If β_{dc} increases to the point where V_{CB} falls below 1 V, the transistor is saturated and linear operation is impossible.

Example 104-1

In Fig. 104-1A, $R_B = 200 \text{ k}\Omega$, $R_C = 2 \text{ k}\Omega$, $R_E = 1 \text{ k}\Omega$, $V_{BB} = 10 \text{ V}$, $V_{CC} = 20 \text{ V}$, $\beta_{dc} = 100$. Calculate the values of I_E, V_E, V_C, V_B, and V_{CB}.

Solution

Emitter current,

$$I_E = \frac{10 \text{ V} - 0.7 \text{ V}}{1 \text{ k}\Omega + \dfrac{200 \text{ k}\Omega}{100}} = \frac{9.3 \text{ V}}{3 \text{ k}\Omega} \qquad (104\text{-}2)$$

$$= 3.1 \text{ mA}$$

Emitter voltage is given by

$$V_E = 3.1 \text{ mA} \times 1 \text{ k}\Omega \qquad (104\text{-}4)$$

$$= +3.1 \text{ V.}$$

Collector voltage is given by

$$V_C \approx 20 \text{ V} - 3.1 \text{ mA} \times 2 \text{ k}\Omega \qquad (104\text{-}5)$$

$$= +13.8 \text{ V.}$$

Base voltage is given by

$$V_B = 3.1 + 0.7 \qquad (104\text{-}6)$$

$$= +3.8 \text{ V.}$$

Collector-to-base voltage is given by

$$V_{CB} = 13.8 - 3.8 \tag{104-7}$$
$$= 10 \text{ V}.$$

Example 104-2

In Fig. 104-1B, $R_B = 100 \text{ k}\Omega$, $R_C = 680 \text{ }\Omega$, $R_E = 470 \text{ }\Omega$, $V_{CC} = 9 \text{ V}$, $\beta_{dc} = 80$. Calculate the values of I_E, V_E, V_C, V_B, V_{CB}.

Solution

Emitter current is given by

$$I_E = \frac{9 \text{ V} - 0.7 \text{ V}}{470 \text{ }\Omega + \dfrac{100 \text{ k}\Omega}{80}} = \frac{8.3 \text{ V}}{1720 \text{ }\Omega} \tag{104-7}$$
$$= 4.83 \text{ mA}.$$

Emitter voltage is given by

$$V_E = 4.83 \text{ mA} \times 470 \text{ }\Omega \tag{104-4}$$
$$= +2.27 \text{ V}.$$

Collector voltage is given by

$$V_C = 9 \text{ V} - 4.83 \text{ mA} \times 680 \text{ }\Omega \tag{104-5}$$
$$= +5.72 \text{ V}.$$

Base voltage is given by

$$V_B = 2.27 + 0.7 \tag{104-6}$$
$$= +2.97 \text{ V}.$$

Collector-to-base voltage is given by

$$V_{CB} = 5.72 - 2.97 \text{ V} \tag{104-7}$$
$$= 2.75 \text{ V}.$$

Example 104-3

In Fig. 104-1C, $R_B = 200 \text{ k}\Omega$, $V_{CC} = +12 \text{ V}$, $R_C = 2.2 \text{ k}\Omega$, $\beta_{dc} = 50$. What is the value of V_{CB}?

Solution

Emitter current is given by

$$I_E = \frac{12 \text{ V} - 0.7 \text{ V}}{200 \text{ k}\Omega/50} = \frac{11.3 \text{ V}}{4 \text{ k}\Omega} \tag{104-8}$$
$$= 2.83 \text{ mA}.$$
$$V_C = 12 \text{ V} - 2.83 \text{ mA} \times 2.2 \text{ k}\Omega \tag{104-9}$$
$$= +5.77 \text{ V}$$
$$V_B = +0.7 \text{ V}$$
$$V_{CB} = 5.77 - 0.7 = 5.07 \text{ V}$$

Notice that if, for example, β_{dc} increased to 200, I_E would be theoretically 11.3 mA and V_C would be negative—an impossible situation. Consequently, the transistor would then have entered the saturated state.

Chapter 105
Voltage divider bias

THE VOLTAGE DIVIDER CIRCUIT IS THE MOST IMPORTANT AND THE MOST commonly used of all the base-biased arrangements. The resistors, $R1$, $R2$, form a voltage divider (Fig. 105-1A) and are connected between the collector supply voltage (V_{CC}) and ground. The voltage drop across $R2$ then provides the necessary forward bias for the emitter-base junction. The inclusion of R_E is essential, because this resistor primarily determines the value of the emitter current, I_E. It follows that the voltage divider arrangement provides a stable value of I_E that is almost independent of any fluctuations in the value of β_{dc}. In addition this type of bias only requires a single voltage source (V_{CC}).

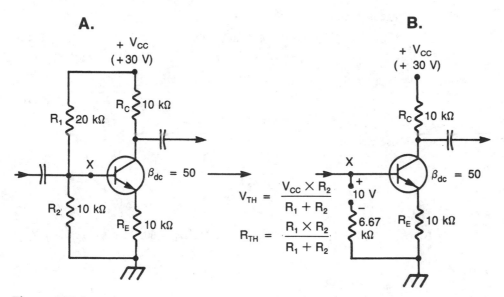

Figure 105-1

MATHEMATICAL DERIVATIONS

Thévenize the circuit of Fig. 105-1A at the point X. The equivalent circuit is then shown in Fig. 105-1B.

Thévenin generator voltage,

$$V_{TH} = V_{CC} \times \frac{R_2}{R_1 + R_2} \text{ volts} \tag{105-1}$$

Thévenin resistance,

$$R_{TH} = \frac{R_1 \times R_2}{R_1 + R_2} \text{ ohms} \qquad (105\text{-}2)$$

Regard V_{TH} as the forward base bias voltage and R_{TH} as the base resistor. Then from equation 104-2,

Emitter current,

$$I_E = \frac{V_{TH} - V_{BE}}{R_E + (R_1 \times R_2)/\beta_{dc}(R_1 + R_2)} \qquad (105\text{-}3)$$

$$= \frac{V_{TH} - 0.7 \text{ V}}{R_E + (R_1 \times R_2)/\beta_{dc}(R_1 + R_2)} \text{ milliamperes}$$

assuming that $V_{BE} = 0.7$ V.

Provided $R_E \gg (R_1 \times R_2)/\beta_{dc}(R_1 + R_2)$,

Emitter current,

$$I_E \approx \frac{V_{TH} - 0.7 \text{ V}}{R_E} \text{ milliamperes} \qquad (105\text{-}4)$$

The emitter current is largely independent of any fluctuations in the value of β_{dc}.

Example 105-1

In Fig. 105-1A, calculate the value of I_E.

Solution

Thévenin voltage,

$$V_{TH} = V_{CC} \times \frac{R_2}{R_1 + R_2} \qquad (105\text{-}1)$$

$$= 30 \text{ V} \times \frac{10 \text{ k}\Omega}{20 \text{ k}\Omega + 10 \text{ k}\Omega}$$

$$= 10 \text{ V}.$$

Thévenin resistance,

$$R_{TH} = \frac{R_1 \times R_2}{R_1 + R_2} \qquad (105\text{-}2)$$

$$= \frac{20 \times 10}{20 + 10}$$

$$= 6.67 \text{ k}\Omega.$$

Emitter current,

$$I_E = \frac{V_{TH} - 0.7 \text{ V}}{R_E + R_{TH}/\beta_{dc}} \qquad (105\text{-}3)$$

$$= \frac{10 \text{ V} - 0.7 \text{ V}}{10 \text{ k}\Omega + 6.67 \text{ k}\Omega/50}$$

359

$$= \frac{9.3 \text{ V}}{10.133 \text{ k}\Omega}$$

$$= 0.918 \text{ mA}.$$

Using the approximate formula,
    Emitter current,

$$I_E \approx \frac{V_{TH} - 0.7 \text{ V}}{R_E} \qquad (105\text{-}4)$$

$$= \frac{9.3 \text{ V}}{10 \text{ k}\Omega}$$

$$= 0.93 \text{ mA}.$$

Chapter 106
Emitter bias

COMPARED WITH THE BASE BIAS METHOD OF CHAPTER 104, THE EMITTER BIAS circuit of Fig. 106-1A is a superior way of establishing the dc conditions in a transistor amplifier. It does however, require the presence of the emitter resistor, R_E, and the additional power supply, V_{EE}. The mathematical derivations will show that the ideal expression for the emitter bias does not contain a β_{dc} term so that the values of the emitter and collector currents tend to be independent of any fluctuations in β_{dc} due to temperature variations.

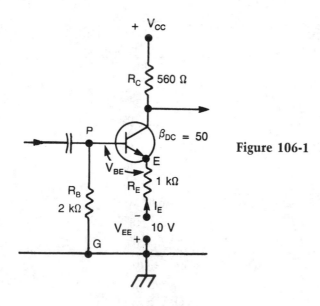

Figure 106-1

MATHEMATICAL DERIVATIONS

For the loop, BEG,

$$\frac{I_E \times R_B}{\beta_{dc}} + V_{BE} + I_E R_E - V_{EE} = 0 \qquad (106\text{-}1)$$

This yields
   Emitter current,

$$I_E = \frac{V_{EE} - V_{BE}}{R_E + R_B/\beta_{dc}} \text{ milliamperes} \qquad (106\text{-}2)$$

For the silicon npn transistor, $V_{BE} \approx 0.7$ V and $V_{EE} \gg V_{BE}$. In addition we can arrange that $R_E \gg R_B/\beta_{dc}$. Then the ideal expression for the emitter current is

Emitter current,

$$I_E \approx \frac{V_{EE}}{R_E} \text{ milliamperes} \qquad (106\text{-}3)$$

This value of I_E is stable since it is independent of any fluctuations in the static current gain, β_{dc}.
Base current,

$$I_B \approx \frac{I_E}{\beta_{dc}} \text{ milliamperes} \qquad (106\text{-}4)$$

   Base voltage,

$$V_B = -I_B R_B \text{ volts} \qquad (106\text{-}5)$$

Example 106-1

In Fig. 106-1, calculate the values of the base and emitter potentials (use the ideal value of the emitter current).

Solution

Emitter current,

$$I_E = \frac{V_{EE}}{R_E} \qquad (106\text{-}3)$$

$$= \frac{10 \text{ V}}{1 \text{ k}\Omega}$$

$$= 10 \text{ mA}.$$

   Base current,

$$I_B \approx \frac{I_E}{\beta_{dc}} \qquad (106\text{-}4)$$

$$= \frac{10 \text{ mA}}{50}$$

$$= 0.2 \text{ mA}$$

Potential at the bias,

$$V_B = -I_B R_B$$ (106-5)
$$= 0.2 \text{ mA} \times 2 \text{ k}\Omega$$
$$= -0.4 \text{ V}.$$

Potential at the emitter,

$$V_E = (-0.4 \text{ V}) + (-0.7 \text{ V})$$
$$= -1.1 \text{ V}.$$

Chapter 107
Collector feedback bias

COLLECTOR FEEDBACK BIAS IS CREATED BY CONNECTING A SINGLE RESISTOR between the collector and the base (Fig. 107-1). The forward base bias is therefore provided by the collector voltage, V_C, and not by the collector supply voltage, $+V_{CC}$. Under amplifier conditions the output signal is taken from the collector so that V_C is a combination of a dc voltage and an ac voltage. Consequently, R_B will not only determine the base/emitter bias but will also provide a path between the output collector signal and the input base signal. Because these two signals are 180° out of phase, the feedback is negative or degenerative. The result of this feedback is a reduction in the gain of the amplifier but the circuit will also have the advantages that are outlined in Chapter 120. The main disadvantage of self bias is its dependency on the value of β_{dc}.

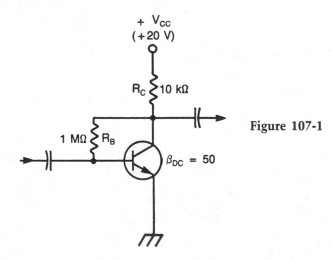

Figure 107-1

MATHEMATICAL DERIVATIONS

From the base bias equation

Emitter current,

$$I_E = \frac{V_C - V_{BE}}{R_B/\beta_{dc}} \text{ milliamperes} \tag{107-1}$$

Collector voltage,

$$V_C = V_{CC} - (I_C + I_B) R_C \tag{107-2}$$

$$= V_{CC} - I_E R_C \text{ volts}$$

From Equations 107-1 and 107-2,
Emitter current,

$$I_E = \frac{V_{CC} - V_{BE}}{R_C + R_B/\beta_{dc}} \text{ milliamperes} \tag{107-3}$$

Assuming $V_{BE} = 0.7$ V,
Emitter current,

$$I_E = \frac{V_{CC} - 0.7 \text{ V}}{R_C + R_B/\beta_{dc}} \text{ milliamperes} \tag{107-4}$$

$$\approx \frac{V_{CC}}{R_C + R_B/\beta_{dc}} \text{ milliamperes} \tag{107-5}$$

if $V_{CC} \gg 0.7$ V.

Example 107-1

In Fig. 107-1, calculate the values of I_E and V_C.

Solution

Emitter current,

$$I_E = \frac{V_{CC} - 0.7 \text{ V}}{R_C + R_B/\beta_{dc}} \tag{107-4}$$

$$= \frac{20 \text{ V} - 0.7 \text{ V}}{10 \text{ k}\Omega + 1 \text{ M}\Omega/50}$$

$$= \frac{19.3 \text{ V}}{10 \text{ k}\Omega + 20 \text{ k}\Omega}$$

$$= 0.64 \text{ mA}$$

Collector voltage,

$$V_C = V_{CC} - I_E R_C \tag{107-2}$$

$$= +20 \text{ V} - 10 \text{ k}\Omega \times 0.64 \text{ mA}$$

$$= +13.6 \text{ V}$$

Chapter 108
The common emitter amplifier

THE BASE-DRIVEN STAGE IS THE MOST FREQUENTLY USED TYPE OF TRANSISTOR amplifier. In Fig. 108-1A, the input signal, v_b, is applied between the base of an npn transistor and ground while the amplified output signal appears between the collector and ground. As far as the signal is concerned, the equivalent circuit of Fig. 108-1B shows that the transistor behaves as a constant current source in association with the emitter diode's ac resistance r_e'.

If the input signal drives the base more positive with respect to ground, the forward voltage across the emitter/base junction will be higher so that the emitter current is increased. This in turn will raise the level of the collector current so that there will be a greater voltage drop across

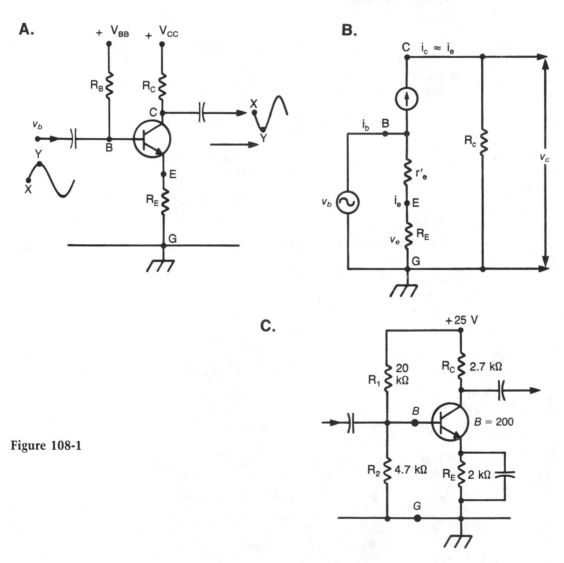

Figure 108-1

the collector load, R_C. As a result the collector voltage, v_C, will be less positive so that the input and output signals are 180° out of phase.

The expressions for such quantities as the voltage gain and the input impedance are shown in the mathematical derivations.

MATHEMATICAL DERIVATIONS

In the loop $BEGB$ (Fig. 108-1B),

$$i_e(r_e' + R_E) = v_b \text{ volts} \tag{108-1}$$

Signal emitter current,

$$i_e = \frac{v_b}{R_E + r_e'} \text{ milliamperes} \tag{108-2}$$

Signal emitter voltage,

$$v_e = i_e R_E \text{ volts} \tag{108-3}$$

Output signal collector voltage,

$$v_c = i_c R_C \tag{108-4}$$
$$\approx i_e R_C \text{ volts}$$

$$\text{Voltage gain, } G_V = \frac{v_c}{v_b} \approx \frac{i_e R_C}{i_e(R_E + r_e')} \tag{108-5}$$
$$= \frac{R_C}{R_E + r_e'}$$

Emitter-base diode's ac resistance

$$r_e' = \frac{\Delta V_{BE}}{\Delta I_E} \text{ ohms} \tag{108-6}$$
$$= \frac{25 \text{ mV}}{I_E}$$

where: I_E = dc level of the emitter current (mA)

Equation 108-6 is obtained by differentiating Shockley's expression for a diode current. Signal base current,

$$i_b = \frac{i_c}{\beta} \approx \frac{i_e}{\beta} \text{ milliamperes} \tag{108-7}$$

where β is the ac current gain from base to collector and is defined by: $\beta = \Delta I_C / \Delta I_B$. For a particular transistor the values of β and β_{dc} are comparable.

Input impedance at the base,

$$z_{in} = \frac{v_b}{i_b} \tag{108-8}$$
$$\approx \frac{i_e(r_e' + R_E)}{i_e/\beta}$$
$$= \beta (r_e' + R_E) \text{ ohms}$$

If the emitter resistor, R_E, is bypassed by a suitable capacitor (Fig. 108-1C), the emitter is at signal ground and the stage can be referred to as a common emitter (CE) amplifier. The voltage gain of such a stage is given by

Voltage gain of the CE amplifier,

$$G_V = \frac{R_C}{r_e'} \qquad \text{(108-9)}$$

Input impedance at the base,

$$z_{in} = \beta r_e' \text{ ohms} \qquad \text{(108-10)}$$

Example 108-1

In Fig. 108-1C, what are the values of the voltage gain from the base to the collector and the input impedance at the base?

Solution

Thévenize the circuit at the points B and G,
Thévenin voltage,

$$V_{TH} = 25 \text{ V} \times \frac{4.7 \text{ k}\Omega}{4.7 \text{ k}\Omega + 20 \text{ k}\Omega}$$

$$= 4.76 \text{ V}$$

Dc level of the emitter current,

$$I_E \approx \frac{4.76 \text{ V}}{2 \text{ k}\Omega}$$

$$= 2.38 \text{ mA}$$

Emitter-base diodes ac resistance,

$$r_e' = \frac{25 \text{ mV}}{I_E} \qquad \text{(108-6)}$$

$$= \frac{25 \text{ mV}}{2.38 \text{ mA}}$$

$$= 10.5 \text{ }\Omega$$

Voltage gain,

$$G_V = \frac{R_C}{r_e'} \qquad \text{(108-9)}$$

$$= \frac{2.7 \text{ k}\Omega}{10.5 \text{ }\Omega}$$

$$\approx 250$$

Input impedance at the base,

$$z_{in} = \beta r_e' \qquad \text{(108-10)}$$

$$= 200 \times 10.5 \text{ }\Omega$$

$$= 2.1 \text{ k}\Omega$$

❧

Chapter 109
The common-base amplifier

IN THE COMMON BASE ARRANGEMENT THE BASE OF THE NPN TRANSISTOR IS grounded and the input signal, v_e, is applied to the emitter (Fig. 109-1A). In the ac equivalent circuit of Fig. 109-1B, the output signal, v_c, appears between the collector and ground.

If the input signal drives the emitter (n region) more positive, the forward voltage on the emitter/base junction is lower so that the emitter current is reduced. This in turn will decrease the collector current so that the voltage drop across the collector load will be less. The collector voltage, v_c, will therefore be more positive so that the input and the output signals are in phase. This is in contrast with the common emitter arrangement (Chapter 108) where the input and output signals are 180° out of phase.

The expressions for the voltage gain and the input impedance of the CB amplifier appear in the mathematical derivations.

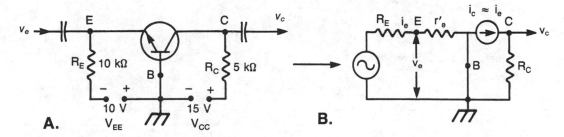

Figure 109-1

MATHEMATICAL DERIVATIONS

In Fig. 109-1B,

Input emitter signal,

$$v_e = i_e r_e' \text{ volts} \tag{109-1}$$

where r_e' is the ac resistance of the emitter-base diode. The value of r_e' is given by 25 mV/I_E (Equation 108-6) where I_E is the dc level of the emitter current.

Output collector signal,

$$v_c \approx i_e R_C \text{ volts} \tag{109-2}$$

Voltage gain from emitter to collector is,

$$G_V = \frac{v_c}{v_e} = \frac{i_e R_C}{i_e r_e'} \tag{109-3}$$

$$= \frac{R_C}{r_e'}$$

The expression for the voltage gain is identical to that of the *CE* arrangement (equation 108-9).

The input impedance at the emitter is given by,

$$z_{in} = \frac{v_e}{i_e} = \frac{i_e r_e'}{i_e} \qquad (109\text{-}4)$$

$$= r_e' \text{ ohms}$$

The input impedance of the CB amplifier is therefore extremely low when compared with the input impedance $(\beta r_e')$ at the base of the CE amplifier. This is the main disadvantage of the CB arrangement whose main use is normally confined to high frequency oscillator circuits.

Example 109-1

In Fig. 109-1A, calculate the values of the voltage gain from the emitter to the collector and the input impedance at the emitter.

Solution

Dc level of the emitter current,

$$I_E \approx \frac{10 \text{ V}}{10 \text{ k}\Omega}$$

$$= 1 \text{ mA}.$$

Ac emitter-base diode resistance,

$$r_e' = \frac{25 \text{ mV}}{I_E} \qquad (109\text{-}1)$$

$$= \frac{25 \text{ mV}}{1 \text{ mA}}$$

$$= 25 \ \Omega.$$

Voltage gain from the emitter to the collector,

$$G_V = \frac{R_c}{r_e'} \qquad (109\text{-}3)$$

$$= \frac{5 \text{ k}\Omega}{25 \ \Omega}$$

$$= 200.$$

Input impedance at the emitter,

$$z_{in} = r_e' \qquad (109\text{-}4)$$

$$= 25 \ \Omega.$$

Chapter 110
The emitter follower

THE EMITTER FOLLOWER IS SOMETIMES REFERRED TO AS THE COMMON collector (CC) stage. In Fig. 110-1A, the input signal is applied between the base and ground while the output signal is taken from the emitter and ground; the collector is normally joined directly to the V_{CC} supply. As will be shown in the mathematical derivations the voltage gain from the base to the emitter is less than unity but the CC arrangement has a high input impedance and a low output impedance. The emitter follower is therefore primarily used for impedance matching by allowing a high impedance source to feed a low impedance load. In this way it behaves as a matching transformer but with the advantages of a better frequency response, greater simplicity and lower cost. This also means that the emitter follower can sometimes provide some degree of isolation between two stages.

Assume that the input signal is driving the p-type base more positive with respect to ground. This increases the forward voltage applied to the emitter-base junction so that the emitter current increases. This increase of current passes through the emitter resistor, R_E; as a result the emitter voltage becomes more positive with respect to ground. The emitter output signal is therefore in phase with the input base signal. In other words the emitter output "follows" the base input. However, the whole of the output signal is applied as negative or degenerative feedback to the input signal. This is the reason for the high input impedance, the low output impedance and the voltage gain of less than unity.

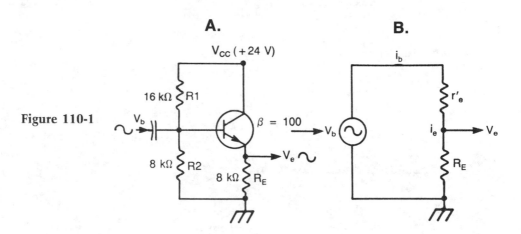

Figure 110-1

MATHEMATICAL DERIVATIONS

In the equivalent circuit of Fig. 110-1B,

Input base signal,

$$v_b = i_e(r_e' + R_E) \text{ volts} \tag{110-1}$$

where: r_e' = ac emitter-base diodes resistance = 25 mV/I_E ohms

Output emitter signal,

$$v_e = i_e \times R_E \text{ volts} \tag{110-2}$$

Voltage gain,

$$G_V = \frac{v_e}{v_b} = \frac{i_e \times R_E}{i_e(r_e' + R_E)} = \frac{R_E}{r_e' + R_E} < 1 \qquad (110\text{-}3)$$

Signal base current,

$$i_b \approx \frac{i_e}{\beta} \text{ milliamperes} \qquad (110\text{-}4)$$

Input impedance at the base,

$$z_{in} = \frac{v_b}{i_b} = \beta \times \frac{v_b}{i_e} \qquad (110\text{-}5)$$

$$= \beta \times (r_e' + R_E) \text{ ohms}$$

Example 110-1

In Fig. 110-1A what are the values of the voltage gain from base to emitter and the input impedance at the base?

Solution

dc level of the emitter current,

$$I_E \approx \frac{24 \text{ V} \times 8 \text{ k}\Omega}{(8 \text{ k}\Omega + 16 \text{ k}\Omega) \times 8 \text{ k}\Omega}$$

$$= 1 \text{ mA.}$$

ac emitter-base diode resistance,

$$r_e' = \frac{25 \text{ mV}}{I_E}$$

$$= \frac{25 \text{ mV}}{1 \text{ mA}}$$

$$= 25 \text{ }\Omega.$$

Voltage gain from the base to the emitter,

$$G_V = \frac{R_E}{r_e' + R_E} \qquad (110\text{-}3)$$

$$= \frac{8 \text{ k}\Omega}{25 \text{ }\Omega + 8 \text{ k}\Omega}$$

$$= 0.997.$$

Input impedance at the base,

$$z_{in} = \beta \times (r_e' + R_E) \qquad (110\text{-}5)$$

$$= 100 \times (25 \text{ }\Omega + 8 \text{ k}\Omega)$$

$$= 802.5 \text{ k}\Omega.$$

Chapter 111
Class-A power amplifier

FIGURE 111-1A ILLUSTRATES A RECEIVER'S OUTPUT STAGE WHOSE LOAD IS A low-impedance loudspeaker. The step-down transformer is used to match the speaker to the required value of the collector load. Under optimum conditions the circuit will deliver audio power to the loudspeaker with the minimum amount of signal distortion.

The collector characteristics of a power transistor are shown in Fig. 111-1B. Initially the quiescent or no-signal conditions are analyzed to determine the dc value, V_{CEQ}, of the collector-emitter voltage. If the resistance of the transformer's primary winding is ignored, a vertical dc load line can be drawn and a convenient Q point can be selected to determine the dc level (I_{CQ}) of the collector current. The effective collector load of the loudspeaker is given by:

$$R_L \times \left(\frac{N_P}{N_S}\right)^2$$

and determines the slope of the ac load line.

When the input signal is applied to the base, the operating conditions move up and down the ac load line. To allow for the greatest input signal, it is important that Q is approximately in the center of this line. Such a condition is known as class-A operation in which the collector current flows throughout the cycle of the base signal voltage. However, the amount of the input signal is limited by saturation on the positive alternation (point Q_1) and by cut-off clipping on the negative alternation (point Q_2). We must also be careful that the transistor's maximum V_{CE} and power

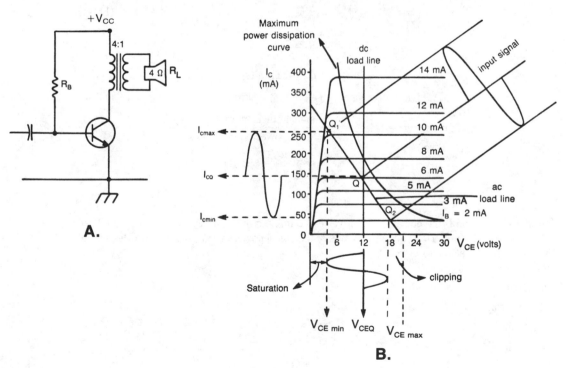

Figure 111-1

ratings are not exceeded at any point on the operating section of the ac load line. This is ensured by keeping the load line below the curve representing the transistor's maximum power rating.

MATHEMATICAL DERIVATIONS

In Fig. 111-1B,

Effective ac collector load,

$$r_c = \left(\frac{N_P}{N_S}\right)^2 \times R_L \text{ ohms} \tag{111-1}$$

The ac load line is derived from the relationship:

$$\frac{\Delta V_{CE}}{\Delta I_C} = r_c \text{ ohms} \tag{111-2}$$

Peak voltage fluctuation at the collector is

$$v_{cpeak} = \frac{V_{CEmax} - V_{CE}}{2} \text{ volts} \tag{111-3}$$

Peak fluctuation in the collector current is

$$i_{cpeak} = \frac{I_{Cmax} - I_{Cmin}}{2} \text{ milliamperes} \tag{111-4}$$

Signal power output,

$$P_0 = \frac{v_{cpeak}}{\sqrt{2}} \times \frac{i_{cpeak}}{\sqrt{2}} \tag{111-5}$$

$$= \frac{v_{cpeak} \times i_{cpeak}}{2}$$

$$= v_{cRMS} \times i_{crms} \text{ milliwatts}$$

$$\text{dc power input} = I_{CQ} \times V_{CC} \text{ milliwatts} \tag{111-6}$$

$$\text{Percentage efficiency, } \eta = \frac{\text{signal power output}}{\text{dc power input}} \times 100\% \tag{111-7}$$

$$= \frac{v_{crms} \times i_{crms}}{I_{CQ} \times V_{CC}} \times 100\%$$

Example 111-1

In Figs. 111-1A and B the dc base current value is 6 mA and the input signal creates a peak base current swing of 4 mA. Determine the values of the signal power output, the dc power input and the percentage efficiency.

Solution

Effective collector load

$$r_c = \left(\frac{N_P}{N_S}\right)^2 \times R_L \tag{111-1}$$

$$= 4^2 \times 4 \ \Omega$$

$$= 64 \ \Omega$$

Ignoring the dc voltage drop across the resistance of the primary winding, Quiescent collector voltage, $V_{CEQ} = 12$ V.

At the Q point the collector current, $I_{CQ} = 140$ mA.

$$\text{Dc power input} = V_{CEQ} \times I_{CQ} \tag{111-6}$$
$$= 12 \text{ V} \times 140 \text{ mA}$$
$$= 1.68 \text{ W}$$

From Fig. 111-1B,

$$V_{CEmax} = 19 \text{ V, } V_{CEmin} = 4 \text{ V, } I_{Cmax} = 260 \text{ mA, } I_{Cmin} = 40 \text{ mA}$$

Signal power output

$$P_0 = \frac{v_{cpeak} \times i_{cpeak}}{2} \tag{111-5}$$
$$= \frac{(19 \text{ V} - 4 \text{ V}) \times (260 \text{ mA} - 40 \text{ mA})}{2 \times 2 \times 2}$$
$$= 0.41 \text{ W}$$

Percentage efficiency

$$\eta = \frac{\text{signal power output}}{\text{dc power input}} \times 100\% \tag{111-7}$$
$$= \frac{0.41 \text{ W}}{1.68 \text{ W}} \times 100\%$$
$$= 24.6\%$$

Chapter 112
The class-B push-pull amplifier

FIGURE 112-1A ILLUSTRATES A CLASS-B PUSH-PULL AUDIO AMPLIFIER. IN THIS figure the input signals to the two bases are *180° out of phase* while the output transformer is connected between the collectors so that the collector currents drive in opposite directions through the primary winding. As a result the input signal to the loudspeaker is determined by the value of $i_{c1} - i_{c2}$.

In the class-A operation of Chapter 111, the Q point was located at the middle of the ac load line so that the collector current flowed continuously throughout the cycle of the input signal. Although the distortion was kept to a minimum, the efficiency of the class-A circuit was only of the order of 20% to 30%. To increase the efficiency up to between 50% and 60% we operate under class-B conditions whereby the Q point is theoretically lowered to the bottom of the ac load line (Fig. 112-1B) so that $V_{CEQ} = V_{CC}$ and $I_{CQ} = 0$; in other words, the forward bias on each transistor is zero. However, there is unacceptable (cross-over) distortion near the cut-off region so that in practice each transistor is provided with a small forward "trickle" bias by the voltage divider, $R1$, $R2$.

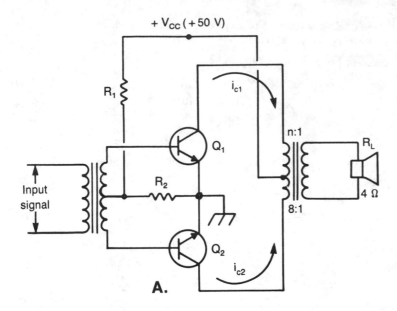

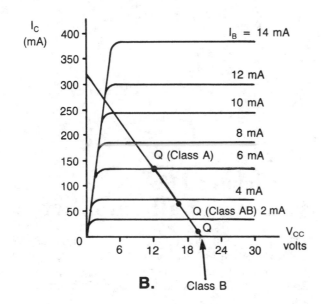

Figure 112-1

As a further compromise between distortion and efficiency we can operate an audio push-pull circuit in class-AB where the corresponding Q point on the ac load line lies half way between the two points representing class-A and class-B (Fig. 112-1B).

Under class-B conditions each transistor conducts for about one half cycle of its input signal and because the two base signals are 180° out of phase, the transistors conduct alternatively; in other words, there is a "cross-over" between the two transistors. The two half cycles then combine to produce the complete sine-wave output that appears across the loudspeaker.

Compared with a single power transistor operated under class-A conditions, the push-pull amplifier offers the following advantages:

1. There is no resultant dc current to cause partial saturation of the output transformer's iron core. If the core were to partially saturate, the output signal would be distorted.
2. Due to nonlinearity in the transistor characteristics, the collector current waveforms cannot be pure sine waves and are therefore, to some extent, distorted. This distortion is in the form of harmonic components, both odd and even. The *even* harmonic components of i_{c1} and i_{c2} are in phase and will therefore cancel in the output transformer. The elimination of *even* harmonic distortion is the major advantage of the push-pull arrangement.
3. The efficiency (50% to 60%) of the class-B push-pull stage is much higher than that of the single transistor class-A amplifier (20% to 30%). Therefore, for a given dc power input, the push-pull stage can provide a greater signal power output.
4. The maximum input signal voltage to the class-B push-pull stage is approximately four times greater than the maximum input signal to the class-A amplifier. All these advantages are achieved at the expense of using a *matched* pair of transistors. If one transistor becomes defective, both should be replaced.

MATHEMATICAL DERIVATIONS

Under quiescent conditions,

$$V_{CEQ} = V_{CC} \text{ volts} \tag{112-1}$$

$$I_{CQ} \approx 0 \tag{112-2}$$

Effective load as seen by each collector,

$$r_c = \left(\frac{n}{2}\right)^2 R_L \text{ ohms}$$

Voltage gain from base to collector is

$$G_V = \frac{v_c}{v_b} = \frac{r_c}{R_e'} \tag{112-4}$$

where R_e' is the emitter-base diode resistance for a large signal input.

Saturation current at the top end of the ac load line is

$$I_{C(\text{sat})} = \frac{V_{CEQ}}{r_c} = \frac{V_{CC}}{r_c} \text{ amperes} \tag{112-5}$$

Maximum load power output is given by
Output power,

$$P_0 = \frac{V_{CEQ} \times I_{C(sat)}}{2} \text{ watts} \tag{112-6}$$

Example 112-1

In Fig. 112-1A, the R_e' of each transistor is 1 Ω. What are the values of the maximum power output and the voltage gain from base to collector?

Solution

Effective load seen by each collector is

$$r_c = \left(\frac{n}{2}\right)^2 \times R_L = 4^2 \times 4 \qquad (112\text{-}3)$$

$$= 64 \ \Omega$$

$$V_{CEQ} = V_{CC} = 50 \ \text{V} \qquad (112\text{-}1)$$

$$I_{C(\text{sat})} = \frac{V_{CEQ}}{r_c} = \frac{50 \ \text{V}}{64 \ \Omega} \qquad (112\text{-}5)$$

$$= 0.78 \ \text{A}$$

Voltage gain from base to collector is

$$G_V = \frac{v_c}{v_b} = \frac{r_c}{R_e{}'} = \frac{64 \ \Omega}{1 \ \Omega} \qquad (112\text{-}4)$$

$$= 64$$

Maximum signal power output is

$$P_0 = \frac{V_{CEQ} \times I_{C(\text{sat})}}{2} \qquad (112\text{-}6)$$

$$= \frac{50 \ \text{V} \times 0.78 \ \text{A}}{2}$$

$$= 19.5 \ \text{W}$$

Chapter 113
The class-C rf power amplifier

FIGURE 113-1 SHOWS A CLASS-C RF POWER AMPLIFIER THAT IS OPERATED AT A high efficiency of 80% or more; in other words, four fifths of the dc power drawn from the V_{CC} supply is converted into the rf signal power output.

The circuit uses signal bias in which base current flows either side of the positive alternation's peak. This base current charges the capacitor, C_B, but when the current ceases, the capacitor discharges through R_B. However, the time constant $R_B C_B$ is high compared with the period of the input signal. As a result the dc voltage across R_B is approximately equal to the peak value of the input signal and applies a *reverse* bias to the emitter/base junction. Care must be taken that this reverse bias does not exceed the emitter-to-base breakdown voltage.

Due to the action of the signal bias, the collector current consists of narrow pulses at the same repetition rate as the frequency of the input signal. When these pulses are applied to the high Q resonant tuned circuit, the result is a circulating current that oscillates between L and C and develops a complete sine wave for the output voltage between the collector and ground. Only a small mean level of collector current is drawn from the V_{CC} supply so that the amplifier's efficiency is high.

Because the collector current waveform is severely distorted, it contains strong components at the harmonic frequencies. It is then possible to tune the LC circuit to a particular harmonic so

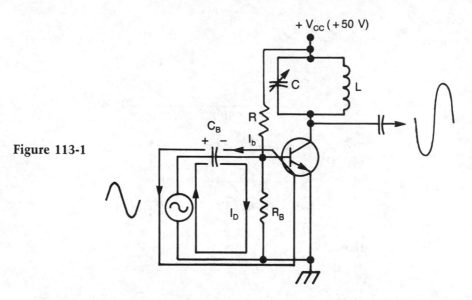

Figure 113-1

that the output frequency is a multiple of the input frequency. Such a stage is referred to as a frequency multiplier such as a doubler, tripler, etc.

MATHEMATICAL DERIVATIONS

Quiescent collector-emitter voltage,

$$V_{CEQ} = V_{CC} \text{ volts} \tag{113-1}$$

Maximum possible voltage swing at the collector extends from the saturated value of V_{CE} to $2 \times V_{CC}$. Knowing the value of $V_{CE(sat)}$ (≈ 1 V), the signal power output is given by:
rf power output,

$$P_0 = \frac{(0.707 \times V_{CC})^2}{r_d} \tag{113-2}$$

$$= \frac{V_{CC}^2}{2\,r_d} \text{ watts}$$

where: r_d = dynamic resistance of the tuned circuit (Ω).

The power dissipation is given by:

$$P_D = \frac{V_{CE(sat)} \times V_{CC}}{2r_d} \tag{113-3}$$

$$= \frac{V_{CE(sat)}}{V_{CC}} \times P_o \text{ watts}$$

Example 113-1

In Fig. 113-1 the input signal has a frequency of 2 MHz and the value of C is 100 pF. Calculate the required values for the inductor L, which has an rf resistance of 30 Ω. Calculate the values of the Q and the dynamic resistance of the LC circuit. Assuming that the complete load line is used, what are the values of the rf power output and the power dissipation ($V_{CE(sat)} \approx 1$ V)?

377

Solution

Inductance,

$$L = \frac{1}{4 \times \pi^2 \times f_r^2 C} \text{ H}$$

$$= \frac{10^6}{4 \times \pi^2 \times (2 \times 10^6)^2 \times 100 \times 10^{-12}} \text{ H}$$

$$= 63.3 \ \mu\text{H}$$

$$Q = \frac{1}{R} \times \sqrt{\frac{L}{C}}$$

$$= \frac{1}{30} \times \sqrt{\frac{63.3 \times 10^{-6}}{100 \times 10^{-12}}}$$

$$= 26.5$$

Dynamic resistance,

$$r_D = \frac{L}{CR}$$

$$= \frac{63.3 \times 10^{-6}}{100 \times 10^{-12} \times 30} \ \Omega$$

$$= 21.1 \ k\Omega$$

Rf power output,

$$P_o = \frac{V_{CC}^2}{2r_d} \tag{113-2}$$

$$= \frac{(50 \text{ V})^2}{2 \times 21.1 \text{ k}\Omega}$$

$$\approx 60 \text{ mW}$$

Power dissipation,

$$P_D = \frac{V_{CE(sat)}}{V_{CC}} \times P_o \tag{113-3}$$

$$= \frac{1 \text{ V}}{50 \text{ V}} \times 60 \text{ mW}$$

$$= 1.2 \text{ mW}$$

Chapter 114
The junction field-effect transistor

THE ACTION OF THE PNP OR NPN BIPOLAR TRANSISTOR INVOLVES TWO TYPES OF charge carrier—the electron and the hole. The FET is a unipolar transistor because its operation requires only one charge carrier that may be either the electron or the hole.

The two common types of FET are the *junction field-effect transistor* (JFET) and the *metal-oxide semiconductor field-effect transistor* (MOSFET); the latter is sometimes referred to as an *insulated-gate field-effect transistor* (IGFET).

The JFET is essentially a doped silicon bar, which is referred to as a channel and behaves as a resistor. The doping may either be p-type or n-type so that we have either a p-channel or an n-channel JFET. At the ends of the channel are two terminals that are referred to as the source and the drain. When a particular drain-source voltage (V_{DS}) is applied between the end terminals, the amount of current flow (I_D) between the source and the drain depends on the channel's resistance. The value of this resistance is controlled by a gate that may either consist of two n-type regions diffused into a p-type channel or two p-type regions diffused into an n-type channel. In either case the two regions are commonly joined to provide a single gate (two separate gates may be used in some mixer circuits). Cutaway views of both types with their schematic symbols are shown in Fig. 114-1A and B. In these schematic symbols the vertical line may be regarded as the channel; the arrow then points towards an n-channel but away from a p-channel. The gate line may either be symmetrically positioned with respect to the source and the drain, or may be drawn closer to the source.

If a reverse bias voltage is applied between the gate and the source of an n channel JFET, depletion layers will surround the two p-regions that form the gate. If the reverse bias is increased, the depletion layers will spread more deeply into the channel until they almost touch. The channel's resistance will then be extremely high so that the gate current is very small.

The reverse biasing of the gate/source junction may be compared to applying a negative voltage to a triode's grid relative to its cathode. Like the tube, the FET is a voltage-controlled device in the sense that only the input voltage to the gate controls the output drain current. This is in contrast with the bipolar transistor where the base/emitter junction is forward biased; the input voltage then controls the input current, which in turn determines the output current.

With reverse biasing of the gate/source junction, very little gate current flows. This means that the input impedance to a JFET is of the order of several megohms; this is a definite advantage of the FET over the bipolar transistor whose input impedance is relatively low. However, when compared with the JFET, the output current of a bipolar transistor is much more sensitive to changes in the input voltage; the result is a lower voltage gain available from the JFET.

In Fig. 114-1C the voltage (V_{DS}) between the source and the drain of the n-channel JFET, is gradually increased from zero; at the same time the voltage between gate and source, V_{GS}, is 0 V, which is referred to as the shorted gate condition. Initially, the available channel is broad so that the drain current, I_D is directly proportional to V_{DS} and rises rapidly as V_{DS} is increased. However, the drain voltage creates a reverse bias on the junction between the channel and the gate. The increase in V_{DS} causes the two depletion regions to widen until finally they almost come into contact. This occurs when V_{DS} equals a value called the *pinch-off voltage*, V_P (Fig. 114-1D), the available channel is then very narrow so that the drain current is limited (or pinched-off). Further raising of V_{DS} above the pinch-off point will only produce a small increase in the drain current. This situation continues until the drain voltage equals $V_{DS(max)}$; at this point an avalanche effect takes place and the JFET breaks down. Over the operating range between V_P and $V_{DS(max)}$, the approximately constant value of the drain current with the shorted gate is referred to as the I_{DSS} (drain-to-source current with shorted gate).

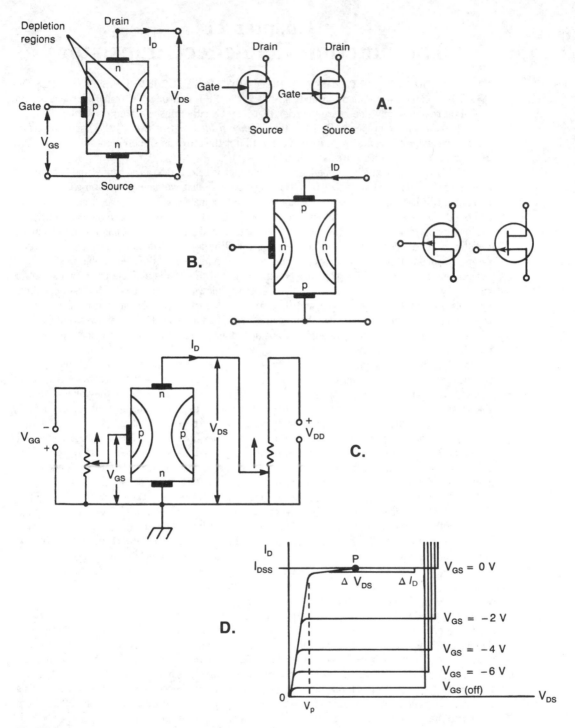

Figure 114-1

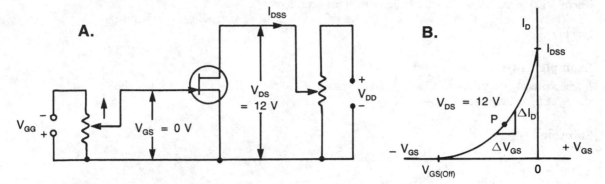

Figure 114-2

For each different negative value of V_{GS}, a different drain current curve can be obtained. This family of curves is illustrated in Fig. 114-1D. Ultimately, V_{GS} can be sufficiently negative so that the drain current is virtually cut off and equal to zero; this value of gate-source voltage is therefore referred to as $V_{GS(off)}$.

At the cutoff condition the depleted layers nearly touch; this also occurred when V_{DS} was equal to V_P. Therefore, $V_{GS(off)}$ has the same value as V_P although $V_{GS(off)}$ is a negative voltage while V_P is positive.

The *transconductance* curve is the graph of the drain current, I_D, plotted against the gate-to-source voltage, V_{GS}, while maintaining the drain-to-source voltage, V_{DS}, at a constant level. For example, in Fig. 114-2A, the drain voltage is set to 12 V while the gate is initially shorted to the source so that $V_{GS} = 0$. The recorded drain current would then equal the value of I_{DSS}. If the reverse gate voltage is now increased from zero, the drain current will fall until ultimately cut-off is reached when $V_{GS} = V_{GS(off)}$.

MATHEMATICAL DERIVATIONS

The shape of the transconductance curve is considered to be a parabola (Fig. 114-2B) so that there is a mathematical relationship between I_D and V_{GS}.

Drain current,

$$I_D = I_{DDS}\left(1 - \frac{V_{GS}}{V_{GS(off)}}\right)^2 \text{ milliamperes} \tag{114-1}$$

In this equation both V_{GS} and $V_{GS(off)}$ are considered to be negative voltages.

In a FET amplifier the control that the gate voltage exercises over the drain current is measured by the transconductance, g_m. At a particular point, P, on the curve, the transconductance is defined by:

$$g_m = \frac{\Delta I_D}{\Delta V_{GS}} \text{ milliamperes per volt} \tag{114-2}$$

and is normally measured in microsiemens (μS). By differentiating the expression for I_D with respect to V_{GS},

Transconductance,

$$g_m = g_{m0}\left(1 - \frac{V_{GS}}{V_{GS(off)}}\right) \text{ microsiemens} \tag{114-3}$$

381

where: g_{m0} = transconductance for the shorted gate condition.

The value of g_{m0} is $-V_{GS(off)}/2\,I_{DSS}$; because $V_{GS(off)}$ is a negative voltage, g_{m0} is a positive quantity.

Example 114-1

A junction field-effect transistor has a pinch-off voltage equal to 4 V. If I_{DSS} = 12 mA, find the values of I_D and g_m when $V_{GS} = -2$ V.

Solution

$$V_{GS(off)} = -V_P$$
$$= -4 \text{ V.}$$

Drain current,

$$I_D = I_{DSS}\left(1 - \frac{V_{GS}}{V_{GS(off)}}\right)^2 \tag{114-1}$$

$$= 12 \times \left(1 - \frac{(-2)}{(-4)}\right)^2$$

$$= 3 \text{ mA.}$$

$$g_{m0} = -\frac{2\,I_{DSS}}{V_{GS(off)}} = \frac{-2 \times 12}{-4} = 6 \text{ mA/V}$$

$$= 6000 \text{ } \mu\text{S.}$$

Transconductance,

$$g_m = g_{m0}\left[1 - \left(\frac{V_{GS}}{V_{GS(off)}}\right)\right] \tag{114-3}$$

$$= 6000\left[1 - \left(\frac{(-2)}{(-4)}\right)\right]$$

$$= 3000 \text{ } \mu\text{S.}$$

Chapter 115
Methods of biasing the JFET

THE METHODS THAT FOLLOW ARE APPLIED TO N-CHANNEL FETS, ALTHOUGH p-channel FETS may be similarly biased by reversing the polarities of the dc supply voltages.

MATHEMATICAL DERIVATIONS
The gate bias

This method is similar to the base bias of the bipolar transistors. In Fig. 115-1A, the drain current is given by:

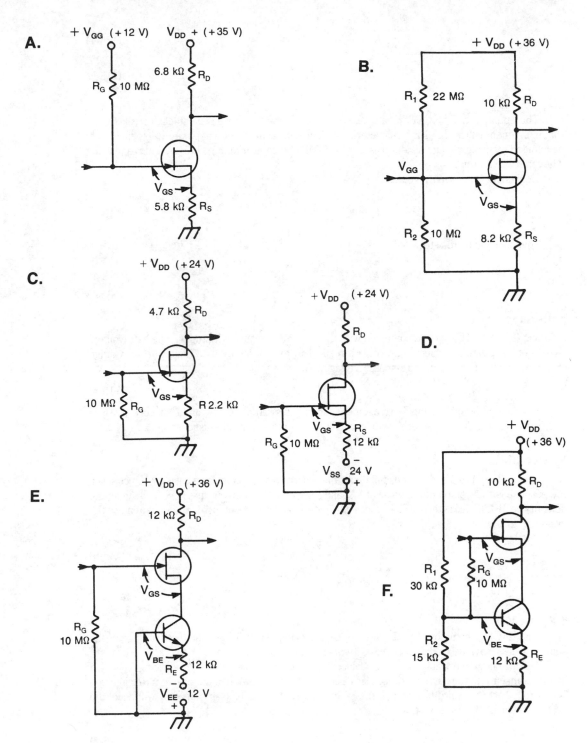

Figure 115-1

383

$$I_D = \frac{V_{GG} - V_{GS}}{R_S} \tag{115-1}$$

$$\approx \frac{V_{GG}}{R_S} \text{ milliamperes} \tag{115-2}$$

if $V_{GG} \gg V_{GS}$.

The equation $I_D \approx V_{GG}/R_S$ means that V_{GG} and R_S may be chosen to establish a value of I_D that is independent of the JFET characteristics. The requirement for a separate V_{GG} supply may be avoided by using voltage divider bias (Fig. 115-1B). By comparing Fig. 115-1A and B and using Thévenin's Theorem,

Equivalent gate supply voltage,

$$V_{GG} = \frac{V_{DD} R_2}{R_1 + R_2} \text{ volts} \tag{115-3}$$

Equivalent gate resistor,

$$R_G = \frac{R_1 R_2}{R_1 + R_2} \text{ ohms} \tag{115-4}$$

Self-bias

Self-bias (Fig. 115-1C) may be compared with the cathode bias used with tubes. The gate voltage is zero but the gate is biased negative with respect to the source by the amount of the voltage drop across R_S. Therefore, gate-source voltage,

$$V_{GS} = I_D R_S \text{ volts} \tag{115-5}$$

Self-bias, although a simple arrangement, does not swamp out V_{GS} and therefore the bias point depends on the characteristics of the JFET; this is its principle disadvantage.

Source bias

Source bias (Fig. 115-1D) is comparable with the emitter bias of a bipolar transistor. Once again the purpose is to swamp out the value of V_{GS} and achieve a drain current that is virtually independent of the JFET characteristics. Because the gate is at ground potential, the equation for the drain current is:

$$I_D = \frac{V_{SS} - V_{GS}}{R_S} \tag{115-6}$$

$$\approx \frac{V_{SS}}{I_S} \text{ milliamperes}$$

if $V_{SS} \gg V_{GS}$.

It is easy to swamp out V_{GS} if a large negative voltage is available. The disadvantage of the source bias is obvious in that it requires two separate supplies.

If a large negative voltage is not available, the problem may be solved by replacing R_S with a bipolar transistor in the current source bias circuit of Fig. 115-1E. When the base/emitter junction is controlled by a forward bias action, the emitter current of the bipolar transistor will equal the drain current of the FET so that:

$$I_D = I_E = \frac{V_{EE} - V_{BE}}{R_E} \tag{115-7}$$

$$\approx \frac{V_{EE}}{R_E} \text{ milliamperes}$$

if $V_{EE} \gg V_{BE}$.

Because $V_{BE} = 0.7$ V and only varies by 0.1 V from one transistor to another, a high drain current can be fixed by using a low voltage source for V_{EE}. If the low negative voltage is not available, the bipolar transistor may be forward biased by the voltage divider circuit of Fig. 115-1F. Then by Thévenin's Theorem,

Drain current,

$$I_D = I_E = \frac{\dfrac{V_{DD} \times R_2}{R_1 + R_2} - V_{BE}}{R_E} \tag{115-8}$$

$$\approx \frac{V_{DD} \times R_2}{(R_1 + R_2)R_E} \text{ milliamperes}$$

provided

$$\frac{V_{DD} \times R_2}{R_1 + R_2} \gg V_{BE}$$

Current-source bias has the advantage of providing the best swamping action; its disadvantage is the requirement for the additional bipolar transistor.

Example 115-1

In Fig. 115-1A calculate the values of I_D and V_{DS} (ignore V_{GS}).

Solution

$$\text{Drain current} = \frac{V_{GG}}{R_S} = \frac{12 \text{ V}}{5.6 \text{ k}\Omega} \tag{115-2}$$

$$= 2.14 \text{ mA.}$$

Drain-source voltage,

$$V_{DS} = 35 \text{ V} - (6.8 \text{ k}\Omega + 5.6 \text{ k}\Omega) \times 2.14 \text{ mA}$$
$$= 8.5 \text{ V, rounded off.}$$

Example 115-2

In Fig. 115-1B calculate the value of I_D if $V_{GS} = -2$ V.

Solution

Effective gate supply voltage is (10 MΩ × 36 V)/(10 MΩ + 22 MΩ) = 11.25 V. (115-3)

Then the drain current,

$$I_D = \frac{11.25 \text{ V} + 2 \text{ V}}{8.2 \text{ k}\Omega} \tag{115-4}$$

$$= 1.6 \text{ mA, rounded off.}$$

Example 115-3

In Fig. 115-1C, $V_{GS} = -2$ V. Calculate the values of I_D and V_D.

Solution

Drain current,

$$I_D = \frac{2 \text{ V}}{2.2 \text{ k}\Omega} \tag{115-5}$$
$$= 0.91 \text{ mA}$$

Drain voltage,

$$V_D = +24 \text{ V} - 0.91 \text{ mA} \times 4.7 \text{ k}\Omega$$
$$= +24 \text{ V} - 4.277 \text{ V}$$
$$= +19.7 \text{ V, rounded off.}$$

Example 115-4

In Fig. 115-1D, $V_{GS} = -2$ V. Calculate the values of I_D and V_{DS}.

Solution

Drain current,

$$I_D = \frac{24 \text{ V} + 2 \text{ V}}{12 \text{ k}\Omega} \tag{115-6}$$
$$= 2.17 \text{ mA.}$$

Drain voltage,

$$V_D = +24 \text{ V} - 2.17 \text{ mA} \times 3.3 \text{ k}\Omega$$
$$= +24 \text{ V} - 7.16 \text{ V}$$
$$= +16.8 \text{ V, rounded off.}$$

Source voltage,

$$V_S = +2 \text{ V.}$$

Drain-to-source voltage,

$$V_{DS} = +16.8 \text{ V} - (+2 \text{ V})$$
$$= 14.8 \text{ V}$$

Example 115-5

In Fig. 115-1E, calculate the values of I_D and V_{DS} (assume $V_{GS} = -2$ V).

Solution

Assuming that the V_{BE} of the bipolar transistor is 0.7 V,

Drain current,

$$I_D = I_E = \frac{12 \text{ V} - 0.7 \text{ V}}{12 \text{ k}\Omega} \tag{115-7}$$

$$= 0.94 \text{ mA}$$

Then $V_E = -0.7$ V.
If $V_{GS} = 2$ V, $V_S = V_C = +2$ V and $V_{CE} = 2.7$ V.
Drain potential, $V_D = +36$ V $- 0.94$ mA $\times 12$ k$\Omega = +24.7$ V.
Drain-to-source voltage, $V_{DS} = 24.7 - 2.0 = 22.7$ V.

Example 115-6

In Fig. 115-1F, calculate the value of V_{DS} (assume $V_{GS} = -2$ V).

Solution

Due to the voltage divider action, the potential at the point A is

$$+36 \text{ V} \times \frac{15 \text{ k}\Omega}{15 \text{ k}\Omega + 30 \text{ k}\Omega}$$

$$= +12 \text{ V}$$

which is equal to V_G. Since $V_{GS} = -2$ V, $V_S = +14$ V.
  The emitter current,

$$I_E = I_D = \frac{12 \text{ V} - 0.7 \text{ V}}{12 \text{ k}\Omega} \tag{115-8}$$

$$= 0.94 \text{ mA}$$

  The drain potential is $+36$ V $- 10$ k$\Omega \times 0.94$ mA $= +26.6$ V and $V_{DS} = 26.6$ V $- 14$ V $= 12.6$ V.

Chapter 116
The JFET amplifier
and the source follower

THE PRINCIPLES OF THE JFET AMPLIFIER ARE SHOWN IN FIG. 116-1A. (Lowercase letters are used to indicate signal values.) The signal, v_g, to be amplified is applied between the gate and the source and produces variations, I_D, in the drain current. The resultant voltage variations across the drain load, r_d, produce voltage variations of opposite polarity between gate and ground. This output signal, v_d, is 180° out of phase with the input signal. The same phase inversion occurs with the common emitter arrangement of a bipolar transistor amplifier, and the grounded cathode tube stage.

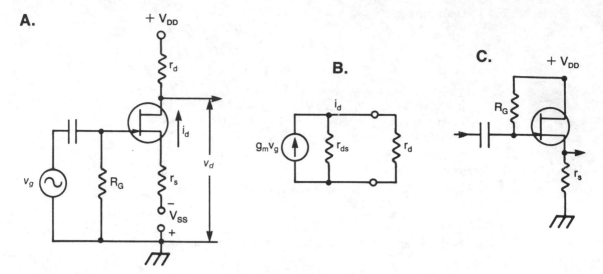

Figure 116-1

MATHEMATICAL DERIVATIONS

Assuming that r_s is bypassed by a capacitor to avoid negative or degenerative feedback, the JFET behaves as a current source of value $g_m v_g$ in parallel with a very high resistance, r_{ds}, which is the reciprocal of the drain characteristics' slope (Fig. 116-1B). Normally r_{ds} exceeds 100 kΩ and may be neglected in the analysis. Then:

Output signal,

$$v_d = g_m v_g \times r_d \text{ volts} \qquad (116\text{-}1)$$

and

$$\text{Voltage gain} = \frac{v_d}{v_g} = g_m r_d \qquad (116\text{-}2)$$

If the source resistance, r_s, is not bypassed, negative feedback is introduced into the circuit and the voltage gain is reduced to:

$$\text{Voltage gain} = \frac{r_d}{r_s + \dfrac{1}{g_m}} \qquad (116\text{-}3)$$

Provided $r_s \gg 1/g_m$,

$$\text{Voltage gain} = \frac{r_d}{r_s} \qquad (116\text{-}4)$$

In the source follower circuit of Fig. 116-1C, $r_d = r_s$ so that

$$\text{Voltage gain} = \frac{v_s}{v_g} = \frac{r_s}{r_s + \dfrac{1}{g_m}} \to 1 \qquad (116\text{-}5)$$

if $r_s \gg 1/g_m$.

388

The low voltage gain is due to the negative feedback developed across r_s. However, this also results in the source follower having an extremely high input impedance but a low output impedance. The same properties are possessed by the cathode follower and common collector circuits.

Example 116-1

In Fig. 116-2, the g_m of the JFET is 6000 μS. Calculate the voltage gain from gate to drain (neglect the value of r_{ds}). If the capacitor, C, is removed, what is the new value of the voltage gain?

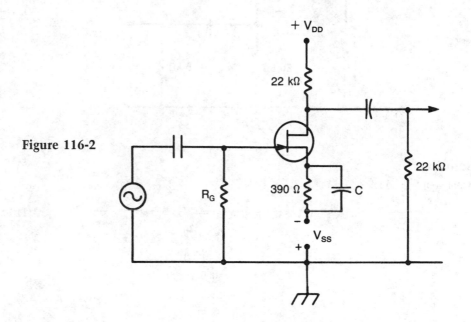

Figure 116-2

Solution

Total drain load, $r_d = 22$ kΩ ∥ 22 k$\Omega = 11$ kΩ.

$$\text{Voltage gain} = g_m r_d \qquad (116\text{-}2)$$
$$= 6000 \ \mu\text{S} \times 11 \ \text{k}\Omega$$
$$= 66.$$

When C is removed,

$$\text{New voltage gain} = \frac{r_d}{r_s + \dfrac{1}{g_m}} \qquad (116\text{-}3)$$

$$= \frac{11000}{390 + 167}$$

$$= 20, \text{ rounded off.}$$

Example 116-2

In Fig. 116-3, calculate the value of the voltage gain from gate to source.

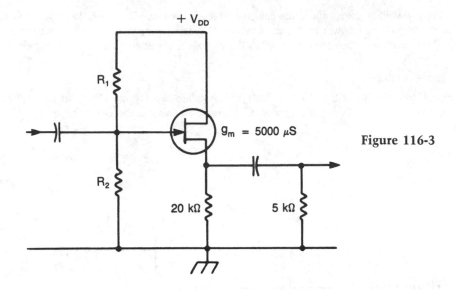

Figure 116-3

Solution

Total source load, $r_s = 20$ kΩ $\parallel 5$ kΩ $= 4$ kΩ.

$$\text{Voltage gain} = \frac{r_s}{r_s + \dfrac{1}{g_m}} \qquad (116\text{-}5)$$

$$= \frac{4000}{4000 + 200}$$

$$= 0.95.$$

Chapter 117
The metal-oxide
semiconductor field-effect transistor

TO AVOID EXCESSIVE GATE CURRENT, IT IS NECESSARY TO APPLY A NEGATIVE bias to the gate of a JFET. However, with a MOSFET the gate current is still virtually zero even if the gate is positive with respect to the source. This is made possible by the MOSFET's construction.

In Fig. 117-1A the n-channel MOSFET has only one p-region which is called the substrate. The substrate may have its own terminal (four-terminal device) or may be internally connected to

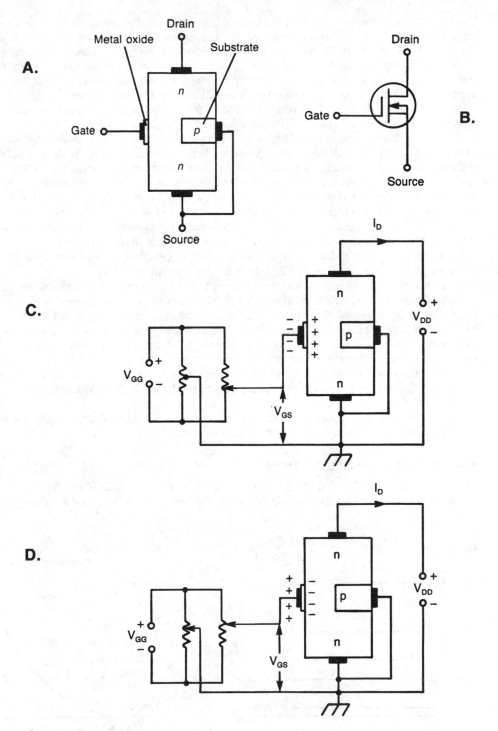

A.

Drain

Metal oxide

Substrate

Gate

n

p

n

Source

B.

Drain

Gate

Source

C.

I_D

n

−
−
−
−

+
+
+
+

p

+
V_{DD}
−

V_{GG} +
−

V_{GS}

n

D.

I_D

n

+
+
+
+

−
−
−
−

p

+
V_{DD}
−

V_{GG} +
−

V_{GS}

n

Figure 117-1

391

the source (three-terminal device); in this chapter it is assumed that the source and the substrate are either internally or externally connected. This is illustrated in the schematic symbol of Fig. 117-1B. As with the JFET, the arrow in the symbol points towards the n-channel. The action of the substrate is to reduce the width of the channel through which the electrons pass from source to drain. On the other side of the narrow channel a thin layer of silicon dioxide (a metal oxide) is deposited and acts as an insulator. A metallic gate is then spread over the opposite surface of the silicon dioxide layer. Because the gate is now insulated from the channel, the device is sometimes referred to as an *insulated-gate FET* (IGFET).

In the operation of the n-channel MOSFET the drain supply voltage, V_{DD}, causes electrons to flow from source to drain through the channel while the gate voltage controls the channel's resistance. If a negative voltage is applied to the gate with respect to the source (Fig. 117-1C), electrostatic induction will cause positive charges to appear in the channel. These charges will be in the form of positive ions that have been created by the repulsion of conduction-band electrons away from the gate; in other words, the number of conduction-band electrons existing in the n-channel has been reduced or depleted. If the gate is made increasingly negative, there will be fewer and fewer conduction-band electrons available until ultimately the MOSFET is cut off. This action is very similar to that of the JFET; because a negative gate causes a depletion of conduction electrons, this manner of operating a MOSFET is called the *depletion mode*.

Because the channel and the gate are insulated from each other, it is possible to apply a positive voltage to the MOSFET gate (Fig. 117-1D). The result will be to induce negative charges into the n-channel; these will be in the form of additional conduction-band electrons that are drawn into the channel by the action of the positive gate. The total number of conduction-band electrons has therefore been increased or enhanced; consequently, this manner of operation for the MOS-FET is called the *enhancement mode*.

Unlike the JFET, the MOSFET may be operated with either a positive or a negative gate voltage; in either mode of operation the input resistance of a MOSFET is very high and is typically of the order of hundreds of GΩ.

Apart from the necessity or reversing the polarity of the drain and gate supply voltages, the operation of p-channel and n-channel MOSFETs is identical.

MATHEMATICAL DERIVATIONS

The appearance of the MOSFET's characteristic curves is similar to those of the JFET and is illustrated in Fig. 117-2A and B. The only difference is the extension of the gate voltage into the

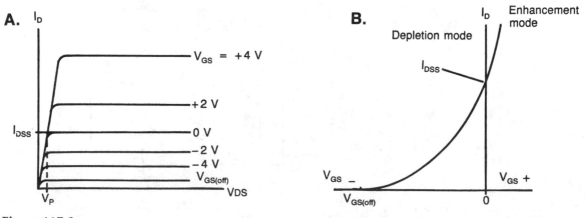

Figure 117-2

positive region of the enhancement mode. The transconductance curve is still a parabola with its equation:

Drain current,

$$I_D = I_{DSS} \left(1 - \frac{V_{GS}}{V_{GS(off)}}\right)^2 \text{ milliamperes} \qquad (117\text{-}1)$$

In this equation $V_{GS(off)}$ is always a negative voltage but V_{GS} may either be positive or negative.

The type of MOSFET, which may either operate in the depletion or the enhancement mode, conducts when $V_{GS} = 0$ with a drain current value equal to I_{DSS}. For this reason the device is called a normally on or depletion-enhancement (DE) MOSFET. Such a MOSFET can be operated with zero bias because the dc operating point can be chosen at $V_{GS} = 0$. When the signal is applied to the gate, the operation of the MOSFET will swing back and forth between the depletion and the enhancement modes.

Example 117-1

A DE MOSFET has values of $I_{DSS} = 10$ mA and $V_{GS(off)} = -4$ V. What are the values of I_D and g_{mO} at (a) $V_{GS} = -2$ V, and (b) $V_{GS} = +1$ V?

Solution

(a) $V_{GS} = -2$ V.

Drain current,

$$I_D = 10\left(1 - \frac{(-2)}{(-4)}\right)^2 = 2.5 \text{ mA} \qquad (117\text{-}1)$$

$$g_{mO} = \frac{-2\,I_{DSS}}{V_{GS(off)}} = \frac{-2 \times 10}{-4}$$

$$= 5 \text{ mS} = 5000 \ \mu\text{S}.$$

Transconductance,

$$g_m = 5000\left(1 - \frac{(-2)}{(-4)}\right)$$

$$= 2500 \ \mu\text{S}.$$

(b) $V_{GS} = +1$ V.

Drain current,

$$I_D = 10\left(1 - \frac{(+1)}{(-4)}\right)^2$$

$$= 10 \times \left(\frac{5}{4}\right)^2$$

$$= 15.6 \text{ mA}.$$

Transconductance,

$$g_m = 5000\left(1 - \frac{(+1)}{(-4)}\right)$$

$$= 6250 \ \mu\text{S}.$$

Chapter 118
The DE MOSFET amplifier

THE GATE OF THE DE MOSFET MAY BE BIASED EITHER POSITIVELY OR negatively with respect to the source. Working in either the depletion mode or the enhancement mode can be provided by both voltage divider and source bias. However, self-bias and current-source bias can only be used to operate in the depletion mode.

Like the JFET amplifier, the DE MOSFET behaves like a constant current source with a value of $g_m v_d$ in parallel with the very high value of r_{ds}. The equations for the signal output and the voltage gain are therefore the same for both the JFET and the MOSFET amplifiers (Chapter 116).

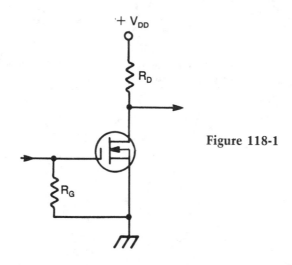

Figure 118-1

MATHEMATICAL DERIVATIONS

The DE MOSFET can be operated with zero bias which is achieved in the circuit of Fig. 118-1. It follows that

Drain-source voltage,

$$V_{DS} = V_{DD} - I_{DSS} R_D \text{ volts} \qquad (118\text{-}1)$$

Provided $V_{DS} > V_p$, the DE MOSFET will automatically operate on the nearly flat section of the $V_{GS} = 0$ V drain curve. This signal may also be directly coupled to the gate without the necessity for a coupling capacitor; this will allow a flat response for the amplifier at low frequencies.

Example 118-1

A DE MOSFET has values of $I_{DSS} = 10$ mA, $V_{GS(off)} = -4$ V and is used in the circuit of Fig. 118-1. If $R_D = 1.5$ kΩ, calculate the values of V_{DS} and the amplifier's voltage gain (neglect any effect of r_{ds}).

Solution

The circuit is operating with zero bias and therefore the drain current, $I_D = I_{DSS} = 10$ mA. It follows that

$$V_{DS} = V_{DD} - I_{DSS}\,R_D \qquad\qquad (118\text{-}1)$$
$$= 24 \text{ V} - 10 \text{ mA} \times 1.5 \text{ k}\Omega$$
$$= 9 \text{ V}.$$

The value of V_{DS} exceeds the pinch-off voltage, $V_P = 4$ V, and therefore the DE MOSFET will operate on the flat section of the drain curve for $V_{GS} = 0$ V.

From Chapter 114,

$$g_{m0} = \frac{-2\,I_{DSS}}{V_{GS(\text{off})}}$$
$$= \frac{-2 \times 10}{-4}$$
$$= 5 \text{ mS} = 5000 \ \mu\text{S}.$$

Voltage gain $= g_{m0} \times r_d$
$$= 5000 \ \mu\text{S} \times 1.5 \text{ k}\Omega$$
$$= 7.5.$$

Chapter 119
The enhancement-only or E MOSFET

IN THIS TYPE OF MOSFET THERE IS NO LONGER A CONTINUOUS N-CHANNEL between the drain and the source. The substrate stretches all the way across the metal oxide layer (Fig. 119-1A) so that no current can flow if the MOSFET is in the depletion mode. When $V_{GS} = 0$ V, there will be a small number of conduction-band electrons created in the substrate by thermal energy but the current flow due to the drain supply is still extremely small. Consequently, the E MOSFET is also referred to as a *normally off* MOSFET. This normally off condition is shown in the schematic symbol (Fig. 119-1B) by the broken line which represents the channel. Because Fig. 119-1A shows an n-channel E MOSFET, the arrow in the schematic symbol points towards the channel; with a p-type MOSFET the arrow would point away from the channel.

To produce an appreciable flow of drain current it is necessary to apply a positive voltage to the gate. If this voltage is low, the charges induced in the substrate are negative ions, which are created by filling holes in the p-substrate with valence electrons. When the positive gate voltage is increased above a certain *minimum* threshold level, ($V_{GS} > V_{GS(\text{th})}$), the additional induced charges are conduction band electrons that exist in a thin n-type inversion layer next to the metal oxide and allow an appreciable flow of electrons from source to drain.

MATHEMATICAL DERIVATIONS

In Fig. 119-1C each drain curve represents a fixed positive value of V_{GS}. For the lowest curve $V_{GS} = +V_{GS(\text{th})}$ and the E MOSFET is virtually cut off.

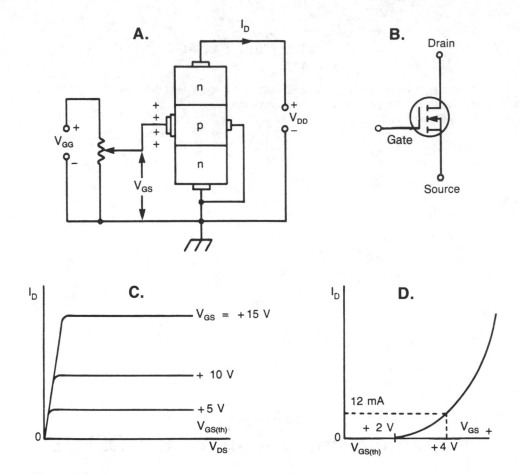

Figure 119-1

The transconductance curve is again parabolic in shape with its vertex at V_{GS} (Fig. 119-1D). The curve's equation is:

Drain current,

$$I_D = k \ (V_{GS} - V_{GS(th)})^2 \text{ milliamperes} \qquad (119\text{-}1)$$

where k is a constant value for a particular MOSFET.

The range of the threshold voltage is typically $+1$ V to $+5$ V. To obtain the value of k, a point on the transconductance curve must be specified. For example, if $V_{GS(th)} = +2$ V and $I_D = 12$ mA when $V_{GS} = +4$ V, it follows from Equation 119-1,

$$12 \text{ mA} = k \ (4 \text{ V} - 2 \text{ V})^2$$

$$k = 0.003.$$

The equation for the transconductance (g_m) is:

$$g_m = 2 \ k \ (V_{GS} - V_{GS(th)}) \text{ siemens} \qquad (119\text{-}2)$$

Example 119-1

For an E MOSFET, $k = 0.001$ and $V_{GS(\text{th})} = +2.5$ V. What are the values of I_D and g_m at the point on the transconductance curve for which $V_{GS} = +5.5$ V?

Solution

Drain current,

$$I_D = k \ (V_{GS} - V_{GS(\text{th})})^2 \tag{119-1}$$
$$= 0.001 \ (5.5 \text{ V} - 2.5 \text{ V})^2 \text{ A}$$
$$= 9 \text{ mA}.$$

Transconductance,

$$g_m = 2 \ k \ (V_{GS} - V_{GS(\text{th})}) \tag{119-2}$$
$$= 2 \times 0.001 \ (5.5 \text{ V} - 2.5 \text{ V}) \text{ S}$$
$$= 6000 \ \mu\text{S}.$$

Chapter 120
The E MOSFET amplifier

E MOSFETS CAN USE VOLTAGE DIVIDER AND SOURCE BIAS BUT NOT SELF AND current-source bias, which can only provide operation in the depletion mode. However, there is another circuit called drain feedback bias, which is only suitable for the enhancement mode (Fig. 120-1A). A high-value resistor of several MΩ is connected between the drain and the gate. Because the voltage drop across R_G due to the gate current is negligible, $V_{DS} = V_{GS} = V_{DD} - I_D R_D$. By arranging that the value of V_{DS} is well above the pinch-off-voltage, the operation of the MOSFET will occur on the nearly flat section of the drain curve. Once I_D, V_{DD} and V_{GS} have been determined, the value of R_D may be chosen to provide the required operating condition.

The action of drain feedback bias for the E MOSFET is similar to the bipolar's collector feedback or self-bias, which compensates for changes that occur in the FET's characteristics. For example if I_D is tending to decrease, both V_{DS} and V_{GS} will increase and this will level off the tendency for I_D to fall. The disadvantage of this type of bias is the degenerative feedback, which occurs from the drain output to the gate input. This has the effect of reducing the amplifier's voltage gain.

Because the gate of an E MOSFET operates with a positive voltage on the gate, it is possible to use direct coupling between the amplifier stages (Fig. 120-1B); not only is the circuitry simple, it has the advantage of an excellent flat response at low frequencies. The positive bias required by one stage is provided by the dc drain voltage of the previous stage.

The equations of the E MOSFET amplifier are the same as those for the JFET and DE MOSFET amplifiers previously discussed in Chapters 116 and 118. For a common source stage the voltage gain from gate to drain is $g_m r_d$.

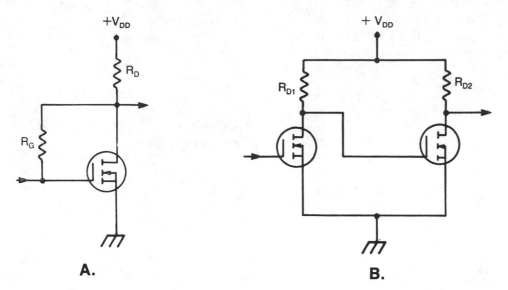

A.   B.

Figure 120-1

Example 120-1

For a particular E MOSFET, $k = 0.001$ and $V_{GS(th)} = +2.5$ V. It is required to bias this MOS-FET to the point where $V_{GS} = +5.5$ V. If the drain feedback circuit of Fig. 120-1A is used, calculate the required value of R_D. Determine the value of the voltage gain from gate to drain.

Solution

Drain current,

$$I_D = 0.001 \ (5.5 \text{ V} - 2.5 \text{ V})^2 \text{ A}$$
$$= 9 \text{ mA}.$$

Transconductance,

$$g_m = 2 \times 0.001 \ (5.5 \text{ V} - 2.5 \text{ V}) \text{ S}$$
$$= 6 \text{ mS}$$
$$= 6000 \ \mu\text{S}.$$

Drain load resistor,

$$R_D = \frac{24 \text{ V} - 5.5 \text{ V}}{9 \text{ mA}}$$
$$\approx 2 \text{ k}\Omega.$$

Voltage gain from gate to drain,

$$G_v = g_m r_d$$
$$= 6000 \ \mu\text{S} \times 2 \text{ k}\Omega$$
$$= 12.$$

Chapter 121
Operational amplifiers

OPERATIONAL AMPLIFIERS ARE DESIGNED TO SIMULATE CERTAIN MATHEMATICAL operations such as addition, subtraction, differentiation, and integration. In Fig. 121-1A, the input signal, V_i, is fed through an impedance, Z_i, to a solid-state amplifier whose voltage gain is $-A$ (the minus sign indicates that the output signal, V_o, is inverted with respect to the input signal such as would occur with some form of common-emitter arrangement). Z_f provides feedback from the output circuit to the input circuit and the combination of Z_f and Z_i may be regarded as a voltage divider between V_o and V_i.

The circuit in Fig. 121-1B is a noninverting operational amplifier whose feedback action is similar to that of the inverting type. Figure 121-1C and D show operational amplifiers that are capable of performing addition and subtraction; in these cases, Z_f and Z_i are resistors.

In the operational amplifier of Fig. 121-1E, C_f is a capacitor and the amplifier is then capable of integration. Owing to the Miller effect, the input capacitance will be equal to $(A + 1)C_f$. If the input signal, V_i, is a step voltage as shown, the voltage at the amplifier input terminal will slowly rise and the output signal, V_o, will be an amplified linear fall, which will represent the result of integrating V_i. If C_f is replaced by a suitable inductor, L_f, the operational amplifier will be capable of differentiation.

MATHEMATICAL DERIVATIONS

Provided A is sufficiently large, the voltage gain of the inverting amplifier with feedback (Fig. 121-1A) is

Voltage gain with feedback,

$$A' = \frac{V_o}{V_i} = -\frac{Z_f}{Z_i} \tag{121-1}$$

For the noninverting operational amplifier of Fig. 121-1B.
Voltage gain with feedback,

$$A' = \frac{V_o}{V_i} = 1 + \frac{Z_f}{Z_i} \tag{121-2}$$

In the summing amplifier (Fig. 121-1C) the gain for each input signal is $-R_f/R_i$.
Output signal,

$$V_o = -R_f \left(\frac{V_{i1}}{R_{i1}} + \frac{V_{i2}}{R_{i2}} + \frac{V_{i3}}{R_{i3}} \right) \text{ volts} \tag{121-3}$$

For the differential amplifier (Fig. 121-1D),
Output signal,

$$V_o = -\frac{R_f}{R_i}(V_{i1} - V_{i2}) \text{ volts} \tag{121-4}$$

Example 121-1

In Fig. 121-1A, Z_f is a 150 kΩ resistor and Z_i is a 10 kΩ resistor. Assuming that A is large, what is the gain of this operational amplifier?

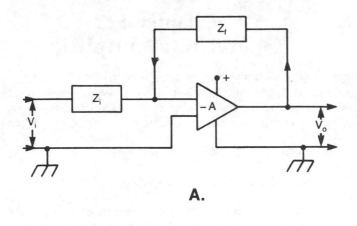

A.

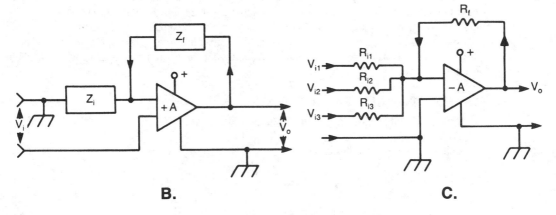

B.

C.

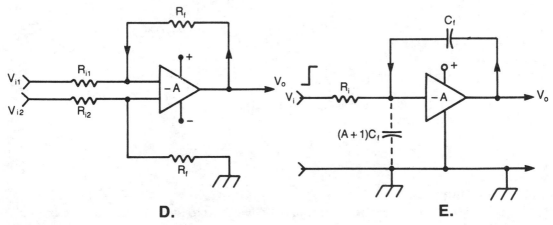

D.

E.

Figure 121-1

Solution

Operational amplifier gain,

$$A' = -\frac{R_f}{R_i} \qquad (121\text{-}1)$$

$$= -\frac{150}{10}$$

$$= -15$$

Example 121-2

In Fig. 121-1B, Z_f is a 150 kΩ resistor and Z_i is a 10 kΩ resistor. Assuming that A is large, calculate the gain of the operational amplifier.

Solution

Operational amplifier gain,

$$A' = 1 + \frac{R_f}{R_i} \qquad (121\text{-}2)$$

$$= 1 + \frac{150}{10}$$

$$= +16$$

Example 121-3

In Fig. 121-1C, V_{i1}, V_{i2}, V_{i3} are step voltages of +2 mV, +5 mV and +7 mV respectively, $R_f = 1$ MΩ and each of the resistors R_{i1}, R_{i2}, and R_{i3} has a value of 0.5 MΩ. Determine the value of the output signal.

Solution

Voltage gain for each signal is $\quad -R_f/R_i = -1$ MΩ/0.5 MΩ

$$= -2.$$

Output signal, $\qquad V_o = -2\,(2 + 5 + 7)$

$$= -28 \text{ mV}$$

Example 121-4

In Fig. 121-1D, $R_f = 120$ kΩ and $R_i = 10$ kΩ. V_{i1} and V_{i2} are step voltages of +15 mV and +10 mV respectively. What is the value of the output signal?

Solution

Voltage gain for each signal is

$$-\frac{R_f}{R_i} = -\frac{120 \text{ kΩ}}{10 \text{ kΩ}}$$

$$= -12$$

Output signal,

$$V_o = -12 \times (15 - 10)$$
$$= -60 \text{ mV}.$$

Chapter 122
Positive or regenerative feedback

POSITIVE FEEDBACK MAY BE USED TO INCREASE THE GAIN OF AN AMPLIFIER circuit. This is illustrated in Fig. 122-1A where V_i is the input signal from the preceding stage. Such a signal would be applied between the base of a transistor (or the control grid of a tube) and ground. The input signal, V_o, appears between collector (or plate) and ground and a fraction, β (beta), of this output signal is then fed back to the input circuit so that this feedback voltage, $+\beta V_o$, is in phase with V_i. β is called the feedback factor, which may either be expressed as a decimal fraction or as a percentage.

In order for the feedback to be positive, there must be a total of 360° phase shift (equivalent to zero phase shift) around the feedback loop base → collector → base (or grid → plate → grid). The total signal voltage when applied between base and emitter (or control grid and cathode) is the sum of the input signal, V_i, and the positive feedback voltage, $+\beta V_o$. The voltage gain (open loop gain) of the active device is A and the mathematical derivations will show that $A' = A/(1 - A\beta)$ where A' is the overall voltage gain with the positive feedback present (closed loop gain). There are then three possible conditions in the circuit:

1. If A and β are chosen so that the value of $A\beta$ is less than 1, then A' is greater than A and the amplifier's gain has been increased as the result of the positive feedback. Such is the case with the so-called *regenerative* amplifier.
2. If $A\beta = 1$ (for example $A = 10$ and $\beta = 0.1$ or 10%), A' is theoretically infinite. This means that the circuit can provide a continuous output without any input signal from the previous stage. This is the condition in a stable oscillator.
3. If $A\beta > 1$, the oscillator is unstable. The output, V_o, then increases which tends to reduce the value of A until the equilibrium condition of $A\beta = 1$ is reached.

The condition for oscillation is therefore $A\beta = 1\angle0°$; this is sometimes referred to as the *Barkhausen* or *Nyquist criterion*. The inclusion of "$\angle0°$" in the polar value of $A\beta$ means that the resultant phase shift around the loop is zero degrees and, consequently, the feedback is positive. By contrast an angle of 180° would indicate that the feedback is negative.

Figure 122-1B shows the principle of positive feedback in an oscillator circuit. If we assume that the input signal is 1 V rms and the voltage gain of the active device is 10, the output signal is 10 V rms; if the feedback factor is 1/10 or 10%, the 10 V output signal will be responsible for creating the 1 V input signal (this does *not* mean that there is only 9 V left of the output signal). This argument sounds rather like the chicken and the egg so the question arises "How does the circuit get started in the first place?" The answer is that all active devices are inherently noisy.

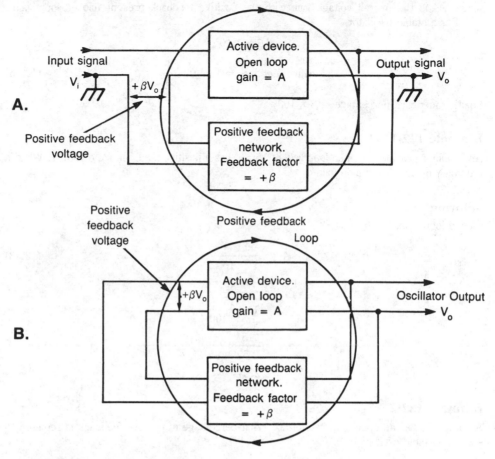

A.

Input signal V_i

$+\beta V_o$

Positive feedback voltage

Active device.
Open loop
gain $= A$

Output signal V_o

Positive feedback
network.
Feedback factor
$= +\beta$

B.

Positive
feedback
voltage

$+\beta V_o$

Positive feedback
Loop

Active device.
Open loop
gain $= A$

Oscillator Output V_o

Positive feedback
network.
Feedback factor
$= +\beta$

Figure 122-1

Because the noise is spread throughout the frequency spectrum, it will contain a component at the frequency of oscillation. This component will trigger the positive feedback network so that the oscillation will increase until the equilibrium condition of $A\beta = 1\angle 0°$ is reached.

MATHEMATICAL DERIVATIONS

For a regenerative amplifier:

Total input signal between base and emitter

$$= V_i + \beta V_o \text{ volts} \tag{122-1}$$

Output signal,

$$V_o = A \times (V_i + \beta V_o) \text{ volts} \tag{122-2}$$

This yields

$$A' = \frac{V_o}{V_i} = \frac{A}{1 - A\beta} \tag{122-3}$$

where A' is the overall voltage gain with the positive feedback present (closed loop gain).
For a stable oscillator,

$$A\beta = 1\angle 0°$$ (122-4)

$$\text{Positive feedback factor, } \beta = \frac{1}{A}$$ (122-5)

Positive feedback percentage $= \beta \times 100\%$

Example 122-1

An amplifier provides an open loop gain of 15 and has a positive feedback factor of 3%. What is the value of the closed loop gain?

Solution

Closed loop gain,

$$A' = \frac{A}{1 - A\beta}$$ (122-3)

$$= \frac{15}{1 - \dfrac{15 \times 3}{100}}$$

$$= \frac{15}{0.55}$$

$$= 27.3$$

Example 122-2

An amplifier has an open loop gain of 50. What percentage of positive feedback is required to sustain a stable oscillation?

Solution

Positive feedback factor,

$$\beta = \frac{1}{A}$$ (122-5)

$$= \frac{1}{50}$$

Positive feedback percentage

$$= \beta \times 100$$ (122-6)

$$= \frac{1}{50} \times 100$$

$$= 2\%.$$

Chapter 123
The Wien bridge oscillator

THIS OSCILLATOR (FIG. 123-1) EMPLOYS TWO COMMON-EMITTER STAGES SO that there is theoretically zero phase shift between a signal voltage on the base of $Q1$ and the output voltage at the collector of $Q2$. The feedback loop is completed by the Wien filter consisting of $R1\ R2\ C1\ C2$ and therefore, in order for the feedback to be positive, the input voltage to the filter V_o, and the output voltage from the filter, V_i, must be in phase.

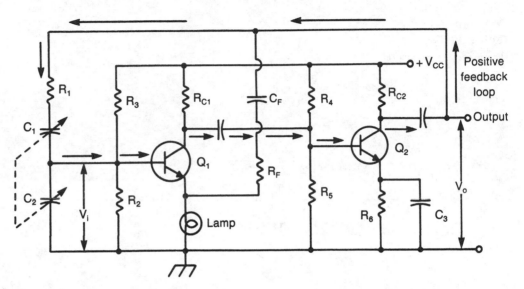

Figure 123-1

MATHEMATICAL DERIVATIONS

The frequency of oscillation is given by:
Frequency of oscillation,

$$f_o = \frac{1}{2\pi\sqrt{R1\ R2\ C1\ C2}}$$ (123-1)

$$= \frac{1}{2\pi RC} \text{ hertz}$$

if $R1 = R2 = R$ ohms and $C1 = C2 = C$ farads.
At the frequency, f_o, the attenuation factor is:
Attenuation factor,

$$\beta = \frac{V_i}{V_o} = \frac{1}{3}$$ (123-2)

The mathematical derivations show that the combined voltage gain of $Q1$ and $Q2$ must be equal to three. This is not practical and, therefore, the oscillator circuit contains negative feedback, provided by R_f and the lamp which form the bridge circuit in conjunction with $R1\ C1\ R2\ C2$.

The combined gain of $Q1$, $Q2$ without the negative feedback can then be high but their gain with feedback will equal three under stable conditions, which can be determined by the operating resistance of the lamp.

Like the RC phase-shift oscillator (Chapter 124), the Wien bridge circuit is especially suitable for the generation of low frequency sine waves with good stability and a lack of harmonic distortion.

Example 123-1

In Fig. 123-1, $C1 = C2 = 8200$ pF, $R1 = R2 = 12$ kΩ. Determine the frequency of the oscillation.

Solution

Frequency of oscillation,

$$f_o = \frac{0.159}{RC} \text{ Hz} \qquad (123\text{-}1)$$

$$= \frac{0.159}{12 \times 10^3 \times 8200 \times 10^{-12}}$$

$$= \frac{1.59 \times 10^{-1}}{1.2 \times 8.2 \times 10^{-5}}$$

$$= \frac{15900}{1.2 \times 8.2} \text{ Hz}$$

$$= 1.62 \text{ kHz, rounded off.}$$

Chapter 124
The RC phase-shift oscillator

THIS TYPE OF OSCILLATOR (FIG. 124-1) IS CAPABLE OF GENERATING A SINE WAVE that is relatively free of harmonic distortion; its frequency output can range from less than 1 Hz to a few hundred kHz.

The circuit uses the common-emitter configuration so that there is 180° phase change from base to collector; the feedback loop is completed by the RC phase shift network that contains a minimum number of three sections (as shown). For the feedback to be positive, the network must therefore provide a further 180° shift (ignoring any effect of the transistor circuitry) and it would appear, at first glance, that each RC section should contribute a 60° shift. However, this simple approach ignores the shunting effect of one section on another. The results of a precise analysis are given in the mathematical derivations.

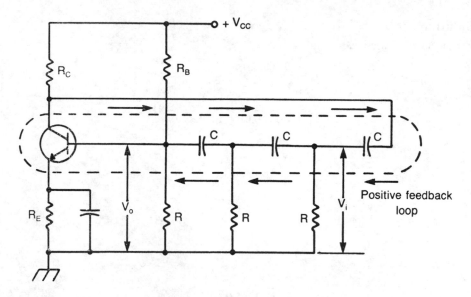

Figure 124-1

MATHEMATICAL DERIVATIONS

Frequency of oscillation,

$$f_o = \frac{1}{2\pi\sqrt{6}RC} = \frac{0.159}{\sqrt{6} \times RC} \text{ Hz} \qquad (124\text{-}1)$$

where R is measured in ohms and C in farads.

The attenuation factor of the network is:

Attenuation factor,

$$\beta = \frac{V_o}{V_i} = \frac{1}{29} \qquad (124\text{-}2)$$

and, therefore, if oscillations are to be sustained, the open-loop voltage gain of the common-emitter amplifier must be 29. Notice that the formula for f_o is inversely proportional to C (and not $\sqrt{C}$ as in the LC oscillator).

If a four-section RC feedback network is used:

Frequency of oscillation,

$$f_o = \frac{1}{2\pi RC} \text{ Hz} \qquad (124\text{-}3)$$

Attenuation factor,

$$\beta = \frac{1}{18.4} \qquad (124\text{-}4)$$

Example 124-1

In Fig. 124-1, $C = 0.01\ \mu\text{F}$, $R = 10\ \text{k}\Omega$. Calculate the frequency of the oscillation.

Solution

Frequency of oscillation,

$$f_o = \frac{0.159}{\sqrt{6}RC} \tag{124-1}$$

$$= \frac{1.59 \times 10^{-1}}{\sqrt{6} \times 0.01 \times 10^{-6} \times 10^4} \text{ Hz}$$

$$= \frac{1590}{\sqrt{6}}$$

$$= 650 \text{ Hz, rounded off.}$$

Chapter 125
The multivibrator

THE MULTIVIBRATOR CIRCUIT OF FIG. 125-1, CONSISTS OF TWO COMMON-emitter stages that are cross-connected for positive feedback. The resultant instability causes the transistors to cut on and off alternatively so that approximate square-wave voltage outputs appear at the collectors; the base waveforms have a sawtooth appearance.

When the transistor $Q1$ has been driven to the cutoff condition as a result of the positive feedback action, the base potential, e_{b1}, is approximately equal to $-V_{CC}$. The capacitor $C2$ will then discharge through R_{B2} and $Q2$ so that e_{b1} will rise towards V_{CC} with a time constant approximately equal to $R_{B2}C_2$.

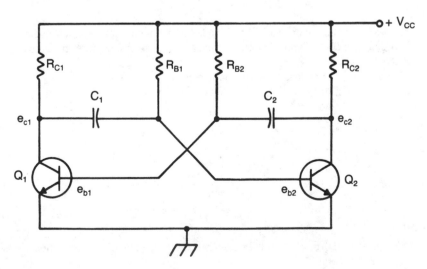

Figure 125-1

When e_{b1} becomes slightly positive, $Q1$ will switch on and the positive feedback action will drive $Q2$ to the cutoff position. Since e_{b1} approximately reaches its halfway mark in rising from $-V_{CC}$ to a slightly positive potential on its way towards $+V_{CC}$, $Q1$ is cut off for a time interval of approximately $0.7R_{B2}C_2$.

MATHEMATICAL DERIVATIONS

$$\text{Multivibrator frequency} = \frac{1}{\text{total period}} \tag{125-1}$$

$$= \frac{1}{0.7(R_{B1}C_1 + R_{B2}C_2)}$$

$$= \frac{1}{1.4R_B C} \text{ hertz}$$

if the multivibrator is symmetrical with $C1 = C2 = C$ farads and $R_{B1} = R_{B2} = R_B$ ohms.

Example 125-1

In Fig. 125-1, $R_{B1} = R_{B2} = 100$ kΩ and $C1 = C2 = 0.01$ μF. What is the approximate frequency of the multivibrator?

Solution

Multivibrator frequency,

$$f_o = \frac{1}{1.4R_B C} \text{ Hz} \tag{125-1}$$

$$= \frac{1}{1.4 \times 0.01 \times 10^{-6} \times 10^5} \text{ Hz}$$

$$= 714 \text{ Hz, rounded off.}$$

4
PART

Tubes and their associated circuits

Chapter 126
The triode tube—static characteristics

THE DIODE, WHICH IS THE SIMPLEST FORM OF THERMIONIC TUBE, IS LIMITED TO a single function, namely rectification (the conversion of an ac voltage to a dc voltage). In the triode or three electrode tube, the flow of the electrons from the heated cathode to the plate is controlled by means of an additional electrode, which is interposed between the cathode and the plate (Fig. 126-1A). This electrode is called the control grid (G1) on account of its form taken in early examples of such tubes. Its modern form is in the shape of a thin spiral of wire or an open mesh. This grid is commonly operated at a negative potential relative to the cathode so that it attracts no electrons to itself and there is no flow of grid current; however it tends to repel those electrons that are being attracted toward the positive plate.

The number of electrons reaching the plate per second is determined mainly by the electric field near the cathode, and hardly at all by the rest of the field in the remaining space between the cathode and the plate. Near the cathode the electrons are travelling slowly compared to those that have already moved some distance toward the plate; therefore the electron density in the inter-electrode space will be high near the cathode but will decrease towards the plate. The total space charge will be concentrated near the cathode because once an electron has left this region, it contributes to the space charge for only a very brief interval of time. Therefore, the space current in the triode is determined by the electric field near the cathode produced by the combined effect of the plate and grid potentials.

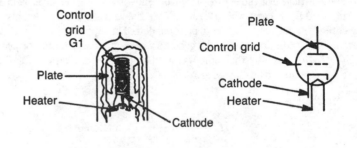

Figure 126-1

A.

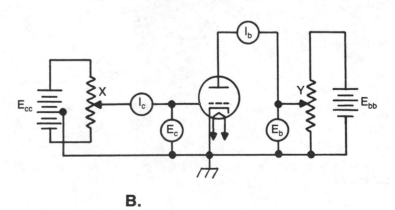

B.

When the grid is made sufficiently negative with respect to the cathode, all of the emitted electrons will be repelled back to the cathode and no plate current will flow. Therefore, the plate current, $I_b = 0$ and the tube is said to be cutoff.

It is clear that the value of the plate current is governed by both the grid voltage and the plate voltage. We can say that the triode is a voltage controlled device like the field-effect transistor (FET). For a particular tube we need to know:

1. The degree of control that the plate voltage exercises over the plate current while maintaining the grid voltage constant.
2. The degree of control that the grid voltage exercises over the plate current while maintaining the plate voltage constant.
3. The relative controls of the grid and the plate in maintaining a constant level of plate current.

This information is discussed in Chapters 127, 128, and 129. It is derived from static characteristic curves that show the interaction between the variables I_b (plate current) E_c (grid voltage) and E_b (plate voltage). The word *static* means that the curves are obtained under controlled laboratory conditions in which two of the three quantities are varied during the experiment but the third quantity is kept constant (Fig. 126-1B). By contrast, under dynamic conditions such as occur in an amplifier circuit, all three variables are changing simultaneously.

MATHEMATICAL DERIVATIONS

For a symmetrical grid structure it can be shown that the electric field near the cathode is proportional to $E_c + E_b/\mu$ where E_c and E_b are the grid and plate potentials as measured relative to the cathode, which is assumed to be grounded. The amplification factor, μ, (Chapter 129) is a "constant" that is determined by the geometry of the tube. The total, or plate current, I_b, varies with $E_c + E_b/\mu$ in exactly the same way as the plate current varies with the plate voltage for a space-charge limited diode. Therefore

Plate current,

$$I_b = K\left(E_c + \frac{E_b}{\mu}\right)^{3/2} \text{ milliamperes} \qquad (126\text{-}1)$$

where K is a constant that is determined by the dimensions of the tube.

This yields

$$I_b = 0$$

when

$$E_c = -\frac{E_b}{\mu} \text{ volts} \qquad (126\text{-}2)$$

Consequently, the value of the cutoff bias on the grid is $-E_b/\mu$ volts.

Example 126-1

A triode has an amplification factor of 40 and its plate voltage is $+240$ V. What is the value of its cutoff bias?

Solution

$$\text{Cutoff bias} = -\frac{E_b}{\mu} \qquad\qquad (126\text{-}2)$$

$$= -\frac{240 \text{ V}}{40}$$

$$= -6 \text{ V}.$$

Chapter 127
Ac and dc plate resistances

THE AC PLATE RESISTANCE, r_p, IS A STATIC PARAMETER THAT MEASURES THE control that the plate voltage (E_b) exercises over the plate current, (I_b). Figure 127-1B shows a typical family of plate characteristic curves (Chapter 126); these are the graphs of the plate current versus the plate voltage and are obtained from the experimental set-up of Fig. 127-1A. Let us assume that the grid voltage, E_c, is 2 V negative with respect to the grounded

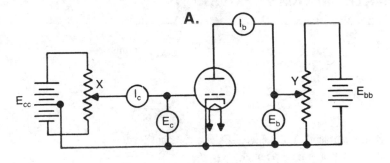

A.

Figure 127-1

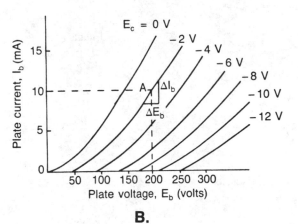

B.

cathode; this is achieved by adjusting the setting of the potentiometer, X. At the same time we adjust the potentiometer, Y, to apply $+200$ V to the plate. From the plate characteristics we see that the corresponding plate current is 10 mA (point A). The dc plate resistance R_p, of the triode is then 200 V/10 mA $= 20$ kΩ; of course this resistance value is not a constant because the triode is a nonlinear device and therefore its dc resistance depends on the particular operating point that you have selected.

Under dynamic (amplifier) conditions the plate voltage and the plate current are continuously changing so that we are more concerned with the (small) change in the plate current that is produced by a corresponding (small) change in the plate voltage. For example, if we increase the plate voltage by 20 V (ΔE_b) to $+220$ V, the plate current might increase from 10 mA to 12 mA and therefore the corresponding change in plate current is $12 - 10 = 2$ mA (ΔI_b). The ac plate resistance, r_p, is then defined as the ratio of $\Delta E_b : \Delta I_b = 20$ V : 2 mA $= 10$ kΩ (the ac plate resistance value is less than the dc resistance value). This means that a 10 V change in the plate voltage produces a 1 mA change in the plate current.

Because a high r_p value means that a large change in the plate voltage only produces a small change in the plate current, the ac plate resistance is an inverse measurement of the control that the plate voltage exercises over the plate current. In strict mathematical terms the ac plate resistance is the reciprocal of the characteristic slope at the operating point, A.

The value of the ac plate resistance is not a constant but depends on the chosen operating point.

MATHEMATICAL DERIVATIONS

dc plate resistance,

$$R_p = \frac{E_b}{I_b} \text{ ohms} \tag{127-1}$$

ac plate resistance,

$$r_p = \frac{\Delta E_b}{\Delta I_b} \text{ ohms} \tag{127-2}$$

while maintaining the value of E_c at a constant level.

At the operating point A,
ac plate resistance,

$$r_p = \left(\frac{\delta E_b}{\delta I_b}\right)_{E_c \text{ constant}} \text{ ohms,} \tag{127-3}$$

From the equation 126-1, $I_b = K(E_c + E_b/\mu)^{3/2}$

Therefore,

$$\left(\frac{\delta I_b}{\delta E_b}\right)_{E_c \text{ constant}} = \frac{3K}{2\mu}\left(E_c + \frac{E_b}{\mu}\right)^{1/2}$$

or

ac plate resistance,

$$r_p = \left(\frac{\delta E_b}{\delta I_b}\right)_{E_c \text{ constant}} \tag{127-4}$$

$$= \frac{2\mu}{3K}\left(E_c + \frac{E_b}{\mu}\right)^{-1/2} \text{ ohms}$$

Equation 127-4 shows that the value of r_p is dependent on the values of E_c and E_b at the chosen operating point.

Example 127-1

On a family of plate characteristics the operating point is specified by $E_c = -3$ V, $I_b = 6$ mA, $E_b = 180$ V. Without altering the grid voltage, the plate voltage is reduced to 160 V and the plate current then falls to 4.5 mA. What are the values of the dc and ac plate resistances?

Solution

dc plate resistance,

$$R_p = \frac{E_b}{I_b} \tag{127-1}$$

$$= \frac{180 \text{ V}}{6 \text{ mA}}$$

$$= 30 \text{ k}\Omega.$$

ac plate resistance,

$$r_p = \frac{\Delta E_b}{\Delta I_b} \tag{127-2}$$

$$= \frac{180 \text{ V} - 160 \text{ V}}{6 \text{ mA} - 4.5 \text{ mA}}$$

$$= \frac{20 \text{ V}}{1.5 \text{ mA}}$$

$$= 13.33 \text{ k}\Omega.$$

Chapter 128
Transconductance, g_m

THE PLATE CURRENT OF A TRIODE TUBE IS SIMULTANEOUSLY DETERMINED BY the voltages that exist on the plate and the grid relative to the cathode. In Chapter 127 we investigated the control that the plate voltage exercised over the plate current; this was governed by the value of the ac plate resistance, r_p. We are now going to explore the degree of control that the grid electrode exercises over the plate current.

In the experimental set-up of Fig. 128-1A, let us adjust the setting of the potentiometer, X, so that the grid voltage, E_c, is 2 V negative with respect to the grounded cathode. At the same time the potentiometer, Y, is adjusted to apply +200 V to the plate. On the plate characteristics of Fig. 127-1B the corresponding plate current is 10 mA and we have arrived at the operating point, A. The same operating point is shown on the transfer characteristic of Fig. 127-1C; this characteristic is the graph of the plate current, I_b, versus the grid voltage, E_c, while maintaining a constant level of the plate voltage, E_b.

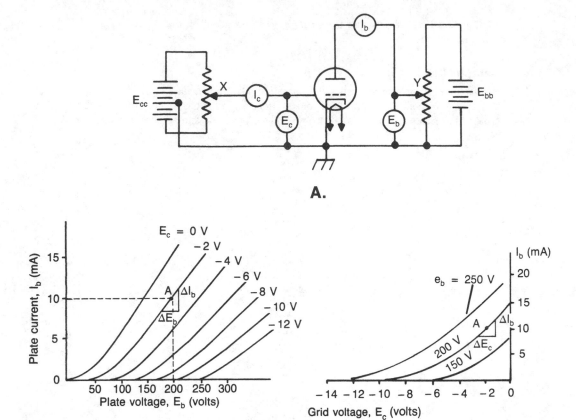

Figure 128-1

Let us now alter the setting of the potentiometer, X, so that the grid is 3 V negative with respect to the cathode; at the same time the potentiometer, Y, is unchanged. The plate current drops from 10 mA to 8 mA (for example) so that a 1-volt *change* on the grid (ΔE_c) is responsible for a 2 mA *change* (ΔI_b) in the value of the plate current. The transconductance, g_m, is then defined as the ratio of $\Delta I_b : \Delta I_c$ while maintaining the value of E_b at a constant level. This ratio of current to voltage is a conductance (letter symbol, G) which is measured in siemens (S).

In our example the value of g_m is 2 ma/V or 2000 μS (however the tube manual still refers to micromhos in which one mho has the same value as one siemens). The transconductance is a direct measure of the control that the grid exercises over the plate current because a high value of g_m means that a small change in the grid voltage produces a shift in the plate current that would otherwise only be achieved by a relatively large change in the plate voltage. In strict mathematical terms the value of g_m is equal to the slope of the transfer characteristic at the operating point, A.

Like the value of the ac plate resistance, the magnitude of the transconductance is not a constant but depends on the chosen operating point.

MATHEMATICAL DERIVATIONS

Transconductance,

$$g_m = \frac{\Delta I_b}{\Delta E_c} \text{ siemens} \qquad (128\text{-}1)$$

while maintaining the value of E_b at a constant level.

Transconductance,

$$g_m = \left(\frac{\delta I_b}{\delta E_c}\right)_{(Eb \text{ constant})} \qquad (128\text{-}2)$$

From Equation 126-1,

$$I_b = K\left(E_c + \frac{E_b}{\mu}\right)^{3/2}$$

therefore

Transconductance,

$$g_m = \left(\frac{\delta I_b}{\delta E_c}\right)_{(Eb \text{ constant})} \qquad (128\text{-}3)$$

$$= \frac{3K}{2}\left(E_c + \frac{E_b}{\mu}\right)^{1/2} \text{ siemens}$$

This shows that the value of g_m depends on the values of E_c and E_b at the chosen operating point.

Example 128-1

On the plate characteristics of a triode, the operating point is specified by $E_c = -3$ V, $I_b = 6$ mA, $E_b = 180$ V. Without changing the plate voltage, the grid voltage is raised to -2 V and the plate current increases to 9 mA. What is the value of the transconductance?

Solution

Change in the plate current,

$$\Delta I_b = 9 - 6$$
$$= 3 \text{ mA.}$$

Change in the grid voltage,

$$\Delta E_c = -2 - (-3)$$
$$= 1 \text{ V.}$$

Transconductance,

$$g_m = \frac{\Delta I_b}{\Delta E_c} \qquad (128\text{-}1)$$

$$= \frac{3 \text{ mA}}{1 \text{ V}}$$

$$= 3 \text{ mA/V}$$

$$= 3000 \ \mu\text{S}$$

Chapter 129
Amplification factor

THE AMPLIFICATION FACTOR (μ) IS A STATIC PARAMETER THAT COMPARES THE relative controls that the plate voltage (E_b) and the grid voltage (E_c) exercises over the plate current (I_b). Its value depends on the geometry of the triode, in particular the size and spacing of the electrodes.

Because the control grid is wound close to the cathode, its effect on the plate current is greater than that of the plate, which is positioned further from the cathode.

Referring to Fig. 129-1A, we will assume that the potentiometer, X, is set so that the grid is 2 V negative with respect to the cathode. At the same time we adjust the potentiometer, Y, to apply +200 V to the plate. From the plate characteristics illustrated in Fig. 129-1B, we can see that the plate current is 10 mA (point A). If the potentiometer, X, is now reset to apply −3 V to the grid, the plate current will be reduced to 8 mA (point B). However, we can restore the current to 10 mA if we increase the plate voltage from +200 V to +220 V (point C). We can deduce that 1 V change on the grid can be compensated by a 20 V change on the plate. This means that the grid is twenty times more effective than the plate in controlling the plate current.

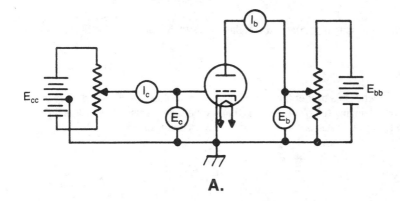

A.

Figure 129-1

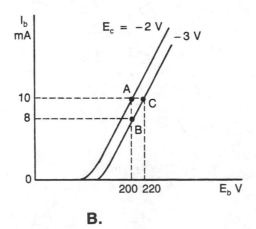

B.

Therefore at the operating point, A, the value of the amplification factor, μ, is 20. Notice that μ has no units because it merely compares two voltage changes.

Triodes may be classified in terms of their μ values. Low-μ triodes have an amplification factor of less than 10 and are primarily power amplifier tubes. By comparison medium-μ ($\mu \approx 20$) and high-μ ($\mu \approx 100$) triodes are used for the voltage amplification of small signals.

MATHEMATICAL DERIVATIONS

Amplification factor,

$$\mu = \frac{\Delta E_b}{\Delta E_c}, \text{ keeping } I_b \text{ constant} \tag{129-1}$$

Transconductance,

$$g_m = \frac{\Delta I_b}{\Delta E_c}, \text{ keeping } E_b \text{ constant} \tag{128-1}$$

Plate resistance,

$$r_p = \frac{\Delta E_b}{\Delta I_b}, \text{ keeping } E_c \text{ constant} \tag{127-1}$$

Then

$$\mu = \frac{\Delta E_b}{\Delta E_c} \tag{129-2}$$

$$= \frac{\Delta E_b}{\Delta I_b} \times \frac{\Delta I_b}{\Delta E_c}$$

$$= r_p \times g_m$$

This relationship is only true provided the values of μ, r_p, and g_m are all measured at the same operating point.

Alternatively, from Equations 127-4 and 128-3,

$$r_p = \frac{2\mu}{3K}\left(E_c + \frac{E_b}{\mu}\right)^{-1/2}$$

and

$$g_m = \frac{3K}{2}\left(E_c + \frac{E_b}{\mu}\right)^{1/2}$$

then,

$$r_p \times g_m = \mu \tag{129-2}$$

Example 129-1

In Fig. 129-1A the potentiometers, X and Y, are set to provide the initial operating point: $E_b = +250$ V, $E_c = -6$ V, $I_b = 8$ mA. When E_c is changed to -4 V, I_b increases to 12 mA. The plate current is restored to 8 mA by reducing the plate voltage to $+200$ V. Calculate the value of the amplification factor.

Solution

Change in the plate voltage, $\Delta E_b = 250 - 200 = 50$ V. Corresponding change in the grid voltage, $\Delta E_c = 6 - 4 = 2$ V.

Amplification factor,

$$\mu = \frac{\Delta E_b}{\Delta E_c} = \frac{50 \text{ V}}{2 \text{ V}} \qquad (129\text{-}2)$$
$$= 25$$

Notice that the value of μ is not a constant, but depends to some degree on the chosen operating point.

Example 129-2

In Example 129-1, what is the value of the grid cutoff bias if $E_b = +250$ V?

Solution

Grid cutoff bias,

$$E_c = -\frac{E_b}{\mu} \qquad (126\text{-}2)$$
$$= -\frac{250 \text{ V}}{25}$$
$$= -10 \text{ V}$$

Example 129-3

At a triode's operating point $r_p = 15$ kΩ and $g_m = 4000$ μS. What is the value of the amplification factor?

Solution

Amplification factor,

$$\mu = r_p \times g_m \qquad (129\text{-}2)$$
$$= 15 \times 10^3 \times 4000 \times 10^{-6}$$
$$= 60.$$

Notice that μ is actually a negative number since, if the grid is made more *negative,* the plate voltage must be made more *positive* in order to maintain the same level of plate current.

Chapter 130
The triode tube as an amplifier

FIGURE 130-1 SHOWS THE TRIODE AS AN AMPLIFIER TUBE IN THE SIMPLEST possible way. The signal to be amplified is some type of alternating voltage, e_{in}, (for example, sine wave, pulse, sawtooth, square wave) which is applied to the control grid. E_{cc} is a steady dc

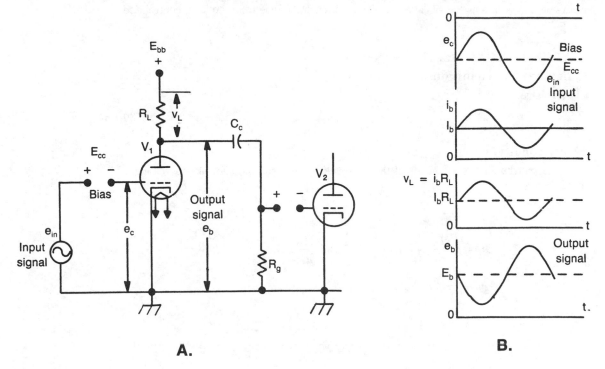

A.

B.

Figure 130-1

voltage that is supplied from a "C" battery and is referred to as the bias. The value of bias is such that, throughout the cycle of the signal, the grid is always negative with respect to the cathode. The plate is maintained at a high positive potential by the "B" battery which provides a voltage, E_{bb}, in series with the tube and the resistor, R_L. Plate current will flow and there will be a voltage drop across the amplifier's load, R_L. Consequently, the voltage at the plate, E_b (with respect to the grounded cathode) will always be less than the value of E_{bb}.

Under quiescent (dc) conditions when no signal is being applied to the grid ($e_{in} = 0$ V), the bias voltage, E_{cc}, will determine the value of the steady plate current, I_b, and the steady plate voltage, E_b. With signal conditions, an alternating voltage is applied to the control grid. This creates a fluctuating plate current, which will now contain an alternating current component, i_b. This component will develop an alternating voltage drop across R_L and because, at all times, the sum of the voltages across the load and the tube must equal the fixed value of E_{bb}, an alternating component, e_b, will appear in the waveform of the plate voltage. This fluctuation in the voltage at the plate is the amplified output signal. To determine the actual voltages and currents in the circuit we combine the dc levels with the signal values by using the principle of superposition (Chapter 29).

Notice that under the dynamic conditions in an amplifier circuit, the three quantities e_c, e_b and i_b are varying simultaneously. This is in contrast with the laboratory conditions that we used when developing the tube's static parameters. In the procedures for calculating these parameters, two of the quantities were varied while the third quantity was kept constant.

When the grid voltage, e_c, is becoming less negative, the plate current, i_b, is increasing and there is a greater voltage drop across the load, R_L. Therefore the plate voltage, e_b, across the tube decreases so that e_c and e_b are 180° out of phase. This is often described as the phase

423

inversion that occurs as the signal is transferred through the tube from the input control grid to the output plate.

MATHEMATICAL DERIVATIONS
Quiescent (dc) conditions

Plate voltage,

$$E_b = E_{bb} - I_b R_L \text{ volts} \tag{130-1}$$

Signal conditions

The plate current, I_b, is a function of the plate voltage, E_b, and the control grid voltage, E_c. Therefore

$$I_b = f(E_b, E_c) \tag{130-2}$$

$$\Delta I_b = \left(\frac{\delta I_b}{\delta E_b}\right)_{(Eb \text{ constant})} \times \Delta E_b + \left(\frac{\delta I_b}{\delta E_c}\right)_{(Eb \text{ constant})} \times \Delta E_c$$

$$i_b = \frac{1}{r_p} \times e_b + g_m \times e_c \text{ amperes}$$

But

$$e_b = -i_b R_L \text{ volts} \tag{130-3}$$

where the negative sign indicates the 180° phase change between the input and output signals. Combining Equations 130-2 and 130-3,

$$i_b = \frac{-i_b R_L}{r_p} + g_m \times e_c$$

This yields

$$i_b = \frac{r_p g_m e_c}{r_p + R_L} = \frac{-\mu e_c}{r_p + R_L} \text{ amperes} \tag{130-4}$$

Equation 130-4 indicates that the amplifier behaves as a constant voltage source whose open circuit output is $-\mu e_c$ and whose internal resistance is r_p (Fig. 130-2).

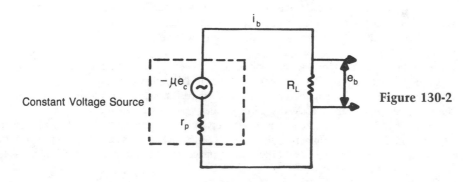

Constant Voltage Source

Figure 130-2

The output signal is

$$e_b = i_b R_L = -\frac{\mu R_L e_c}{r_p + R_L} \text{ volts} \tag{130-5}$$

Voltage gain,

$$G_v = \frac{\text{output signal}}{\text{input signal}} = \frac{e_b}{e_c} = \frac{\mu R_L}{r_p + R_L} \tag{130-6}$$

The negative sign in the expression G_v is dropped because μ itself is actually a negative number. Normally the value of R_L is about five times greater than the value of r_p so that the voltage gain approaches the level of μ.

It is common practice to connect the output signal to the next stage for further amplification. This is achieved by a coupling capacitor, C_c, which blocks the dc level of the plate voltage from being applied to the grid of the following stage. If this capacitor shorts, the tube V2 is saturated and the output signal of the next stage is reduced to a very low level.

The use of the plate load resistor and the coupling capacitor is referred to as RC coupling. However, the following grid resistor, R_g, is effectively in parallel with the plate load, R_L; this reduces the effective load so that the voltage gain is lowered to a value given by

Voltage gain,

$$G_v = \frac{\mu \times R_L \| R_g}{r_p + R_L \| R_g} \tag{130-7}$$

Example 130-1

A triode amplifier has an ac plate resistance of 5 kΩ, an amplification factor of 25 and a plate load resistor of 33 kΩ. What is the voltage gain of the amplifier? If the amplifier is now connected to the next stage whose grid resistor has a value of 100 kΩ, what is the new value of the voltage gain?

Solution

Voltage gain of the amplifier,

$$G_v = \frac{\mu R_L}{r_p + R_L} \tag{130-6}$$

$$= \frac{25 \times 33 \text{ k}\Omega}{5 \text{ k}\Omega + 33 \text{ k}\Omega}$$

$$= \frac{25 \times 33}{38}$$

$$= 21.7$$

When the amplifier is connected to the next stage,

$$\text{Effective load} = 33 \text{ k}\Omega \| 100 \text{ k}\Omega$$

$$= \frac{33 \times 100}{100 + 33}$$

$$= \frac{33 \times 100}{133}$$

$$= 24.8 \text{ k}\Omega.$$

New voltage gain,

$$G_v = \frac{25 \times 24.8 \text{ k}\Omega}{5 \text{ k}\Omega + 24.8 \text{ k}\Omega}$$

$$= \frac{25 \times 24.8}{29.8}$$

$$= 20.8$$

As an example (Fig. 130-1B), let us assume that $E_{cc} = -5$ V, $I_b = 8$ mA, $R_L = 10$ kΩ and $E_{bb} = +250$ V. Notice that capital letters represent the steady quiescent values while lowercase letters are used for the signal values. The dc voltage drop across the load is 8 mA $\times$ 10 kΩ = 80 V and therefore $E_b = +250 - 80 = +170$ V (with respect to the grounded cathode).

When a sine-wave voltage, e_{in}, of 1.5 V peak is applied to the control grid, the triode's assumed parameters are such that the fluctuation in the plate current waveform has a peak value of 3 mA. The plate current therefore varies between $8 + 3 = 11$ mA and $8 - 3 = 5$ mA; these extremes correspond to plate voltages of $+250$ V $- (11$ mA $\times 10$ k$\Omega) = +140$ V and $+250$ V $- (5$ mA $\times 10$ k$\Omega) = +200$ V. The plate voltage waveform therefore consists of a $+(200 + 140)/2$ V $= +170$ Vdc level together with an alternating component (output signal) of $(200 - 140)/2 = 30$ V peak.

The input signal has a peak value of 1.5 V and the voltage gain, G_v, of the amplifier is therefore 30 V/1.5 V = 20. Because G_v is a voltage ratio, it has no units. The waveforms of e_c, i_b and e_b are shown in Fig. 130-1B.

Chapter 131
Dynamic characteristics

THE MUTUAL CHARACTERISTICS SO FAR CONSIDERED HAVE SHOWN HOW I_b varies with E_c provided E_b is kept constant. Similarly, in the case of the plate characteristics, there was the proviso that E_c be kept constant. These characteristics are known as the "static characteristics," and give certain information about the tube itself, making possible the choice of suitable tubes and suitable working conditions for any particular purpose.

If a tube is connected in a particular circuit, as in Fig. 131-1A with a plate load resistance, R_L, then if the potential on the grid is varied, the potential on the plate is also varied. This did not, however, occur under the conditions when the static mutual characteristics were plotted with a constant plate voltage. When the plate current changes, so does the voltage drop across R_L, and because the plate voltage is applied from a source of constant voltage E_{bb}, the plate voltage E_b will change. To get a true picture of what is happening, a set of characteristics is required giving the variations of I_b with E_c subject to the simultaneous and consequent variations of E_b, the extent of which will vary with the load resistance. Such a family of characteristics is called a set of "dynamic" mutual characteristics, and would be characteristic not of the tube itself, but of the tube when connected to a particular value of the plate load. This would appear to necessitate a set of dynamic characteristics for every value of load resistance; fortunately the dynamic characteristics corresponding to any particular value of load resistance can be deduced from the static characteristics as will be shown by an example. For this reason, only the static characteristics are found on tube data sheets.

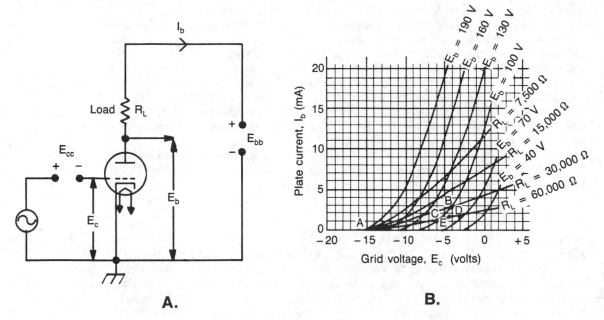

Figure 131-1

Consider the case where the available plate supply voltage, E_{bb}, 190 volts and the tube's static mutual characteristics are as illustrated in Fig. 131-1B. For simplicity, assume a purely resistive load of 30,000 Ω.

Now when $I_b = 0$, there will be no potential drop across the plate load, and $E_b = 190$ volts. From the static characteristic corresponding to 190 volts, it is seen that $I_b = 0$ corresponds to $E_c = -15$ volts. The point A is therefore on the dynamic characteristic.

When $I_b = 3$ mA, the potential drop across the resistive load will be 90 volts, so that $E_b = 100$ volts; the point B, corresponding to $E_b = 100$ volts and $I_b = 3$ mA, will therefore lie on the dynamic characteristic. In the same way by assuming other values for I_b, the dynamic characteristic corresponding to a resistive plate load of 30,000 ohms (or for any other value of resistance) may be plotted completely. A number of these dynamic characteristics are shown in Fig. 131-1B. These facts are at once apparent:

(a) The characteristics are for the most part very straight except for slight curvature near the cutoff region.

(b) The higher the load resistance, the less is the slope of the dynamic characteristic, and vice versa.

(c) The smaller the load resistance, the more nearly does the dynamic characteristic coincide with the static characteristic for $E_b = 190$ volts, and the greater is the curvature at the lower end.

After the dynamic characteristic corresponding to an available plate supply voltage of 190 V and a load resistance of 30,000 ohms has been deduced, the operating point on that characteristic (say, $E_c = -4.5$ volts, corresponding to a plate current of 2.8 mA) may be chosen. Now, suppose that an alternating voltage is applied to the grid in addition to the steady bias of -4.5 volts, and let the peak value of the signal be 1.5 volts; then the grid potential will vary between -6 volts and -3 volts, and it is now apparent that for the given load of 30,000 Ω, the variations in the plate current will all lie on the dynamic mutual characteristic corresponding to that value of load resistance. Therefore I_b will vary between 2.3 mA and 3.3 mA; that is, a total variation of 1 mA, or 0.5 mA on either side of the "no signal" or quiescent current of 2.8 mA. The change in the plate current will be proportional to the change in the grid voltage, and an undistorted signal will result provided that the dynamic characteristic is straight throughout the range of the variation of the grid voltage. For this reason the operating point is chosen in the center of the straight portion of the characteristic lying in the range of negative values of grid potential. This allows the maximum voltage signal to be applied to the grid without causing distortion; for, generally speaking, the voltage on the grid, relative to the cathode, must always be sufficiently negative to prevent the flow of grid current, and yet, on the other hand, not so negative as to cause operation over the lower curved portion of the dynamic characteristic.

With a signal of peak voltage 1.5 volts on the grid, an alternating anode current of peak value 0.5 mA flows in the load resistance of 30,000 Ω, thereby developing a peak voltage of 15 volts across the load. The voltage gain is therefore 15 V/1.5 V = 10.

THE LOAD LINE

Corresponding to the dynamic mutual characteristics, are the "load lines" of the family of the static plate characteristics. Figure 131-2 shows the static plate characteristics for the same triode tube together with the load lines for the various values of the plate current and the plate voltage; this implies a minimum level of distortion. In choosing a load, a value must be selected such that the corresponding load line makes equal intercepts on the plate characteristics. The selection of a load line making equal intercepts is equivalent to choosing a dynamic mutual characteristic with a straight portion.

From Fig. 131-1A, it is clear that $E_b = E_{bb} - I_b R_L$. If R_L and E_b are constants, the two variables I_b and E_b may be plotted in the form of a graph. This is a straight line which will clearly pass through the points given by (1) $E_b = E_{bb}$, $I_b = 0$ and (2) $E_b = 0$ when $I_b = E_{bb}/R_L$. This line is called the "load line" for the particular load considered and the E_{bb} supply available. Consider in particular the load line for $R_L = 30,000$ Ω and $E_b = 190$ volts. Assume that $E_c = -4.5$ volts. The -4.5 volts plate characteristic E_c meets the 30,000 Ω load line at the point F, corresponding

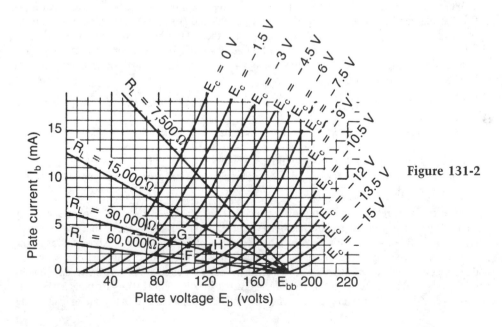

Figure 131-2

428

to $E_b = 106$ volts, $I_b = 2.8$ mA. This is taken as the operating point for $E_c = -4.5$ volts; then the corresponding plate current is 2.8 mA. Now let a signal of peak value 1.5 volts be applied to the grid so that E_c will vary between -3 volts and -6 volts (points G and H). From the points of intersection of the corresponding characteristics with the 30,000 Ω load line, it can be seen that I_b varies between 3.3 mA and 2.3 mA and E_b between 91 and 121 volts. Consequently for equal swings of grid voltage about the standing bias, there are approximately equal swings in the values of the plate voltage which has an amplification of 10. This is known as the *voltage gain, G_v*, whose value depends on the value of the load resistance.

Assume that the load is now changed to 60,000 Ω. Choosing again $E_c = -4.5$ volts, it is seen that a 1.5-volt peak signal will cause an alternating plate current of peak value 0.275 mA (points C and D, Fig. 131-1B) about the steady value of 1.60 mA (point E). This gives a peak alternating voltage of 16.5 volts across the 60,000 Ω load so that the voltage gain increases to 11.

The higher the load resistance, the higher is the voltage gain; but the voltage gain is always less than the amplification factor (μ) of the tube as derived from its static characteristics; in our example the amplification factor of the tube is 11.5.

Just as the static transfer and plate characteristics are exactly equivalent as far as imparting information about the tube itself, the dynamic mutual characteristic and the load line are equivalent ways of expressing the behavior of the tube with a given resistive plate load. It is, however, somewhat easier to detect inequality of the intercepts on the load line than to detect a slight curvature of the dynamic characteristic; therefore the load line method of choosing operating conditions is the one usually employed.

Example 131-1

Referring to Fig. 131-1B, $E_{bb} = 190$ V and $R_L = 15$ kΩ. If $E_{cc} = -6$ V, what is the dc level of the plate current? A signal whose rms value is 1.414 V is applied to the control grid. What are the values of the amplifier's voltage gain and the amount of dc power drawn from the E_{bb} supply?

Solution

From the dynamic mutual characteristic corresponding to $R_L = 15$ kΩ, $I_b = 4$ mA if $E_{cc} = -6$ V. A sine-wave signal whose rms value is 1.414 V, has a peak value of 1.414 V $\times$ 1.414 = 2 V. The grid voltage, E_c, therefore swings between $-6 + (-2) = -8$ V and $-6 - (-2) = -4$ V.

Plate current corresponding to $E_c = -8$ V is 2.8 mA. Plate current corresponding to $E_c = -4$ V is 5.2 mA. Peak value of the plate current fluctuation is 4 mA $-$ 2.8 mA = 1.2 mA (or 5.2 mA $-$ 4 mA = 1.2 mA). Peak value of the output signal = 1.2 mA $\times$ 15 kΩ = 18 V. Amplifier's voltage gain = 18 V/2 V = 9. Dc power drawn from the E_{bb} supply = 4 mA $\times$ 190 V = 760 mW.

Example 131-2

Referring to Fig. 131-2, $E_{bb} = 190$ V and $R_L = 15$ kΩ. If $E_c = -6$ V, what is the dc level of the plate current? A sine-wave signal whose rms value is 1.414 V is applied to the control grid. What is the (rms) value of the output signal voltage? Calculate the amplifier's voltage gain.

Solution

From Fig. 131-2, the dc level of the plate current is 6 mA, and the corresponding plate voltage is 134 V. The peak value of the signal input is 1.414 V $\times$ 1.414 = 2 V. The grid voltage, E_c, therefore swings between $-6 + (-2) = -8$ V and $-6 - (-2) = -4$ V. Plate voltage corresponding to $E_c = -8$ V is 152 V. Plate voltage corresponding to $E_c = -4$ V is 116 V. Peak value of the plate voltage fluctuation is 152 V $-$ 134 V = 18 V (or 134 V $-$ 116 V = 18 V) and therefore the rms value of the output signal is 18 V × 0.707 = 12.73 V. Amplifier's voltage gain = 18 V/ 2 V = 9.

These results are the same as those obtained in Example 131-1.

Chapter 132
Types of bias

THE BIAS IS A DC VOLTAGE THAT IS APPLIED IN SERIES WITH THE SIGNAL between the control grid and the cathode. The polarity of this voltage is such as to make the grid negative with respect to the cathode and thereby establish the operating point on the dynamic transfer characteristic (Fig. 132-1A). The position of this point determines the amplifier's class of operation. The various classes of operation are as follows:

1. **Class A.** This requires that the bias point is approximately half-way between $E_c = 0$ V and the cutoff point ($E_c = -E_{bb}/\mu$ V). For this reason class-A operation is sometimes referred to as midpoint bias, which requires a value of approximately $-E_{bb}/(2\mu)$ volts. This type of operation ensures minimum distortion of the output signal waveform and is therefore commonly used in audio amplifiers for speech and music. However, the corresponding dc level of plate current is high so that the plate efficiency

$$\frac{\text{signal power output from the plate}}{\text{dc power input from the } E_{bb} \text{ supply}} \times 100\%$$

is low and is typically less than 25%. This is of little importance when the signal levels are small such as in the audio stages immediately following a microphone or in the rf voltage amplifier of a receiver.

2. **Class B.** The triode is biased to the projected cutoff point so that the plate current only flows during the positive half cycle in the input signal; this means that there is considerable distortion in the output signal waveform. However, the dc level of the plate current is low and therefore it is possible to achieve practical plate efficiencies of 50% to 60%. This type of operation is used in audio push-pull stages that employ two matched tubes; the circuit arrangement is such that the distortion associated with one tube is cancelled by an opposite distortion created by the other tube. Class B is also employed in rf linear stages that are capable of amplifying amplitude modulated (AM) signals. Because distortion is not a major factor, the amplitude of the signal is as large as possible and, at its positive peak, the grid is driven positive with respect to the cathode so that grid current flows.

3. **Classes AB-1 and AB-2.** The bias point lies approximately halfway between class-A and class-B operation. The distortion is therefore more severe when compared with class-A operation but it is possible to achieve plate efficiencies of the order of 35%. One of the main uses of class-AB is in audio push-pull stages so that if there were a cascaded series of such amplifiers, the early ones would be biased in class A, subsequent stages would operate in class AB, and the final stage would be a class-B amplifier. When compar-

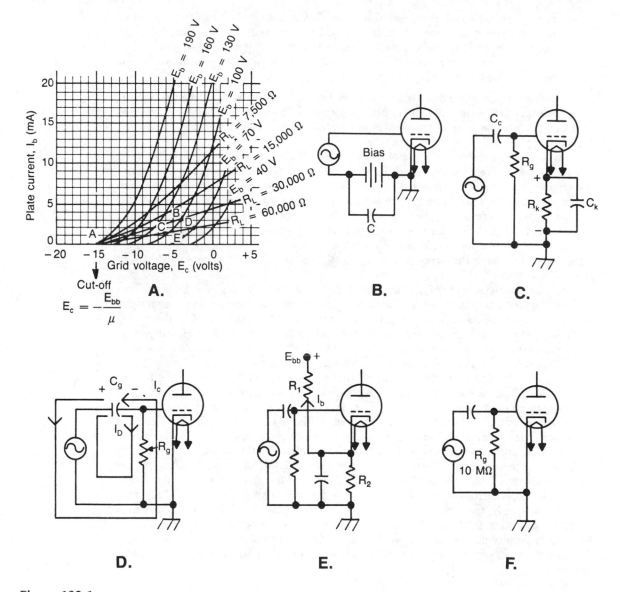

Figure 132-1

ing class AB-2 with class AB-1, class AB-2 allows a larger input signal so that at its positive peak, grid current flows. This allows a greater signal power output but at the expense of increased distortion.

4. **Class C.** The bias point is typically between two and four times the cutoff value so that the plate current only flows for about one quarter to one third of the input cycle. The distortion is very severe but the plate efficiency is as high as 85%. However, this distortion is not a problem because class C is used in rf power amplifiers that have tank circuits as their loads. The tank circuit may then be tuned to resonance at the input signal frequency and unwanted frequencies will then be eliminated. However, in frequency multi-

plier stages such as doublers and triplers, the tank circuit may be tuned to the required harmonic. The doubler stage requires a bias of five or ten times the cutoff value, while the bias for a tripler stage is ten to twenty times the cutoff value; these frequency multipliers therefore have a lower plate efficiency (about 50%) than that of the standard class-C power amplifier (up to 85%).

In previous chapters the bias has been provided by a "C" battery; such a solution is cumbersome and expensive because the battery must periodically be replaced. The battery is bypassed by a capacitor so that none of the signal is lost across the battery's internal resistance (Fig. 132-1B).

An alternative solution is the use of cathode bias, which consists of the resistor, R_k (Fig. 132-1C). The dc current passes through R_k and develops the bias voltage between the cathode and ground. The direction of the electron flow is from ground to the cathode which therefore carries a positive potential. Because there is no dc voltage drop across the grid resistor, R_g, the grid is negative with respect to the cathode by the amount of the cathode bias. For example, if the amount of cathode bias required is -5 V (grid relative to cathode) and the corresponding plate current is 2.5 mA, the required value of R_k is 5 V/2.5 mA = 2 kΩ. Because the amount of the cathode bias depends on the dc plate current level, this type of bias is only suitable for class-A or class-AB operation.

Because the signal component of the plate current also passes through R_k, there will be a signal voltage across the bias resistor. This signal voltage is 180° out of phase with the input signal and therefore represents negative or degenerative feedback. The result is to reduce the amplifier's voltage gain. To prevent this degeneration we can include a bypass or decoupling capacitor, C_k, which offers a low reactance (compared with the value of R_k) to the frequencies contained in the input signal. Taking music as an example, the frequency range will extend from below 100 Hz to several kHz. A common criterion is to arrange that the reactance of C_k does not exceed 1/10th of the value of R_k at the *lowest* frequency to be bypassed. For audio amplifiers the values of C_k are normally several microfarads while for receiver rf amplifiers, C_k is 0.01 μF or less.

Grid-leak bias (Fig. 132-1D) is commonly used in class-C rf power amplifier stages and in oscillators. The amount of the bias depends on the amplitude of the input signal that causes grid current to flow at the peak of its positive half cycle. This grid current then charges C_g towards the peak value of the input signal. When the grid current ceases, C_g discharges slowly through R_g but their time constant is high compared with the period of the input signal. As a result of this action, there is developed across R_g a dc bias level that is determined by the magnitude of the input signal; consequently, this type of bias is sometimes referred to as "signal" bias. If the drive from the preceeding stage fails, no signal bias will be generated, so that some rf amplifiers include a fixed safety battery bias or cathode bias in order to limit the plate current to a "safe" value. Provided the input signal is sufficiently large, this form of bias can be used to provide class-C operation.

Bleeder bias (Fig. 132-1E) uses a voltage divider ($R1$, $R2$) which is connected between E_{bb} and ground. Ignoring the flow of the cathode current through $R2$, the positive potential on the cathode due to the bleeder current, I_b, is $+E_{bb} \times R_2/(R_1 + R_2)$ volts. Theoretically, this form of bias could provide any class of operation but in practice the amount of bias is limited by the maximum dc voltage that can be safely applied between the cathode and the heater.

Contact bias is possible with certain amplifier tubes. The control grid requires a narrow pitch for its spiral of wire and is also wound very close to the cathode. As a result there is an accumulation of the space charge around the control grid which carries a bias of approximately -0.5 V to -1.0 V. To maintain this bias (Fig. 132-1F), most of the space charge must not be allowed to leak away to ground through the grid resistor, R_g. This is achieved by increasing the value of R_g from a typical value of 470 kΩ to about 10 MΩ. Clearly, contact bias is only suitable for small signals and is rarely used nowadays.

MATHEMATICAL DERIVATIONS
Cathode bias
Value of cathode bias resistor,

$$R_k = \frac{E_c}{I_k} \text{ ohms} \qquad (132\text{-}1)$$

where: E_c = the value of the required bias (V)

I_k = dc level of the cathode current (A)

Reactance of the cathode bypass capacitor, C_k, is one-tenth the value of the cathode bias resistor. Therefore

$$\frac{R_k}{10} = \frac{1}{2\pi f C_k}$$

This yields

$$C_k = \frac{10}{2\pi f R_k} \text{ F} = \frac{10^7}{2\pi f R_k} \,\mu\text{F} \qquad (132\text{-}2)$$

where: f = lowest frequency to be bypassed.

Grid-leak (signal) bias
Value of the grid leak or signal bias,

$$E_c = -I_{c1} \times R_g \text{ volts} \qquad (132\text{-}3)$$

where: R_g = total of all the resistances associated with the grid current (Ω)

I_{c1} = dc level of the control grid current (A).

Bleeder bias
Value of the bleeder bias,

$$E_c = E_{bb} \times \frac{R_2}{R_1 + R_2} \text{ volts} \qquad (132\text{-}4)$$

This formula neglects the flow of the cathode current through the resistor, $R1$.

Example 132-1
A triode is used in an audio amplifier stage that covers the frequency range of 50 Hz to 15 kHz. The dc level of the plate current is 7 mA and the cathode bias resistor has a value of 1.2 kΩ. What is the amount of the cathode bias and what is the required value for the cathode bypass capacitor?

Solution
Amount of the cathode bias,

$$E_c = I_b \times R_k \qquad (132\text{-}1)$$
$$= 7 \text{ mA} \times 1.2 \text{ k}\Omega$$
$$= 8.4 \text{ V.}$$

433

Value of the cathode bypass capacitor,

$$C_k = \frac{10^7}{2\pi f R_k} \ \mu F \qquad\qquad (132\text{-}2)$$

$$= \frac{10^7}{2 \times \pi \times 50 \times 1.2 \times 10^3}\mu F$$

$$= 21 \ \mu F.$$

A 20 μF, 10 WVdc capacitor would be adequate.

Example 132-3

An rf amplifier employs grid-leak bias with a grid resistor of 47 kΩ. If the level of the control grid current is 1.2 mA, calculate the value of the bias. What is the approximate rms value of the input sine-wave signal?

Solution

Value of the bias,

$$E_c = -I_{c1} \times R_g \qquad\qquad (132\text{-}2)$$

$$= -1.2 \ \text{mA} \times 47 \ \text{k}\Omega$$

$$= -56.4 \ \text{V.}$$

The level of the bias is approximately equal to the peak value of the signal. Therefore the rms value of the input signal,

$$e_{\text{rms}} \approx 0.707 \times 56.4 \ \text{V}$$

$$= 40 \ \text{V.}$$

Chapter 133
The development of the amplifier tube

WITHIN THE TRIODE ARE THE CATHODE, THE CONTROL GRID, AND THE PLATE. These electrodes represent conducting surfaces that are separated by the vacuum dielectric. Consequently, there are three interelectrode capacitances whose values are normally of the order of a few picofarads. Referring to Fig. 133-1A, these capacitances are:

1. The grid-cathode capacitance, C_{gk}, which is effectively in parallel with the input signal circuit and is called the tube's input capacitance.
2. The plate-cathode capacitance, C_{pk}, which is effectively in parallel with R_L as far as the signal is concerned and is called the tube's output capacitance.
3. The plate-grid capacitance, C_{pg}, which allows the output signal to drive a current *back* into the input circuit.

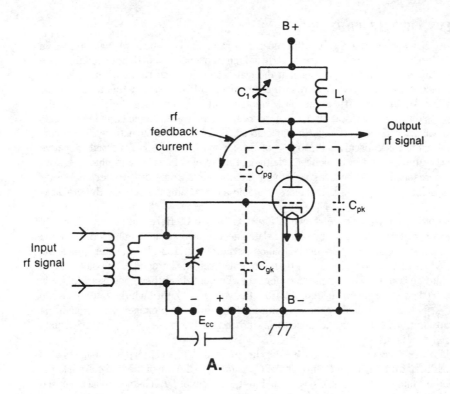

A.

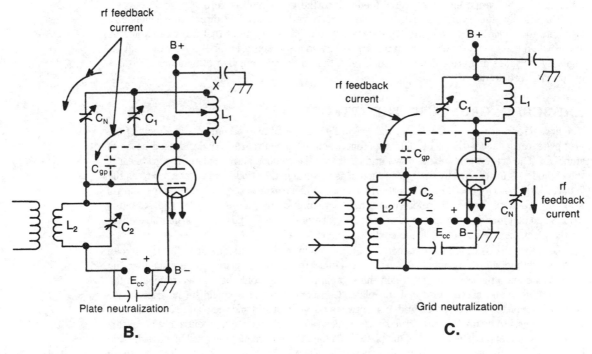

Plate neutralization

B.

Grid neutralization

C.

Figure 133-1

INSTABILITY OF THE TRIODE

Of these three capacitances the C_{pg} has the most important effect, which we will discuss in greater detail. Clearly the C_{pg} provides, particularly at high frequencies, a path between the plate and the grid circuits so that the output signal at the plate can drive a current back into the grid circuit and develop a *feedback* voltage across the impedance of this circuit; this is commonly referred to as the *Miller effect*. If this feedback voltage is of sufficient magnitude and has the correct phase (positive feedback), the circuit may cease to function as an amplifier and become an oscillator. This will occur if the plate load (the tank circuit $L1 \, C1$) behaves inductively. However, if the plate tank circuit is capacitive, the phase of the feedback voltage is reversed (negative feedback) and the amplifier's gain is reduced. Neither of these effects will occur when the plate tank circuit is at resonance and behaves resistively but this circuit may drift to become either inductive or capacitive. The triode is therefore unstable as an rf amplifier unless the circuit is neutralized.

Typical neutralization circuits are shown in Fig. 133-1B and C. Here the purpose is not to eliminate the feedback through the C_{gp} but to cancel its effect by an opposite feedback through the neutralizing capacitor, C_N. In the *plate* neutralization circuit of Fig. 133-1B, a center tap on the *plate* tank coil, $L1$, is connected to $B+$ and is therefore effectively at rf ground. The points, X and Y, are at opposite ends of the coil $L1$ and therefore the rf potentials at these points are equal in magnitude but 180° out of phase. The feedback from the voltage at Y through the interelectrode capacitance, C_{gp}, is then cancelled by the opposite feedback from the voltage at X through the neutralizing capacitor, C_N. During the neutralizing procedure the value of C_N is varied until complete cancellation is achieved.

In the grid neutralizing circuit of Fig. 133-1C, the *grid* coil, $L2$, is center-tapped and is at rf ground. The feedback from the point P through the C_{gp} is taken to the top of the grid coil while the feedback from the same point through C_N is applied to the bottom of $L2$. The neutralizing capacitor may then be varied until the overall feedback to the grid circuit is zero.

Historically, the evolution of the various screen grid tubes are a direct consequence of attempts to reduce the plate-control grid capacitance to such an extent that the tube would be stable when used in an rf amplifier circuit. However, some screen-grid tubes have other advantages over triodes and are commonly employed where a voltage amplifier at audio or radio frequencies is required.

THE SCREEN GRID TUBE: THE TETRODE

The first screen-grid tube, or tetrode, was originally introduced to overcome the ill-effects of the grid-plate capacitance; these become apparent when an unneutralized triode is used as an rf amplifier. The screen-grid tube has two grids between the cathode and the plate; the grid nearer the cathode performs exactly the same function as the grid in the triode and is referred to as the "control grid" ($G1$) while the additional grid acts as an electrostatic screen between the control grid and the plate, and is therefore called the "screen-grid" or "screen" ($G2$). The screen is maintained at a high positive potential approaching that of the plate, and has a considerable effect on the electron stream between the control grid and the plate.

Consider the electric field between the electrodes in terms of "its flux lines." If the screen were a solid metal plate maintained at a potential equal to that of the plate, the flux lines leaving the cathode and grid would terminate on the screen, and there would be no electric field in the space between the screen grid and the plate. Consequently, there would be no capacitance between the plate and the screen grid, nor between the control grid and the plate. Now, consider the screen grid to be made in the form of a close mesh and maintained at a potential not necessarily equal to, but approaching, that of the plate. This time the screening effect will be considerable but not perfect, though with a fine mesh structure it will be practically so. The result is that there

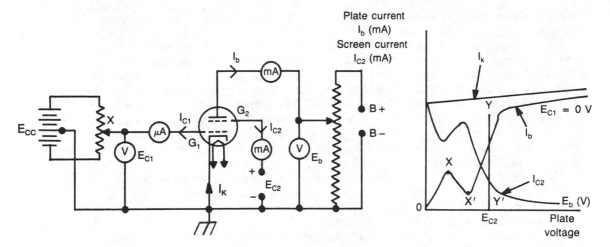

Figure 133-2

will be capacitance between the pairs of electrodes, control grid and screen grid, screen grid and plate, plate and control grid though the grid-plate capacitance will be very much reduced from that in the triode. In commercial types of screen grid tubes the residual control grid-plate capacitance varies from 0.001 pF to 0.02 pF, as compared with 2 pF to 8 pF for a triode.

Figure 133-2 shows a typical plate characteristic for a tetrode, drawn under conditions of constant control grid voltage (E_{c1}) and constant screen voltage (E_{c2}). When the plate potential is zero, all the emitted electrons are attracted to the screen, giving a fairly high screen current (I_{c2}); the plate current (I_b) will be zero. If, now, the plate potential is increased, some of the electrons passing through the mesh of the screen are carried on by their momentum and come under the influence of the plate, to which they are attracted, so that the plate current will increase with a greater plate potential. Due to the shielding effect of the screen, however, the potential of the plate will have very little effect on the electric field in the vicinity of the cathode, and an increase in plate potential will not appreciably increase the total space or cathode current. Any increase in plate current will therefore be at the expense of the screen current.

As the plate potential increases, so also will the velocity of the electrons on their arrival at the plate. One effect of bombarding the plate with fast moving electrons is that other electrons may be ejected by the force of impact. The quantity of these ejected electrons (or *secondary electrons,* as they are usually called) will vary with the material of the plate and the velocity of the electrons reaching the plate from the cathode *(primary electrons).* In certain circumstances as many as ten secondary electrons may be liberated by one fast-moving primary electron. This phenomenon of "secondary emission" also occurred in the diode and the triode, but in those cases the secondary electrons were attracted back into the plate surface and had no effect on the tube. With the tetrode, however, the velocity of the primary electrons is sufficiently high to cause secondary emission while the plate is at a lower potential than the screen grid. It follows that there is a tendency for the screen to collect these secondary electrons emitted from the plate. The result is an increase in the screen current at the expense of the plate current. A further increase in the plate potential will increase the velocity of the primary electrons and therefore increase the emission of the secondary electrons. If the screen is still at a higher potential than the plate, it will collect practically all of these slow-moving secondary electrons, with the result that the plate current will actually decrease with a greater plate potential.

This state of affairs is represented by the portion of the plate characteristic *XX'* (Fig. 133-2). Under the operating conditions that control this portion of the characteristic, the tube behaves as a negative resistance device, because a decrease in the plate voltage causes an increase in the plate current.

If the plate potential is still further increased, the majority of the secondary electrons will no longer be attracted to the screen, but more and more will be drawn back into the plate and the plate current will once more increase with a greater plate potential (at the expense of a decreasing screen current). The portion of the tetrode characteristic that is useful for most purposes is that portion well to the right of the vertical line YY' in Fig. 133-2. In this region the curve becomes practically straight, and the plate current is nearly independent of the plate voltage indicating a very high value for the ac plate resistance, r_p. The effect of the grid, however, is practically the same as if the screen and the plate together form the collecting electrode—that is to say, the transconductance is of the same order as for a triode.

The required value of the screen voltage is commonly obtained by connecting the screen to $B+$ by means of a dropping resistor of suitable value or a potentiometer, as shown in Fig. 133-3A and B; the value of R_{sg} may be calculated from the fact that the screen current is normally about one fifth to one tenth of the value of the plate current.

With an alternating signal applied between the cathode and the grid, there will be fluctuations in the screen current, just as there are fluctuations in the plate current. The effect of the fluctuating screen voltage is overcome by connecting the screen grid to the cathode through a capacitor, C_{sg}. This capacitor represents a negligible reactance at high frequencies, and the screen grid and the cathode will be virtually at the same alternating potential. There will be no coupling between the plate and the control grid circuits, apart from the very small residual grid-plate capacitance.

Due to the restriction on the working part of the characteristic imposed by secondary emission, the screen grid tetrode is of little or no use as a power amplifier, and its use as a voltage amplifier is limited, because it can handle only a very small input signal. These tubes are virtually obsolete.

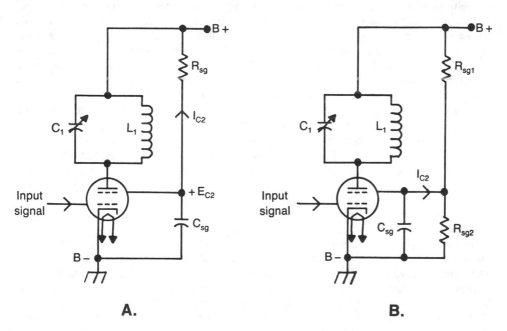

Figure 133-3

438

THE PENTODE TUBE

One method of reducing secondary emission effect is the introduction of an additional electrode, in the form of a third grid, which is positioned between the screen and the plate. This third grid ($G3$) is called the *suppressor,* and the resulting five-electrode tube is referred to as a pentode. The suppressor is given a negative potential relative to the plate and the screen grid, and this prevents the low-velocity secondary electrons from reaching the screen. At the same time the suppressor is usually built of open-mesh wire so that it does not interfere appreciably with the passage of the high-velocity primary electrons towards the plate. The suppressor grid is usually connected to the cathode, but, because other connections may be needed, the lead to the suppressor grid is usually brought out of the tube to a separate pin and the connection made externally. In certain cases where a pentode is suitable only as a power amplifier the connection is made internally.

Figure 133-4A illustrates the pentode's plate characteristics, which do not contain the negative resistance section associated with the tetrode. The transfer characteristics are displayed in Fig. 133-4B and the approximate values of the tube's parameters are $r_p = 1.5$ MΩ, $g_m = 2400$ μS and $\mu = 3600$. With such high values of r_p and μ, it is preferable to regard the tube in the equivalent amplifier circuit as a constant current generator (Fig. 133-4C).

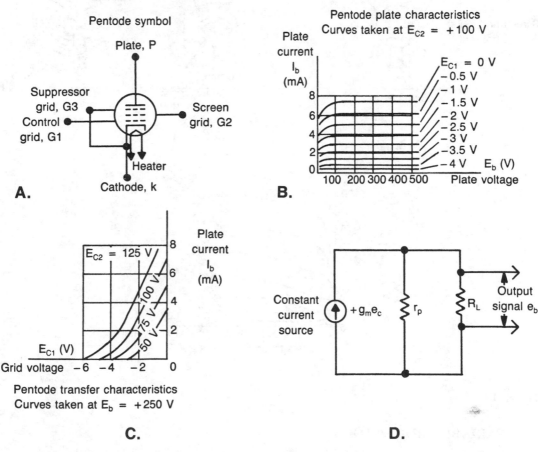

Figure 133-4

439

THE BEAM POWER TUBE

A second method of reducing the effects of secondary emission, is to include additional electrodes between the screen grid and the plate. These electrodes are connected to the cathode inside the tube and will repel the electron stream. The electrodes are arranged so that they concentrate the electron stream into a comparatively narrow beam and for this reason are usually referred to as "beam-forming plates" (Fig. 133-5A). The concentration of the electrons into this beam, combined with a large distance between the screen grid and the plate, gives an intensified space-charge effect in the screen grid-plate space that will repel the secondary electrons back into the plate's surface. The screen current is made small by having an open-meshed screen, together with optical alignment of the control grid and screen grid. Such a tube is referred to as a "beam power tube" and its plate characteristics are shown in Fig. 133-5B.

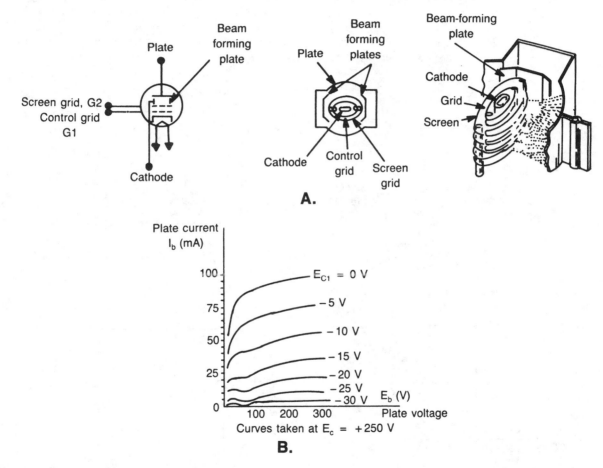

Figure 133-5

THE VARIABLE MU-PENTODE

The *variable mu-pentode* is a pentode in which the control grid is made to have an asymmetrical structure. This is normally done by making the pitch of the control grid vary along its length, the mesh of the grid being closer at the ends rather than the center. The result is that various parts of

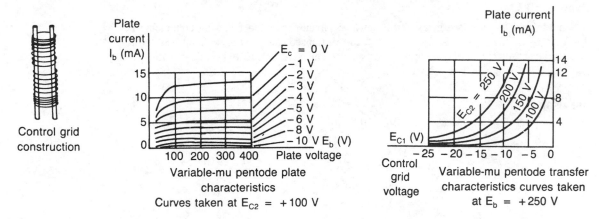

Figure 133-6

Control grid construction

Variable-mu pentode plate characteristics
Curves taken at $E_{C2} = +100$ V

Variable-mu pentode transfer characteristics curves taken at $E_b = +250$ V

the tube cut-off with different grid bias voltages, so that the overall cut-off comes gradually rather than abruptly.

Figure 133-6 shows a set of transfer characteristics for a typical variable mu-pentode, plotted for a constant plate voltage, but a variable screen voltage. With curved characteristics the value of the g_m will depend on the chosen bias point. Because the voltage gain, $G_v = g_m R_L$, this will enable the voltage gain of an amplifier to be varied over a wide range by changing the bias voltage on the control grid.

MATHEMATICAL DERIVATIONS

Screen grid voltage,

$$E_{c2} = E_{bb} - I_{c2} \times R_{SG} \text{ volts} \tag{133-1}$$

Cathode current,

$$I_K = I_b + I_{c2}(+ I_{c1}) \text{ milliamperes} \tag{133-2}$$

The control grid current, I_{c1}, would be present in a power amplifier but not in a voltage amplifier.

In Fig. 133-4D

Output signal,

$$e_b = g_m e_c \times (r_p \| R_L) \tag{133-3}$$

$$= \frac{g_m e_c}{\dfrac{1}{r_p} + \dfrac{1}{R_L}}$$

$$\approx g_m R_L e_c \text{ volts if } r_p \gg R_L$$

Voltage gain of the pentode amplifier is

$$G_v = \frac{e_b}{e_c} = g_m R_L \tag{133-4}$$

Example 133-1

A pentode tube is operated in a voltage amplifier circuit under the following conditions: $I_b = 3$ mA, $R_L = 33$ kΩ, $R_{SG} = 120$ kΩ, $E_{bb} = +250$ V, transconductance $g_m = 2700$ μS. If the dc level of the screen current is one-fifth of the plate current value, calculate the values of the screen potential and the amplifier's voltage gain.

Solution

Dc level of screen current, $I_{c2} = 3$ mA/5 $= 0.6$ mA.

Screen grid voltage,

$$E_{c2} - E_{bb} - I_{c2} \times R_{SG} \tag{133-1}$$
$$= 250 \text{ V} - 0.6 \text{ mA} \times 120 \text{ kΩ}$$
$$= +178 \text{ V}.$$

Voltage gain,

$$G_v = g_m R_L \tag{133-4}$$
$$= 2700 \times 10^{-6} \times 33 \times 10^3$$
$$= 89.$$

Chapter 134
Negative feedback

THE INTRODUCTION OF NEGATIVE FEEDBACK IN AN AUDIO AMPLIFIER REQUIRES that a fraction, β of the output signal is fed back in opposition to the input signal. In contrast with the disadvantage of reducing the amplifier gain, negative or degenerative feedback provides the following features:

1. Stabilization of the amplifier voltage gain against changes in the parameters of the active device such as a transistor or tube.
2. Reduction in the amplitude distortion caused by nonlinearity in the characteristics of the active device.
3. Reduction in frequency and phase distortion produced by the junction and stray capacitances.
4. Reduction in noise.
5. Changes in the amplifier's input and output impedances.

Note that the second, third, and fourth advantages refer to the distortion and the noise created within the amplifier itself. Negative feedback has no effect on the noise and the distortion that are fed in from the previous stage.

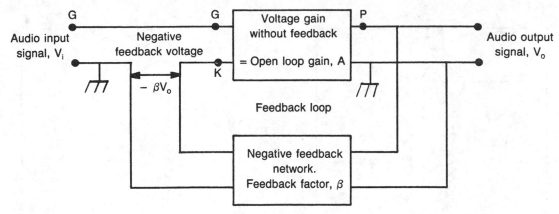

Figure 134-1

MATHEMATICAL DERIVATIONS

Voltage gain with negative feedback

In the block diagram of Fig. 134-1, the signal applied between the grid and the cathode is the sum of the input audio signal, V_i, between the control grid and ground, and the negative feedback voltage, $-\beta V_o$. Therefore,

$$\text{Signal between the grid and the cathode} = V_i - \beta V_o \text{ volts} \qquad (134\text{-}1)$$

where: V_o = output audio signal.

Output audio signal,

$$V_o = A \times (\text{signal between } G \text{ and } K) \qquad (134\text{-}2)$$
$$= A(V_i - \beta V_o) \text{ volts}$$

where: A = open loop gain.

Amplifier gain with negative feedback,

$$A' = \frac{V_o}{V_i} = \frac{A}{1 + A\beta} \qquad (134\text{-}3)$$

If $A\beta$ is appreciably greater than 1,

$$A' \rightarrow \frac{A}{A\beta} = \frac{1}{\beta} \qquad (134\text{-}4)$$

which is independent of A; consequently the value of A' will be little affected by any changes in the parameters of the active device.

Because

$$A' = \frac{A}{1 + A\beta'} \qquad (134\text{-}5)$$
$$A = \frac{A'}{1 - A'\beta}$$

443

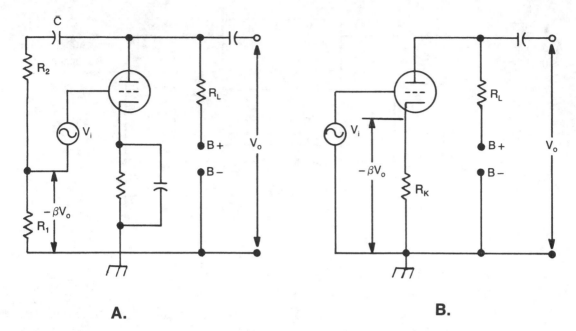

A.   **B.**

Figure 134-2

and

$$\beta = \frac{1}{A'} - \frac{1}{A} = \frac{A - A'}{AA'} \qquad (134\text{-}6)$$

The feedback factor, β, may either be expressed as a fraction or as a percentage.

The two basic negative feedback circuits are shown in Fig. 134-2A and B. Figure 134-2A represents voltage negative feedback in which the audio signal voltage, V_o, is divided between $R1$ and $R2$ (C is only a dc blocking capacitor). The voltage $-\beta V_o$, which is developed across $R1$, is applied as degenerative feedback to the input signal, V_i.

The feedback factor is given by
Feedback factor,

$$\beta = \frac{R_1}{R_1 + R_2} \qquad (134\text{-}7)$$

The voltage gain of the amplifier with feedback is:

$$A' = \frac{A}{1 + A\beta} \qquad (134\text{-}8)$$

Open loop gain,

$$A = \frac{\mu R_L}{r_p + R_L}$$

This expression for the voltage gain without feedback, A, applies when the value of $R1 + R2$ is sufficiently large so that the equivalent value of the plate load resistance is not appreciably affected by the shunting action of the two resistors.

444

Current feedback, as shown in Fig. 134-2B, uses the signal component of the plate current to develop the degenerative feedback voltage across the cathode resistor, P_K; β is then equal to R_K/R_L and the voltage gain of the amplifier with feedback,

$$A' = \frac{A}{1 + A\beta}$$

where

$$A = \frac{\mu R_L}{r_p + R_K + R_L} \tag{134-9}$$

The last two equations yield

$$A' = \frac{\mu R_L}{r_p + R_L + R_K(1 + \mu)} \tag{134-10}$$

Example 134-1

The voltage gain of an audio amplifier without negative feedback is 35. If 20% degenerative feedback is introduced into the circuit, calculate the value of the amplifier gain with feedback.

Solution

Because $A = 35$ and $\beta = 20/100 = 0.2$, the gain with feedback is:

$$A' = \frac{A}{1 + A\beta}$$

$$= \frac{35}{1 + (35 \times 0.2)}$$

$$= \frac{35}{1 + 7} = \frac{35}{8}$$

$$= 4.4, \text{ rounded off.}$$

Example 134-2

The voltage gain of an audio amplifier with degenerative feedback is 8. If the negative feedback factor is 1/10, what is the amplifier gain without feedback?

Solution

Because $A' = 8$ and $\beta = 1/10$, the gain without feedback is:

$$A = \frac{A'}{1 - A'\beta} \tag{134-5}$$

$$= \frac{8}{1 - (8 \times 0.1)} = \frac{8}{0.2}$$

$$= 40.$$

Example 134-3

The voltage gain of an audio amplifier without feedback is 25. When degenerative feedback is introduced into the circuit, the voltage gain falls to 10. What is the feedback percentage?

Solution

Because $A = 25$ and $A' = 10$, the feedback factor is:

$$\beta = \frac{A - A'}{AA'} \tag{134-6}$$

$$= \frac{25 - 10}{25 \times 10}$$

$$= \frac{15}{250}$$

$$\text{Feedback percentage} = \frac{15 \times 100}{250} - 6\%.$$

Example 134-4

In Fig. 134-3, what is the negative feedback percentage? If the triode's amplification factor is 24 and its ac plate resistance is 11 kΩ, what is the voltage gain with feedback?

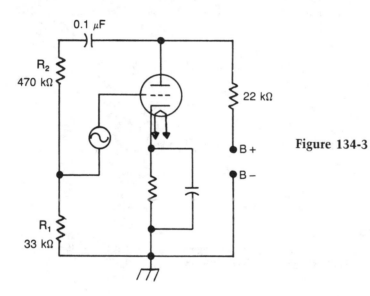

Figure 134-3

Solution

Feedback factor,

$$\beta = \frac{R_1}{R_1 + R_2} \tag{134-7}$$

$$= \frac{33 \text{ k}\Omega}{470 \text{ k}\Omega + 33 \text{ k}\Omega}$$

$$= \frac{33}{503}$$

$$\text{Feedback Percentage} = \frac{33 \times 100}{503} = 6.6\%, \text{ rounded off.}$$

Voltage gain without feedback,

$$A = \frac{\mu R_L}{r_p + R_L}$$

$$= \frac{24 \times 22}{11 + 22}$$

$$= 16$$

Voltage gain with feedback,

$$A' = \frac{16}{1 + (16 \times 0.066)}$$

$$= 7.9, \text{ rounded off.}$$

Chapter 135
The grounded grid triode circuit and the cathode follower

IN PREVIOUS CHAPTERS WE HAVE DISCUSSED THE OPERATION OF GROUNDED cathode amplifiers in which the input signal is applied between the control grid and ground and the output signal appears between the plate and ground. However, there are two other possible arrangements:

1. The grounded grid amplifier. In this circuit the input signal is applied between cathode and ground. The grid is grounded as far as the signal is concerned and the output signal appears between the plate and ground.
2. The cathode follower or grounded plate current. In this arrangement the input signal is applied between the grid and ground. The plate is grounded as far as the signal is concerned but is still connected to the $B+$ supply. The output signal appears between the cathode and ground.

THE GROUNDED GRID TRIODE CIRCUIT

This circuit (Fig. 135-1A and B) is commonly used for rf amplification in the VHF and UHF bands; one example is the first stage of a TV receiver. Its advantages are twofold:

1. It contributes less noise than an amplifier using a conventional screen grid, which suffers from partition noise due to the random effect in the division of the electron stream between the screen grid and the plate circuits.
2. The circuit does not require neutralization because the grounded grid provides an electrostatic screen between the output plate circuit and the input cathode circuit. This is particularly advantageous at high frequencies where the adjustment of the neutralizing circuit is

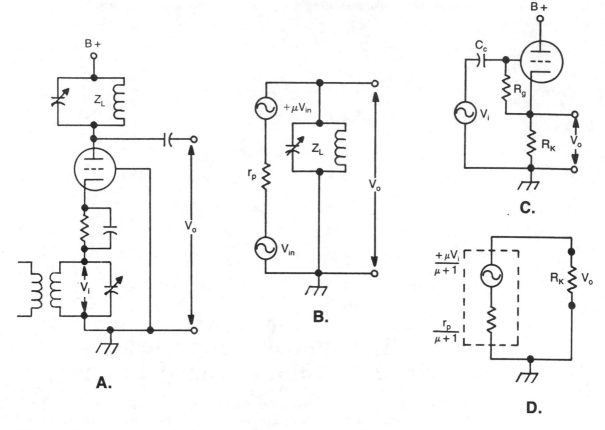

Figure 135-1

more critical. In addition the elimination of the neutralizing capacitor reduces the input and output capacitances of the stage. Unlike the conventional grounded-cathode amplifier, there is no phase change between the input signal at the cathode and the output signal on the plate. Both of these signal voltages are 180° out of phase with the signal component of the plate current.

The value of the stage's input impedance is much lower than that of the conventional grounded-cathode amplifier discussed earlier. This low input impedance is the main disadvantage of the grounded-grid circuit because it limits the amplification that can be obtained.

THE CATHODE FOLLOWER CIRCUIT

The cathode follower circuit is shown in Fig. 135-1C, and is an example of voltage negative feedback with a feedback factor, β, of unity. The features of the circuit are:

1. There is no phase change between the cathode-ground output signal and the grid-ground input signal (the cathode output "follows" the grid input).
2. The voltage gain of the circuit is less than unity. This voltage gain is given by:

Voltage gain with feedback,

$$A' = \frac{A}{A + 1}$$

where

Voltage gain without feedback,

$$A = \frac{\mu R_k}{r_p + R_k},$$

assuming no load is connected across R_k.

3. A higher input impedance and a lower output impedance as compared with a conventional grounded-cathode amplifier.

This circuit is an impedance matching device and may be used to match a high impedance source to a low impedance load. The cathode follower can therefore provide some degree of isolation between a stage with a low input impedance and a preceeding stage with a high output impedance.

MATHEMATICAL DERIVATIONS
In the equivalent circuit of Fig. 135-1B,

Plate signal current,

$$i_b = \frac{(\mu + 1)V_i}{r_p + Z_L} \text{ amperes} \tag{135-1}$$

Output signal voltage,

$$V_o = i_b Z_L = \frac{(\mu + 1)Z_L}{r_p + Z_L} \times V_i \text{ volts} \tag{135-2}$$

Voltage gain,

$$G_v = \frac{V_o}{V_i} = \frac{(\mu + 1)Z_L}{r_p + Z_L} \tag{135-3}$$

Input impedance,

$$Z_{in} = \frac{V_i}{i_b} = \frac{r_p + Z_L}{\mu + 1} \text{ ohms} \tag{135-4}$$

In the circuits of Fig. 135-1C and D,

Input impedance of the cathode follower circuit, Z_{in}, consists of the parallel combination of:

$$(A + 1)R_g, \frac{C_{gk}}{1 + A}$$

and C_{gp}, where $A = \mu R_k/(r_p + R_k)$ and C_{gk} C_{gp}, C_{pk} are the interelectrode capacitances.

Output impedance, Z_o, consists of the parallel combination of R_k, $r_p/(\mu + 1)$ and C_{pk}. These values of the input and output impedances can be compared with the equivalent expressions for the grounded cathode arrangement:

Input impedance, Z_{in}, of the grounded cathode circuit is the result of combining R_g, C_{gk} and $(A + 1)C_{pg}$ in parallel where $A = \mu R_L/(r_p + R_L)$.

Output impedance, Z_o, of the grounded cathode circuit is equal to the parallel combination of R_L, r_p and C_{pk}.

Example 135-1

The plate load of a grounded-grid triode amplifier is a tank circuit with a dynamic resistance of 20 kΩ. The tube has an amplification factor, μ, of 15 and an ac plate resistance, r_p, of 8 kΩ. Calculate the values of the stage's voltage gain and its input resistance.

Solution

Voltage gain,

$$G_v = \frac{(\mu + 1)Z_L}{r_p + Z_L} \tag{135-3}$$

$$= \frac{16 \times 20}{8 + 20}$$

$$= \frac{16 \times 20}{28}$$

$$= 11.$$

Input resistance,

$$R_{in} = \frac{r_p + Z_L}{\mu + 1} \tag{135-4}$$

$$= \frac{8 + 20}{16}$$

$$= \frac{28}{16}$$

$$= 1.8 \text{ k}\Omega.$$

Example 135-2

In Fig. 135-1C, $R_k = 10$ kΩ, $R_g = 1$ MΩ, $r_p = 8$ kΩ, and $\mu = 20$. What is the voltage gain of the cathode follower? Calculate the values of the stage's input and output resistances.

Solution

Voltage gain without feedback,

$$A = \frac{\mu R_k}{r_p + R_k}$$

$$= \frac{20 \times 10 \text{ k}\Omega}{8 \text{ k}\Omega + 10 \text{ k}\Omega}$$

$$= 11.1$$

Voltage gain with feedback,

$$A' = \frac{A}{A + 1}$$

$$= \frac{11.1}{11.1 + 1}$$

$$= \frac{11.1}{12.1}$$

$$= 0.92$$

Input resistance,

$$R_{in} = (A + 1)R_g$$

$$= (11.1 + 1) \times 1 \text{ M}\Omega$$

$$= 12.1 \text{ M}\Omega.$$

The output resistance consists of R_k(10 kΩ) in parallel with $r_p/(\mu + 1) = 8/(20 + 1)$ kΩ = 381 Ω.

Output resistance,

$$R_o = 381 \text{ }\Omega\|10000 \text{ }\Omega$$

$$= \frac{381 \times 10000}{381 + 10000}$$

$$= 367 \text{ }\Omega.$$

5
PART

Principles of radio communications

Chapter 136
Amplitude modulation—
percentage of modulation

MODULATION MEANS CONTROLLING SOME FEATURE OF THE RF CARRIER BY THE audio signal in such a way that the receiver is able to reproduce the information (speech, music, etc.). Because the carrier may be regarded as a high frequency sine wave, it only possesses three features that may be modulated—amplitude, frequency, and phase (such quantities as period and wavelength are directly related to the frequency). In amplitude modulation, the instantaneous amplitude (voltage) of the rf wave is linearly related to the instantaneous magnitude (voltage) of the af signal while the rate of amplitude variation is equal to the modulating frequency.

In Fig. 136-1A, the modulating signal is a single sine wave or test tone, which is used to amplitude modulate an rf carrier. The extent of modulation is measured by the percentage modulation as defined by: $(E_{max} - E_o) \times 100/E_o\%$. This expression is for positive peaks (upward modulation) where E_o is the amplitude of the unmodulated carrier. The percentage modulation on negative (downward modulation) peaks is $(E_o - E_{min}) \times 100/E_o\%$ (Fig. 136-1B).

For a variety of reasons the modulation envelope for a symmetrical tone contain assymmetrical distortion that is referred to as carrier shift. In negative carrier shift the troughs of the modulation envelope are greater than the peaks and this will cause a decrease (compared with the unmodulated value) in the dc plate current reading of a plate modulated class-C stage. Similarly, with positive carrier shift, the envelope's peaks exceed the troughs and the plate current will increase.

Increasing the amplitude of the test tone raises the percentage modulation, improves the signal-to-noise ratio at the receiver, and results in a higher audio power output. However, if the modulation percentage is increased to 100% and beyond (overmodulation) the modulation envelope is severely distorted and this distortion will ultimately appear in the output from the loudspeaker. Overmodulation also results in the generation of spurious sidebands that will cause

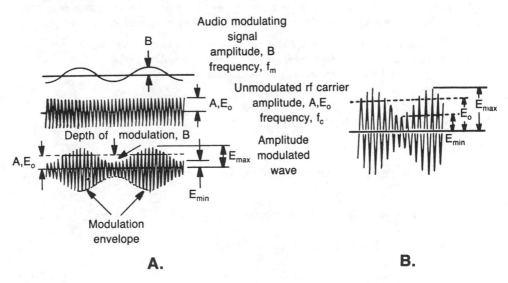

Figure 136-1

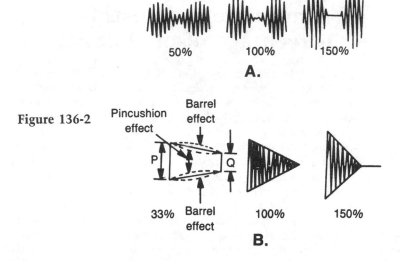

Figure 136-2

interference to adjacent channels by heterodyning with the sidebands of those channels. Figure 136-2A shows the AM waveforms for various modulation percentages.

Other than measuring the percentage modulation from the AM waveform, it is common practice to use a trapezoidal display on an oscilloscope. This is achieved by feeding the AM signal to the vertical deflecting (Y) plates and the audio signal to the horizontal deflecting plates. In addition to measuring percentage modulation, this method will clearly indicate whether there exists the required linear relationship between the instantaneous amplitude of the rf carrier and the instantaneous magnitude of the tone signal. Nonlinearity will appear either as a "barrel" or a "pincushion effect" on the trapezoidal display (Fig. 136-2B). The percentage of modulation is given by $[(P - Q)/(P + Q)] \times 100\%$ so that the 50% modulation, $P = 3Q$ (*not* 2Q). It should be emphasized that when AM is being used to transmit speech or music, the percentage of modulation varies from instant to instant. Permitted values may then be quoted for:

1. The minimum percentage modulation on average peaks of the audio signal.
2. The maximum modulation percentage on negative modulation peaks.
3. The maximum modulation on positive modulation peaks.
4. The maximum carrier shift.

MATHEMATICAL DERIVATIONS
In Fig. 136-1

$$\text{Percentage modulation on positive peaks} = \frac{E_{max} - E_o}{E_o} \times 100\% \qquad (136\text{-}1)$$

$$\text{Percentage modulation on negative peaks} = \frac{E_o - E_{min}}{E_o} \times 100\% \qquad (136\text{-}2)$$

If the modulation is symmetrical about the E_o level,

$$E_o = \frac{E_{max} + E_{min}}{2} \text{ volts} \qquad (136\text{-}3)$$

Equations 136-1, 136-3 yield

$$\text{Percentage modulation} = \frac{E_{max} - \dfrac{(E_{max} + E_{min})}{2}}{E_o} \times 100\% \qquad (136\text{-}4)$$

$$= \frac{E_{max} - E_{min}}{2E_o} \times 100\%$$

$$= \frac{E_{max} - E_{min}}{E_{max} + E_{min}} \times 100\% \qquad (136\text{-}5)$$

Degree of modulation,

$$m = \frac{\text{percentage modulation}}{100} \qquad (136\text{-}6)$$

$$= \frac{(A + B) - (A - B)}{(A + B) + (A - B)}$$

$$= \frac{B}{A}$$

Percentage of carrier shift =

$$\frac{\text{dc level (modulated)} - \text{dc level (unmodulated)}}{\text{dc level (unmodulated)}} \times 100\% \qquad (136\text{-}7)$$

Trapezoidal display:

$$\text{Percentage of modulation} = \frac{P - Q}{P + Q} \times 100\% \qquad (136\text{-}8)$$

Example 136-1

In Fig. 136-1, it is assumed that the modulation envelope is symmetrical about the E_o level. If $E_{max} = 800$ V and $E_o = 500$ V, what is the value of the percentage of modulation? Calculate the value of the instantaneous peak power if the AM signal is developed across an effective load of 100 Ω.

Solution

$$\text{Percentage modulation} = \frac{(E_{max} - E_o)}{E_o} \times 100 \qquad (136\text{-}1)$$

$$= \frac{800 - 500}{500} \times 100$$

$$= 60\%.$$

This instantaneous peak voltage, E_{max}, is 800 V and therefore the peak power is:

$$\frac{E_{max}^{\,2}}{R} = \frac{(800 \text{ V})^2}{100 \ \Omega}$$

$$= 6400 \text{ W}.$$

Note that the unmodulated carrier power is $(500 \text{ V})^2/100 \ \Omega = 2500$ W and the instantaneous minimum power is $(200 \text{ V})^2/100 \ \Omega = 400$ W.

The next chapter will show that the average power in this AM signal is 2950 W, which is different from the arithmetic mean of $(6400 + 400)/2 = 3400$ W. This difference is due to the fact that the instantaneous power curve is not symmetrical about the unmodulated carrier power level.

Example 136-2

A trapezoidal display (Fig. 136-2) is used to measure the percentage modulation of the AM signal in Example 136-1. If $P = 3$ cm, calculate the value of Q.

Solution

Because the percentage of modulation is 60%,

$$\frac{P - Q}{P + Q} \times 100 = 60 \tag{136-8}$$

Therefore:

$$\frac{3 - Q}{3 + Q} = 0.6$$

or

$$Q = \frac{1.2}{1.6} = 0.75 \text{ cm.}$$

Example 136-3

When unmodulated, the dc plate current in the final stage of an AM broadcast is 5.4 A. When modulated by a symmetrical audio tone, the plate current falls to 5.2 A. What is the percentage of the negative carrier shift?

Solution

$$\text{Percentage of negative carrier shift} = \frac{5.2 - 5.4}{5.4} \times 100\% \tag{136-7}$$

$$= -3.7\%.$$

Chapter 137
AM sidebands and bandwidth

FIGURE 137-1 ILLUSTRATES THE RESULT OF AMPLITUDE MODULATING AN RF carrier (amplitude A, frequency f_c) by an audio test tone (amplitude B, frequency f_m). As previously discussed, the degree of modulation, $m = B/A$ (Equation 136-6). In the mathematical

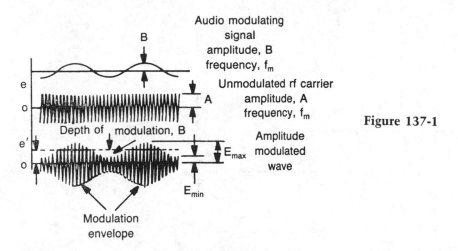

Figure 137-1

analysis shown below, it is revealed that the AM (voltage) signal is entirely rf in nature (no audio voltage is present) and contains the following three components:

1. An upper sideband component whose amplitude is $B/2 = mA/2$ and whose frequency is $f_c + f_m$.
2. A carrier component whose amplitude is A and whose frequency is f_c.
3. A lower sideband component whose amplitude is $B/2 = mA/2$ and whose frequency is $f_c - f_m$.

When these three rf components are literally added together, the resultant waveform is that of the AM signal.

As an example, let a 100 V, 1 MHz carrier be 50% amplitude modulated by a 1 kHz test tone. The act of modulation creates the sidebands that contain the signal's intelligence so that the AM signal is composed of (1) the upper sideband with an amplitude of $0.5 \times 100/2 = 25$ V at a frequency of 1 MHz + 1 kHz = 1001 kHz, (2) the carrier component with an amplitude of 100 V at a frequency of 1 MHz = 1000 kHz and (3) the lower sideband whose amplitude is 25 V at a frequency of 1 MHz − 1 kHz = 999 kHz. The instantaneous maximum voltage is 100 + 25 + 25 = 150 V while the minimum voltage is 100 − 25 − 25 = 50 V. The spectrum analyzer display of this AM signal is shown in Fig. 137-2.

The AM signal is accommodated within a certain bandwidth that is defined as the frequency difference between the upper and lower sidebands. In our example the bandwidth required is 1001 − 999 = 2 kHz, which is equal to twice the frequency of the modulating tone. When the signal is speech or music that contains many instantaneous frequencies of varying amplitudes, each audio component will produce a pair of sidebands, and the bandwidth occupied by the AM signal will be the frequency difference between the highest upper sideband and the lowest lower sideband transmitted. Consequently, the bandwidth is equal to twice the *highest* audio modulating frequency.

In addition to the waveform (Fig. 137-1) and spectrum (Fig. 137-2) representations, the AM signal may be shown in terms of phasor diagrams (Fig. 137-3).

The corresponding times in the waveform and the phasor diagrams are indicated by the points X, Y, and Z.

Each sideband component contains a certain rf power that is determined by the degree of modulation, m, and the unmodulated carrier power, P_c. The expressions for the sideband and carrier powers are shown in the mathematical derivations.

459

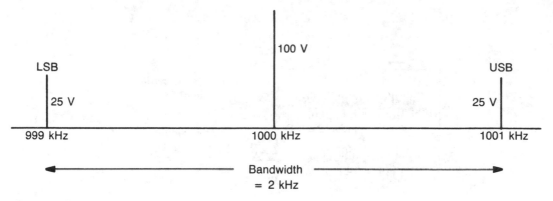

Figure 137-2

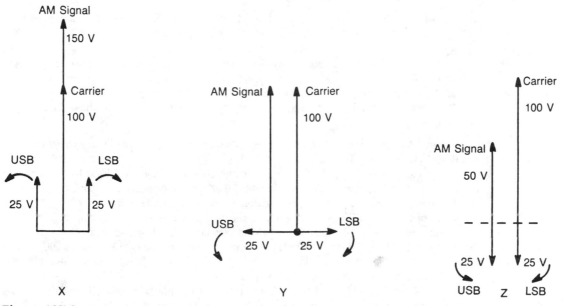

X   Y

Figure 137-3

MATHEMATICAL DERIVATIONS

The voltage equation of the unmodulated carrier wave is:

$$e = A \sin 2\pi f_c t \text{ volts} \tag{137-1}$$

The expression for the modulation envelope of the AM signal is $A + B \sin 2\pi f_m t$. Consequently, the voltage equation of the AM signal is:

$$e' = (A + B \sin 2\pi f_m t) \sin 2\pi f_c t.$$
$$= A(1 + m \sin 2\pi f_m t) \sin 2\pi f_c t.$$
$$= A \sin 2\pi f_c t + mA \sin 2\pi f_m t \sin 2\pi f_c t.$$

$$= A \sin 2\pi f_c t + \frac{mA}{2} \cos 2\pi(f_c - f_m)t - \frac{mA}{2} \cos 2\pi(f_c + f_m) \text{ volts}$$

Carrier Component   Lower Sideband   Upper Sideband

$$\text{Maximum instantaneous voltage} = A(1 + m) \text{ volts} \tag{137-2}$$

$$\text{Minimum instantaneous voltage} = A(1 - m) \text{ volts} \tag{137-3}$$

$$\text{Bandwidth} = (f_c + f_m) - (f_c - f_m) = 2 f_m \text{ hertz} \tag{137-4}$$

Let the carrier and the sidebands be associated with the same resistive load, R. Converting the peak values to their rms values,

Carrier power,

$$P_c = \left(\frac{A}{\sqrt{2}}\right)^2 \Big/ R = \frac{A^2}{2R} \text{ watts} \tag{137-5}$$

Upper sideband power,

$$P_{USB} = m^2 \frac{\left(\frac{A}{\sqrt{2}}\right)^2}{4R} = \frac{m^2 A^2}{8R} \text{ watts} \tag{137-6}$$

Lower sideband power,

$$P_{LSB} = \frac{m^2 A^2}{8R} \text{ watts} \tag{137-7}$$

Total sideband power,

$$P_{SB} = 2 \times \frac{m^2 A^2}{8R} = \frac{m^2 A^2}{4R} \tag{137-8}$$

$$= \frac{1}{2} m^2 P_c \text{ watts}$$

Peak instantaneous power,

$$P_{max} = \frac{A^2(1 + m)^2}{2R} \text{ watts} \tag{137-9}$$

Minimum instantaneous power,

$$P_{min} = \frac{A^2(1 - m)^2}{2R} \text{ watts} \tag{137-10}$$

Total rf power (carrier and sidebands),

$$P_T = P_c + \frac{1}{2} m^2 P_c \tag{137-11}$$

$$= P_c\left(1 + \frac{1}{2} m^2\right) \text{ watts}$$

In the particular case of 100% modulation, $m = 1$ so that the total sideband power $= 0.5 P_c$ and the total rf power is $1.5 P_c$. Consequently, one third of the total rf power is concentrated in the sidebands and two thirds in the carrier.

Example 137-1

A 2 kW, 1.3 MHz carrier is 80% amplitude modulated by a 2 kHz test tone. Calculate the power in each of the sidebands and their frequencies. What is the bandwidth of the AM signal?

Solution

Frequency of the upper sideband,

$$f_{USB} = 1.3 \text{ MHz} + 2 \text{ kHz}$$
$$= 1302 \text{ kHz}.$$

Frequency of the lower sideband,

$$f_{LSB} = 1.3 \text{ MHz} - 2 \text{ kHz}$$
$$= 1298 \text{ kHz}.$$

Bandwidth,

$$f_{USB} - f_{LSB} = 2 \times 2 \text{ kHz} = 4 \text{ kHz}. \tag{137-4}$$

Total sideband power,

$$P_{SB} = \frac{1}{2} \times (0.8)^2 \times 2000 \tag{137-8}$$
$$= 640 \text{ W}.$$

Power in each sideband,

$$\frac{P_{SB}}{2} = \frac{640}{2} \tag{137-6, 137-7}$$
$$= 320 \text{ W}.$$

Example 137-2

A 1 kW carrier is 60% modulated by a 1 kHz test tone. If the modulation is increased to 80%, what is the percentage increase in the total sideband power?

Solution

Total sideband power, 60% modulation,

$$P_{SB} = \frac{1}{2} \times (0.6)^2 \times 1000 \tag{137-8}$$
$$= 180 \text{ W}$$

Total sideband power, 80% modulation,

$$P_{SB} = \frac{1}{2} \times (0.8)^2 \times 1000$$
$$= 320 \text{ W}$$

Percentage increase in the total sideband power:

$$\frac{320 - 180}{180} \times 100 = \frac{1400}{18}$$
$$= 78\%, \text{ rounded off.}$$

Chapter 138
Percentage changes
in the antenna current
and the sideband power
due to amplitude modulation

THE ACT OF AMPLITUDE MODULATION CREATES THE ADDITIONAL SIDEBAND power and therefore the total rf power associated with the antenna is greater under modulated conditions. The higher the degree of modulation, the greater is the amount of the sideband power.

The antenna current is directly proportional to the square root of the total rf power and will increase with the degree of modulation. Problems involving the percentage changes in the antenna current and the sideband power may be solved by assuming convenient and compatible values for the unmodulated carrier power, the antenna current, and the effective antenna resistance. A suitable set of values is:

Unmodulated carrier power = 100 W
Unmodulated antenna current = 1 A
Effective antenna resistance = 100 Ω

As an example, if the modulation percentage produced by a single tone is 50%, the total sideband power is $1/2 \times 0.5^2 \times 100$ W = 25 W and the total rf power is 100 + 25 W = 125 W. The new antenna current is $\sqrt{125 \text{ W}/100 \text{ }\Omega} = \sqrt{1.225} = 1.1068$ A. If the percentage modulation is now increased to 100%, the new sideband power = $1/2 \times 1^2 \times 100$ W = 50 W, the new total rf power is 100 + 50 = 150 W and the new antenna current is $\sqrt{150 \text{ W}/100 \text{ }\Omega} = \sqrt{1.5} = 1.2247$ A. Consequently, when compared with the unmodulated condition, the antenna current increases by 10.68% for 50% modulation and 22.47% for 100% modulation. If the percentage modulation is increased from 50% to 100%, the sideband power increases by:

$$\frac{50 \text{ W} - 25 \text{ W}}{25 \text{ W}} \times 100\% = 100\%$$

and the percentage increase in the antenna current is:

$$\frac{1.2247 - 1.1068}{1.1068} = 10.62\%$$

MATHEMATICAL DERIVATIONS

With double-sideband amplitude modulation produced by a single audio tone (Fig. 138-1),

$$\text{Total sideband power, } P_{SB} = 0.5 \ m^2 P_c \text{ watts}$$

$$\text{Total rf power} = P_c(1 + 0.5 \ m^2) \text{ watts} \tag{138-1}$$

If I_A is the unmodulated antenna current and $I_A{}'$ is the value of the modulated antenna current,

$$I_A{}' = I_A \times \sqrt{\frac{P_c(1 + 0.5 \ m^2)}{P_c}} \tag{138-2}$$

$$= I_A \times \sqrt{1 + 0.5 \ m^2} \text{ amperes}$$

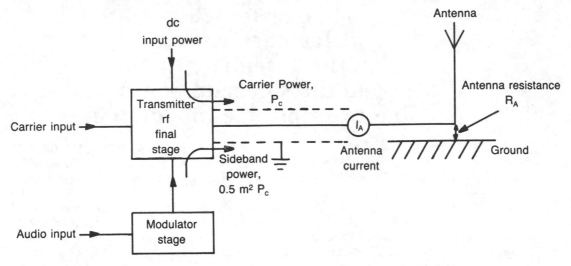

Figure 138-1

$$\text{Increase in antenna current} = I_A{}' - I_A \qquad (138\text{-}3)$$
$$= I_A \times (\sqrt{1 + 0.5\ m^2} - 1)\ \text{amperes}$$

$$\% \text{ increase in antenna current} = \left(\frac{I_A{}' - I_A}{I_A}\right) \times 100\% \qquad (138\text{-}4)$$
$$= (\sqrt{1 + 0.5\ m^2} - 1) \times 100\%$$

If the degree of the amplitude modulation produced by a single tone is increased from m_1 to m_2,

$$\text{Increase in antenna current} = I_A[\sqrt{1 + 0.5\ m_2{}^2} - \sqrt{1 + 0.5\ m_1{}^2}]$$

$$\% \text{ increase in antenna current} = \frac{I_A(\sqrt{1 + 0.5\ m_2{}^2} - \sqrt{1 + 0.5\ m_1{}^2})}{I_A\sqrt{1 + 0.5\ m_1{}^2}} \times 100 \quad (138\text{-}5)$$
$$= \left(\sqrt{\frac{2 + m_2{}^2}{2 + m_1{}^2}} - 1\right) \times 100\%$$

Similarly, if the degree of modulation is decreased from m_2 to m_1, the percentage decrease in the antenna current is given by:

$$\left(1 - \sqrt{\frac{2 + m_1{}^2}{2 + m_2{}^2}}\right) \times 100\% \qquad (138\text{-}6)$$

If the degree of modulation is increased from m_1 to m_2,

$$\% \text{ increase in } P_{SB} = \left(\frac{0.5\ m_2{}^2 P_c - 0.5\ m_1{}^2 P_c}{0.5\ m_1{}^2 P_c}\right) \times 100\% \qquad (138\text{-}7)$$
$$= \left(\frac{m_2{}^2 - m_1{}^2}{m_1{}^2}\right) \times 100\%$$
$$= \left(\frac{m_2{}^2}{m_1{}^2} - 1\right) \times 100\%$$

If the degree of modulation is decreased from m_2 to m_1, the percentage decrease in the total sideband power is

$$\% \text{ decrease in } P_{SB} = \left(\frac{m_2{}^2 - m_1{}^2}{m_2{}^2} \right) \times 100\% \qquad (138\text{-}8)$$

$$= \left(1 - \frac{m_1{}^2}{m_2{}^2} \right) \times 100\%$$

Example 138-1

An rf carrier is 70% amplitude modulated (double-sideband) by a single tone. What is the percentage increase in the antenna current as compared with its unmodulated value?

Solution

$$\% \text{ increase in antenna current} = (\sqrt{1 + 0.5m^2} - 1) \times 100\% \qquad (138\text{-}4)$$

$$= (\sqrt{1 + 0.5 \times 0.7^2} - 1) \times 100$$

$$= 11.6\%.$$

Example 138-2

An rf carrier is double-sideband amplitude modulated by a single tone. If the degree of modulation is increased from 0.50 to 0.80, what are the percentage increases in the antenna current and the sideband power?

Solution

$$\% \text{ increase in antenna current} = \left(\sqrt{\frac{2 + m_2{}^2}{2 + m_1{}^2}} - 1 \right) \times 100\% \qquad (138\text{-}5)$$

$$= \left(\sqrt{\frac{2 + 0.64}{2 + 0.25}} - 1 \right) \times 100$$

$$= \left(\sqrt{\frac{2.64}{2.25}} - 1 \right) \times 100$$

$$= 8.32\%$$

$$\% \text{ increase in } P_{SB} = \frac{m_2{}^2 - m_1{}^2}{m_1{}^2} \times 100\% \qquad (138\text{-}7)$$

$$= \frac{0.8^2 - 0.5^2}{0.5^2} \times 100$$

$$= \frac{0.64 - 0.25}{0.25} \times 100$$

$$= \frac{0.39}{0.25} \times 100$$

$$= 156\%.$$

Chapter 139
Plate modulation

IT IS APPROPRIATE TO CONSIDER A TUBE VERSION OF THIS CIRCUIT BECAUSE modulation primarily occurs in the final stage of a commercial AM transmitter. The rf power level of this stage is normally in the order of 10 kilowatts or more and no solid-state devices are capable of operating at this level.

The plate modulated rf power amplifier of Fig. 139-1 is a class-C amplifier in which the audio signal output is superimposed on the dc plate supply voltage, E_{bb}; this total voltage is then applied to the plate circuit and corresponds to the required modulation envelope of the AM signal.

The level of the carrier signal input, the value of the control grid bias and the impedance of the plate load tank circuit may be adjusted so that the instantaneous amplitude of the AM signal output is approximately equal to the instantaneous voltage applied to the plate circuit. This will provide the necessary linear relationship between the instantaneous amplitude of the AM signal and the instantaneous magnitude of the audio signal.

Modulation may occur in the plate circuit of the transmitter final rf stage (high level modulation) or in an earlier stage where the modulator will be required to provide less audio power. However, if low level modulation is used, the following stages cannot be operated under class-C

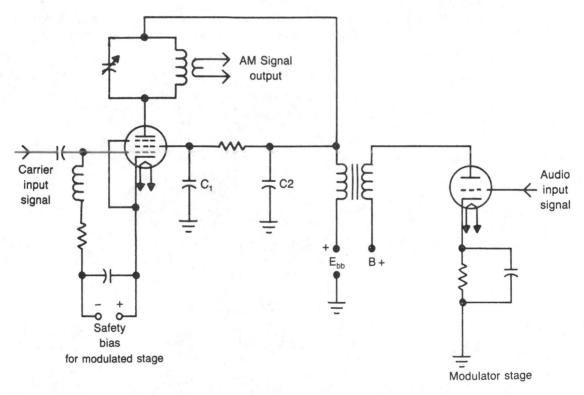

Figure 139-1

conditions for high plate efficiency or to provide frequency multiplication but must be linear class-A or class-B amplifiers.

If a beam power tube is used for the rf stage, both the plate and the screen grid are modulated by the audio signal ($C1$ and $C2$ are rf but not af bypass capacitors). By modulating the screen grid, excessive screen current is prevented at the trough of the modulation cycle.

In plate modulation, the power in the rf sidebands is obtained from the modulator stage while the carrier power is derived from the E_{bb} supply. The modulator is therefore required to deliver to the rf stage the amount of power that is sufficient to generate the sidebands while taking into account the percentage value of the class-C plate efficiency.

MATHEMATICAL DERIVATIONS

If the audio signal is a single tone and the required degree of modulation is m, the total double sideband power is $0.5\ m^2 P_c$ where P_c is the amount of power developed in the plate circuit. Then carrier power,

$$P_c = \frac{E_{bb} \times I_b \times \eta_R}{100} \text{ watts} \tag{139-1}$$

where: η_R = plate efficiency percentage of the final stage.

E_{bb} = plate supply voltage of the rf final stage (V).

I_b = dc level of plate current for the rf final stage (A).

The audio power output required from the modulator stage is

$$\text{Modulator power output} = \frac{0.5\ m^2 P_c}{\eta_R} \times 100 = \frac{50\ m^2 P_c}{\eta_R} \text{ watts} \tag{139-2}$$

The modulator stage of Fig. 139-1 is a single tube audio power amplifier that will operate under class-A conditions with a relatively low plate efficiency, η_M.

$$\text{dc plate input power to the modulator stage} = \frac{\text{audio power input}}{\eta_M} \times 100 \tag{139-3}$$

$$= \frac{50\ m^2 P_c}{\eta_R \times \eta_M} \times 100$$

$$= \frac{5000\ m^2 P_c}{\eta_R \times \eta_M}$$

$$= \frac{50\ m^2 \times E_{bb} \times I_b}{\eta_M} \text{ watts}$$

$$\text{Plate dissipation of the modulated stage} = (\text{Modulator's dc input power}) \times \left(1 - \frac{\eta_M}{100}\right) \tag{139-4}$$

$$= \left(\frac{100}{\eta_M} - 1\right) \times \frac{50\ m^2 P_c}{\eta_R} \text{ watts}$$

The effective load connected across the secondary of the output modulator's transformer is E_{bb}/I_b ohms. The turns ratio of the transformer must be used to match this effective load to the optimum plate load required by the modulator tube. For example, if the modulator tube is a triode, the optimum plate load is $2r_p$, where r_p is the triode's ac plate resistance.

Example 139-1

The plate modulated final rf stage of an AM transmitter has a dc supply voltage of 800 V and a dc plate current level of 1.5 A. The stage is operated in class-C with a plate efficiency of 75%, and is 80% modulated by a single tone. The modulator stage uses two triodes in parallel and has a plate efficiency of 30%. What are (a) the total power in the sidebands, (b) the modulator output power, (c) the required dc power input to the modulator, (d) the required turns ratio for the modulator output transformer if each triode modulator tube has an ac plate resistance of 800 Ω, and (e) the plate dissipation in each of the triodes?

Solution

(a) Dc plate input power = $E_{bb} \times I_b$ = 800 × 1.5 = 1200 W. The rf carrier power developed in the plate circuit of the final rf stage is given by:

$$\text{Carrier power, } P_c = 1200 \times \frac{75}{100}$$

$$= 900 \text{ W}$$

Total double sideband power,

$$P_{SB} = 0.5 \, m^2 P_c$$

$$= 0.5 \times (0.8)^2 \times 900$$

$$= 288 \text{ W.}$$

(b) The modulator output must provide the sideband power, taking into account the class-C plate efficiency of the final rf stage.

$$\text{Modulator output power} = \frac{P_{SB}}{\eta_R} \times 100 = \frac{288}{75} \times 100 \qquad (139\text{-}2)$$

$$= 384 \text{ W}$$

(c) The plate efficiency of the modulator stage is 30%. The dc plate power input to the modulator is:

$$\frac{\text{Modulator output power}}{\eta_M} \times 100 = \frac{384}{30} \times 100 \qquad (139\text{-}3)$$

$$= 1280 \text{ W}$$

Alternatively, the dc plate power input to the modulator is:

$$\frac{50 \, m^2 E_{bb} I_b}{\eta_M} = \frac{50 \times (0.8)^2 \times 800 \times 1.5}{30} = 1280 \text{ W}$$

$$= 1.28 \text{ kW}$$

(d) The effective load on the transformer secondary is:

$$\frac{E_{bb}}{I_b} = \frac{800 \text{ V}}{1.5 \text{ A}}$$

$$= 533 \ \Omega$$

Effective r_p of the two triodes in parallel = 800/2 = 400 Ω.

Optimum load required by the triodes = 2 × 400 = 800 Ω. Required turns ratio for the modulator transformer = $\sqrt{800/533}$ = 1.23:1.

(e) The total plate dissipation of the modulator stage is

$$1.28 \times \frac{70}{100} = 0.896 \text{ kW} \qquad (139\text{-}4)$$

$$= 896 \text{ W}$$

The plate dissipation in each triode is

$$\frac{896}{2} = 450 \text{ W, rounded off.}$$

Note that if the triode had been connected in push-pull and not in parallel, the plate dissipation in each triode would still have been half of the total plate dissipation in the modulator stage.

Chapter 140
Frequency and phase modulation

WITH FREQUENCY MODULATION, THE INSTANTANEOUS FREQUENCY OF THE RF wave is varied in accordance with the modulating signal while the amplitude of the rf wave is kept constant. The rf power output and the antenna current of the FM transmitter are therefore independent of the modulation.

FREQUENCY MODULATION

In the FM wave (Fig. 140-1), the instantaneous amount of the *frequency shift* or deviation away from its average unmodulated value is linearly related to the instantaneous *magnitude* (voltage) of the modulating signal while the rate at which the frequency deviation (f_d) occurs, is equal to the modulating frequency. Note carefully that the amount of the frequency shift in the rf wave is independent of the modulating frequency. Therefore, if they are of the same amplitude, modulating tones of 200 Hz and 400 Hz will provide the same amount of frequency shift in the FM wave. However, it must be pointed out that in FM and TV broadcast transmitters, the higher audio frequencies above 800 Hz are progressively accentuated (preemphasized) in order to improve their signal-to-noise ratio at the receiver. The degree of preemphasis is measured by the time constant of the RC circuit whose audio output increases with frequency. This time constant is specified by the FCC as 75 microseconds. In order to restore the total balance, the receiver discriminator output is fed to an RC deemphasis circuit with the same time constant.

When 100% modulation of an FM wave occurs, the amount of frequency shift reaches the maximum value allowed for the particular communications system. This maximum value is called the frequency deviation which, in the FM broadcast service (emission designation F3E), is a shift of 75 kHz on either side of the unmodulated carrier frequency. The output frequency swing for 100% modulation is therefore ±75 kHz. However, in the aural transmitter of a TV broadcast station, 100% modulation corresponds to a frequency deviation of 25 kHz while in the Public Safety Radio Services, the frequency deviation is only 5 kHz. The percentage modulation and the amount of the frequency shift are directly proportional so that 40% modulation in the FM broadcast service would correspond to an output frequency swing of ±40 × 75/100 = ±30 kHz.

For FM the degree of modulation is defined by:

$$\text{Modulation index, } m = \frac{f_d}{f_m}$$

where: f_m = modulating frequency

f_d = frequency deviation

During the transmission of speech and music, the value of the instantaneous modulation index can vary from less than 1 to over 100. However, for a particular system there is a certain modulation index value, which is called the *deviation ratio* and is determined by:

$$\text{Deviation ratio} = \frac{f_d \text{ corresponding to 100\% Modulation}}{\text{Highest transmitted value of } f_m}$$

For commercial FM broadcast, the transmitted audio range is 50 Hz to 15 kHz. Since the frequency deviation value for 100% modulation is 75 kHz, the deviation ratio is 75 kHz/15 kHz = 5.

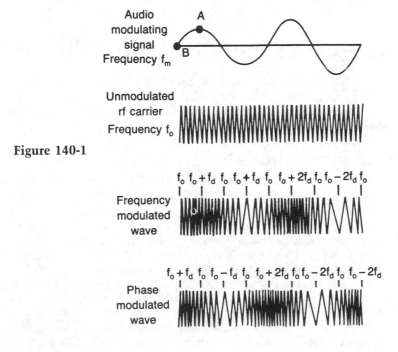

Figure 140-1

470

PHASE MODULATION

In phase modulation (Fig. 140-1) the instantaneous phase of the rf carrier wave is varied in accordance with the modulating signal while the amplitude of the rf wave is kept constant. The instantaneous amount of the phase shift away from its unmodulated value is linearly related to the instantaneous magnitude (voltage) of the modulating signal. Because a rate of change of phase is equivalent to a shift in frequency, the PM wave is similar in appearance to the FM wave and, in fact, they cannot be distinguished except by reference to the modulating signal. The important differences are:

1. With PM, the output frequency swing is proportional to the product of the amplitude and the frequency of the modulating signal, while in FM, the swing is proportional to the amplitude alone. Therefore, if the carrier is being modulated by a single tone and the tone amplitude and the frequency are both doubled, the output frequency swing in the PM signal would be quadrupled (neglecting any effect of preemphasis).
2. Relative to the cycle of the modulating signal, the instantaneous maximum (and/or minimum) frequency positions are 90° apart in the FM and PM waves. In Fig. 140-1, point A in the modulating signal corresponds to the instantaneous maximum frequency in the FM wave. For the PM wave the maximum frequency occurs at point B which is one-fourth of a cycle or 90° from point A.

The main importance of PM is its use in the indirect method of creating FM. For reasons of frequency stability (and to avoid the requirement for an AFC system) it is common practice in many FM transmitters to use a master crystal oscillator whose frequency is fixed and which therefore cannot be directly modulated. The modulated stage comes after the crystal oscillator and consists of a circuit that actually creates phase modulation. However, before reaching the modulated stage, the audio signal is passed through a correction network (not to be confused with the preemphasis circuit) whose voltage output is inversely proportional to frequency and which introduces an approximate 90° phase shift. Then as far as the undistorted audio signal *input to* the correction network is concerned, the output from the modulated stage is the required FM signal, but with regard to the *output from* the correction network, the modulated stage produces PM.

MATHEMATICAL DERIVATIONS

The equation of an FM wave is

$$f(t) = A \cos [\omega_o t + m \sin \omega_m t] \tag{140-1}$$

where: A = peak value of the unmodulated carrier voltage (V)

ω_o = angular frequency of the carrier (rad/s)

ω_m = angular frequency of the audio modulating tone (rad/s)

m = modulation index

Then,

$$f(t) = A \cos (\omega_o t + m \sin \omega_m t) \tag{140-2}$$
$$= A \cos \omega_o t \cos (m \sin \omega_m t) - A \sin \omega_o t \sin (m \sin \omega_m t)$$

Now,

$$\cos (m \sin \omega_m t) = J_o(m) + 2[J_2(m) \cos 2\omega_m t + J_4(m) \cos 4\omega_m t + . . .] \tag{140-3}$$

and,

$$\sin (m \sin \omega_m t) = 2[J_1(m) \sin \omega_m t + J_3(m) \sin 3\omega_m t + \ldots] \qquad (140\text{-}4)$$

where: $J_n(m)$ = Bessel function coefficients

$\qquad n = 0, 1, 2, 3, \ldots$ etc.

Substituting Equations 140-3 and 140-4 in the Equation 140-2,

$$f(t) = A[J_o(m) \cos \omega_o t + J_1(m)[\cos (\omega_o + \omega_m)t - \cos (\omega_o - \omega_m)t] \qquad (140\text{-}5)$$
$$+ J_2(m)[\cos (\omega_o + 2\omega_m)t + \cos (\omega_o - 2\omega_m)t]$$
$$+ J_3(m)[\cos (\omega_o + 3\omega_m)t - \cos (\omega_o - 3\omega_m)t]$$
$$+ J_4(m)[\cos (\omega_o + 4\omega_m)t + \cos (\omega_o - 4\omega_m)t], \ldots \text{ etc.}$$

Remembering that the modulating signal is only a single sine wave, the most outstanding conclusion drawn from Equation 140-5 is that the modulated carrier consists of an infinite number of sidebands above and below the carrier frequency. In practice, the sidebands are limited to those greater than 1% of the *unmodulated carrier voltage*.

As an example let us assume that an unmodulated carrier has a frequency of 100 MHz and a peak value of 1 V. The modulating signal is a tone of 15 kHz and produces a modulation index of 5, so the frequency deviation is $5 \times 15 = 75$ kHz.

From a table of Bessel Functions:

$J_0(5) = -0.178$
$J_1(5) = -0.328$
$J_2(5) = +0.047$
$J_3(5) = +0.365$
$J_4(5) = 0.391$ } significant sidebands
$J_5(5) = 0.261$
$J_6(5) = 0.131$
$J_7(5) = 0.053$
$J_8(5) = 0.018$ } significant sideband limit
$J_9(5) = 0.005$ insignificant sideband

It can be seen that there are eight pairs of significant sidebands, so that the effective bandwidth is $15 \times 8 \times 2$ kHz = 240 kHz. The bandwidth for commercial FM broadcast is 200 kHz.

Table 140-1 has been compiled from Bessel Function values. It shows the number of pairs of significant sidebands for values of the modulation index, m.

Table 140-1

Modulation index m	Number of pairs of sidebands
1	3
2	4
3	6
4	7
5	8
6	9
7	10
8	$m + 4$

Note: Once $J_0(x)$ and $J_1(x)$ have been computed, the remainder can be calculated from the relationship:

$$J_{n+1}(x) + J_{n-1}(x) = \frac{2n}{x} J_n(x) \tag{140-6}$$

For example if $n = 3$ and $x = 5$,

$$J_4(5) + J_2(5) = 0.047 + 0.391 = 0.438$$

$$\frac{2 \times 3}{5} J_3(5) = 1.2 \times 0.365 = 0.438.$$

Example 140-1

For test purposes, a 10 kW FM broadcast transmitter is 80% modulated by a 2 kHz tone. Including the carrier and sidebands, what is the total power in the FM signal? Calculate the value of the output frequency swing.

Solution

Because the rf power in an FM signal is independent of the modulation, the total power remains at 10 kW. In the FM broadcast service, 100% modulation corresponds to a frequency deviation of 75 kHz. The frequency swing for 80% modulation is $\pm(80/100) \times 75 = \pm 60$ kHz.

Example 140-2

The aural transmitter of a TV broadcast station is 60% modulated by a 500 Hz test tone. The amplitude and the frequency of the tone are both halved. What is the output frequency swing in the FM signal?

Solution

In the aural transmitter of a TV broadcast station, 100% modulation corresponds to a frequency deviation of 25 kHz. Because the percentage modulation is directly proportional to the tone amplitude (the effect of the preemphasis may be neglected at 500 Hz and 250 Hz) but is independent of its frequency, the new modulation frequency will be 30%, which will correspond to an output frequency swing of:

$$\pm \frac{30}{100} \times 25 = \pm 7.5 \text{ kHz}$$

Example 140-3

A 50 kW FM broadcast station is instantaneously 40% modulated by an audio frequency of 5 kHz. What is (a) the value of the modulation index and (b) the bandwidth of the FM signal?

Solution

(a) For an FM broadcast station, 40% modulation corresponds to the frequency shift of:

$$\frac{40}{100} \times 75 = 30 \text{ kHz}$$

The modulation index is

$$\frac{30 \text{ kHz}}{5 \text{ kHz}} = 6$$

(b) From Table 140-1, the number of significant sidebands on either side of the carrier is 9. Since adjacent sidebands are separated by the audio frequency of 5 kHz, the bandwidth is $2 \times 9 \times 5 = 90$ kHz.

Example 140-4

An FM transmitter is operating on 102 MHz and the oscillator stage is being directly modulated by a 3 kHz test tone so that its frequency swing is ± 2 kHz. The oscillator stage generates 4250 kHz when unmodulated. If 100% modulation corresponds to a frequency swing of 75 kHz, what is the percentage modulation produced by the test tone?

Solution

The total product factor of the frequency multiplier stages is 120 MHz/4250 kHz = 24. The output frequency swing = $\pm 2 \times 24 = 48$ kHz and the percentage modulation is $(48/75) \times 100 = 64\%$.

Example 140-5

When an FM broadcast transmitter is modulated by a 450 Hz tone, the output frequency swing is ± 30 kHz. If the tone frequency is lowered to 225 Hz but the amplitude is unchanged, what is the new percentage of modulation?

Solution

For tones of 450 Hz and 225 Hz the effect of preemphasis may be neglected. Therefore the change in the tone frequency does not alter the modulation percentage, which remains at $30/75 \times 100 = 40\%$.

Chapter 141
Calculation of operating power
by the direct and indirect methods

THE OPERATING POWER OF A TRANSMITTER IS THE ACTUAL POWER SUPPLIED TO the antenna system and is not the same as the authorized power. The difference between the authorized and operating power is associated with the station's power tolerance. For example, if an AM broadcast station has an authorized power of 10 kW, its power tolerance extends from $+5\%$ to -10% and therefore the operating power may legitimately lie between 9 kW and 10.5 kW. The operating power may be calculated by either the direct method or the indirect method.

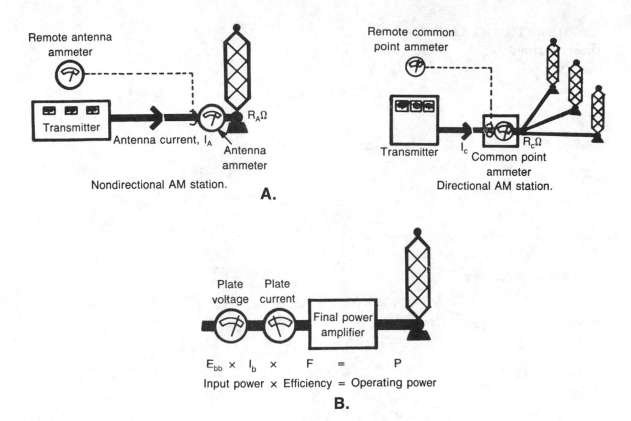

A.

Remote antenna ammeter

Transmitter

Antenna current, I_A

Antenna ammeter

$R_A \Omega$

Nondirectional AM station.

Remote common point ammeter

Transmitter

I_c Common point ammeter

$R_c \Omega$

Directional AM station.

Plate voltage   Plate current   Final power amplifier

$E_{bb} \times I_b \times F = P$

Input power $\times$ Efficiency = Operating power

B.

Figure 141-1

THE DIRECT METHOD

Using the direct method (Fig. 141-1A) a transmitter's operating power is determined from the rf current (I_A) delivered by the transmitter to the antenna and the antenna's effective load at resonance (R_A). The operating power in watts is then given by the expression (antenna current)$^2 \times$ antenna resistance.

Directional AM broadcast stations employ multiple radiating elements. The operating power of these stations, as determined by the direct method, is equal to the product of the resistance common to all the antenna towers (the common point resistance, R_c) and the square of the current common to all of the antenna towers (common point current, I_c).

THE INDIRECT METHOD

The indirect method (Fig. 141-1B) involves the plate input power to the final rf stage and an efficiency factor, F, which is normally expressed as a decimal fraction but may be given as a percentage. The operating power is then calculated from the value of the dc plate supply voltage $\times$ dc level of the plate current $\times$ efficiency factor.

For a standard AM broadcast station, the efficiency factor depends on the method of modulation (plate, grid, or low level) and the maximum rated carrier power output. By contrast the value of the efficiency factor for FM stations is normally supplied by the manufacturer.

MATHEMATICAL DERIVATIONS
Direct method
Operating power,

$$P = I_A{}^2R_A \text{ watts} \tag{141-1}$$

This yields,

$$I_A = \sqrt{\frac{P}{R_A}} \text{ amperes} \tag{141-2}$$

and,

$$R_A = \frac{P}{I_A{}^2} \text{ ohms} \tag{142-3}$$

where: I_A = antenna current in amperes (A)

R_A = effective antenna resistance (Ω)

For a directional AM station,

$$\text{Operating power} = I_c{}^2R_c \text{ watts} \tag{141-4}$$

where: I_c = common point current (A)

R_c = common point antenna resistance (Ω).

Indirect method
Operating power,

$$P = E_{bb} \times I_b \times F \text{ watts} \tag{141-5}$$

where: E_{bb} = plate supply voltage (V)

I_b = dc level of the plate current (A)

Example 141-1
The rf current delivered to an antenna is 8 A and the antenna's effective resistance is 125 Ω. What is the value of the operating power as calculated by the direct method?

Solution

$$\text{Operating power} = I_A{}^2 \times R_A \text{ watts} \tag{141-1}$$
$$= (8A)^2 \times 125 \ \Omega$$
$$= 8000 \text{ W}$$
$$= 8 \text{ kW}.$$

Example 141-2
The final rf stage of a standard broadcast station has a plate supply voltage of 2500 V and a dc level of plate current equal to 1.5 A. The stage uses plate modulation with an efficiency factor of 0.7. What is the value of the operating power, as determined by the indirect method?

Solution

$$\text{Operating power} = E_{bb} \times I_b \times F \text{ watts} \qquad (141\text{-}5)$$
$$= 2500 \text{ V} \times 1.5 \text{ A} \times 0.7$$
$$= 2625 \text{ W}.$$

Chapter 142
The field strength of an antenna —effective radiated power

THE ELECTRIC FIELD STRENGTH (FIELD INTENSITY), ε, AT A PARTICULAR POSITION from a transmitting antenna is measured in millivolts or microvolts per meter and determines the signal voltage induced in a receiving antenna. The field strength is directly proportional to the antenna current and to the square root of the radiated power but is inversely proportional to the distance from the transmitting antenna. For example, a thin center-fed dipole has a radiation resistance of 73.2 ohms so that if the rf power delivered to the antenna is 1 kilowatt, the antenna current is $\sqrt{1000 \text{ W}/73.2 \text{ }\Omega} = 3.696$ A. The field strength at a position of one mile or 1609.3 meters from the antenna is 60×3.696 A/1609.3 m $= 0.1378$ volts per meter or 137.8 millivolts per meter. This value is used as the FCC standard to determine the gain of more sophisticated antennas.

In displaying the radiation pattern of a transmitting antenna, positions of equal field strength are joined together to form a particular field strength contour (Fig. 142-1).

If parasitic elements such as reflectors and directors are attached to a simple dipole (or more antennas are added to produce an array), the radiated rf power is concentrated in particular directions and the antenna gain is increased. This gain can either be considered in terms of the field strength or the rf power.

From measurements taken at a particular position, the field gain is the ratio of the field strength produced by the complex antenna system to the field strength created by a simple ideal dipole (assuming that this dipole is fed with the same rf power as the complex antenna). Because the square of the field intensity is directly proportional to the radiated power, the power gain of the antenna is equal to the square of the field gain and is either expressed as a power ratio or in decibels.

The effective radiated power (ERP) in a particular direction is calculated from the product of the rf power delivered to the antenna and its power gain ratio.

MATHEMATICAL DERIVATIONS
In Fig. 142-1,

Electric field strength, ε

$$\varepsilon = \frac{60 \times I_A}{d} \text{ volts per meter} \qquad (142\text{-}1)$$

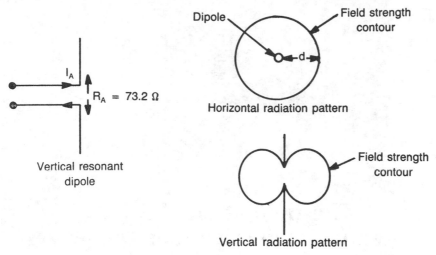

Figure 142-1

where: I_A = current at the antenna's feedpoint (A)

        d = distance from the transmitting antenna (m)

Antenna current,

$$I_A = \sqrt{\frac{P_A}{R_A}} \text{ amperes} \tag{142-2}$$

where: P_A = antenna power (W)

        R_A = antenna's radiation resistance (Ω)

Effective radiated power,

ERP = (rf power delivered to the antenna) × (antenna power gain ratio) watts   (142-3)

Antenna power gain = 10 log (antenna power gain ratio) decibels   (142-4)

= 20 log (antenna field gain ratio) decibels

Example 142-1

At a position 3 miles from a transmitting antenna, the field strength is 650 μV/m. If the antenna current is doubled, what is the new field strength (a) at the same position and (b) at a position 5 miles from the antenna?

Solution

(a) The field strength is directly proportional to the antenna current. The new field strength is therefore 2 × 650 = 1300 μV per meter = 1.3 mV per meter.

(b) The field strength is inversely proportional to the distance from the transmitting antenna. Therefore, the new field strength at a position 5 miles from the transmitting antenna is 1300 × 3/5 = 780 μV per meter.

Example 142-2

The rf power input to a transmitting antenna is 8.7 kW. If the antenna power gain is 8 dB, what is the value of the effective radiated power?

Solution

Antenna power gain ratio = inv log (8/10) = 6.309.

Effective radiated power = 8.7 × 6.309 = 54.9 kW.

Chapter 143
The single-sideband system— frequency conversion in the superheterodyne receiver

THE SIDEBANDS IN AN AM WAVE REPRESENT THE AUDIO INFORMATION AND IN order to convey this information, it is not essential that the carrier and both sets of sidebands are transmitted.

THE SINGLE-SIDEBAND SYSTEM

There are various types of AM emission and each type must be considered separately as regards the power content of the sidebands and the bandwidth. For example in the single-sideband suppressed carrier (SSSC) system of Fig. 143-1, designated J3E (formerly A3J), all of the rf power is concentrated in one set of sidebands (either the upper set or the lower set). This improves the signal-to-noise ratio at the receiver as well as reducing the bandwidth required. However, when compared with double-sideband operation (A3E), the SSSC system demands a higher degree of frequency stability for the carrier generated in the transmitter's oscillator and also requires that a carrier component be reinserted at the receiver.

The advantage of power saving with the SSSC system lies in the fact that the whole of the rf carrier power can be concentrated in one set of sidebands rather than being distributed over the carrier and two sets of sidebands; this is the reason for the improvement in the signal-to-noise ratio at the receiver.

For example, if a 100 W carrier is 100% (A3E) modulated by a single tone, the power in each of the sidebands is 25 W. If the whole of the 100 W unmodulated carrier power is concentrated in one sideband, the power advantage is equivalent to:

$$10 \log_{10} \left(\frac{100 \text{ W}}{25 \text{ W}} \right) = 10 \log_{10} 4$$

$$= 6 \text{ dB}$$

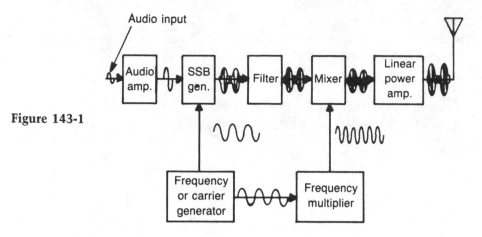

Audio input

Figure 143-1

This power advantage will result in the improved signal-to-noise ratio at the receiver.
Other single-sideband emission designations are:

- R3E   Single-sideband with reduced (or pilot) carrier.
- H3E   Single-sideband with full carrier power.

MATHEMATICAL DERIVATIONS

Let the unmodulated carrier power by P_c watts and let the carrier be 100% double sideband modulated by a single audio tone.

Total double sideband power

$$P_{SB} = \frac{1}{2} \times m^2 \times P_c \text{ watts}$$

$$= \frac{1}{2} \times 1^2 \times P_c$$

$$= P_c/2 \text{ watts.}$$

Total rf power with double sideband operation $= P_c + P_c/2 = 3P_c/2$ watts.

Power in one sideband $= P_c/4$ watts.

% power saving with the SSSC system $= \dfrac{3P_c/2 - P_c/4}{3P_c/2} \times 100\%$

$$= \frac{6-1}{6} \times 100$$

$$= 83\frac{1}{3}\%$$

Example 143-1

The unmodulated carrier of a 1 MHz transmitter is 5 kW. The carrier is then 80% double-sideband amplitude modulated by a 600 Hz tone. Calculate the value of the total rf power. If the carrier and one sideband are now suppressed what is the percentage saving in the transmitter's rf power?

480

Solution

Double sideband

Total sideband power,

$$P_{SB} = \frac{1}{2} \times m^2 \times P_c$$

$$= \frac{1}{2} \times 0.8^2 \times 5$$

$$= \frac{1}{2} \times 0.64 \times 5$$

$$= 1.6 \text{ kW}$$

Total rf power = $5 + 1.6 = 6.6$ kW.

Single sideband

Total power in the single sideband,

$$\frac{P_{SB}}{2} = 1.6 \text{ kW}/2$$

$$= 800 \text{ W}.$$

$$\% \text{ saving in total rf power} = \frac{6.6 - 0.8}{6.6} \times 100$$

$$= 88\%.$$

FREQUENCY CONVERSION IN THE SUPERHETERODYNE RECEIVER

Figure 143-2 represents the block diagram of the initial stages in either an AM or an FM super-heterodyne broadcast receiver. The purpose of the mixer and the local oscillator is to achieve frequency conversion. In this process, the receiver is tuned to a particular signal and the carrier frequency is then converted to a fixed intermediate frequency (i-f) which is 455 kHz in the AM broadcast band. Therefore, for all incoming carrier frequencies, the i-f stages always operate on the same fixed intermediate frequency and may be designed to provide high selectivity and sensitivity.

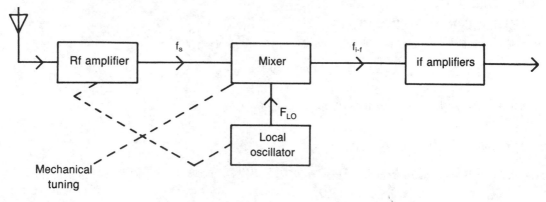

Figure 143-2

As indicated in Fig. 143-2 the rf tuned circuits of the initial amplifier, the mixer and the local oscillator, are mechanically ganged so that as the rf amplifier and the input circuit of the mixer are tuned to the desired signal frequency, the local oscillator stage generates its own frequency and produces a continuous rf output that differs in frequency from the incoming carrier by the value of the intermediate frequency (i-f).

The output of the oscillator and the incoming signal are both fed to the mixer stage, which contains a nonlinear device. In general, when two signals of different frequencies are mixed together (this mixing process is sometimes called *heterodyning,* or *beating*) the result contains a component at their difference frequency.

The difference frequency is equal to the intermediate frequency. The mixer output circuit may therefore be tuned to the intermediate frequency and this will automatically eliminate the unwanted components created by the mixing process. During the frequency conversion, the modulation originally carried by the wanted signal is transferred to the intermediate frequency component in the mixer's output. However, if the local oscillator frequency is greater than the incoming signal frequency (known as *tracking above*) the upper sidebands carried by the incoming carrier frequency will become the lower sidebands in the intermediate frequency amplifiers and vice-versa.

In an AM broadcast receiver, the local oscillator always tracks above in order to reduce the physical size required for the local oscillator tuning capacitor. However, in TV and FM broadcast receivers, which operate in the VHF and UHF bands, the local oscillator may track either above or below.

IMAGE CHANNEL

Consider that a superheterodyne receiver is tuned to a carrier frequency of 640 kHz in the AM broadcast band. Because the i-f is 455 kHz, the local oscillator will generate 640 + 455 = 1095 kHz. Assume that there exists an unwanted signal whose carrier frequency is 1095 + 455 = 1550 kHz (which also lies within the AM broadcast band of 535 kHz to 1605 kHz). If the signal is sufficiently strong to reach the mixer stage, it will beat with the local oscillator output, and its carrier frequency will also be converted to the i-f. Once this has occurred, the i-f stages will be unable to separate the wanted and the unwanted signals; both will be detected and both will be heard on the loudspeaker. The unwanted signal of 1550 kHz is an example of an image channel because its frequency is the image of the wanted signal in the "mirror" of the local oscillator frequency (the frequency of the wanted signal is the same amount below the local oscillator frequency as the image channel is above). The frequency of an image channel therefore differs from the wanted carrier frequency by an amount equal to twice the intermediate frequency.

MATHEMATICAL DERIVATIONS

If the local oscillator "tracks above" the incoming signal frequency,

$$f_{LO} = f_S + f_{i\text{-}f}, \; f_S = f_{LO} - f_{i\text{-}f}, \; f_{i\text{-}f} \qquad (143\text{-}1)$$
$$= f_{LO} - f_S$$

If the local oscillator "tracks below" the incoming frequency:

$$f_{LO} = f_S - f_{i\text{-}f}, \; f_S = f_{LO} + f_{i\text{-}f}, \; f_{i\text{-}f} \qquad (143\text{-}2)$$
$$= f_S - f_{LO}$$

where: f_S = wanted signal frequency (kHz)

f_{LO} = local oscillator frequency (kHz)

$f_{i\text{-}f}$ = intermediate frqeuency (kHz)

The image channel frequency is given by:
Tracking *above,*

$$\text{Image frequency} = f_S + (2 \times f_{i\text{-}f}) \qquad (143\text{-}3)$$
$$= f_{LO} + f_{i\text{-}f}$$

Tracking *below,*

$$\text{Image frequency} = f_S - (2 \times f_{i\text{-}f}) \qquad (143\text{-}4)$$
$$= f_{LO} - f_{i\text{-}f}$$

Example 143-1

A superheterodyne AM broadcast receiver with an i-f of 455 kHz is tuned to receive a station on 670 kHz. What are the frequencies of the local oscillator and a possible image channel?

Solution

In an AM broadcast receiver, the frequency of the local oscillator is tracking above the wanted signal frequency.

Local oscillator frequency,

$$f_{LO} = f_S + f_{i\text{-}f} \qquad (143\text{-}1)$$
$$= 670 + 455$$
$$= 1125 \text{ kHz.}$$

$$\text{Image channel frequency} = f_S + (2 \times f_{i\text{-}f}) \qquad (143\text{-}3)$$
$$= 670 + 2 \times 455$$
$$= 1580 \text{ kHz.}$$

Chapter 144
The crystal and its
temperature coefficient

WHEN A CRYSTAL SUCH AS QUARTZ VIBRATES, AN ALTERNATING VOLTAGE appears between a pair of opposite faces; this is called the *piezoelectric effect.* The fundamental frequency of vibration for a particular crystal is inversely proportional to the crystal's thickness and also depends on the type of cut. The crystal may be compared with a LCR circuit, with the inductance, capacitance, and resistance related to equivalent physical properties of the crystal. When used in an oscillator circuit, the crystal is placed between metal surfaces (the holder) to which the positive feedback voltage is applied. Electrically, the holder may be regarded as a capacitance (typically a few pF) in parallel with the LCR circuit of the crystal. This equivalent circuit of the crystal and its holder is shown in Fig. 144-1A.

Although a crystal's fundamental frequency is primarily determined by the thickness of the cut, the frequency may be varied by connecting a small trimmer capacitor across the crystal and its holder. In addition, the vibration of the crystal is complex and contains overtone frequencies; these are virtually harmonics of the fundamental frequency and it is mainly the odd overtones that are produced. The upper limit of the fundamental frequency is about 20 to 25 MHz and is determined by the minimum thickness to which a crystal may be cut without the danger of fracturing.

The Q of a crystal is normally several thousand (compared with 300 or less for a conventional LCR circuit). This is one important reason for the high degree of frequency obtainable from a crystal oscillator. Another reason is the crystal's low temperature coefficient.

A change in crystal temperature can cause a change in resonant frequency. The crystal's temperature coefficient may be used to calculate the amount of frequency shift that can be expected from a given change in temperature. This coefficient is measured in Hz per MHz per degree celsius (parts per million per degree celsius) and may be either positive or negative. For example, if a crystal is marked as $-30/10^6/°C$, it indicates that if the temperature increases by one degree celsius, the decrease (because of the negative coefficient) will be 30 Hz for every megahertz of the crystal's operating frequency. Accordingly, the change in the crystal's operating

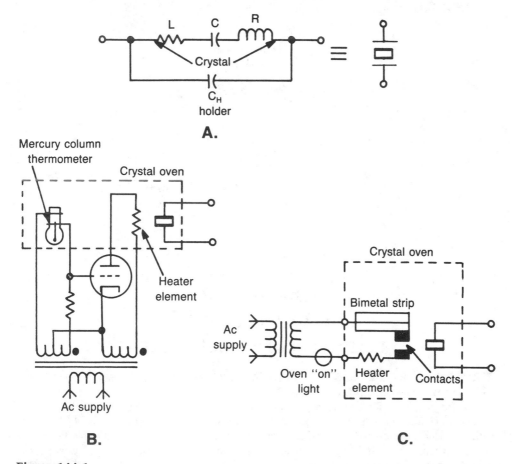

Figure 144-1

484

frequency is Δf = value of the temperature coefficient (either positive or negative) × crystal frequency in MHz × °C temperature change, which is positive if the temperature is increasing and negative if the temperature is decreasing. Note that if the temperature coefficient is negative and the temperature is decreasing, the result will be an increase in the frequency.

Because the frequency change, Δf, occurs at the crystal oscillator stage, the change in the output frequency of the final rf stage will be Δf × the total product factor of the frequency multiplier stages (if any).

To improve the frequency stability, the crystal may be enclosed in a thermostatically controlled oven. For broadcast stations the FCC has specified certain temperature tolerances at the crystal position within the oven. If an X-cut or Y-cut crystal is used, the permitted tolerance is only ±0.1°C, but with a low temperature-coefficient crystal, the tolerance is ±1.0°C.

The form of thermostat may either be a mercury thermometer type or the less sensitive, but simpler and cheaper, thermocouple variety (Fig. 144-1B and C). In Fig. 144-1B any increase of temperature above the tolerance limit will cause the triode to be permanently cut off so that the heater element will carry no current. If the temperature in Fig. 144-1C falls below the tolerance limit, the bimetal strip will cause the contacts to close and complete the circuit for the heater element.

When using crystals, precautions must be taken to ensure that excessive voltage is not developed across the crystal, because this may cause the crystal to fracture. If the crystal becomes dirty, it must be removed from its holder, held by the edges (not the face) and then washed in either alcohol or soap and water.

MATHEMATICAL DERIVATIONS

If the crystal is operating at a frequency of f MHz and the temperature coefficients is Δf Hz/MHz/°C (Δf is either positive or negative),

$$\text{Frequency shift at the oscillator stage} = \Delta f \times f \times \Delta T \text{ Hz} \qquad (144\text{-}1)$$

where ΔT = temperature change (°C).

If the frequency multiplier stages have a product factor, m,

$$\text{Output frequency shift} = m \times (\Delta f \times f \times \Delta T) \text{ Hz} \qquad (144\text{-}2)$$

Example 144-1

A Y-cut crystal has an operating frequency of 3750 kHz at a temperature of 20° C. If the crystal temperature coefficient is $-15/10^6/°C$, what is the operating frequency at 25° C?

Solution

Because 3750 kHz = 3.75 MHz, the change in frequency is given by:

$$\Delta f = -15 \times 3.75 \times (25 - 20)$$
$$= -281.25 \text{ Hz}$$

The operating frequency at 25° C = 3750 kHz − 281.25 Hz = 3749.719 kHz, rounded off.

Example 144-2

A transmitter has an operating frequency of 22 MHz and uses a 2725 kHz crystal enclosed in an oven with the temperature at 65° C. If the crystal temperature coefficient is $+5/10^6/°C$, and the oven temperature decreases to 63° C, what is the transmitter's output frequency?

Solution

The total product factor of the frequency multiplier stages is 22 MHz/2725 kHz = 8. Change in the crystal operating frequency is $\Delta f = +5 \times 2.725 \times (63 - 65) = -27.25$ Hz. Change in the transmitter output frequency is $8 \times \Delta f = 8 \times (-27.25) = -218$ Hz. Transmitter output frequency is 22 MHz − 218 Hz = 21999.782 kHz.

Chapter 145
The frequency monitor

THE FREQUENCY TOLERANCE OF AN AM BROADCAST STATION IS ±20 HZ. THE FCC requires that every broadcast station has a means of determining the number of Hz that the actual radiated carrier frequency differs from the assigned value. Such a device has to be approved by the FCC and is known as a frequency monitor. This auxiliary unit contains a crystal-controlled oscillator that is entirely separate from the AM transmitter. The amount by which the carrier frequency is too high or too low is shown on a frequency deviation indicator (Fig. 145-1A), that must be checked at intervals during the broadcast day and the reading entered in the transmitter log.

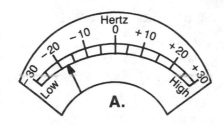

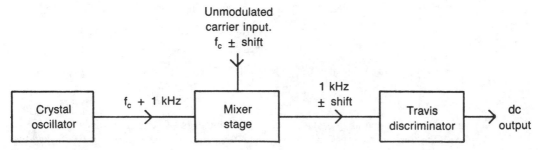

Figure 145-1

486

The basic principle of the frequency monitor is shown in the block diagram of Fig. 145-1B. A crystal oscillator with a thermostatically controlled oven produces a stable frequency that differs from the authorized carrier frequency by 1 kHz. The output of the crystal oscillator is fed to a mixer stage that is also supplied with a sample of the carrier voltage, derived from an unmodulated stage of the transmitter. The frequency of the mixer's output to the discriminator circuit is the difference of 1 kHz ± any shift of the carrier frequency from its assigned value. The discriminator is of the tuned high (1050 Hz)/tuned low (950 Hz) or Travis variety so that the magnitude and polarity of its dc output is determined by the instantaneous shift in the carrier frequency. This dc output is then passed to the monitor's indicating device with its center zero scale.

Periodically, the accuracy of the frequency monitor is measured. An external check is made of the radiated carrier frequency and the result is compared with the simultaneous reading of the frequency monitor. The error in the frequency monitor can then be corrected by adjusting the monitor oscillator frequency (although the oscillator is crystal-controlled, the frequency can be varied a few Hz by adjusting the value of a trimmer capacitor across the crystal).

MATHEMATICAL DERIVATIONS

The error in the frequency monitor is equal to the deviation (positive for "high," negative for "low") from the assigned frequency as recorded by the frequency monitor minus the deviation (positive for "high," negative for "low") as indicated by the external frequency check. (145-1)

Example 145-1

An external frequency check of an AM broadcast station carrier reveals that at a particular time the carrier frequency was 8 Hz low. The transmitter log showed for the same time, the carrier frequency was 10 Hz high. What was the error in the station frequency monitor?

Solution

$$\text{Error in the frequency monitor} = +10 \text{ Hz} - (-8 \text{ Hz})$$
$$= +18 \text{ Hz or 18 Hz high.}$$

Example 145-2

At a particular time the log of an AM broadcast station showed that the carrier frequency was 10 Hz low. An external frequency check made at the same time, indicated that the carrier frequency was 7 Hz low. What was the frequency monitor reading?

Solution

The error in the frequency monitor was $-10 \text{ Hz} - (-7 \text{ Hz}) = -3 \text{ Hz}$, or the frequency monitor was reading 3 Hz low.

Chapter 146
Television broadcast frequencies

THE COMPOSITE VIDEO SIGNAL OF A TELEVISION BROADCAST CONTAINS THE picture information together with the blanking, synchronizing, and equalizing pulses. When the composite signal amplitude-modulates the rf carrier, a white picture element produces minimum modulation and a black element creates maximum modulation. This system is called negative transmission, which has the advantages of providing greater power efficiency and reducing the effects of noise.

The modulation percentages produced by the various parts of the composite signal are:

1. Synchronizing and equalizing pulses, 100%.
2. Blanking pulses, 75 ± 2.5%.
3. Black level, 70%, approximately.
4. White level, 12.5 ± 2.5%.

The high video frequencies, extending up to 4 MHz, represent the fine detail in the picture to be transmitted. If a double-sideband system were used, the total bandwidth required would be 8 MHz, which would exceed the allocated channel width of 6 MHz. To reduce the bandwidth the double-sideband system is used for video frequencies up to 750 kHz, but only the upper sidebands are transmitted for the remaining frequencies. The result is vestigial sideband transmission (C3F), which is produced by the filter system following the rf power amplifier.

The modulated signal is finally radiated as a horizontal polarized wave from a VHF/UHF antenna, generally of the turnstile variety. The use of a horizontal antenna system eliminates some of the effects of noise, which is primarily vertically polarized.

The sound, or aural, transmitter employs frequency modulation, with 100% modulation corresponding to a frequency deviation of 25 kHz. The transmitted audio range is 50 Hz to 15 kHz so

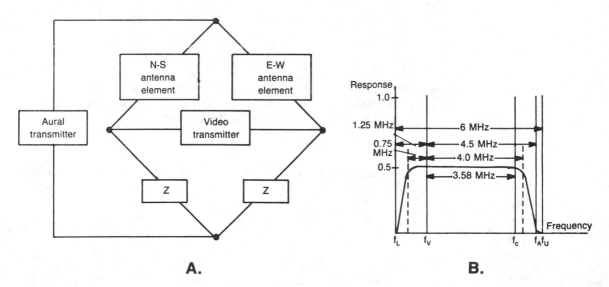

A.   **B.**

Figure 146-1

that the deviation ratio is 20 kHz/15 kHz = 1.667. Preemphasis and corresponding deemphasis are used, with a time constant of 75 microseconds. A common antenna system is used for both the AM video signal and the FM aural signal, so it is necessary to avoid interaction between the two transmitters. This is achieved by means of a diplexer (Fig. 146-1A), which is essentially a balanced bridge. The north-south and east-west antenna elements are two arms of the bridge. It is balanced by equal impedances in the other arms. The output of the video transmitter is fed to one pair of opposite corners of the bridge while the aural transmitter is connected to the other pair. With the bridge balanced, both transmitters will activate the antenna elements, but neither will interact with the other.

MATHEMATICAL DERIVATIONS

The video carrier frequency, f_V, has a tolerance of ± 1 kHz and is located 1.25 MHz above the channel's lower frequency limit, f_L, so that

$$f_V = (f_L + 1.25) \text{ MHz} \tag{146-1}$$
$$= (f_U - 4.75) \text{ MHz}$$

where f_U is the channel's upper frequency limit. These frequencies are shown in Fig. 146-1B, which represent a channel's response curve. The aural carrier frequency, f_A, also has a frequency tolerance of ± 1 kHz and is positioned 0.25 MHz below the channel's upper frequency limit. Therefore

$$f_A = (f_U - 0.25) \text{ MHz} \tag{146-2}$$
$$= (f_L + 5.75) \text{ MHz}$$
$$= (f_V + 4.5) \text{ MHz}$$

As a result of the frequency conversion process in the superheterodyne TV receiver, the video and aural intermediate frequencies are respectively 45.75 MHz and 41.25 MHz; their frequency separation is still 4.5 MHz. These values are independent of the channel selected.

In a color transmission the positioning of the video and aural carriers is the same as in the monochrome transmission, but the color information (chrominance signal) is used to amplitude modulate a chrominance subcarrier with its frequency, f_C, located 3.579545 MHZ $\pm$ 10 Hz (commonly rounded off to 3.58 MHz) above the video carrier frequency, f_v. Then

$$f_C = (f_v + 3.579545) \text{ MHz} \tag{146-3}$$
$$= (f_L + 4.829545) \text{ MHz}$$
$$= (f_U - 1.170455) \text{ MHz}$$
$$= (f_a - 0.920455) \text{ MHz}$$

Example 146-1

The width of TV channel 4 extends from 66 to 72 MHz. What are the frequencies of the video carrier, the aural carrier and the chrominance subcarrier?

Solution

Video carrier frequency,

$$f_v = f_L + 1.25 \tag{146-1}$$
$$= 66 + 1.25$$
$$= 67.25 \text{ MHz.}$$

Aural carrier frequency,

$$f_a = f_U - 0.25 \tag{146-2}$$
$$= 72 - 0.25$$
$$= 71.75 \text{ MHz.}$$

Chrominance subcarrier frequency,

$$f_C = f_L + 4.83 \tag{146-3}$$
$$= 66 + 4.83$$
$$= 70.83 \text{ MHz.}$$

Note: The TV receiver's local oscillator frequency for Channel 4 is $67.25 + 45.75 = 113$ MHz.

Chapter 147
Distributed constants
of transmission lines

IN PREVIOUS CHAPTERS WE HAVE BEEN CONCERNED WITH SO-CALLED "LUMPED" circuitry where the three electrical properties (resistance, inductance, capacitance) were related to specific components. For example, we regarded a resistor as being a "lump" of resistance and considered all connecting wires to be perfect conductors. By contrast, the resistance, R, associated with a two-wire transmission line is not concentrated into a "lump" but is "distributed" along the entire length of the line. The distributed constant of resistance is therefore measured in the basic unit of ohms per meter, rather than ohms.

Because the straight wires of the parallel line are conducting surfaces separated by an insulator or dielectric, the line will possess the distributed constants of self-inductance, L (henrys per meter) and capacitance, C (farads per meter). In addition no insulator is perfect and consequently there is another distributed constant, G (siemens per meter), which is the leakage conductance between the wires. These four distributed constants are illustrated in Fig. 147-1A; their order of values in a practical line are R mΩ/m, L μH/m, C pF/m, and G nS/m.

In physical terms the properties of resistance and leakage conductance will relate to the line's power loss in the form of heat dissipated and will therefore govern the degree of attenuation, measured in decibels per meter (dB/m). The self-inductance results in a magnetic field surrounding the wires while the capacitance means that an electric field exists between the wires (Fig. 147-1B); these L and C properties determine the line's behavior in relation to the frequency.

The two-wire ribbon line that connects an antenna to its TV receiver is often referred to as a 300 Ω line. But what is the meaning of the 300 Ω? Certainly we cannot find the value with an ohmmeter so we are led to the conclusion that it only appears under working conditions when the

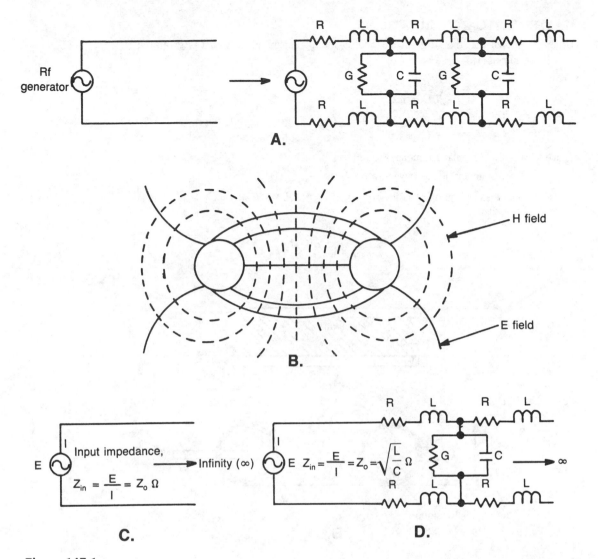

Figure 147-1

line is being used to convey rf power. In fact the 300 Ω value refers to the line's surge character-istic impedance whose letter symbol is Z_o. The surge impedance is theoretically defined as the input at the rf generator to an infinite length of the line (Fig. 147-1C and D). In the equivalent circuit of Fig. 147-1D, the C and G line constants will complete a path for current to flow so that an effective current, I, will be drawn from the rf generator whose effective output voltage is E. Then the input impedance is

$$Z_{in} = Z_o = \frac{E_{rms}}{I_{rms}} \ \Omega.$$

The mathematical derivations show that an rf line behaves resistively and has a surge impedance of $\sqrt{L/C}$ Ω; this value must also depend on the lines' physical construction.

MATHEMATICAL DERIVATIONS

It may be shown that the surge impedance, Z_o, is a complex quantity that is given by:

Surge impedance,

$$Z_o = \sqrt{\frac{R + j\omega L}{G + j\omega C}} \text{ ohms} \tag{147-1}$$

$$= \sqrt[4]{\frac{R^2 + \omega^2 L^2}{G^2 + \omega^2 C^2}} \;\; \angle \left[\frac{1}{2} \left(\text{inv tan } \frac{\omega L}{R} - \text{inv tan } \frac{\omega C}{G} \right) \right] \text{ ohms}$$

where: $\omega = 2\pi f$, angular frequency (rad/s)

f = frequency of the rf generator (Hz).

For a low-loss operating at radio frequencies (rf), $\omega L \gg R$ and $\omega C \gg G$. Then,

Surge impedance,

$$Z_o \to \sqrt{\frac{L}{C}} \; \angle 0° \; \Omega \tag{147-2}$$

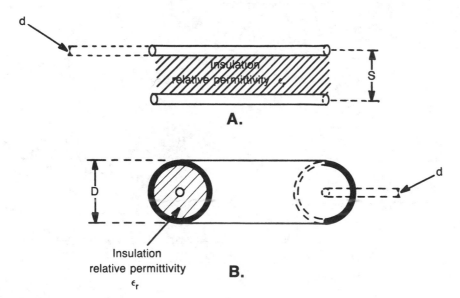

Figure 147-2

Two-wire line or twin lead (Fig. 147-2A)

Surge impedance,

$$Z_o = \frac{276}{\sqrt{\epsilon_r}} \log \left(\frac{2S}{d} \right) \text{ ohms} \tag{147-3}$$

where: S = spacing between the conductors (m)

d = diameter of each conductor (m)

ϵ_r = relative permittivity of the dielectric.

492

Coaxial cable (Fig. 147-2B)

Surge impedance,

$$Z_o = \frac{138}{\sqrt{\epsilon_r}} \log \left(\frac{D}{d}\right) \text{ ohms} \tag{147-4}$$

where: D = inner diameter of the outer conductor (m)

  d = outer diameter of the inner conductor (m)

Example 147-1

Each conductor of a twin-lead transmission line has a radius of 1.5 mm and the spacing between the centers of the conductors is 0.8 cm. If the dielectric is air, what is the value of the line's surge impedance?

Solution

Surge impedance,

$$Z_o = \frac{276}{\sqrt{\epsilon_r}} \log \left(\frac{2S}{d}\right) \tag{147-3}$$

$$= \frac{276}{\sqrt{1}} \log \left(\frac{2 \times 0.8 \times 10^{-2} \text{ m}}{2 \times 1.5 \times 10^{-3} \text{ m}}\right)$$

$$= 276 \log 5.333$$

$$\approx 200 \ \Omega.$$

Example 147-2

The outer conductor of a coaxial cable has an inner diameter of 1.75 cm while the inner conductor has an outer diameter of 2.5 mm. If teflon (relative permittivity = 2.1) is used as the insulator between the conductors, what is the value of the cable's surge impedance?

Solution

Surge impedance,

$$Z_o = \frac{138}{\sqrt{\epsilon_r}} \log \left(\frac{D}{d}\right) \tag{147-4}$$

$$= \frac{138}{\sqrt{2.1}} \log \left(\frac{1.75 \times 10^{-2} \text{ m}}{2.5 \times 10^{-3} \text{ m}}\right)$$

$$= \frac{138}{\sqrt{2.1}} \log 7$$

$$\approx 81 \ \Omega.$$

Example 147-3

The primary constants of an rf transmission line are $R = 50$ mΩ/m, $L = 1.6$ μH/m, $C = 7.5$ pF/m and $G = 10$ nS/m. If the frequency of the generator feeding the line is 300 MHz, what is the value of the line's surge impedance?

Solution

Surge impedance,

$$Z_o = \sqrt{\frac{L}{C}}$$

(147-2)

$$= \sqrt{\frac{1.6 \times 10^{-6}}{7.5 \times 10^{-12}}}$$

$$= 462 \ \Omega.$$

Chapter 148
The matched line

REFERRING TO FIG. 148-1A, LET US CONSIDER THE CONDITIONS THAT EXIST AT the position, X, Y, on the infinite line. Because there is still an infinite length to the right of X, Y, the input impedance at this position looking down the line will be equal to the value of Z_o. Consequently, if the section of the line to the right of X, Y is removed and replaced by a resistive load whose value in ohms is the same as that of the surge impedance, it will appear to the generator as if it is still connected to an infinite line and the input impedance at the generator will remain equal to Z_o.

When a line is terminated by a resistive load of value Z_o, the line is said to be "matched" to the load. Under matched conditions the line is most effective in conveying rf power from the generator (for example, a transmitter) to the load such as an antenna.

Let us examine in detail what happens on a matched line. Travelling sine waves of voltage and current start out from the rf generator and move down the line in phase. Due to the small amount of attenuation present on the line, the effective (rms) values of the voltage and current decay slightly but at all positions $V_{rms}/I_{rms} = Z_o$ (Fig. 148-1B).

On arrival at the termination, the power contained in the voltage and current waves is completely absorbed by the load. It is important to realize that power is being conveyed down the line and is *not* being dissipated and lost as heat in the surge impedance.

In previous chapters we have assumed that in a series circuit consisting of a source, a two-wire line and a load, the current was instantaneously the same throughout the circuit. This can only be regarded as true provided the distances involved in the circuit are small compared with the wavelength of the output from the source. The wavelength whose letter symbol is the Greek lambda, λ, is defined as the distance between two consecutive identical states in the path of the wave. For example, the wavelength of a wave in water is the distance between two neighboring crests (or troughs). On a matched transmission line the wavelength is the distance between two adjacent positions where identical voltage (and current) conditions occur instantaneously. Travelling or progressive waves exist on the line because time is involved in propagating rf energy from the source to the load.

In free space the velocity of an electromagnetic wave (radio wave) is a constant that is approximately equal to 3×10^8 m/s (the speed of light). However, on a transmission line the velocity is the speed at which the voltage and current waves, as well as the electric and magnetic fields, travel towards the matched load. This velocity is always less than the speed of light, c. The

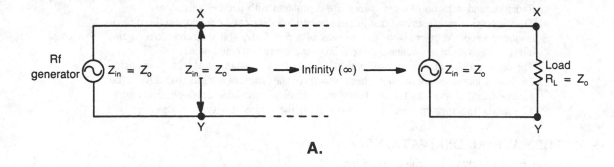

A.

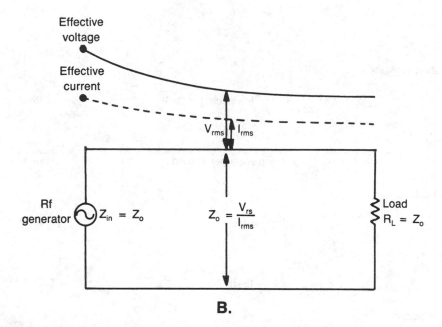

B.

Figure 148-1

ratio of the velocity on the line to the velocity in free space is called the velocity factor, δ. The value of δ varies from 0.66 for certain types of coaxial cable to 0.975 for an air insulated two-wire line. It follows that because the velocity is the product of the frequency and the wavelength, the wavelength on the line is shorter than the wavelength in free space.

The features of the matched line may be summarized as follows:

1. Travelling waves of the voltage and current move down the line in phase and their power is completely absorbed by the load.
2. The ratio of the effective voltage to the effective current is constant over the entire line and is equal to the surge impedance, Z_o.
3. The input impedance at the generator is equal to the surge impedance and is independent of the line's length.
4. The power losses on the line are subdivided into:

a. Radiation and induction losses, which are a problem with parallel wire lines.
b. The dielectric hysteresis loss that increases with frequency and depends on the type of insulator used. At microwave frequencies of a few GHz, the dielectric loss is the ultimate reason for abandoning coaxial lines and using waveguides instead.
c. The copper loss that is associated with the conductor's resistance. At high frequencies this loss is increased by the skin effect, which confines most of the electron flow to the surface (skin) of a conductor and therefore reduces the available cross-sectional area. The larger the surface area of the conductors, the less is this type of loss.

MATHEMATICAL DERIVATIONS

The power conveyed down the line is given by:

Conveyed power,

$$P = V_{rms} \times I_{rms} \tag{148-1}$$
$$= I^2_{rms} \times Z_o$$
$$= V^2_{rms}/Z_o \text{ watts}$$

Attenuation constant,

$$\alpha = 9.8 \left(\frac{R}{2Z_o} + \frac{GZ_o}{2} \right) \text{ decibels per meter} \tag{148-2}$$

Phase velocity,

$$v = \frac{\lambda}{T} = f \times \lambda \text{ meters per second} \tag{148-3}$$

Wavelength,

$$\lambda = \frac{v}{f} \text{ meters} \tag{148-4}$$

Frequency,

$$f = \frac{v}{\lambda} \text{ hertz} \tag{148-5}$$

where: T = period(s)

On a transmission line let us consider the generator's output as the reference voltage. At a position A, which is a distance of $\lambda/4$ from the generator, the instantaneous voltage will lag by 90° on this reference level while at distances of $\lambda/2$ and $3\lambda/4$, (positions B and C), the voltages will be respectively 180° out of phase and 270° lagging (90° leading). A particular phase condition will then travel down at a speed, which is called the phase velocity, v; this movement of a phase condition as illustrated in Fig. 148-2. Consequently, over a distance of 1 meter there will be an angular difference which is measured by the phase shift constant, β. The unit of β is the radian per meter and because there must be a difference of 360° (2π radians) over a distance of λ meters,

Phase shift constant,

$$\beta = \frac{2\pi}{\lambda} \text{ radians per meter} \tag{148-6}$$

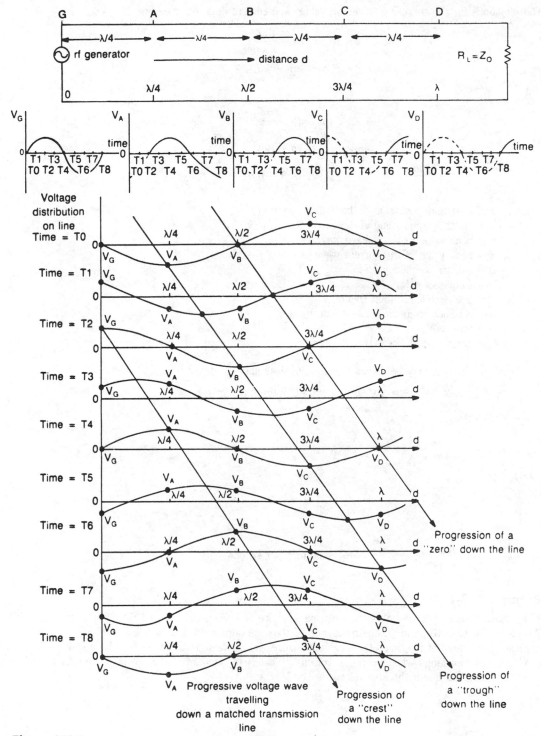

Figure 148-2

The equations for the travelling waves of voltage and current on a loss-free line are

$$v = E \sin\left(\omega t - \frac{2\pi d}{\lambda}\right)$$ (148-7)

$$= E \sin(2\pi ft - \beta d)$$

and

$$i = I \sin\left(\omega t - \frac{2\pi d}{\lambda}\right)$$ (148-8)

$$= I \sin(2\pi ft - \beta d)$$

$$= \frac{E}{Z_o} \sin(2\pi ft - \beta d)$$

where: v = instantaneous value of the voltage wave (V)
i = instantaneous value of the current wave (A)
E = peak value of the voltage wave (V)
I = peak value of the current wave (A)
ω = angular frequency (rad/s)
f = frequency (Hz)
β = phase shift constant (rad/m)
d = distance from the generator (m)
t = time (s)
Z_o = surge impedance (Ω).

Mathematically it may be shown that in a low-loss line,

$$\beta = \omega\sqrt{LC} = 2\pi f\sqrt{LC} \text{ radians per meter}$$ (148-9)

Therefore,

Phase velocity,

$$v = f \times \lambda$$ (148-10)

$$= \frac{2\pi f}{\beta}$$

$$= \frac{\omega}{\beta}$$

$$= \frac{1}{\sqrt{LC}} \text{ meters per second}$$

Example 148-1

The primary constants of an rf transmission line are $R = 50$ mΩ/m, $L = 1.6$ μH/m, $C = 7.5$ pF/m and $G = 10$ nS/m. If the frequency of the generator feeding in a line is 300 MHz, calculate the values of the line's (a) surge impedance, (b) attenuation constant, (c) phase-shift constant, (d) wavelength, and (e) phase velocity. What is the amount of the power conveyed down this matched line if the load voltage is 160 V rms?

Solution

(a) Surge impedance,

$$Z_o = \sqrt{\frac{L}{C}} = \sqrt{\frac{1.6 \times 10^{-6} \text{ H}}{7.5 \times 10^{-12} \text{ F}}} = 462 \; \Omega$$

(b) Attenuation constant,

$$\alpha = \frac{R}{2Z_o} + \frac{GZ_o}{2} \qquad (148\text{-}2)$$

$$= \frac{50 \times 10^{-2}}{2 \times 462} + \frac{10 \times 10^{-9} \times 462}{2}$$

$$= 54 \times 10^{-6} + 2.3 \times 10^{-6}$$

$$= 56.3 \times 10^{-6} \text{ nepers per meter}$$

$$= 490 \times 10^{-6} \text{ dB/m}$$

(c) Phase shift constant,

$$\beta = \omega\sqrt{LC} \qquad (148\text{-}9)$$

$$= 2 \times \pi \times 300 \times 10^{-6} \times \sqrt{1.6 \times 10^{-6} \times 7.5 \times 10^{-12}}$$

$$= 6.53 \text{ rad/m}$$

(d) Line's wavelength,

$$\lambda = \frac{2\pi}{\beta} \qquad (148\text{-}10)$$

$$= \frac{2\pi}{6.53}$$

$$= 0.96 \text{ m}$$

(e) Phase velocity,

$$v = f \times \lambda \qquad (148\text{-}3)$$

$$= 300 \times 10^6 \times 0.96$$

$$= 288 \times 10^6 \text{ m/s.}$$

Power conveyed to the load,

$$P = V^2_{rms}/Z_o \qquad (148\text{-}1)$$

$$= \frac{(160 \text{ V})^2}{462\Omega}$$

$$= 55 \text{ W}$$

Chapter 149
The unmatched line—
reflection coefficient

I F AN RF LINE IS TERMINATED BY A RESISTIVE LOAD THAT IS NOT EQUAL TO THE
surge impedance, the generator will still send voltage and current waves down the line in phase
but their power will only be partially absorbed by the load. A certain fraction of the voltage and
current waves that arrive (are incident) at the load, will be reflected back towards the generator.
At any position on the line the instantaneous voltage (or current) will be the resultant of the
incident and reflected voltage (or current) waves; these combine to produce so-called standing
waves. In the extreme cases of open and short circuited lines, no power can be absorbed by the
"load" and total reflection will occur.

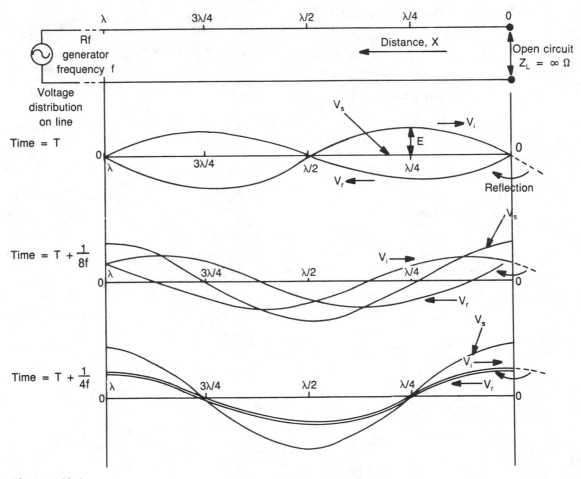

Figure 149-1

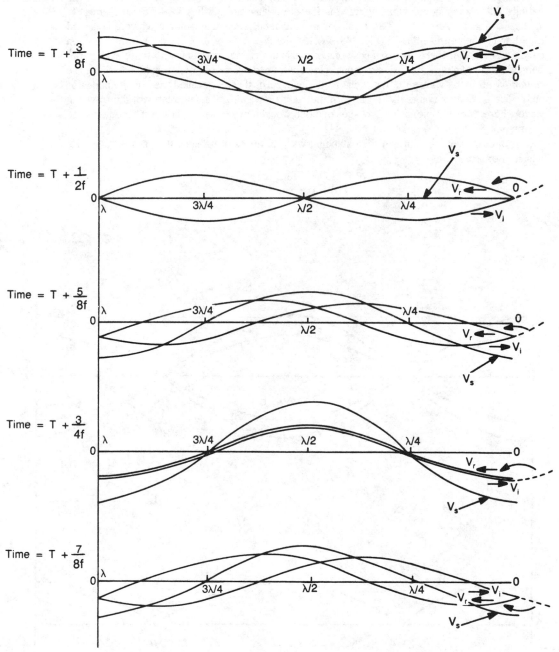

Figure 149-1 (cont.)

As the incident voltage wave moves down the line towards the open circuit, we can derive the changes in the instantaneous standing wave for each subsequent interval of one eighth of the period. The instantaneous standing wave is therefore illustrated in Fig. 149-1 for the time, T, $T + 1/8f$, $T + 1/4f$, . . . $T + 7/8f$. Finally all instantaneous voltage standing waves are collected together in one representation (Fig. 149-2A). At the distances of 0, $\lambda/2$, and λ from the open circuit termination, there are *fixed* positions of maximum voltage variation with time; such variations are called voltage antinodes or loops. However, at the distances of $\lambda/4$ and $3\lambda/4$, there are *stationary* positions where the voltage is at all times zero; these are called the voltage nodes or nulls. The stationary points that occur with the standing wave, are in contrast with the traveling wave where, for example, a zero voltage condition would move down the line at a speed equal to the phase velocity.

It is customary to represent the standing wave in terms of its effective or rms value, V_s, which is expressed by

$$2 E \cos \frac{2\pi x}{\lambda}$$

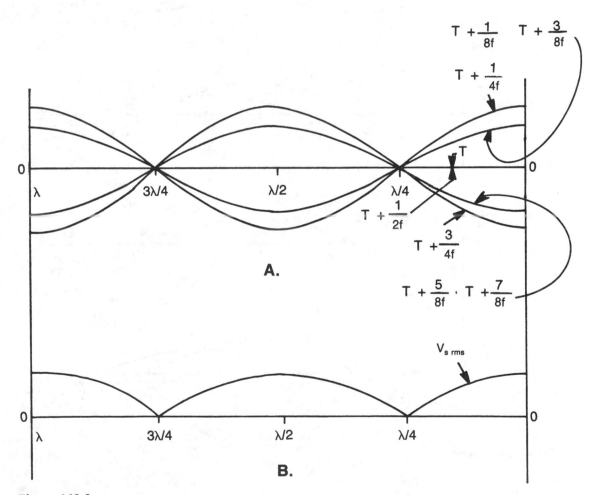

Figure 149-2

502

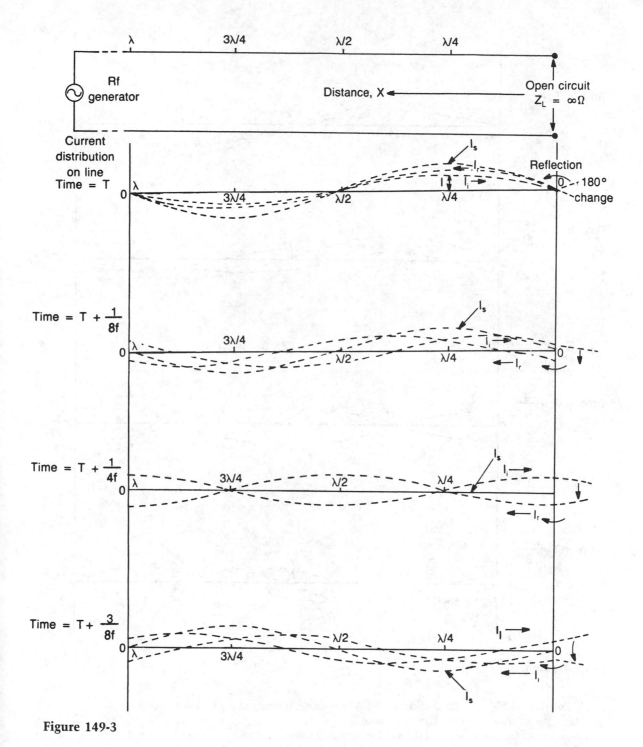

Figure 149-3

503

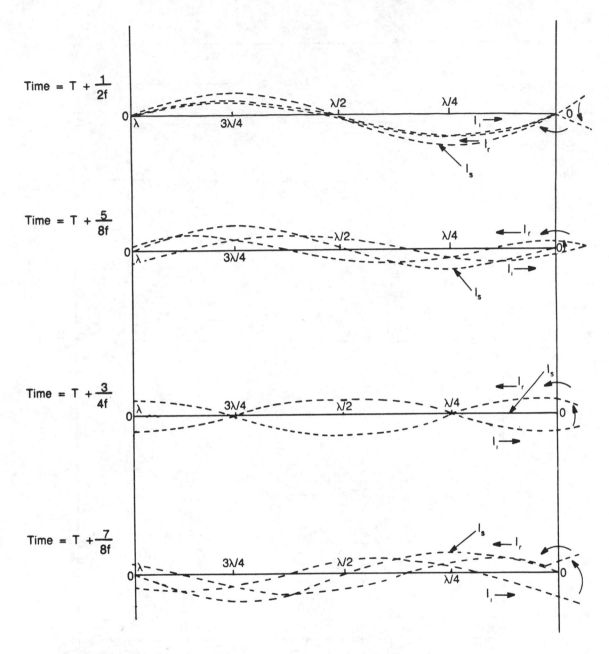

Figure 149-2 (cont.)

and is illustrated in Fig. 149-2B. Notice that we only consider the magnitude of this expression and that no positive or negative sign is involved.

Now let us examine the current (broken line) standing wave distribution on an open circuited line. At the open circuit termination the circuit must reverse its direction and therefore we have an "out of phase" reflection in which the wave is extended beyond the open circuit, then phase shifted by 180° and afterwards folded back to provide the reflected wave. Incident and reflected waves are then combined to produce the current standing wave at times, T, $T + 1/8f$, $T + 1/4f$, . . . $T + 7/8f$ (Fig. 149-3). Finally, all eight instantaneous current standing waves are collected together in one presentation (Fig. 149-4).

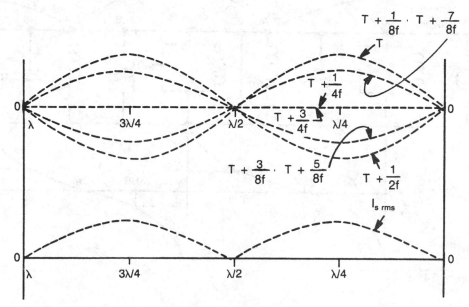

Figure 149-4

The variation of the impedance along the open-circuited line is shown in Fig. 149-5. One is struck with the similarity to the behavior of series and parallel LC circuits. Remember that the impedance response of the series circuit changes from capacitive through resistive (low value) to inductive. By contrast the parallel circuit has an impedance response that varies from inductive through resistive (high value) to capacitive. These equivalent LC circuits have been included in Fig. 149-5. Because the line exhibits resonant properties it may be referred to as *tuned;* by contrast the matched line is called "flat, untuned or nonresonant".

We will now examine the effect of terminating a loss-less line with a resistive load that is not equal to the surge impedance. The amount of reflection will be reduced so that the nodes and antinodes will be less pronounced. Neglecting any losses on the line, the effective voltage and current distribution for the three possible cases of a resistive load, namely $R_L > Z_o$, $R_L = Z_o$ and $R_L < Z_o$, are shown in Fig. 149-6.

MATHEMATICAL DERIVATIONS

It is important to realize that a standing wave represents a sinusoidal distribution both with time and distance. The incident traveling wave can be represented by

$$v_i = E \sin \left(2\pi f t + \frac{2\pi x}{\lambda} \right) \text{ volts} \tag{149-1}$$

where: v_i = instantaneous value of the incident voltage wave (V)
E = peak value of the incident voltage wave (V)

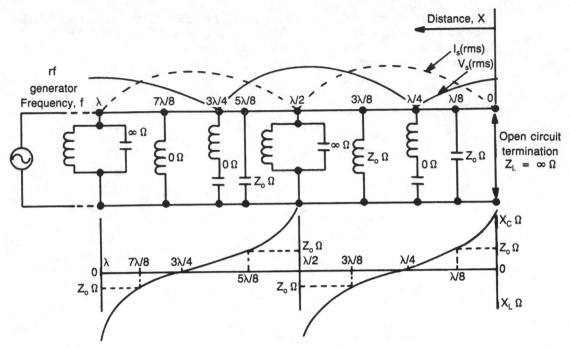

Figure 149-5

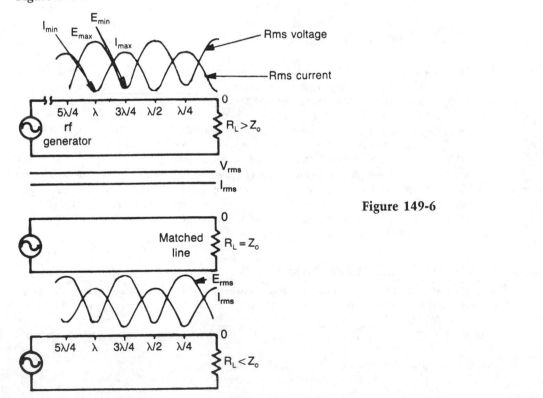

Figure 149-6

f = frequency (Hz)

x = distance measured from the open circuit termination (m)

λ = wavelength existing on line (m)

The equation of the reflected wave is

$$v_r = E \sin \left(2\pi ft - \frac{2\pi x}{\lambda} \right) \text{ volts} \qquad (149\text{-}2)$$

where: v_r = instantaneous value of the reflected voltage wave (V)

The instantaneous standing wave voltage, v_s, is

$$v_s = v_i + v_r = E \left[\sin \left(2\pi ft + \frac{2\pi x}{\lambda} \right) + \sin \left(2\pi ft - \frac{2\pi x}{\lambda} \right) \right] \qquad (149\text{-}3)$$

$$= 2 E \sin 2\pi ft \cos \frac{2\pi x}{\lambda} \text{ volts}$$

$\left(\begin{array}{c} \text{Sinusoidal distribution} \\ \text{with time} \end{array} \right)$ $\left(\begin{array}{c} \text{Sinusoidal distribution} \\ \text{with distance} \end{array} \right)$

The incident current wave that is in phase with the incident voltage, is presented by:

$$i_i = I \sin \left(2\pi ft + \frac{2\pi x}{\lambda} \right)$$

$$= \frac{E}{Z_o} \sin \left(2\pi ft + \frac{2\pi x}{\lambda} \right) \text{ amperes} \qquad (149\text{-}4)$$

where: i_i = instantaneous value of the incident current wave (A)

I = peak value of the incident current wave (A)

The equation of the reflected current wave is:

$$i_r = -I \sin \left(2\pi ft - \frac{2\pi x}{\lambda} \right) \qquad (149\text{-}5)$$

$$= -\frac{E}{Z_o} \sin \left(2\pi ft - \frac{2\pi x}{\lambda} \right) \text{ amperes}$$

where the minus sign preceding I indicates the "out of phase" reflection, and i_r = instantaneous value of the reflected current wave (A).

The instantaneous standing wave current, i_s, is

$$i_s = i_i + i_r \qquad (149\text{-}6)$$

$$= I \left[\sin \left(2\pi ft + \frac{2\pi x}{\lambda} \right) - \sin \left(2\pi ft - \frac{2\pi x}{\lambda} \right) \right]$$

$$= \frac{2E}{Z_o} \cos 2\pi ft \sin \frac{2\pi x}{\lambda} \text{ amperes}$$

Notice that, because of the "sin $2\pi ft$" and "cos $2\pi ft$" terms, v_s and i_s are 90° out of phase in terms of time. The rms value of the current standing wave, I_s, has a magnitude of

$$\frac{2E}{Z_o} \sin \frac{2\pi x}{\lambda},$$

which is illustrated in Fig. 149-4. Current nulls exist at distances of zero, $\lambda/2$ and λ from the open circuit termination, while current antinodes occur at distances of $\lambda/4$ and $3\lambda/4$.

The magnitude of the impedance, Z, at any position is given by,

$$Z = \frac{v_s}{i_s} \tag{149-7}$$

$$= \frac{2E \cos \dfrac{2\pi x}{\lambda}}{\dfrac{2E}{Z_o} \sin \dfrac{2\pi x}{\lambda}}$$

$$= Z_o \cot \frac{2\pi x}{\lambda} \ \text{ohms}$$

Because v_s and i_s are 90° out of phase, the nature of this impedance must be reactive. But at what position on the line is the impedance capacitive and at what positions is it inductive? Let us start by considering the impedance at the position which is at the distance of $\lambda/8$ from the open circuit termination (Fig. 149-7). From this position the values of the instantaneous voltage and current standing waves are plotted versus a time scale that contains T, $T + 1/4f$, $T + 1/2f$, $T + 3/4f$, $T + 1/f$. It is clear that i_s leads v_s by 90° so that the impedance is capacitive. Physically, over the last eighth of a wavelength the voltage distribution, starting from a maximum at the open circuited end, is greater than the current distribution. The electric field associated with the distributed capacitance will dominate the magnetic field produced by the distributed inductance. If $x = \lambda/8$,

$$\cot \frac{2\pi x}{\lambda} = \cot \frac{\pi}{4} = \cot 45° = 1$$

and therefore the input impedance to a $\lambda/8$ line is a capacitive reactance whose value is equal to Z_o. A similar analysis for the open circuited $3\lambda/8$ line will show that i_s lags v_s and that the input impedance is an inductive reactance of value Z_o. At the $\lambda/4$ position there is a voltage null and a current antinode. Theoretically, the impedance is zero but on a practical line the impedance would be equivalent to a low value of resistance. The conditions are reversed for the $\lambda/2$ position where there is a voltage null and a current antinode. The impedance is theoretically infinite but in practice is equivalent to a high value of resistance (Fig. 149-5).

The fraction of the incident voltage and current reflected at the load is called the reflection coefficient whose letter symbol is the Greek, rho, ρ. If $R_L > Z_o$, the effective standing wave voltage at the load is

$$V_L = V_i + V_r = V_i + \rho V_i = V_i(1 + \rho) \ \text{volts} \tag{149-8}$$

where: V_i = effective value of the incident voltage wave.
  V_r = effective value of the reflected voltage wave.

The effective standing wave current through the load is

$$I_L = I_i + I_r = I_i - \rho I_i = I_i(1 - \rho) \ \text{amperes} \tag{149-9}$$

The negative sign indicates the "out of phase" reflection of the current wave compared with the "in phase" reflection of the voltage wave.

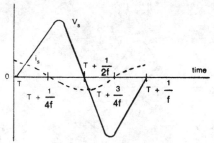

Distance of λ/16 from open circuited end. Impedance is a high value of capacitive reactance.

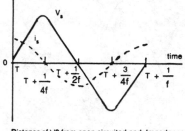

Distance of λ/8 from open circuited end. Impedance is a capacitive reactance whose value is equal to the lines surge impedance, Z_o.

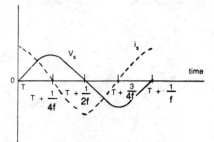

Distance of 3λ/16 from open circuited end. Impedance is a low value of capacitive reactance.

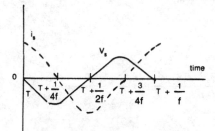

Distance of 5λ/16 from open circuited end. Impedance is a low value of inductive reactance.

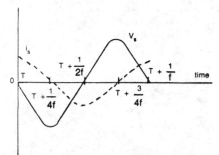

Distance of 3λ/8 from open circuited end. Impedance is an inductive reactance whose value is equal to the line's surge impedance, Z_o.

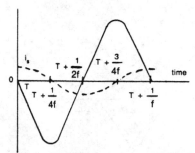

Distance of 7λ/16 from open circuited end. Impedance is a high value of inductive reactance.

Figure 149-7

Then

$$R_L = \frac{V_L}{I_L} = \frac{V_i(1 + \rho)}{I_i(1 - \rho)} \tag{149-10}$$

$$= \frac{Z_o(1 + \rho)}{1 - \rho} \text{ ohms}$$

This yields,

$$\rho = \frac{R_L - Z_o}{R_L + Z_o} \tag{149-11}$$

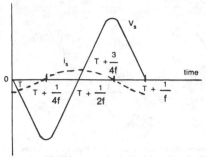

Distance of 9λ/16 from open circuited end. Impedance is a high value of capacitive reactance.

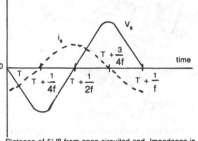

Distance of 5λ/8 from open circuited end. Impedance is a capacitive reactance whose value is equal to the line's surge impedance, Z_o.

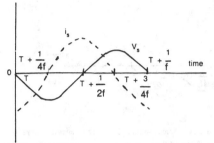

Distance of 11λ/16 from open circuited end. Impedance is a low value of capacitive reactance.

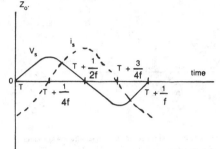

Distance of 13λ/16 from open circuited end. Impedance is a low value of inductive reactance.

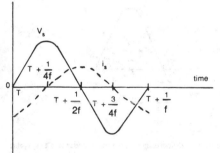

Distance of 7λ/8 from open circuited end. Impedance is an inductive reactance equal in value to the line's surge impedance, Z_o.

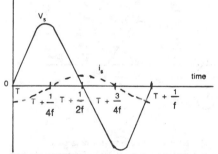

Distance of 15λ/16 from open circuited end. Impedance is a high value of inductive reactance.

if: $R_L = \infty$ (open circuited line), $\rho = +1$
$R_L = Z_o$ (matched line), $\rho = 0$
$R_L = 0$ (short circuited line), $\rho = -1 = 1\angle 180°$

The negative sign indicates that when $R_L < Z_o$, the current wave is reflected "in phase" but the voltage reflection is "out of phase." These results are only true for purely resistive loads.

If the load is a general impedance possessing both resistance and reactance, ρ is not just a number but is a complex quantity with a magnitude P (capital rho), which ranges in value between 0 and 1, and a phase angle, θ, whose value may lie between $+180°$ and $-180°$.

Because the reflected voltage is P times the incident voltage and the reflected current is P times the incident current, it follows that the reflected power, P_r, is P^2 times the incident power, P_i.

Then the power absorbed by the load, P_L is

$$P_L = P_i - P_r = (1 - P^2) P_i \qquad (149\text{-}12)$$

$$= \frac{1 - P^2}{P^2} P_r$$

$$= \frac{4S}{(1 + S)^2} P_i \text{ watts}$$

where: S = the voltage standing wave ratio (see Chapter 150).

Example 149-1

A transmission line with an air dielectric and a surge impedance of 50 Ω is terminated by a load consisting of 40 Ω resistance in series with 65 Ω capacitive reactance. Calculate the value of the reflection coefficient. When an rf generator delivers 120 W to the line which is assumed to be lossless, what are the amounts of the reflected power and the power absorbed by the load? If the wavelength on the line is 3 m, what is the distance between the adjacent voltage nulls?

Solution

Load impedance,

$$Z_L = 40 - j65 \ \Omega.$$

Reflection coefficient,

$$\rho = \frac{Z_L - Z_o}{Z_L + Z_o} \qquad (149\text{-}11)$$

$$= \frac{(40 - j65) - 50}{50 + (40 - j65)}$$

$$= \frac{-10 - j65}{90 - j65}$$

$$= \frac{65.76\angle -98.75°}{111.01\angle -35.84°}$$

$$= 0.6\angle -63°$$

Magnitude of reflection coefficient, P = 0.6
Reflected power,

$$P_r = P^2 \times P_i$$

$$= (0.6)^2 \times 120$$

$$= 43.2 \text{ W.}$$

Power absorbed by the load,

$$P_L = P_i - P_r$$

$$= 120 \times 43.2$$

$$= 76.8 \text{ W.}$$

Distance between two adjacent voltage nulls = $\lambda/2$
$$= 3/2$$
$$= 1.5 \text{ m}$$

Chapter 150
Voltage standing-wave ratio

REFERRING TO FIG. 150-1, THE EFFECTIVE VALUE OF A VOLTAGE ANTINODE IS E_{max} while the effective value of an adjacent voltage node is E_{min}. The degree of standing waves is measured by the voltage standing wave ratio (VSWR) whose letter symbol is S. The VSWR is defined as the ratio $E_{max}:E_{min}$, which is also equal in magnitude to $I_{max}:I_{min}$. E_{max} is the result of an in-phase condition between incident and reflected voltages while E_{min} is the result of a 180° out-of-phase situation.

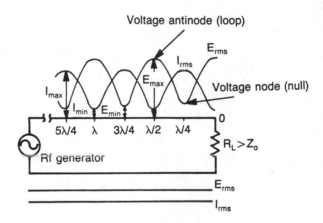

Figure 150-1

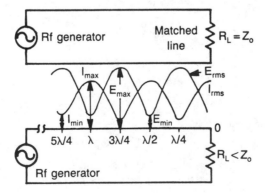

The presence of standing waves on a practical transmission line has the following disadvantages:

1. The incident power reaching the termination is not fully absorbed by the load. The difference between the incident power and the load power is the power reflected back towards the generator.
2. The voltage antinodes may break down the insulation (dielectric) between the conductors and cause arc-over.
3. Because power losses are proportional to the square of the voltage and the square of the current, the attenuation of the practical line increases due to the presence of the standing waves.
4. The input impedance at the generator is a totally unknown quantity that varies with the length of the line and the frequency of the generator.

MATHEMATICAL DERIVATIONS

Maximum effective voltage,

$$E_{max} = V_i + V_r = V_i(1 + P) \text{ volts} \tag{150-1}$$

Minimum effective voltage,

$$E_{min} = V_i - V_r = V_i(1 - P) \text{ volts} \tag{150-2}$$

Then
   VSWR,

$$S = \frac{E_{max}}{E_{min}} = \frac{1 + P}{1 - P} \tag{150-3}$$

If $P = 0$ (matched line), $S = 1$
If $P = 1$ (open or short-circuited line), $S = \infty$

The VSWR is just a number whose value will range from 1 to ∞. On a practical matched system a value of S, which is less than 1.2, is normally regarded as acceptable. It is sometimes preferable to measure the VSWR in decibels, in which case VSWR = 20 log S dB.

Equation 150-3 yields

$$P = \frac{S - 1}{S + 1} \tag{150-4}$$

If a resistive load, R_L, is greater than Z_o the magnitude of the reflection coefficient is given by:

$$P = \frac{R_L - Z_o}{R_L + Z_o} \tag{150-5}$$

Combining Equations 150-4 and 150-5,

$$S = \frac{1 + P}{1 - P} = \frac{R_L}{Z_o} \tag{150-6}$$

If R_L is less than Z_o,

$$P = \frac{Z_o - R_L}{Z_0 + R_L}$$

and

$$S = \frac{Z_o}{R_L} \qquad (150\text{-}7)$$

At the position where the voltage antinodes and the current nodes coincide, the impedance of the line is a maximum and is resistive.

Maximum impedance,

$$Z_{max} = \frac{E_{max}}{I_{min}} = \frac{E_i(1 + P)}{I_i(1 - P)} \qquad (150\text{-}8)$$

$$= Z_o\frac{(1 + P)}{(1 - P)} = SZ_0 \text{ ohms}$$

At the voltage node position, there is a minimum resistive impedance which is given by:

$$Z_{min} = \frac{E_{min}}{I_{max}} = \frac{E_i(1 - P)}{I_i(1 + P)} = Z_o \times \frac{(1 - P)}{(1 + P)} = \frac{Z_o}{S} \text{ ohms.} \qquad (150\text{-}9)$$

Example 150-1

An rf generator delivers 100 W to a 50 Ω loss-free line. If the line is terminated by a 40 Ω resistive load, what are the values of the standing wave ratio, the reflection coefficient, the reflected power and the load power?

Solution

Voltage standing wave ratio,

$$S = \frac{Z_o}{R_L} \qquad (150\text{-}6)$$

$$= \frac{50}{40}$$

$$= 1.25$$

Reflection coefficient,

$$P = \frac{S - 1}{S + 1} \qquad (150\text{-}4)$$

$$= \frac{1.25 - 1}{1.25 + 1}$$

$$= 0.11$$

Reflected power,

$$P_r = P^2 \times P_i$$

$$= (0.11)^2 \times 100$$

$$= 1.2 \text{ W}$$

Load power,

$$P_L = P_i - P_r$$
$$= 100 - 1.2$$
$$= 98.8 \text{ W}$$

Chapter 151
The quarter-wave line

WHEN A QUARTER-WAVE LINE IS SHORTENED AT ONE END AND IS EXCITED TO resonance by the source frequency applied at the other end, standing waves of voltage and current appear on the line. At the short circuit the voltage is zero and the current is a maximum. At the input end the current is nearly zero and the voltage is at its peak. The input impedance to the resonant quarter-wave line is extremely high and therefore this section behaves as an insulator. Figure 151-1A shows a quarter-wave section of line acting as a standoff insulator (support stub) for a two-wire transmission line. At the particular frequency that makes the section a quarter-wavelength line, the stub acts as a highly efficient insulator. However, at higher frequencies the stub is no longer resonant and will behave as a capacitor; such behavior causes a mismatch and standing waves will appear on the two-wire line. At the second harmonic frequency the stub is $\lambda/2$ long and because there is no impedance transformation over the distance of one-half wavelength, the stub will place a short circuit across the main line. No power at the second harmonic frequency can then be transferred down the line. In this way a section of line which is $\lambda/4$ long at a fundamental frequency and is shorted at one end, will have the effect of filtering out all even harmonics.

The quarter-wave section may also be used to match two nonresonant lines with different surge impedances (see mathematical derivations).

MATHEMATICAL DERIVATIONS

When a $\lambda/4$ line of surge impedance, Z_o, is terminated by a load, Z_L, the input impedance, Z_{in}, is given by

$$Z_{in} = \frac{Z_o{}^2}{Z_L} \text{ ohms} \tag{151-1}$$

This yields,

$$Z_o = \sqrt{Z_{in} \times Z_L} \text{ ohms} \tag{151-2}$$

This formula can be used to derive some results that we have already obtained in previous chapters. For example,

1. If $Z_L = \infty \ \Omega$ (an open circuit), $Z_{in} = 0 \ \Omega$ (a short circuit).
2. If $Z_L = Z_0 \ \Omega$ (matched conditions), $Z_{in} = Z_o \ \Omega$.
3. If $Z_L = 0 \ \Omega$ (a short circuit), $Z_{in} = \infty \ \Omega$ (an open circuit).

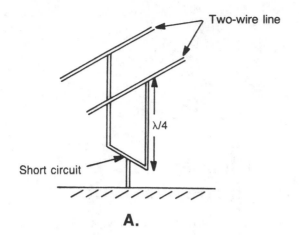

Two-wire line

λ/4

Short circuit

A.

Figure 151-1

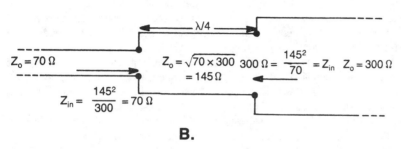

$Z_o = 70 \, \Omega$

$\xrightarrow{\quad\quad\quad} $

$Z_{in} = \dfrac{145^2}{300} = 70 \, \Omega$

λ/4

$Z_o = \sqrt{70 \times 300}$
$= 145 \, \Omega$

$300 \, \Omega = \dfrac{145^2}{70} = Z_{in}$ $Z_o = 300 \, \Omega$

B.

Example 151-1

In Fig. 151-1B, what is the required value for the surge impedance of the λ/4 matching section?

Solution

It is required to match 70 Ω and 300 Ω nonresonant lines. The λ/4 matching section has a surge impedance given by

Surge impedance,

$$Z_o = \sqrt{Z_{in} \times Z_L} \qquad (151\text{-}1)$$
$$= \sqrt{70 \times 300}$$
$$= 145 \, \Omega.$$

The effective load on the 300 Ω line as seen from the generator is $145^2/70 = 300$ Ω, while the 70 Ω line is effectively connected to a $145^2/300 = 70$ Ω termination. Both nonresonant lines are therefore matched and the standing waves only exist on the λ/4 section.

Chapter 152
The Hertz antenna

A N ANTENNA IS DEFINED AS AN EFFICIENT RADIATOR OF ELECTROMAGNETIC energy (radio waves) into free space. The same principles apply to both transmitting and receiving antennas although the rf power levels for the two antennas are completely different. The purpose of a transmitting antenna is to radiate as much rf power as possible either in all directions (omnidirectional antenna) or in a specified direction (directional antenna). By contrast the receiving antenna is used to intercept an rf signal voltage that is sufficiently large compared to the noise existing within the receiver's bandwidth.

Referring to transmission lines, the main disadvantage of the twin-lead is its radiation loss. However, provided the separation between the leads is short compared with the wavelength, the

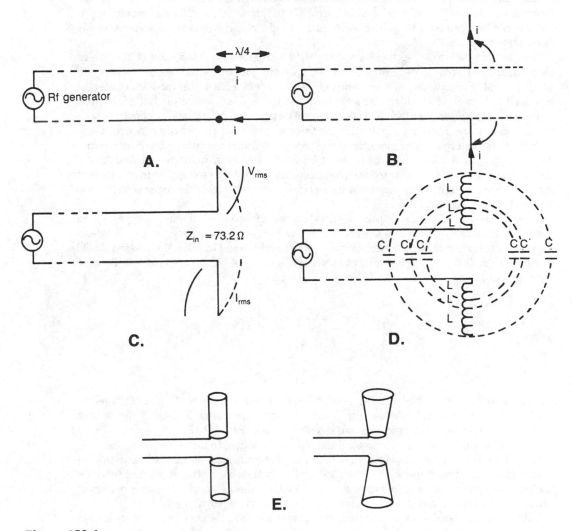

Figure 152-1

radiation loss is small because the fields associated with the two conductors will tend to cancel out. Referring to Fig. 152-1A, the equal currents, i, which exist over the last resonant quarter wavelength of an open circuited transmission line, will intantaneously flow in opposite directions so that their resultant magnetic field is weak. However, if each λ/4 conductor is twisted back through 90° (Fig. 152-1B) the currents, i, are now instantaneously in the same direction so that the surrounding magnetic field is strong.

In addition to the currents, the conductors carry standing-wave voltage distributions with their associated electric fields. The standing-wave voltage and current distributions over the complete half wavelength are shown in Fig. 152-1C. At the center feedpoint the effective voltage is at its minimum level while the effective current value has its maximum value. The effective voltage distribution is drawn on opposite sides of the two sections to indicate that these sections instantaneously carry opposite polarities. In other words, when a particular point in the top section carries a positive voltage with respect to ground, the corresponding point in the bottom section has a negative voltage. The distributed inductance and capacitance (Fig. 152-1D) together form the equivalent of a series resonant LC circuit. Notice, however, that the distributed capacitance exists between the two quarter-wave sections and that the ground is not involved in this distribution.

By bending the two λ/4 sections outward by 90° we have formed the half-wave (λ/2) dipole or Hertz (Heinrich Hertz, 1857–1894) antenna. This antenna will be resonant at the frequency to which it is cut. For example, at a frequency of 100 MHz, which lies within the FM commercial broadcast band of 88 to 108 MHz, the length required for the Hertz antenna is 1.43 m. Such an antenna would be made from two thin conducting rods, each 0.716 m long, and positioned remote from ground. At the frequency of 100 MHz, the antenna behaves as a series resonant circuit with a Q of approximately 10. Therefore the thin dipole is capable of operating effectively within a narrow range that is centered on the resonant frequency. However, we must remember that if the operating frequency is below the resonant frequency, the antenna will appear to be too short and will behave capacitively. Likewise at frequencies above resonance, the antenna will be too long and will be inductive.

If it is required to operate a dipole over a wide range of frequencies, it is necessary to "broad band" the antenna by lowering its Q without changing its resonant frequency. As an example, if we want to operate a dipole satisfactorily over the range of 125 to 175 MHz, the antenna should be cut to the midfrequency of 150 MHz and should possess a Q of $150/(175 - 125) = 3$. Because the resonant frequency of a series LC circuit is given by

$$\frac{1}{2\pi\sqrt{LC}}$$

while:

$$Q = \frac{1}{R}\sqrt{\frac{L}{C}}$$

we must lower the distributed inductance, L, and increase the capacitance, C. The solution is to shorten the antenna while at the same time increasing its surface area. This gives rise to such "broad band" shapes as the cylindrical and biconical dipoles (Fig. 152-1E).

When the Hertz dipole is resonant and the rf power is applied to the center of the antenna, the input impedance at the feedpoint is a low resistance that mathematically may be shown to have a value of 73.2 Ω (for this reason we often speak of a "70 Ω dipole"). This is referred to as the "radiation resistance" of the dipole and is the ohmic load that the half-wave antenna represents at resonance. To achieve a matched condition the dipole should be fed with a 70 Ω line.

As we move from the center of the antenna towards the ends, the impedance increases from approximately 35 Ω (balanced either side with respect to ground) to about 2500 Ω (not infinity

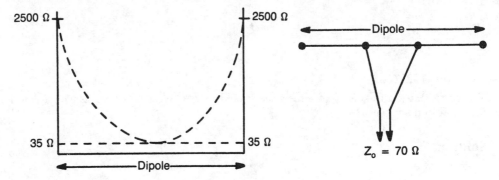

Figure 152-2

because of the "end" capacitance effects). It is therefore possible to select points on the antenna where the impedance can be matched by a gradual taper to a line whose Z_o is not 70 Ω. Such an arrangement is known as a delta feed (Fig. 152-2).

MATHEMATICAL DERIVATIONS

$$\text{Electrical wavelength} = \frac{300}{f} \text{ meters} = \frac{984}{f} \text{ feet} \qquad (152\text{-}1)$$

This leads to:

$$\text{Electrical half wavelength} = \frac{150}{f} \text{ meters} = \frac{468}{f} \text{ feet} \qquad (152\text{-}2)$$

where: f = frequency (MHz).

The voltage and current waves on an antenna travel at a speed that is typically 5% slower than the velocity of light. Therefore the *physical* half wavelength to which the Hertz antenna should be cut, is shorter than the electrical half wavelength.

$$\text{Physical half wavelength} \qquad (152\text{-}3)$$

$$l = \frac{143}{f(\text{MHz})} \text{ meters} = \frac{468}{f(\text{MHz})} \text{ feet}$$

The rf power, P, at the feedpoint is given by

$$P = I_A{}^2 \times R_A \text{ watts} \qquad (152\text{-}4)$$

where: I_A = effective rf current at the center of the antenna (A).

R_A = radiation resistance (Ω).

Example 152-1

What is the physical length of a dipole that is resonant at a frequency of 3 GHz in the microwave region?

Solution

Physical half wavelength,

$$l = \frac{143}{3000 \text{ MHz}} \text{ m} \qquad (152\text{-}3)$$
$$= 4.67 \text{ cm.}$$

Example 152-2

The current at the frequency of a resonant dipole is 3.2 A. What is the amount of rf power delivered to the antenna?

Solution

Antenna power,

$$P = I_A{}^2 \times R_A \qquad (152\text{-}4)$$
$$= (3.2 \text{ A})^2 \times 73.2 \ \Omega$$
$$= 750 \text{ W.}$$

Chapter 153
The Marconi antenna

AT AN OPERATING FREQUENCY OF 2 MHZ THE REQUIRED PHYSICAL LENGTH FOR a resonant hertz dipole is 143.1/2 = 71.6 m. It is difficult to position a vertical antenna of this size so that it is remote from the ground. However, the dipole could be in the form of a long wire antenna, slung horizontally between two towers from which the antenna is insulated. Clearly the vertical Hertz antenna is not a practical proposition at low frequencies and is replaced by the λ/4 unipole or Marconi antenna. The Marconi type may be regarded as a Hertz dipole in which the lower half is replaced by a nonradiating ground image antenna (Fig. 153-1A). What we are saying is that, unlike the Hertz antenna, ground is an integral part of the Marconi antenna system. This also means that the Hertz dipole is balanced with both λ/4 sections mounted remote from ground while the Marconi unipole is unbalanced because part of the antenna is ground itself. In the same way the twin lead with neither conductor grounded is a balanced transmission line, while a coaxial cable is unbalanced with its outer conductor grounded at intervals along its length.

The distributed inductance of the Marconi antenna is associated with the vertical λ/4 rod while the distributed capacitance exists between the rod and ground (Fig. 153-1B). It is therefore impossible to operate with a horizontal Marconi antenna; this compares with the Hertz dipole, which may be mounted either vertically or horizontally. Consequently, all Marconi antennas radiate only vertically polarized waves.

The effective voltage and current distribution on the resonant Marconi antenna is comparable with the distribution on the upper half of the Hertz dipole (Fig. 153-1C). If the λ/4 antenna is end fed by an unbalanced coaxial cable, the radiation resistance is 73.2/2 = 36.6 Ω, which is the effective resistive load of a resonant Marconi antenna. If the operating frequency is below the resonant value, the unipole is too short and behaves capacitively. However, the antenna may be tuned to resonance by adding an inductor in series (Fig. 153-1D). When operating above the

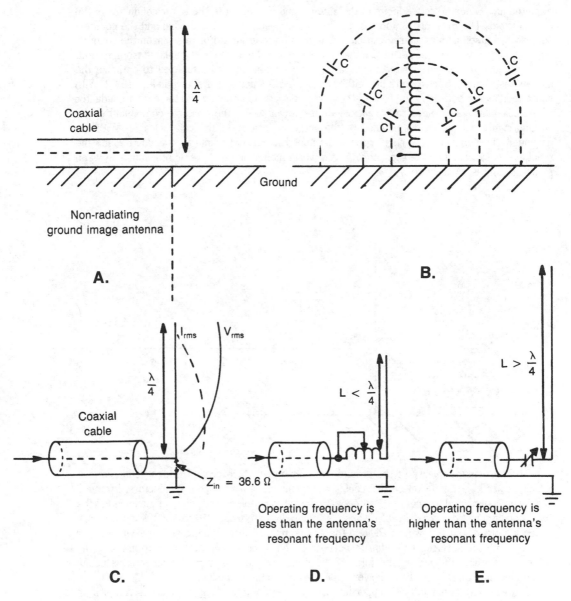

Figure 153-1

resonant frequency the antenna is too long, behaves inductively and may be tuned to resonance with the aid of a series capacitor (Fig. 153-1E).

One type of practical Marconi antenna is the vertical whip (a rod that is flexible to a limited extent). As an example, an end-fed whip antenna is commonly mounted on the top of an automobile so that the roof may act as the required ground plane. Should the antenna be positioned in the vicinity of the rear bumper, most of the ground plane will be provided by the road surface. If it is required to operate a whip antenna at the top of a building, it is necessary to provide the antenna with an apparent ground or counterpoise. This normally consists of a wire structure mounted just beneath the feedpoint of the antenna and connected to the outer conductor of the coaxial cable. The main distribution capacitance then exists between the whip and the counterpoise, which itself acts as a large capacitance in relation to ground. For this reason the counterpoise should be normally larger in size than the antenna and must be well insulated from ground.

Another type of vertical radiator is a steel tower which may be tapered to optimize the current distribution. Such towers may either be end-fed or shunt-fed as shown in Fig. 153-2. With shunt-feeding the bottom of the tower is connected directly to ground while the center conductor of the coaxial cable is joined to a point, P, on the tower where the resistive component of the antenna's impedance can be matched to the surge impedance of the coaxial cable. With the grounded end behaving as a short circuit, the impedance at the point, P, is inductive and the reactance is cancelled by the capacitor, C. It is worth mentioning that the dc resistance between P and ground is virtually zero.

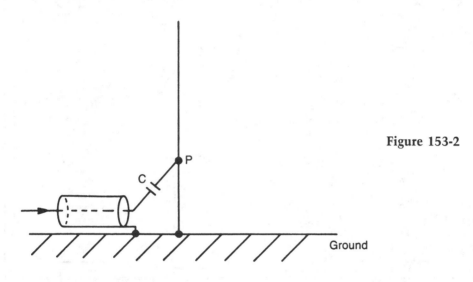

Figure 153-2

To improve the quality of the ground and therefore reduce ground losses, it is common practice to add a ground system that consists of a number of bare copper conductors, arranged radially and connected to a center point beneath the antenna; the center point is then joined to the outer conductor of the coaxial cable. These ground radials are from $\lambda/10$ to $\lambda/2$ in length and are buried a short distance down; in addition the ground may be further improved by laying copper mats beneath the surface. The ultimate purpose of the radial ground system is to improve the antenna efficiency, defined as the ratio of the rf power radiated from the antenna as useful electromagnetic energy to the rf power applied to the antenna feedpoint.

A tower may either be self-supporting or supported by guy wires. Such wires can pick up some of the live transmitted radiation; re-radiation from the guy wires will then occur and this will interfere with the EM wave transmitted from the antenna. To reduce the re-radiation effect to a minimum, an insulator is connected to each end of the guy wire and then intermediate insulators are included with a spacing of about a tenth of the wavelength apart. These insulators are normally of the porcelain "egg" type so that if one of the insulators breaks, the guy wire still provides mechanical support.

If at low frequencies, the vertical antenna is too short for mechanical reasons, the amount of useful vertically polarized radiation can be increased by using "top loading", examples of which are shown in Fig. 153-3. Because the current is zero at the open end of the antenna, the inverted "L"

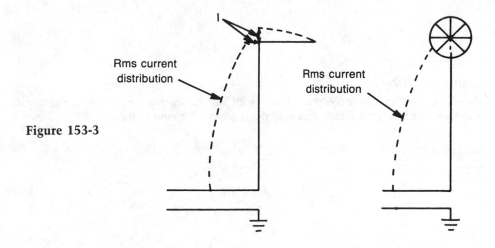

Figure 153-3

arrangement will have the result of increasing the effective current distributed over the antenna's vertical section. More practical to construct is the metallic spoked wheel that will increase the amount of the distributed capacitance to ground and will therefore lower the antenna's resonant frequency to the required value.

MATHEMATICAL DERIVATIONS

Electrical quarter wavelength,

$$\lambda/4 = \frac{75.2}{f} \text{ meters} = \frac{246}{f} \text{ feet} \tag{153-1}$$

Physical length of the Marconi antenna

$$l = \frac{71.6}{f} \text{ meters} = \frac{234}{f} \text{ feet} \tag{153-2}$$

where: f = frequency (MHz).

$$\text{Radiation resistance of the Marconi antenna} = 36.6 \text{ ohms} \tag{153-3}$$

Power delivered to end-fed resonant Marconi antenna,

$$P_A = I_A^2 R_A \text{ watts} \tag{153-4}$$

where: I_A = effective current at the antenna's feedpoint (A)

R_A = radiation resistance (Ω).

Example 153-1

What is the physical length of a Marconi antenna that is resonant at 7.5 MHz?

Solution

$$\text{Physical length} = \frac{71.6}{f} \tag{153-2}$$

$$= \frac{71.6}{7.5}$$
$$= 9.54 \text{ m.}$$

Example 153-2

The effective current at the feedpoint of a resonant Marconi antenna is 8.7 A. Assuming that the antenna is end-fed, what is the amount of the rf power delivered to the antenna?

Solution

Antenna power,

$$P_A = I_A{}^2 \times R_A \text{ W} \tag{153-4}$$
$$= (8.7 \ A)^2 \times 36.6 \ \Omega$$
$$= 2770 \text{ W.}$$

Chapter 154
Microwave antenna
with the parabolic reflector

$\mathbf{A}$T MICROWAVE FREQUENCIES IT IS POSSIBLE TO DESIGN HIGHLY DIRECTIVE antenna systems of reasonable size. One of the most common systems employs a paraboloid as a "dish" reflector with a flared waveguide termination placed at the focus (Fig. 154-1A). If the paraboloid were infinitely large, the result of the reflection would be to produce a unidirectional beam with no spreading. However, with a paraboloid of practical dimensions there is some degree of spreading and the radiation pattern then contains a major lobe (and a number of minor lobes) as illustrated in Fig. 154-1B. The full lobe is, of course, three dimensional so that we are only showing the pattern in one plane.

Referring to the narrow major lobe produced by the antenna with the paraboloid reflector, the beamwidth in the horizontal plane is measured by the angle $P_1 OP_2$ where P_1 and P_2 are the half-power points. In radar systems typical beamwidths are of the order of 1° to 5°. Due to the construction of the paraboloid the horizontal and vertical beamwidths are not necessarily the same. For a microwave link the normal beamwidth is 2°; with such narrow beamwidths the power gains of the antenna systems are extremely high.

For a paraboloid reflector to be effective, the value of d must be greater than ten wavelengths. Consequently, such reflectors would not be practical in the VHF band due to their excessive size.

MATHEMATICAL DERIVATIONS

The beamwidth is given by
Beamwidth angle,

$$\theta = \frac{70\lambda}{d} \text{ degrees} \tag{154-1}$$

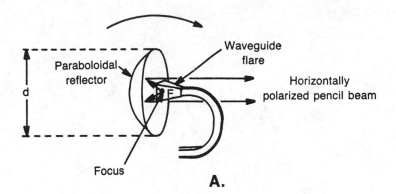

Figure 154-1

A.

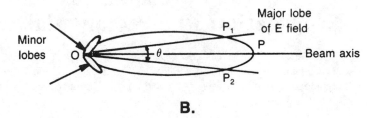

B.

With respect to the half-wave dipole standard, the power gain, G_p is given by

$$G_p = 6\left(\frac{d}{\lambda}\right)^2 \qquad (154\text{-}2)$$

where: λ = wavelength (m)

d = diameter of paraboloid (m).

Example 154-1

An X-band microwave antenna system is operating at 10 GHz. If the parabolic reflector has a diameter of 1 m, what are the values of the beamwidth and the antenna power gain?

Solution

Wavelength,

$$\lambda = \frac{30}{10 \text{ GHz}}$$

$$= 3 \text{ cm}$$

Beamwidth angle,

$$\theta = \frac{70\lambda}{d} \qquad (154\text{-}1)$$

$$= \frac{70 \times 0.03 \text{ m}}{1 \text{ m}}$$

$$= 2.1°$$

Power gain,

$$G_p = 6\left(\frac{d}{\lambda}\right)^2 \tag{154-2}$$

$$= 6 \times \left(\frac{1\ m}{0.03\ m}\right)^2$$

$$= 6667.$$

Chapter 155
Propagation in a rectangular waveguide

OUR FIRST QUESTION IS: "DOES THE ENERGY MOVE STRAIGHT DOWN A rectangular waveguide as a TEM wave?" The answer is "no." If we enclosed a rectangular waveguide around a TEM wave, we could not obey the necessary boundary conditions. These conditions are:

1. There can be no E-field component that is parallel to and existing at an inner surface of the waveguide.
2. There can be no H-field component that is perpendicular to an inner surface of a waveguide.

In fact the electromagnetic energy progresses down the guide by a series of reflections off the internal surface of the narrow dimension (Fig. 155-1A). At each reflection the angles (θ) of incidence and reflection are equal.

Let us consider a TEM wave that approaches a plane (flat metal surface) at an angle θ (Fig. 155-1B). The dark lines with their arrows represent the directions of the incident and reflected wavefronts, which are moving with the free space velocity (virtually at the speed of light). The full lines are $\lambda/2$ apart in free space and represent the incident H-flux lines. The broken lines are also $\lambda/2$ apart and are used to show the reflected H-field. The symbols for the incident and reflected E-fields are respectively "$\otimes$, $\odot$" and "$\times$, $\cdot$." These incident and reflected waves are of the same amplitude but have a phase reversal of $180°$.

The fields of the incident and reflected waves are superimposed and must be combined to produce the resultant E- and H-field patterns. At the metal surface itself, the phase reversal causes cancellation between the incident and the reflected E-fields so that there are no resultant lines that are parallel to the surface.

When the incident and the reflected waves are combined, the resulting pattern of the E- and H-fields between the lines XX' and YY' is exactly the same as we can derive from the support stub analogy. We can therefore infer that the EM wave must progress down the guide by a series of reflections off the narrow dimension.

So far, so good. Now we must find out the factors that determine the angle of incidence and the velocity with which the energy progresses down the guide. To do this we extract the triangle LMN from Fig. 155-1B and display its magnified form in Fig. 155-1C. As the wavefront moves from P to N (a distance of $\lambda/2$), the field pattern progresses a greater distance from M to N. The distance MN is a half wavelength of the field pattern as it exists inside the waveguide and is

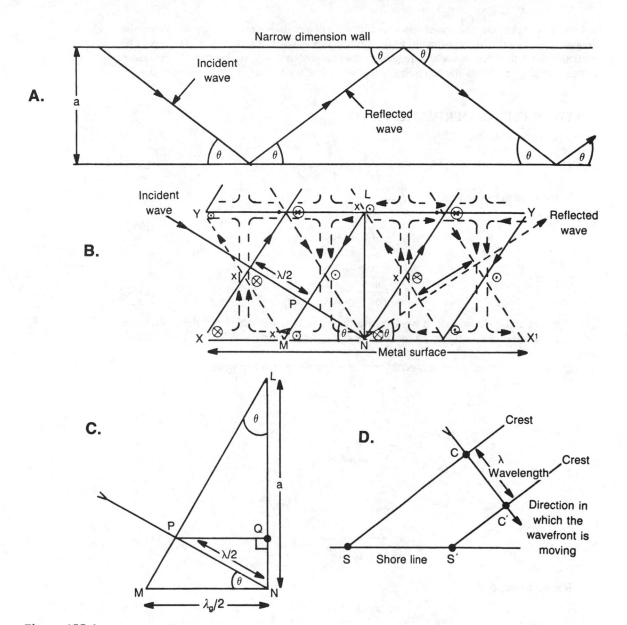

Figure 155-1

termed $\lambda_g/2$ where λ_g is the guide wavelength. For an analogy, think of sea waves approaching the shore line at an angle (Fig. 155-1D). As the wavefront moves from one crest to the next the distance CC' (one wavelength), the pattern of the crests at the shore line covers the greater distance SS'.

For a given rectangular waveguide, the wide "a" and the narrow "b" dimensions are fixed. If the frequency is lowered, the wavelength is longer and the value of λ is increased. In the limiting cut-off condition, the cutoff wavelength equals $2a$ and $\theta = 90°$; the wave will then bounce back and forth between the narrow dimensions and will not progress down the guide. When the frequency is raised, the wavelength decreases and the value of θ is lowered. This could be continued indefinitely; however at a frequency of approximately $1.9\, f_c$, the narrow "b" dimension comes into play and limits the top frequency at which the waveguide can successfully operate.

MATHEMATICAL DERIVATIONS

In the right-angled triangle LPN,

$$\sin \theta = \frac{PN}{LN} = \frac{\lambda/2}{a} = \frac{\lambda}{2a} \qquad (155\text{-}1)$$

In the right-angled triangle MPN,

$$\cos \theta = \frac{PN}{MN} = \frac{\lambda/2}{\lambda_g/2} = \frac{\lambda}{\lambda_g} \qquad (155\text{-}2)$$

Therefore

$$\tan \theta = \frac{\sin \theta}{\cos \theta} = \frac{\lambda/2a}{\lambda/\lambda_g} = \frac{\lambda_g}{2a} \qquad (155\text{-}3)$$

Equation 155-3 shows that, if we decrease the frequency, θ is raised and the guide wavelength, λ_g, is increased. In the cutoff condition, $\theta = 90°$ and the guide wavelength, λ_g, is infinitely long.

Using the trigonometrical relationship $\sin^2\theta + \cos^2\theta = 1$, we can write

$$\left(\frac{\lambda}{2a}\right)^2 + \left(\frac{\lambda}{\lambda_g}\right)^2 = 1$$

This yields,

$$\lambda = \frac{\lambda_g}{\sqrt{1 + \left(\frac{\lambda_g}{2a}\right)^2}} \qquad (155\text{-}4)$$

and,

$$\lambda_g = \frac{\lambda}{\sqrt{1 - \left(\frac{\lambda}{2a}\right)^2}} \qquad (155\text{-}5)$$

Remembering that,

$$\lambda = c/f \text{ and } \lambda_c = 2a = c/f_c, \qquad (155\text{-}6)$$

$$\lambda_g = \frac{\lambda}{\sqrt{1 - \left(\frac{f_c}{f}\right)^2}}$$

where λ_c and f_c are respectively the cutoff wavelength and the cutoff frequency.

In the time that the wavefront moves from P to N at the speed of light c, the field pattern progresses from M to N (a distance of $\lambda_g/2$) with the *phase* velocity, v_ϕ. During the same time interval the electromagnetic energy has moved down the guide a distance PQ. This physical movement of energy takes place at the *group* velocity, v_g.

Because

$$MN = PN \sec \theta \text{ and } PQ = PN \cos \theta,$$

Phase velocity,

$$v_\phi = \frac{c}{\cos \theta} = c \sec \theta \qquad (155\text{-}7)$$

and

Group velocity,

$$v_g = c \cos \theta \qquad (155\text{-}8)$$

Therefore,

$$v_\phi \times v_g = \frac{c}{\cos \theta} \times c \cos \theta$$

$$= c^2$$

where v_ϕ, v_g, and c are all measured in meters per second.

At the cutoff condition the group velocity is zero and the phase velocity is infinite. This result does not contravene any physical laws because the phase velocity involves the *apparent* movement of a field pattern and not the movement of any physical quantity. As the frequency is raised, v_ϕ decreases and v_g increases and both these velocities approach the velocity of light, c. For an X-band waveguide typical graphs of v_g and v_ϕ versus frequency are shown in Fig. 155-2.

We are left to consider the waveguide's phase-shift constant, β. A phase shift of 2π radians occurs over a distance equal to the guide wavelength, λ_g. Therefore

Phase shift constant,

$$\beta = \frac{2\pi}{\lambda_g} \qquad (155\text{-}9)$$

where: β = phase shift constant (rad/m)

   λ_g = guide wavelength (m).

Example 155-1

A practical rectangular waveguide is designed to operate over the X-band, 8.2 to 12.4 GHz. The inner guide dimensions are 2.286 cm and 1.143 cm. If the transmitted frequency is 10 GHz, find the values of the free space wavelength, the cutoff wavelength, the cutoff frequency, the angle of incidence, the guide wavelength, the phase velocity, the group velocity, and the phase-shift constant.

Solution

Free space wavelength,

$$\lambda = \frac{c}{f}$$

$$= \frac{3 \times 10^8 \text{ m/s}}{10 \times 10^9 \text{ Hz}}$$

$$= 0.03 \text{ m}$$

$$= 3 \text{ cm.}$$

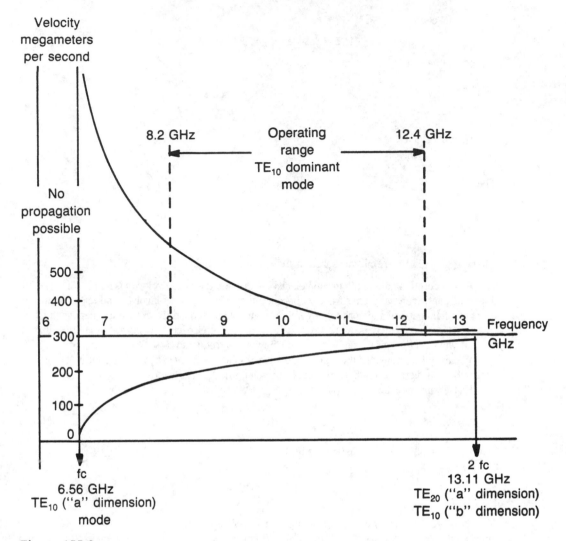

Figure 155-2

Cutoff wavelength,

$$\lambda_c = 2a = 2 \times 2.286$$
$$= 4.572 \text{ cm}$$

Cutoff frequency,

$$f_c = \frac{c}{\lambda_c}$$
$$= \frac{3 \times 10^8 \text{ m/s}}{4.572 \times 10^{-2} \text{ m}} \text{ Hz}$$
$$= 6.562 \text{ GHz}$$

Angle of incidence,

$$\theta = \text{inv sin } \frac{\lambda}{2a} \tag{155-1}$$

$$= \text{inv sin } \frac{3}{4.572}$$

$$= 41.01°$$

Guide wavelength,

$$\lambda_g = \frac{\lambda}{\cos \theta} \tag{155-2}$$

$$= \frac{3}{\cos 41.01°}$$

$$= 3.98 \text{ cm}$$

Phase velocity,

$$v_\phi = \frac{c}{\cos \theta} \tag{155-7}$$

$$= \frac{300}{\cos 41.01°}$$

$$= 397.6 \text{ megameters per second.}$$

Group velocity,

$$v_g = c \cos \theta \tag{155-8}$$

$$= 300 \cos 41.01°$$

$$= 226.4 \text{ megameters per second.}$$

Phase shift constant,

$$\beta = \frac{2\pi}{\lambda_g} \tag{155-9}$$

$$= \frac{2\pi}{3.98 \times 10^{-2} \text{ m}}$$

$$= 159 \text{ radians per meter.}$$

Chapter 156
Parameters of a pulsed radar set

THE SENSITIVITY OR THE MAXIMUM DETECTION RANGE OF A PULSED RADAR SET is proportional to the following expression:

$$\sqrt[4]{\frac{P_t \times G_t \times A_r \times \delta}{P_r}}$$

where: P_t = transmitter peak power output (W)

G_t = gain or directivity of the transmitting antenna.

A_r = effective area of the receiving antenna.

δ = a factor governed by the reflecting properties of the target, namely size, conductivity and inclination.

P_r = the minimum detectable received signal power which must exceed the receiver noise level.

Therefore the maximum range of the radar set may be increased by:

1. Raising the transmitter power output. Bearing in mind that the maximum range is directly proportional to the fourth root of the transmitter power, it would be necessary to increase the transmitted power by sixteen times in order to double the range.
2. Reducing the beam width.
3. Reducing the receiver noise, primarily by the design of the first stage in the receiver.

The main parameters associated with a pulsed radar set (Fig. 156-1) are:

Pulse Shape. Pulses of various shapes may be used in radar systems but for practical reasons, the rectangular shape is usually chosen.

Pulse Duration (Width or Length). This is the duration of the rectangular pulse and is normally measured in microseconds. For most radar sets the pulse duration lies between 0.1 microsecond and 10 microseconds. The pulse width is a factor in determining the minimum range obtainable because the echo pulse and the transmitted pulse must not overlap.

Increasing the pulse duration will increase the energy content of the pulse and will increase

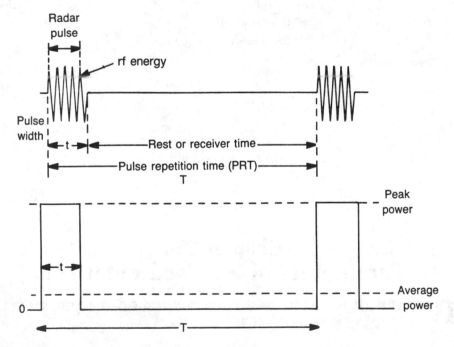

Figure 156-1

the maximum range. However, those targets whose difference in range corresponds to the pulse duration, cannot be distinguished on the display; for example, a pulse duration of 0.1 microsecond will cover approximately 160 yards of range.

Pulse Repetition Time (or Period). This is the time interval between the leading edges of successive rectangular pulses. It is a factor in determining the maximum range obtainable because the echoes from a particular pulse must be received before the next pulse is transmitted. However, for range reliability, the pulse repetition time must be kept sufficiently short to allow a number of pulses to be received from a particular target.

Pulse Repetition Frequency (or Rate). This parameter is also known as the pulse frequency and is the reciprocal of the pulse repetition time. The values of the pulse frequency generally lie between 250 Hz and 5000 Hz. The higher the pulse frequency, the greater is the intensity of the echoes on the display. The result will be an improvement in the target definition.

Peak Power. The peak power is the average power transmitted during the duration of the pulse. This may vary from several kilowatts to a few megawatts. It has been quoted that the maximum range available is directly proportional to the fourth root of the peak power so that to triple the range, would require the transmitter power to be increased by eighty-one times.

Average Power. This is the mean power over the pulse repetition period. It follows that the ratio of the average power to the peak power is the same as the ratio of the pulse duration to the pulse repetition time.

Both of these ratios are equal to the duty cycle, which has no units. The value of the duty cycle typically ranges from 0.01 to 0.0001.

Power Gain of the Antenna. The maximum range is proportional to the square root of the antenna's power gain. Increasing the range by raising the power gain reduces the beamwidth and improves the bearing resolution.

Antenna Rotation Rate. This rate is limited by the number of received pulses required for each target. The higher the antenna rotation rate, the less will be the intensity of the echoes on the display.

Radio Frequency. The efficiency of surface detection increases with the value of the radio frequency. Moreover, as the radio frequency is increased, the sizes of the antenna and the waveguide elements decrease. The parabolic reflectors used with the centimeter wavelengths can provide very narrow pencil beams that are especially suitable for accurate bearing measurements. The beamwidth is proportional to the ratio of the wavelength to the diameter of the parabolic reflector. It is defined as the angle between those two directions in which the received signal power falls to half of the maximum value, which occurs along the axis of the main beam.

The radio frequencies allocated to marine radar sets are in the gigahertz range, which corresponds to wavelengths which are of the order of centimeters.

MATHEMATICAL DERIVATIONS

$$\frac{\text{Average power}}{\text{Peak power}} = \frac{\text{pulse duration}}{\text{pulse repetition time}} = \text{duty cycle} \qquad (156\text{-}1)$$

$$\text{Beamwidth} = \frac{70\lambda}{d} \text{ degrees} \qquad (156\text{-}2)$$

where: λ = wavelength (m)

d = diameter of paraboloid reflector (m)

Frequency,

$$f = \frac{30}{\lambda(\text{cm})} \text{ gigahertz} \qquad (156\text{-}3)$$

Wavelength,

$$\lambda = \frac{30}{f(\text{GHz})} \text{ centimeters} \qquad (156\text{-}4)$$

$$\text{Pulse duration} = \frac{\text{average power} \times \text{pulse repetition time}}{\text{peak power}} \qquad (156\text{-}5)$$

$$= \frac{\text{average power}}{\text{peak power} \times \text{pulse frequency}}$$

$$= \text{duty cycle} \times \text{pulse repetition time}$$

$$= \frac{\text{duty cycle}}{\text{pulse frequency}} \text{ seconds}$$

$$\text{Pulse repetition time} = \frac{\text{pulse duration} \times \text{peak power}}{\text{average power}} \qquad (156\text{-}6)$$

$$= \frac{\text{pulse duration}}{\text{duty cycle}} \text{ seconds}$$

$$\text{Pulse repetition frequency} = \frac{\text{average power}}{\text{pulse duration} \times \text{peak power}} \qquad (156\text{-}7)$$

$$= \frac{\text{duty cycle}}{\text{pulse duration}} \text{ hertz}$$

$$\text{Average power} = \frac{\text{peak power} \times \text{pulse duration}}{\text{pulse repetition time}} \qquad (156\text{-}8)$$

$$= \text{peak power} \times \text{pulse duration} \times \text{pulse frequency}$$

$$= \text{peak power} \times \text{duty cycle watts}$$

$$\text{Peak power} = \frac{\text{average power} \times \text{pulse repetition time}}{\text{pulse duration}} \qquad (156\text{-}9)$$

$$= \frac{\text{average power}}{\text{pulse duration} \times \text{pulse frequency}}$$

$$= \frac{\text{average power}}{\text{duty cycle}} \text{ watts}$$

$$\text{Duty cycle} = \text{pulse duration} \times \text{pulse frequency} \qquad (156\text{-}10)$$

$$= \frac{\text{average power}}{\text{peak power}}$$

Example 156-1

A radar set has a peak power of 0.5 MW and a pulse duration of 2 μs. If the pulse frequency is 500 Hz, what are the values of the average power and the duty cycle?

Solution

$$\text{Average power} = \text{peak power} \times \text{pulse duration} \times \text{pulse frequency} \qquad (156\text{-}8)$$

$$= 0.5 \times 10^6 \times 2 \times 10^{-6} \times 500$$

$$= 500 \text{ W.}$$

$$\text{Duty cycle} = \text{pulse duration} \times \text{pulse frequency} \qquad (156\text{-}10)$$

$$= 2 \times 10^{-6} \times 500 = 1 \times 10^{-3} = 0.001$$

Note that the duty cycle is also given by:

$$\text{Duty cycle} = \frac{\text{average power}}{\text{peak power}} \qquad (156\text{-}1)$$

$$= \frac{500}{0.5 \times 10^6}$$

$$= 0.001$$

Example 156-2

A radar set has a peak power output of 750 kW and a duty cycle of 0.0005. What is the average power?

Solution

$$\text{Average power} = \text{peak power} \times \text{duty cycle} \qquad (156\text{-}8)$$

$$= 750 \times 10^3 \times 0.0005$$

$$= 375 \text{ W.}$$

Example 156-3

The wavelength of a radar transmission in free space is 9.6 centimeters. What is the corresponding radio frequency?

Solution

$$\text{Radio frequency in GHz} = \frac{30}{\text{wavelength in centimeters}} \qquad (156\text{-}3)$$

$$= \frac{30}{9.6}$$

$$= 3.125 \text{ GHz}$$

Example 156-4

The peak power of a radar pulse is 1 MW and its average power is 500 W. If the pulse duration is 2 μs, what are the values of the pulse repetition time and the pulse frequency?

Solution

$$\text{Pulse repetition time} = \frac{\text{pulse duration} \times \text{peak power}}{\text{average power}} \qquad (156\text{-}6)$$

$$= \frac{2 \times 10^{-6} \times 10^6}{500}$$

$$= \frac{1}{250} \text{ s}$$

$$= 4000 \ \mu\text{s.}$$

$$\text{Pulse frequency} = \frac{1}{\text{pulse repetition time}}$$

$$= \frac{1}{4000 \times 10^{-6}}$$

$$= 250 \text{ Hz.}$$

Example 156-5

The length (duration) of a radar pulse is 5 μs, the peak power is 800 kW and the average power is 2000 W. What is the value of the pulse frequency?

Solution

$$\text{Pulse frequency} = \frac{\text{average power}}{\text{pulse duration} \times \text{peak power}} \qquad (156\text{-}7)$$

$$= \frac{2000}{5 \times 10^{-6} \times 800 \times 10^{3}}$$

$$= 500 \text{ Hz.}$$

Example 156-6

The pulse repetition time of a radar set is 2 ms and the duty cycle is 5×10^{-4}. What is the pulse duration?

Solution

$$\text{Pulse duration} = \text{duty cycle} \times \text{pulse repetition time} \qquad (156\text{-}5)$$

$$= 5 \times 10^{-4} \times 2 \times 10^{-3} \text{ s}$$

$$= 1 \ \mu\text{s.}$$

Example 156-7

The average power of a radar pulse is 1 kW, the pulse width is 2 μs and the pulse frequency is 400 Hz. What is the pulse's peak power?

Solution

$$\text{Peak power} = \frac{\text{average power}}{\text{pulse duration} \times \text{pulse frequency}} \qquad (156\text{-}9)$$

$$= \frac{1000}{2 \times 10^{-6} \times 400} \text{ W}$$

$$= 1.25 \text{ MW.}$$

Example 156-8

The peak power of a pulse radar set is increased by 50%. Assuming that all other parameters remain the same, what will be the increase in the maximum range?

Solution

The maximum range is proportional to the fourth root of the transmitter peak power output. The maximum range is therefore multiplied by a factor of $\sqrt[4]{1.5} = 1.1068$. Since $\sqrt[4]{1.5}$ is approximately 1.11, the range is increased by 11%.

Chapter 157
Radar pulse duration and the discharge line

THE PULSE DURATION (LENGTH, WIDTH) IS DETERMINED IN THE MODULATOR unit (Fig. 157-1), which commonly uses an artificial discharge line consisting of a number of L,C sections. The total discharge line is then equal to the duration of the negative pulse applied to the cathode of the magnetron.

MATHEMATICAL DERIVATIONS

$$\text{Pulse duration} = 2\,N \times \sqrt{LC} \text{ seconds} \qquad (157\text{-}1)$$

where: N = number of L,C sections.

L = inductance of each section (H).

C = capacitance of each section (F).

Example 157-1

A modulator unit has an artificial discharge line of six sections, each of which contains an inductance of 6.5 μH and a capacitance of 4000 pF. What is the value of the pulse duration?

Solution

$$\text{Pulse duration, } t = 2\,N \times \sqrt{LC} \text{ seconds} \qquad (157\text{-}1)$$
$$= 2 \times 6 \times \sqrt{6.5 \times 10^{-6} \times 4000 \times 10^{-12}} \text{ s}$$
$$= 1.9 \ \mu\text{s}.$$

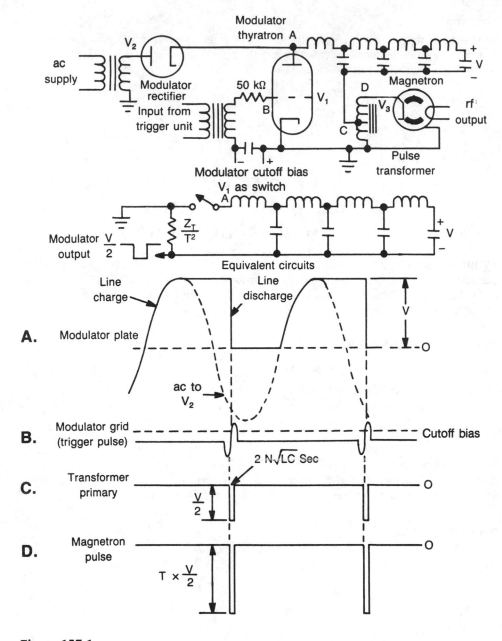

Figure 157-1

Chapter 158
Bandwidth of the radar receiver—intermediate frequency and video amplifier stages

To ACHIEVE ACCURATE RANGING, IT IS NECESSARY TO PRESERVE THE SHAPE OF the rectangular pulse and, in particular, its leading edge. This requires a wide bandwidth for the radar receiver's intermediate amplifier stages in order to pass the numerous harmonics associated with a short duration pulse (Chapter 100).

Because a wide bandwidth is required, the gain of each i-f amplifier stage is low so that the number of such stages in a radar receiver may be five or more. (Fig. 158-1.)

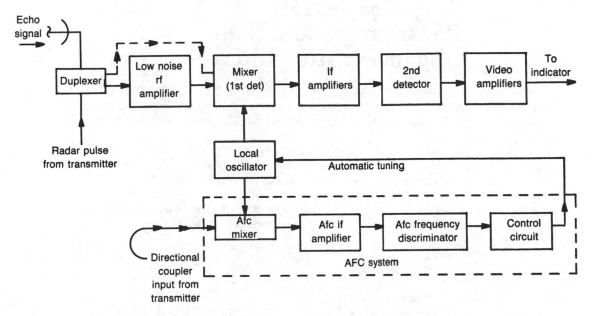

Figure 158-1

MATHEMATICAL DERIVATIONS

$$\text{Bandwidth} = \frac{2}{\text{pulse duration (microseconds)}} \text{ MHz} \qquad (158\text{-}1)$$

$$\text{Pulse duration} = \frac{2}{\text{bandwidth (MHz)}} \text{ microseconds} \qquad (158\text{-}2)$$

Example 158-1

A marine radar set has a pulse duration of 0.5 μs. What is the required bandwidth for the receiver's i-f stages?

Solution

$$\text{Required bandwidth} = \frac{2}{\text{pulse duration } (\mu s)} \qquad (158\text{-}1)$$

$$= \frac{2}{0.5}$$

$$= 4 \text{ MHz}.$$

Chapter 159
Radar ranges and their corresponding time intervals

A RADAR PULSE IS AN ELECTROMAGNETIC WAVE THAT TRAVELS WITH THE velocity of light in air. At standard temperature and pressure this velocity is 2.997×10^8 meters per second or 161700 nautical miles per second where 1 nautical mile ≈ 6080 feet ≈ 2027 yards.

The time taken for a radar pulse to travel one nautical mile is 6.184 microseconds. Therefore, the total time interval for a target range of 1 nautical mile is the total time for the pulse to reach the target and the echo to return to the receiver; this total time will be $2 \times 6.184 = 12.368$ microseconds (Fig. 159-1).

The distance traveled by a radar pulse in 1 microsecond = 0.161711 nautical miles, which is approximately 328 yards. Therefore, a pulse duration of 1 microsecond will cover a range interval of 328/2 = 164 yards.

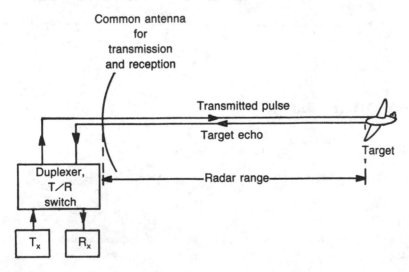

Figure 159-1

MATHEMATICAL DERIVATIONS

$$\text{Radar range in nautical miles} = \frac{\text{total round-trip time interval } (\mu s)}{12.37} \qquad (159\text{-}1)$$

Total time interval for the pulse (μs) = 12.37 × radar range in nautical miles (159-2)
to travel to the target and back.

$$\frac{\text{Distance to the target}}{\text{in nautical miles}} = \frac{\text{time for transmitted pulse to reach the target } (\mu s)}{6.18} \qquad (159\text{-}3)$$

Time taken by the transmitted pulse to reach the target (μs) (159-4)
=6.18 × (distance to the target in nautical miles)

Total distance travelled by the pulse to the target (159-5)
and back to the receiver = 2 × radar range

Example 159-1

The range of a target is 8.7 nautical miles. What is the time in microseconds for the transmitted pulse to reach the target?

Solution

The time taken for the transmitted pulse to reach the target is 6.18 × 8.7 = 53.8 microseconds.

Example 159-2

A radar pulse travels to the target and back to the receiver in a total time interval of 83 microseconds. What is the range of the target in nautical miles?

Solution

Range of the target = 83/12.37 = 6.7 nautical miles.

Chapter 160
Frequency modulated and continuous wave radar systems

APART FROM THE USE OF SHORT DURATION RF PULSES, THERE ARE TWO other possible radar systems. In FM radar the rf wave is modulated by a sawtooth. The difference between the instantaneous frequencies being transmitted and received will then be a direct measure of the reflecting object's range. The most common use of this system is in the radio altimeter where the reflecting object is the earth as seen from the aircraft. The instantaneous frequency will be an accurate measurement of the vertical distance between the ground and the aircraft.

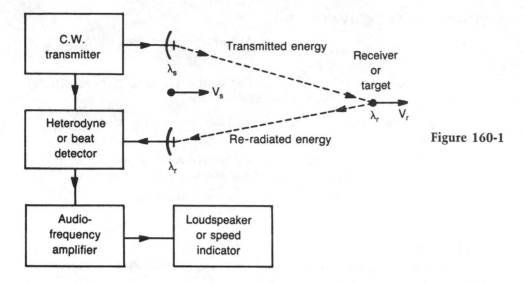

Figure 160-1

In a CW radar system the velocity of a target but not its range may be determined. The principle is based on the Doppler effect in which the frequency of the signal radiated from the reflecting object is shifted when the object is moving relative to the receiver. An example of such a system is a police radar speed-trap (Fig. 160-1).

MATHEMATICAL DERIVATIONS

Let the source be moving with a component of velocity $+v_s$ in the direction of propagation; the result will be to produce a change in the wavelength and hence a shift in the frequency at the receiver. If λ_s is the wavelength and t is the period ($= 1/f_s$) of the radiation from the source when stationary:

New wavelength,

$$\lambda_r = \lambda_s - v_s \times t \text{ meters} \tag{160-1}$$

Velocity of propagation $= c = f_s\lambda_s$

$$= f_r(\lambda_s - v_s \times t) \text{ meters per second} \tag{160-2}$$

where f_r is the received frequency, λ_r is the received wavelength, and c is the velocity of an electromagnetic wave (300,000,000 meters per second or 984,000,000 feet per second). Then:

Received frequency,

$$f_r = \frac{c}{\lambda_s - v_s \times t} = \frac{c}{\lambda_s - \dfrac{v_s}{f_s}} \tag{160-3}$$

$$= f_s \times \left(\frac{c}{c - v_s}\right) \text{ hertz}$$

If the source is stationary but the receiver is moving with a component of velocity $+v_r$ in the direction of propagation, there is, as regards to the receiver, an apparent change in the velocity of propagation.

Apparent change in the velocity of propagation $= c - v_r$

$$= \lambda_s \times f_r$$

This yields
Received frequency,

$$f_r = \frac{c - v_r}{\lambda_s} = f_s \times \frac{(c - v_r)}{c} \text{ hertz} \tag{160-4}$$

If the source and the receiver are both moving,
Received frequency,

$$f_r = f_s \times \left(\frac{c - v_r}{c - v_s} \right) \text{ hertz} \tag{160-5}$$

The Doppler frequency shift $= f_r - f_s \tag{160-6}$

$$= f_s \times \left(\frac{c - v_r}{c - v_s} - 1 \right)$$

$$= f_s \times \left(\frac{v_s - v_r}{c - v_s} \right) \longrightarrow f_s \frac{(v_s - v_r)}{c} \text{ hertz}$$

since c is very much greater than v_s.

The frequency shift depends on the source frequency and the relative velocity between the source and the receiver in the direction of propagation. If the relative velocity is 1 mile per hour and $f_s = 1$ GHz.

$$\text{Doppler frequency shift} = \frac{10^9 \times 5280}{9.84 \times 10^8 \times 3600} \text{ Hz} \tag{160-7}$$

$$\approx 1.5 \text{ Hz per mph per GHz}$$

For a CW radar set, the two-way propagation shift is $2 \times 1.5 = 3.0$ Hz per mph per GHz.

The same principles apply to a sonar set whose radiated acoustic pulse travels with a velocity of approximately 5000 ft/sec in seawater. If the relative velocity between the source and the receiver is 1 knot (6080 ft/hr) and $f_s = 1$ kHz.

$$\text{Doppler shift} = \frac{1000 \text{ Hz} \times 6080 \text{ ft/hr}}{3600 \text{ s/hr} \times 5000 \text{ ft/s}} = 0.34 \text{ Hz} \tag{160-8}$$

For two-way propagation,

$$\text{Doppler shift} = 2 \times 0.34 \approx 7 \text{ Hz per knot per kHz} \tag{160-9}$$

Example 160-1

A CW radar set is operating on a frequency of 10 GHz. If the target is moving in the direction of propagation with a velocity of 60 mph, calculate the amount of the Doppler shift at the receiver.

Solution

$$\text{Two-way Doppler shift} = 3.0 \times 10 \times 60 \tag{160-7}$$

$$= 1800 \text{ Hz.}$$

Example 160-2

A sonar set is operating on a frequency of 9 kHz. If the relative velocity between the frigate and a submarine in the direction of propagation is 18 knots, what is the amount of the two-way Doppler shift?

Solution

$$\text{Two-way Doppler shift} = 0.7 \times 18 \times 9 \qquad (160\text{-}9)$$
$$= 110 \text{ Hz.}$$

6
PART

Mathematics for electronics

Chapter 161
Angles and their measurement

DEFINITION OF AN ANGLE

IT IS IMPOSSIBLE TO SOLVE MANY PROBLEMS IN ELECTRICITY AND ELECTRONICS without a knowledge of angles. For example, in ac analysis two alternating quantities can be separated by a certain phase difference, this difference is normally measured in terms of an angle.

Consider two straight but very thin rods or lines OA, OB (Fig. 161-1A), pivoted at O. Holding OB fixed, turn OA in an counterclockwise direction; OA then describes a positive angle which is measured by the amount of turning through the arc of the circle, AB. At the center of the circle the point O is the *vertex* of the angle, which measures the separation between the rods OA, and OB, and is denoted by $\angle AOB$, or $\angle O$.

The angle described by a complete revolution of OA is arbitrarily divided into 360 equal parts, which are called *degrees* (symbol °).

TYPES OF ANGLES

Acute angle

An acute angle (Fig. 161-1B) involves a rotation that is less than one-quarter of a revolution. The value of the angle then lies between 0° and 90°. Therefore, $0° \angle ABC < 90°$.

Right angle

The right angle (Fig. 161-1C) is related to one-quarter of a revolution, and therefore its value is 90°. When $\angle ABC = 90°$, lines AB and BC are said to be *perpendicular*.

If the line BD is drawn,

$$\angle ABD + \angle DBC = \angle ABC = 90°$$

When the sum of two angles is 90°, the angles are said to be *complementary;* it follows that $\angle ABD$ and $\angle DBC$ are complementary angles. As an example, 20° and 70° are complementary angles because $20° + 70° = 90°$.

Obtuse angle

An obtuse angle (Fig. 161-1D) requires more than one-quarter revolution but less than one-half revolution. The value of an obtuse angle is then greater than 90° but less than 180°. Therefore, $90° < \angle ABC < 180°$.

Straight angle

The straight angle (Fig. 161-1E) is related to one-half of a revolution and consequently its value is 180° ($\angle ABC = 180°$).

If the line BD is drawn,

$$\angle ABD + \angle DBC = 180°$$

When the sum of two angles is 180°, the angles are said to be *supplementary;* it follows that $\angle ABD$ and $\angle DBC$ are supplementary angles. For example, 143° and 37° are supplementary angles because $143° + 37° = 180°$.

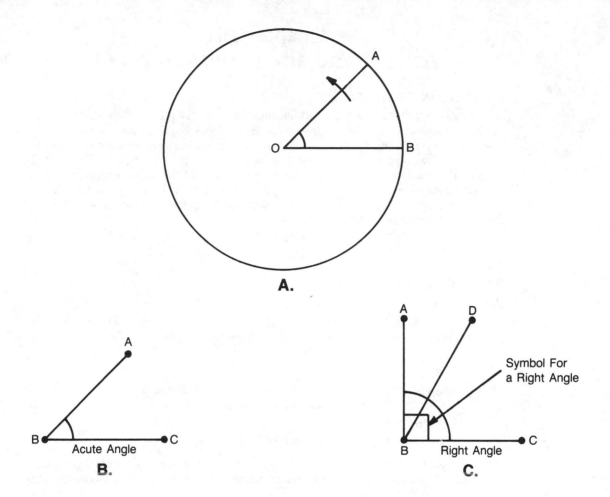

A.

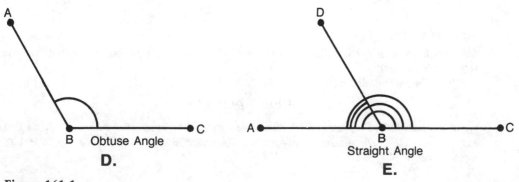

B.

Acute Angle

C.

Symbol For
a Right Angle

Right Angle

D.

Obtuse Angle

E.

Straight Angle

Figure 161-1

548

MEASUREMENT OF ANGLES

Angles of less than one degree can be measured in terms of a decimal fraction of a degree. For example,

$$0.375° = \frac{375°}{1000} = \frac{3°}{8}.$$

Alternatively, the degree can be subdivided into 60 minutes and each minute into 60 seconds. Degrees, minutes, and seconds are respectively denoted by the symbols °, ' and ". Then:

$$\frac{3°}{8} = \frac{3}{8} \times 60' = 22.5' = 22'30''.$$

Scientific calculators normally have mode positions, that are shown as DEG. (degrees), GRAD. (gradients), and RAD. (radians). The gradient (g) is based on the metric system and is defined by $100\ g = 1$ right angle = 90° or $1\ g = 0.9°$. So far it does not appear that the gradient is in common use.

The *radian* is the scientific method of measuring angles; for example, an angle derived from solving differential calculus equations is automatically measured in radians. But what is the radian and how is it related to the degree? In Fig. 161-2A, we measure the length of each arc between P_1 and Q_1, P_2, and Q_2, and so on. We also measure the length of each corresponding radius and then divide each length of arc by that of its radius. In the particular case when the lengths of the arc and the radius are equal, the angle POQ is equal to 1 radian. If we measure this angle with a protractor, we find that the value of the angle is slightly greater than 57°.

MATHEMATICAL DERIVATIONS

If Fig. 161-2A is repeated for any angle, the ratio of arc to radius is constant. This ratio can therefore be taken as a measure of the angle. Such a system of measurement is known as *circular measurement* (^{c}). In Fig. 161-2B the lengths of the arc and the radius are respectively l and r; the value of the angle, θ, in radians is then given by

$$\text{Angle } \theta^c = \frac{l}{r} \text{ radians} \tag{161-1}$$

For the complete circumference (Fig. 161-2C), the angle at the center of the circle is 360° and the length of the perimeter is $2\pi r$ where the circular constant, $\pi = 3.1415926 \ldots .$

Therefore:

$$360° = \frac{2\pi r^c}{r}$$

$$= 2\pi \text{ radians}$$

This yields:

$$1 \text{ radian} = \frac{360°}{2\pi} = \frac{180°}{\pi} = 57.2958$$

$$\approx 57°14'45''$$

Conversion from decimal degrees to radians

$$\theta° = \frac{\pi^c \times \theta°}{180°} \text{ radians} \tag{161-2}$$

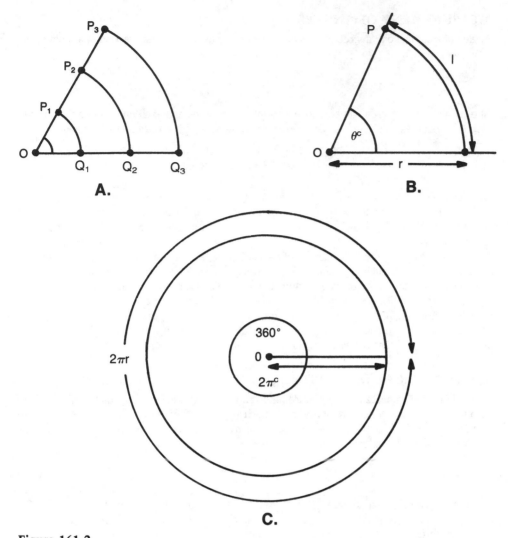

Figure 161-2

Conversion from radians to decimal degrees

$$\theta^c = \frac{180° \times \theta^c}{\pi^c} \text{ degrees} \qquad (161\text{-}3)$$

ANGULAR FREQUENCY (VELOCITY)

In Fig. 161-1A let line OA rotate through an angle of θ radians in a time of t seconds. Then the angular velocity, ω, is given by:

$$\text{Angular velocity, } \omega = \frac{\theta}{t} \text{ radians per second} \qquad (161\text{-}4)$$

This equation should be compared with the following relationship:

$$\text{Linear velocity, } v = \frac{\text{Distance, } d}{\text{Time, } t}$$

Equation 161-4 yields:

$$\text{Angle } \theta = \omega t \text{ radians} \tag{161-5}$$

Let line *OA* be a phasor, which is used to represent a sinewave voltage (or current) whose frequency is f Hz. Then the line will rotate at f revolutions per second, and each revolution is equivalent to an angular sweep of 2π radians. The angular frequency is therefore given by:

$$\text{Angular frequency, } \omega = 2\pi f \text{ rad/s} \tag{161-6}$$

Example 161-1

Convert (a) 137.526° to degrees, minutes, seconds, and radians.
       (b) 73°51′37″ into decimal degrees and radians.
       (c) 4.579ᶜ into decimal degrees and degrees, minutes, and seconds.

Solution

(a) Using a scientific calculator,

Entry, Key	Display
0.526° ⊗	0.526°
60 ⊜	31.56′
⊖ 31 ⊜	0.56′
⊗ 60 ⊜	33.6″

Therefore:

$$137.526° = 137°31′33.6″$$

$$137.526° = \frac{\pi \times 137.526^c}{180} \tag{161-2}$$

$$= 2.4003^c$$

Entry, Key	Display
137.526° ⊗	137.526°
π ⊜ ⊘	432.051
180 ⊜	2.4003ᶜ

(b)

Entry, Key	Display
37 ÷	37″
60 = +	0.616′
51 = ÷	51.616′
60 =	0.8603°

Therefore:

$$73°51'37'' = 73.8603°$$

and,

$$73.8603° = \frac{\pi \times 73.8603^c}{180}$$ (161-2)

$$= 1.289^c$$

(c)

$$4.579^c = \frac{180 \times 4.579°}{\pi}$$ (161-3)

$$= 262.357°$$

Entry, Key	Display
4.579^c ×	4.579^c
180 = ÷	824.22
π =	262.357°

Entry, Key	Display
0.357° ×	0.357°
60 = −	21.442′
21 = ×	0.442′
60 =	26.5″

Therefore:

$$4.579^c = 262.357° = 262°21'26.5''$$

Example 161-2

An AM broadcast station is operating on a frequency of 980 kHz. What is the value of its sine wave's angular frequency?

Solution

Angular frequency,

$$\omega = 2\pi f \qquad \qquad (161\text{-}6)$$
$$= 2 \times \pi \times 980 \times 10^3$$
$$= 6.16 \times 10^6 \text{ rad/s}$$

Chapter 162
The theorem of Pythagoras

A PLANE FIGURE THAT IS BOUNDED BY THREE STRAIGHT LINES IS CALLED A *triangle*. Examples of various triangles are shown in Figs. 162-1A,B,C. Each of these triangles has three sides, which are denoted by the lowercase letters a, b, c. By contrast the capital letters A, B, C represent the three interior angles. Notice that in each triangle side a is opposite or facing angle A; side b is facing angle B, and so on.

Figure 162-1A illustrates an acute-angled triangle in which all of the three interior angles are acute angles. Figure 162-1B shows an obtuse-angled triangle in which one of the interior angles, namely B, is obtuse. Figure 162-1C represents a right-angled triangle in which one of the interior angles, namely C, is a right angle (90°).

All triangles possess the following two properties:

1. The sum of the lengths of any two of the sides must exceed the length of the third side. It follows that it is impossible to construct a triangle whose sides are 3, 4, and 8 inches because (3 + 4) is not greater than 8.
2. The sum of the three interior angles of a triangle is 180° or π radians. Therefore:

$$A + B + C = 180°$$

For the right-angled triangle of Fig. 162-1C, $C = 90°$. Therefore $A + B = 90°$, so that the angles A, B are complementary. The longest side, c, which is facing the right angle, is called the *hypotenuse*.

A famous theorem that applies only to a right-angled triangle was discovered by Pythagoras, who lived in Sicily between the years 570 B.C. and 500 B.C. Formally stated the theorem is as follows: "In a right-angled triangle, the square on the hypotenuse is equal to the sum of the squares on the sides containing the right angle."

In Fig. 162-2, the theorem means that if the area (a^2) of the square *BCHG* is added to the area (b^2) of the square *ACJI*, the result is the area (c^2) of the square *ABEF*; in equation form, $a^2 + b^2 = c^2$, or $c = \sqrt{a^2 + b^2}$. A proof of Pythagoras' theorem is shown in the mathematical derivations.

The theorem of Pythagoras is a vital tool in ac circuit analysis. For example if the effective voltages across a resistor and an inductor in series are respectively 6 V and 8 V, the total voltage across this combination is not 14 V but is, in fact:

$$\sqrt{6^2 + 8^2} = \sqrt{36 + 64} = \sqrt{100} = 10 \text{ V}$$

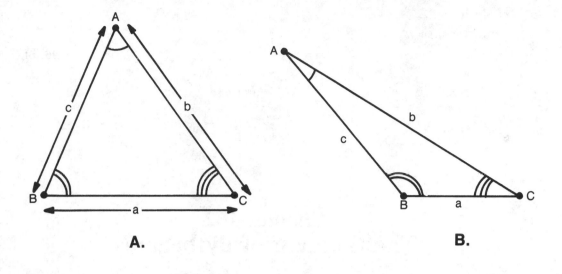

A.

B.

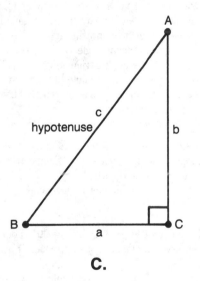

C.

Figure 162-1

(See Chapter 70.) Similarly, if the values of an ac circuit's true power and reactive power are known, it is necessary to combine these values by the Pythagorean equation in order to determine the apparent power (see Chapters 70 through 77).

MATHEMATICAL DERIVATIONS

In Fig. 162-3, the triangles ABC, $A'B'C'$ are equiangular in the sense that $\angle A = \angle A'$, $\angle B = \angle B'$, $\angle C = \angle C'$. Such triangles are said to be *similar*.

In similar triangles the corresponding sides are in proportion. Therefore:

$$\frac{a}{a'} = \frac{b}{b'} = \frac{c}{c'} \tag{162-1}$$

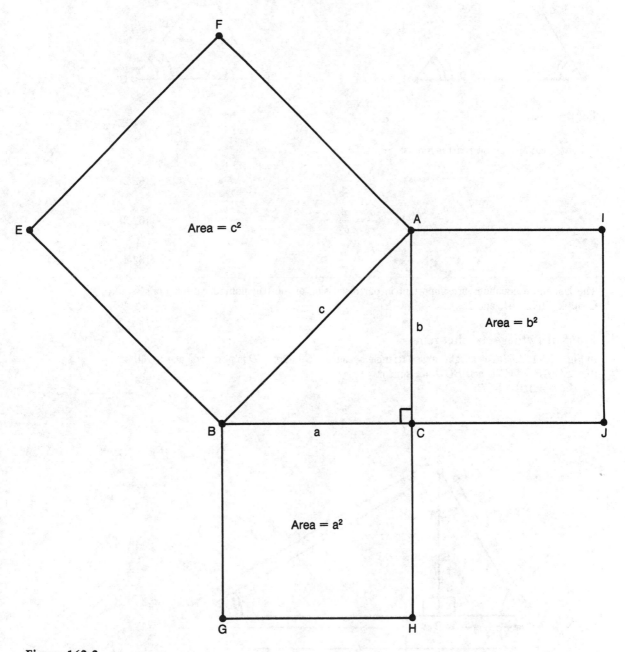

Figure 162-2

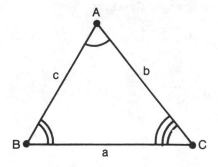

Figure 162-3

These equations can be rearranged as:

$$\frac{b}{c} = \frac{b'}{c'} \qquad (162\text{-}2)$$

$$\frac{a}{c} = \frac{a'}{c'} \qquad (162\text{-}3)$$

$$\frac{b}{a} = \frac{b'}{a'} \qquad (162\text{-}4)$$

The last three equations are important in our discussion of the trigonometrical functions (see Chapters 163, 164, and 165).

Proof of Pythagoras' theorem

In Fig. 162-4, $\triangle ABC$ is a right-angled triangle because $\angle C = 90°$. CD is perpendicular to AB so that $\triangle$'s ABC, ACD, and BCD are similar.

In $\triangle$'s ABC, BCD:

$$\frac{x}{a} = \frac{a}{c}$$

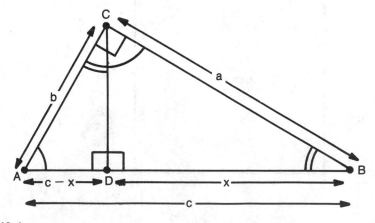

Figure 162-4

556

or,

$$a^2 = cx \qquad (162\text{-}5)$$

In △'s ABC, ACD:

$$\frac{c - x}{b} = \frac{b}{c}$$

or,

$$c^2 - cx = b^2 \qquad (162\text{-}6)$$

Substituting $cx = a^2$ in Equation 162-6,

$$c^2 - a^2 = b^2$$

or,

$$a^2 + b^2 = c^2 \qquad (162\text{-}7)$$

Equation 162-7 is the required Pythagorean relationship. Other forms of the equation are:

$$c = \sqrt{a^2 + b^2} \qquad (162\text{-}8)$$

$$a^2 = c^2 - b^2 \qquad (162\text{-}9)$$

$$a = \sqrt{c^2 - b^2} \qquad (162\text{-}10)$$

$$b^2 = c^2 - a^2 \qquad (162\text{-}11)$$

$$b = \sqrt{c^2 - a^2} \qquad (162\text{-}12)$$

Note that there are certain right-angled triangles in which the lengths of the sides are whole numbers. Three examples are:

(1) $a = 3$, $b = 4$, $c = 5$, since $3^2 + 4^2 = 5^2$
(2) $a = 5$, $b = 12$, $c = 13$, since $5^2 + 12^2 = 13^2$ and
(3) $a = 8$, $b = 15$, $c = 17$, since $8^2 + 15^2 = 17^2$.

Example 162-1

A 1.8 kΩ resistor is connected in series with an inductor whose reactance is 2.3 kΩ. Determine the value of the total impedance, Z (Chapter 70).

Solution

Total impedance

$$Z = \sqrt{R^2 + X_L^2} \qquad (162\text{-}8)$$
$$= \sqrt{1.8^2 + 2.3^2}$$

Entry, Key			Display
1.8	x^2	$+$	3.24
2.3	x^2		5.29
		$=$	8.53
		$\sqrt{x}$	2.92

The impedance is 2.92 kΩ.

Example 162-2

In an ac circuit, the apparent power is 25.6 mVA and the reactive power is 12.8 mVA. Determine the value of the true power.

Solution

$$\text{True Power} = \sqrt{(\text{Apparent Power})^2 - (\text{Reactive power})^2} \qquad (162\text{-}10)$$
$$= \sqrt{25.6^2 - 17.8^2}$$

Entry, Key			Display
25.6	x^2	$-$	655.36
17.8	x^2		316.84
	$=$		338.52
	$\sqrt{x}$		18.4

The true power is 18.4 mW.

Chapter 163
The sine function

FIGURE 163-1 SHOWS THREE RIGHT-ANGLED TRIANGLES THAT ARE SIMILAR because angle B is common to all three triangles and consequently $\angle A = \angle A' = \angle A''$. It follows from Equation 162-2 that:

$$\frac{b}{c} = \frac{b'}{c'} = \frac{b''}{c''}$$

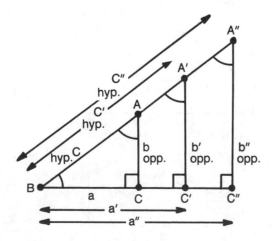

Figure 163-1

558

The values of each of these ratios is determined in some way by the value of angle B. In mathematical language the value of the ratio $b:c$ is a function of angle B since if the value of B is changed, the value of the ratio $b:c$ will alter; in equation form:

$$f(B) = \frac{b}{c}$$

The ratio $b:c$ involves the circular trigonometrical function, sine, which is commonly abbreviated sin.

Therefore:

$$\sin B = \frac{b}{c}.$$

A knowledge of the sine function is important in the ac analysis of series and parallel LCR circuits and also in the use of the j operator.

MATHEMATICAL DERIVATIONS

In Fig. 163-2 side b is facing or *opposite* (abbreviated opp) to angle B, while side a is next or *adjacent* (abbreviated adj) to angle B. It can be argued that side c is also adjacent to the angle B, but side c has already been identified as the *hypotenuse* (abbreviated hyp).

The sine of angle B is defined by:

$$\sin B = \frac{\text{opp}}{\text{hyp}} \qquad (163\text{-}1)$$

$$= \frac{b}{c}$$

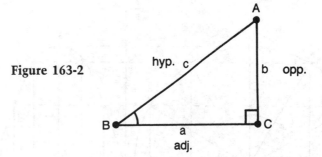

Figure 163-2

As an example, $\sin 30° = 0.5$ so that the length of b is one-half the length of c. But if you know that the value of $\sin B$ is 0.5, how do you obtain the value of 30° on the calculator? If $\sin B = 0.5$,

$$\left. \begin{array}{l} \text{Angle } B = \text{inv sin } 0.5 \\ \text{or arc sin } 0.5 \\ \text{or } \sin^{-1} 0.5 \end{array} \right\} = 30°$$

The terms *inverse (inv) sin*, *arc sin*, and *sin*$^{-1}$ have the same meaning, namely "the value of the angle whose sine value is . . . ." These three key combinations are used on a variety of scientific calculators. Using the inverse, and sin keys,

$$\text{Angle, } B = \text{inv sin } \frac{b}{c} \qquad (163\text{-}2)$$

Because $\sin B = \dfrac{b}{c}$,

$$\text{Side } b = c \sin B \qquad (163\text{-}3)$$

and

$$\text{Side } c = \frac{b}{\sin B} \qquad (163\text{-}4)$$

As always, the choice of which equation to use depends on the information given and the quantity to be found.

The sine is a circular function whose value can be obtained with the aid of measurements taken from a circle. Figure 163-3 shows a circle of unit radius. As line OP rotates in the counter-clockwise or positive direction,

$$\sin \theta = \frac{PN}{OP} = \frac{PN}{1}$$
$$= PN$$

Therefore the length of PN is a direct measure of the value of $\sin \theta$. In this way we can obtain the $\sin \theta$ vs θ graph, which is the waveform of a sinewave voltage (or current). For such a voltage the instantaneous value is:

$$e = E_{max} \sin \theta$$
$$= E_{max} \sin \omega t \qquad (161\text{-}5, \ 163\text{-}5)$$
$$= E_{max} \sin 2\pi ft \qquad (161\text{-}6, \ 163\text{-}6)$$

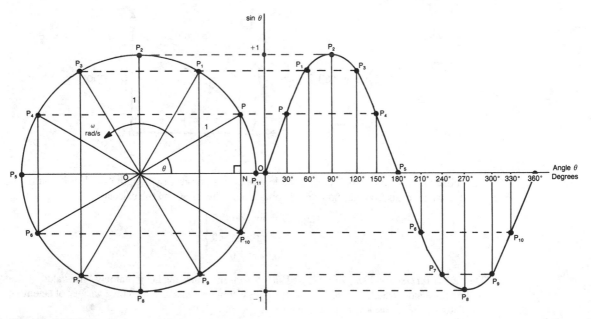

Figure 163-3

Note that the value of sin θ lies between $+1$ and -1.

Example 163-1

In Fig. 163-2, $b = 16.7$, and $c = 23.5$. Find the values of B, A, and a.

Solution

$$\text{Angle } B = \text{inv sin } \frac{b}{c} \qquad (163\text{-}2)$$

$$= \text{inv sin } \frac{16.7}{23.5}$$

Entry, Key		Display
16.7	$\div$	16.7
23.5	$=$	0.7106
inv	sin	45.287°

The value of angle B is 45.3°, rounded off. The angles A, B are complementary.

Therefore:

$$\text{Angle } A = 90° - B$$

$$= 90° - 45.3°$$

$$= 44.7°$$

$$\text{Side } a = \sqrt{c^2 - b^2} \qquad (162\text{-}10)$$

$$= \sqrt{23.5^2 - 16.7^2}$$

$$= 16.5$$

Example 163-2

In Fig. 163-2, $B = 75.8°$, and $c = 112.8$. Find the values of b, A, and a.

Solution

$$\text{Side } b = c \sin B \qquad (163\text{-}3)$$

$$= 112.8 \sin 75.8°$$

Entry, Key		Display
112.8	$\times$	112.8
75.8	sin	0.9694
	$=$	109.35

The length of the side b is 109.35, rounded off.

$$\text{Angle } A = 90° - B$$
$$= 90° - 75.8°$$
$$= 14.2°$$
$$\text{Side } a = \sqrt{c^2 - b^2} \qquad\qquad (162\text{-}10)$$
$$= \sqrt{112.8^2 - 109.35^2}$$
$$= 27.68$$

Example 163-3

In Fig. 163-2, $B = 18.6°$, and $b = 69.3$. Find the values of c, A, and a.

Solution

$$\text{Hypotenuse } c = \frac{b}{\sin B} \qquad\qquad (163\text{-}4)$$
$$= \frac{69.3}{\sin 18.6°}$$

Entry, Key		Display
69.3	$\div$	69.3
18.6°	sin	0.31896
	=	217.3

The length of the hypotenuse, c, is 217.3, rounded off.

$$\text{Angle } A = 90° - B$$
$$= 90° - 18.6°$$
$$= 71.4°$$
$$\text{Side } a = \sqrt{c^2 - b^2} \qquad\qquad (162\text{-}10)$$
$$= \sqrt{217.3^2 - 69.3^2}$$
$$= 206.0, \text{ rounded off.}$$

Chapter 164
The cosine function

FIGURE 164-1 SHOWS THREE RIGHT-ANGLED TRIANGLES THAT ARE SIMILAR because angle B is common to all three triangles and consequently $\angle A = \angle A' = \angle A''$. It follows from Equation 162-3 that

$$\frac{a}{c} = \frac{a'}{c'} = \frac{a''}{c''}$$

The value of each of these ratios is determined in some way by the value of the angle B. In mathematical language the value of the ratio $a:c$ is a function of angle B since if the value of B is changed, the value of the ratio $a:c$ will alter; in equation form: $f(B) = a/c$. The ratio $a:c$ involves the circular trigonometrical function, cosine, which is commonly abbreviated to cos.

Therefore:

$$\cos B = \frac{a}{c}$$

A knowledge of the cosine function is important in the ac analysis of series and parallel LCR circuits and also in the use of the j operator. Moreover if the source voltage and the source current differ in phase, the power factor of the circuit is equal to the cosine of the phase angle (see Chapters 70 thru 77).

Figure 164-1

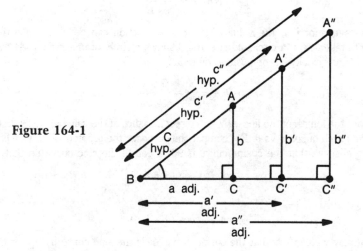

MATHEMATICAL DERIVATIONS

In Fig. 164-2 the cosine of the angle, B is defined by:

$$\cos B = \frac{\text{adj}}{\text{hyp}} \qquad (164\text{-}1)$$

$$= \frac{a}{c}$$

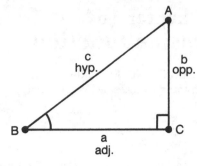

Figure 164-2

Then

$$\text{Angle } B = \text{inv cos } \frac{a}{c} \qquad (164\text{-}2)$$

Since:

$$\cos B = \frac{a}{c}$$

$$\text{Side } a = c \cos B \qquad (164\text{-}3)$$

and,

$$\text{Side } c = \frac{a}{\cos B} \qquad (164\text{-}4)$$

Like the sine function, the value of the cosine function can be obtained with the aid of measurements taken from a circle. Figure 164-3 shows a circle of unit radius. At the line, OP, rotates in a counterclockwise or positive direction,

$$\cos \theta = \frac{ON}{OP} = \frac{ON}{1} = ON$$

Therefore the length of the line, ON, is a direct measure of the value of cos θ. In this way we can obtain the graph of cos θ vs θ. For comparison purposes the graph of sin θ vs θ has also been included. We observe that the cosine curve is 90° ahead of the sine curve so that:

$$\cos \theta = \sin (\theta + 90°) \qquad (164\text{-}5)$$

and

$$\sin \theta = \cos (\theta - 90°)$$

Alternatively we can say that the cosine curve leads the sine curve by 90° or that the sine curve lags the cosine curve by 90°. Like the sine function, the value of the cosine function lies between +1 and −1.

Relationships between the sine and cosine functions

In Fig. 164-2,

$$\cos B = \frac{a}{c}$$

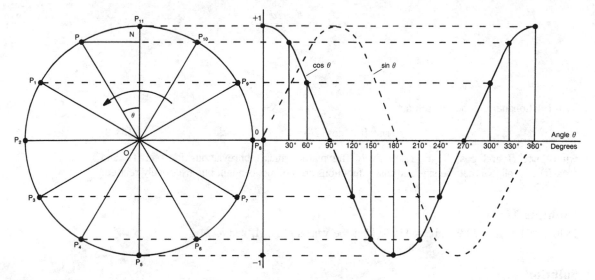

Figure 164-3

With reference to the angle A, the side opposite the angle is a.

Therefore:

$$\sin A = \frac{a}{c}$$

and

$$\cos B = \sin A \qquad (164\text{-}6)$$
$$= \sin (90° - B)$$
$$\sin A = \cos B \qquad (164\text{-}7)$$
$$= \cos (90° - A)$$

The cosine of an angle is equal to the sine of its complement; the sine of an angle is equal to the cosine of its complement. As examples, $\cos 40° = \sin 50° = 0.7660$ and $\sin 20° = \cos 70° = 0.3420$.

In Fig. 164-2,

$$\sin B = \frac{b}{c}$$

and,

$$\cos B = \frac{a}{c}$$

Then:

$$(\sin B)^2 + (\cos B)^2 = \frac{b^2}{c^2} + \frac{a^2}{c^2} \qquad (164\text{-}8)$$

565

$$= \frac{b^2 + a^2}{c^2}$$

$$= \frac{c^2}{c^2}$$

$$= 1$$

This relationship can be written as:

$$\sin^2 B + \cos^2 B = 1 \qquad (164\text{-}9)$$

where $\sin^2 B$ and $\cos^2 B$ are respectively the mathematical abbreviations for $(\sin B)^2$ and $(\cos B)^2$. It follows that the sine and cosine functions are not independent but are closely related.

Example 164-1

In Fig. 164-1, $a = 73.9$ and $c = 112.5$. Find the values of B, A, and b.

Solution

$$\text{Angle } B = \text{inv cos } \frac{a}{c} \qquad (164\text{-}2)$$

$$= \text{inv cos } \frac{73.9}{112.5}$$

Entry, Key	Display
73.9 $\div$	73.9
112.5 $=$	0.6568
$\boxed{\text{inv}}$ $\boxed{\text{cos}}$	48.9°

The value of the angle B is 48.9°.

$$\text{Angle } A = 90° - 48.9°$$

$$= 41.1°.$$

$$\text{Side } b = \sqrt{112.5^2 - 73.9^2}$$

$$= 84.8$$

Example 164-2

In Fig. 164-2, $B = 12.7°$ and $c = 0.763$. Find the values of a, A, and b.

Solution

$$\text{Side } a = c \cos B \qquad (164\text{-}3)$$

$$= 0.763 \cos 12.7°.$$

Entry, Key		Display
0.763	$\times$	0.763
12.7°	cos	0.9755
	$=$	0.744

The length of side a is 0.744.

$$\text{Angle } A = 90° - 12.7°$$
$$= 77.3°$$
$$\text{Side } b = \sqrt{0.763^2 - 0.744^2}$$
$$= 0.169$$

Example 164-3

In Fig. 164-2, $B = 74.8°$, and $a = 27.3$. Find the values of c, A, and b.

Solution

$$\text{Hypotenuse } c = \frac{a}{\cos A} \qquad\qquad (164\text{-}4)$$

$$= \frac{27.3}{\cos 74.8°}$$

Entry, Key		Display
27.3	$\div$	27.3
74.8°	cos	0.2622
	$=$	104.1

The length of the hypotenuse, c, is 104.1.

$$\text{Angle } A = 90° - 74.8°$$
$$= 15.2°$$
$$\text{Side } b = \sqrt{104.1^2 - 27.3^2}$$
$$= 100.5.$$

Example 164-4

In an ac circuit, the source voltage and the source current differ in phase by 40°. What is the value of the circuit's power factor?

Solution

$$\text{Power Factor} = \cos \phi = \cos 40°$$
$$= 0.766.$$

Chapter 165
The tangent function

FIGURE 165-1 SHOWS THREE RIGHT-ANGLED TRIANGLES THAT ARE SIMILAR because angle B is common to all three and consequently $\angle A = \angle A' = \angle A''$. It follows that:

$$\frac{b}{a} = \frac{b'}{a'} = \frac{b''}{a''}$$

The value of each of these ratios is determined in some way by the value of angle B. In mathematical language the value of ratio $b:a$ is a function of angle B. If the value of B is changed, the value of the ratio $b:a$ will alter; in equation form $f(B) = b/a$. The ratio $b:a$ involves the circular trigonometric function, tangent, commonly abbreviated tan. Therefore $\tan B = b/a$.

A knowledge of the tangent function is especially important in the use of operator j and is also required in the ac analysis of various LCR circuits.

MATHEMATICAL DERIVATIONS

In Fig. 165-2 the tangent of angle B is defined by:

$$\tan B = \frac{\text{opp}}{\text{adj}} = \frac{b}{a} \tag{165-1}$$

then

$$\text{Angle } B = \text{inv tan } \frac{b}{a} \tag{165-2}$$

Since

$$\tan B = \frac{b}{a}, \tag{165-3}$$

$$\text{Side } b = a \tan B$$

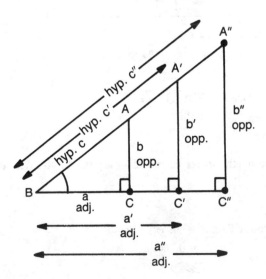

Figure 165-1

568

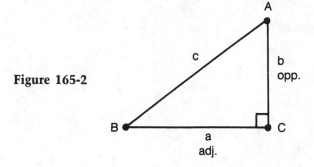

Figure 165-2

and

$$\text{Side } a = \frac{b}{\tan B} \tag{165-4}$$

Like the sine and cosine functions, the value of the tangent function can be obtained with the aid of measurements taken from a circle. Figure 165-3 shows a circle of unit radius; at point Q, a tangent QT is drawn so that lines OQ and TQ are perpendicular to each other (a tangent to a circle touches or grazes the circle at a particular point). As the line OP rotates in the counterclockwise or positive direction,

$$\tan \theta = \frac{TQ}{OQ} = \frac{TQ}{1} = TQ \tag{165-5}$$

Therefore the length of line TQ is a direct measure of the value of $\tan \theta$. In this way we can obtain the graph of $\tan \theta$ versus θ. Note that the value of the tangent function lies between $+\infty$ and $-\infty$.

In Fig. 165-3,

$$\tan \theta = \frac{TQ}{OQ} = \frac{TQ}{OP}$$

$$\text{Angle, } \theta = \frac{\text{Arc } PQ}{OP} \text{ radians}$$

and

$$\sin \theta = \frac{PN}{OP}$$

Since $TQ > \text{Arc } PQ > PN$,

$$\tan \theta > \theta > \sin \theta.$$

For small angles of less than $10°$, TQ, arc PQ, and arc OP are approximately equal and therefore:

$$\tan \theta \approx \theta^c \approx \sin \theta \tag{165-6}$$

As an example, $\sin 5° = 0.0872$, $5° = 0.0873^c$, and $\tan 5° = 0.0875$, rounded off.

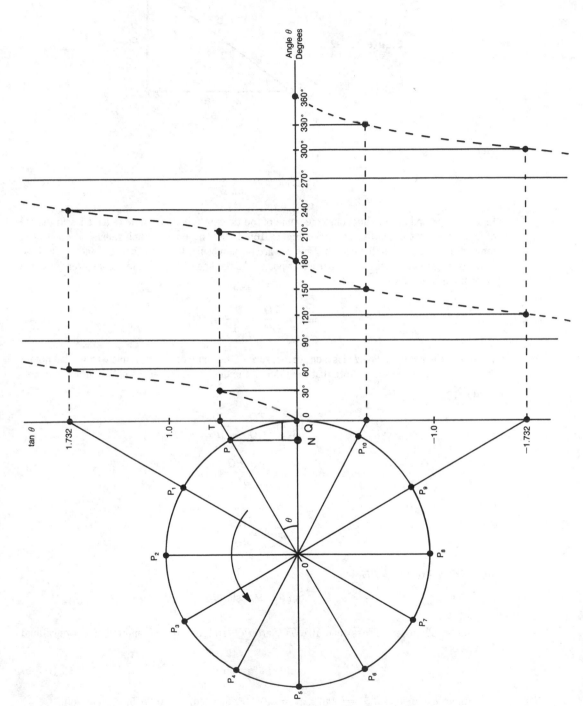

Figure 165-3.

Relationship among sin, cos, and tan

In Fig. 165-2,

$$\tan B = \frac{b}{a} = \frac{b/c}{a/c}$$

(165-7)

$$= \frac{\sin B}{\cos B}$$

This relationship is illustrated in Table 165-1.

Table 165-1.

Angle	Sin	Cos	Tan
0°	0	1	0
30°	0.5	0.866	0.577
45°	0.707	0.707	1
60°	0.866	0.5	1.732
90°	1	0	∞

Example 165-1

In Fig. 165-2, $b = 72.5$, $a = 98.7$. Find the values of B, A and c.

Solution

$$\text{Angle } B = \text{inv tan } \frac{b}{a}$$

(165-2)

$$= \text{inv tan } \frac{72.5}{98.7}$$

Entry, Key		Display
72.5	÷	72.5
98.7	=	0.7345
inv	tan	36.3°

The value of angle B is 36.3°.

$$\text{Angle } A = 90° - 36.3°$$

(162-8)

$$= 53.7°$$

$$\text{Hypotenuse } c = \sqrt{72.5^2 + 98.7^2}$$

$$= 122.5$$

Example 165-2

In Fig. 165-2, $a = 0.654$, and $B = 82.3°$. Find the value of b, A, and c.

Solution

$$\text{Side } b = a \tan B \qquad\qquad (165\text{-}3)$$
$$= 0.654 \tan 82.3°$$

Entry, Key		Display
0.654	$\boxed{\times}$	0.654
82.3°	$\boxed{\tan}$	7.396
	$\boxed{=}$	4.837

The value of side b is 4.837.

$$\text{Angle } A = 90° - 82.3°$$
$$= 7.7°$$
$$\text{Hypotenuse } c = \sqrt{0.654^2 + 4.837^2} \qquad\qquad (165\text{-}8)$$
$$= 4.881$$

Example 165-3

In Fig. 165-2, $b = 116.4$, and $B = 11.7°$. Find the values of a, A, and c.

Solution

$$\text{Side } a = \frac{b}{\tan B} \qquad\qquad (165\text{-}4)$$
$$= \frac{116.4}{\tan 11.7°}$$

Entry, Key		Display
116.4	$\boxed{\div}$	116.4
11.7°	$\boxed{\tan}$	0.2071
	$\boxed{=}$	562.1

The value of side a is 562.1.

$$\text{Angle } A = 90° - 11.7°$$
$$= 78.3°$$
$$\text{Hypotenuse } c = \sqrt{116.4^2 + 562.1^2} \qquad\qquad (162\text{-}8)$$
$$= 574.0$$

Chapter 166
Cosecant, secant, and cotangent

THE SIN, COS, AND TAN FUNCTIONS ARE CONVENIENT MULTIPLIERS FOR converting from one side of a right-angled triangle into another side. However none of these three functions can convert either of the sides forming the right-angle into the hypotenuse. For example, in Fig. 166-1:

$$c = \frac{b}{\sin B} = b \times \frac{1}{\sin B},$$

$$c = \frac{a}{\sin A} = a \times \frac{1}{\sin A}$$

and

$$c = \frac{a}{\cos B} = a \times \frac{1}{\cos B}$$

$$c = \frac{b}{\cos A} = b \times \frac{1}{\cos A}$$

In some problems it is more convenient to have simple multipliers to convert either side a or b into the hypotenuse. In order to do this, we need trigonometrical functions that are the reciprocals of the sine and cosine functions. The reciprocal of sine θ, $1/\sin \theta$, is termed the *cosecant* of θ, where θ is any angle. The reciprocal of cosine θ, $1/\cos \theta$, is called the secant of θ; similarly the reciprocal of tangent θ, $1/\tan \theta$, is the cotangent of θ.

The cosecant (csc), secant (sec), and cotangent (cot) functions are less frequently encountered than the other three functions and do not normally appear as keys on a scientific calculator. If, for example, the value of cosecant 50° is required, the value of sine 50° is found first and the reciprocal key is used to obtain the cosecant of 50°.

The equations of electronics and communications theory rarely involve the reciprocal trigonometrical functions; however, chapter 149 contains the equation:

$$Z = Z_0 \cot \frac{2\pi x}{\lambda},$$

which determined the variation of the impedance along an open-circuited transmission line.

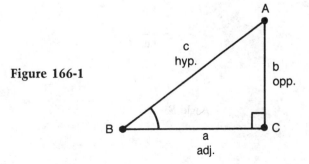

Figure 166-1

MATHEMATICAL DERIVATIONS

In Fig. 166-1, the cosecant of B, is defined by:

$$\csc B = \frac{\text{hypotenuse}}{\text{opposite}} \qquad (166\text{-}1)$$

$$= \frac{c}{b}$$

$$= \frac{1}{\sin B}$$

$$\text{Angle } B = \text{inv csc } \frac{c}{b} \qquad (166\text{-}2)$$

$$\text{Hypotenuse } c = b \csc B \qquad (166\text{-}3)$$

$$\text{Side } b = \frac{c}{\csc B} \qquad (166\text{-}4)$$

The secant of B is defined by:

$$\sec B = \frac{\text{hypotenuse}}{\text{adjacent}} \qquad (166\text{-}5)$$

$$= \frac{c}{a}$$

$$= \frac{1}{\cos B}$$

$$\text{Angle } B = \text{inv sec } \frac{c}{a} \qquad (166\text{-}6)$$

$$\text{Hypotenuse } c = a \sec B \qquad (166\text{-}7)$$

$$\text{Side } a = \frac{c}{\sec B} \qquad (166\text{-}8)$$

The cotangent of B is defined by:

$$\cot B = \frac{\text{adjacent}}{\text{opposite}} \qquad (166\text{-}9)$$

$$= \frac{a}{b}$$

$$= \frac{1}{\tan B}$$

$$= \frac{\cos B}{\sin B}$$

$$\text{Angle } B = \text{inv cot } \frac{a}{b} \qquad (166\text{-}10)$$

$$\text{Side } a = b \cot B \qquad (166\text{-}11)$$

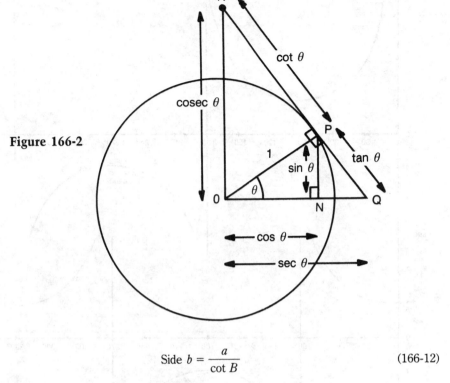

Figure 166-2

$$\text{Side } b = \frac{a}{\cot B} \tag{166-12}$$

Figure 166-2 shows a circle of unit radius with a tangent constructed at point P.

In $\triangle OPN$,

$$\sin \theta = \frac{PN}{OP} = \frac{PN}{1} = PN$$

$$\cos \theta = \frac{ON}{OP} = \frac{ON}{1} = ON$$

In $\triangle OPQ$,

$$\tan \theta = \frac{PQ}{OP} = \frac{PQ}{1} = PQ$$

$$\sec \theta = \frac{OQ}{OP} = \frac{OQ}{1} = OQ$$

In $\triangle OPR$,

$$\cot \theta = \frac{PR}{OP} = \frac{PR}{1} = PR$$

$$\csc \theta = \frac{OR}{OP} = \frac{OR}{1} = OR$$

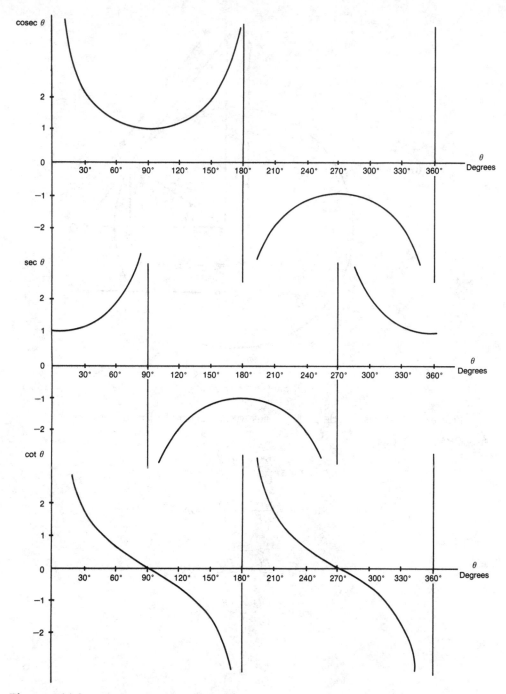

Figure 166-3

Therefore the values of all six trigonometrical functions can be obtained with the aid of measurements taken from the circle of unit radius. The graphs of csc θ, sec θ, and cot θ versus θ, can then be plotted as shown in Fig. 166-3.

Relationships involving csc, sec, and cot

From Equation 164-9:

$$\sin^2 \theta + \cos^2 \theta = 1$$

Therefore:

$$\frac{\sin^2 \theta}{\sin^2 \theta} + \frac{\cos^2 \theta}{\sin^2 \theta} = \frac{1}{\sin^2 \theta}$$

$$1 + \cot^2 \theta = \csc^2 \theta \qquad (166\text{-}13)$$

This relationship is also apparent from $\triangle ORP$, where

$$OP^2 + PR^2 = OR^2 \text{ (see Fig. 166-2)}.$$

From equation 164-9,

$$\cos^2 \theta + \sin^2 \theta = 1$$

Therefore:

$$\frac{\cos^2 \theta}{\cos^2 \theta} + \frac{\sin^2 \theta}{\cos^2 \theta} = \frac{1}{\cos^2 \theta}$$

$$1 + \tan^2 \theta = \sec^2 \theta \qquad (166\text{-}14)$$

This relationship is also apparent from $\triangle OPQ$, where:

$$OP^2 + PQ^2 = OQ^2$$

(see Fig. 166-2).

Example 166-1

Find the values of (a) csc 42.3°, (b) sec 37.4°, and (c) cot 18.6°.

Solution

$$\text{(a) csc } 42.3° = \frac{1}{\sin 42.3°}$$

Entry, Key		Display
42.3°	sin	0.6730
	1/x	1.486

The value of the csc 42.3° = 1.486, rounded off.

$$\text{(b) sec } 37.4° = \frac{1}{\cos 37.4°}$$

Entry, Key	Display
37.4° (cos)	0.7944
(1/x)	1.259

The value of sec 37.4° is 1.259.

$$(c) \ \cot 18.6° = \frac{1}{\tan 18.6°}$$

Entry, Key	Display
18.6° (tan)	0.3365
(1/x)	2.971

The value of cot 18.6° is 2.971.

Example 166-2

In Fig. 166-1, $B = 42.3°$, and $b = 13.2$. Find the value of c.

Solution

$$\text{Hypotenuse } c = b \csc B \qquad \qquad (166\text{-}3)$$
$$= 13.2 \csc 42.3°$$

Entry, Key	Display
13.2 (×)	13.2
1.486 (=)	19.6

The value of hypotenuse c is 19.6.

Example 166-3

In Fig. 166-1, $B = 37.4°$, and $a = 2.78$. Find the value of c.

Solution

$$\text{Hypotenuse } c = a \sec B \qquad \qquad (166\text{-}7)$$

Entry, Key	Display
2.78 (×)	2.78
1.259 (=)	3.50

The value of c is 3.50.

Example 166-4

In Fig. 166-1, $B = 18.6°$, and $b = 117.6$. Find the value of a.

Solution

$$a = b \cot B \qquad\qquad (166\text{-}11)$$
$$= 117.6 \cot 18.6°.$$

Entry, Key	Display
117.6° $\boxed{\times}$	117.6
2.971 $\boxed{=}$	349.4

The value of side a is 349.4.

Chapter 167
Trigonometrical functions
for angles of any magnitude

WE WILL USE FIG. 167-1 TO OBTAIN THE VALUES OF THE TRIGONOMETRICAL functions for an angle of any magnitude. For positive angles, start at the horizontal reference line, OX, and rotate the line OP counterclockwise through the desired angle. This might involve a number of complete rotations but eventually OP will lie in one of the four quadrants. As examples, OP_1 is in the first quadrant while OP_2, OP_3, and OP_4 are, respectively, in the second, third, and fourth quadrants. Perpendiculars are then dropped from points P_1, P_2, P_3, and P_4 onto the horizontal line $X'OX$ so that the right-angled triangles OP_1N_1, OP_2N_2, OP_3N_3, and OP_4N_4 are created. The same procedure is used for negative angles except that the rotation from the reference line, OX, is in the clockwise direction.

In defining the trigonometrical functions for angles of any size, it is necessary to consider the directions in which the perpendiculars (P_1N_1, P_2N_2, P_3N_3, P_4N_4) and the lines (ON_1, ON_2, ON_3, ON_4) are measured. All distances measured horizontally from left to right starting at point O are considered to be positive, while those measured in the opposite direction are taken as negative. Therefore, ON_1 and ON_4 are positive, but ON_2 and ON_3 are negative. Similarly, all distances measured vertically upward from $X'OX$ are positive; whereas all distances measured vertically downward are negative. It follows that P_1N_1 and P_2N_2 are positive but P_3N_3 and P_4N_4 are negative. By contrast, lines OP_1, OP_2, OP_3 and OP_4 are always considered to be positive.

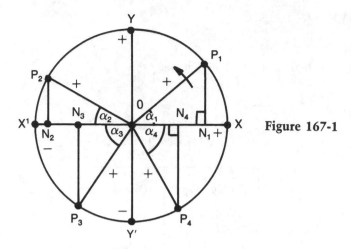

Figure 167-1

The magnitude of the trigonometrical function for any angle is equal to the value of the corresponding function for the acute angle formed by the line OP and the horizontal line $X'OX$. Suppose that for any angle, the line OP after its rotation reaches the position OP_1 in the first quadrant; then:

$$\sin \theta = \frac{(+)P_1N_1}{(+)OP_1} = +\frac{P_1N_1}{OP_1}$$

$$= \sin \alpha_1$$

$$\cos \theta = \frac{(+)ON_1}{(+)OP_1} = +\frac{ON_1}{OP_1}$$

$$= \cos \alpha_1$$

$$\tan \theta = \frac{(+)P_1N_1}{(+)ON_1} = +\frac{P_1N_1}{ON_1}$$

$$= \tan \alpha_1$$

The sine, cosine, and tangent functions all have positive values in the first quadrant.

If OP reaches the position of OP_2 in the second quadrant, then:

$$\sin \theta = \frac{(+)P_2N_2}{(+)OP_2} = +\frac{P_2N_2}{OP_2}$$

$$= \sin \alpha_2$$

$$\cos \theta = \frac{(-)ON_2}{(+)OP_2} = -\frac{ON_2}{OP_2}$$

$$= -\cos \alpha_2$$

$$\tan \theta = \frac{(+)P_2N_2}{(-)ON_2} = -\frac{P_2N_2}{ON_2}$$

$$= -\tan \alpha_2$$

The sine function has a positive value in the second quadrant, but the values of the cosine and tangent functions are negative.

If OP reaches the position of OP_3 in the third quadrant, then:

$$\sin \theta = \frac{(-)P_3N_3}{(+)OP_3} = -\frac{P_3N_3}{OP_3}$$

$$= -\sin \alpha_3$$

$$\cos \theta = \frac{(-)ON_3}{(+)OP_3} = -\frac{ON_3}{OP_3}$$

$$= -\cos \alpha_3$$

$$\tan \theta = \frac{(-)P_3N_3}{(-)ON_3} = +\frac{P_3N_3}{ON_3}$$

$$= \tan \alpha_3$$

The tangent function has a positive value in the third quadrant, but the values of the sine and cosine functions are negative.

If OP reaches the position of OP_4 in the fourth quadrant, then:

$$\sin \theta = \frac{(-)P_4N_4}{(+)OP_4} = -\frac{P_4N_4}{OP_4}$$

$$= -\sin \alpha_4$$

$$\cos \theta = \frac{(+)ON_4}{(+)OP_4} = \frac{ON_4}{OP_4}$$

$$= \cos \alpha_4$$

$$\tan \theta = \frac{(-)P_4N_4}{(+)ON_4} = -\frac{P_4N_4}{ON_4}$$

$$= -\tan \alpha_4$$

The cosine function has a positive value in the fourth quadrant, but the values of the sine and tangent functions are negative. The positive and negative quadrant signs for the six trigonometrical functions are summarized in Fig. 167-2.

MATHEMATICAL DERIVATIONS

If $\sin \theta = \sin \alpha$, where θ is an angle of any size and α is an acute angle, then:

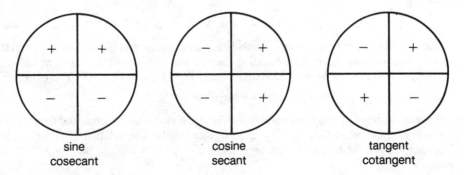

Figure 167-2

$$\text{Angle } \theta = 2n\pi + \alpha \text{ radians} \qquad (167\text{-}1)$$
$$= 360n + \alpha \text{ degrees}$$

or

$$\theta = (2n + 1)\pi - \alpha \text{ radians} \qquad (167\text{-}2)$$
$$= (2n + 1)180 - \alpha \text{ degrees}$$

where n is any positive or negative integer (whole number) or zero.

Therefore if $\alpha = 65°$, sin 115°, sin 425°, sin 475°, sin 785°, sin 835°, and so on all have the same value as sin 65°: 0.9063.

If $\cos \theta = \cos \alpha$, where θ is an angle of any size and α is an acute angle, then:

$$\text{Angle } \theta = 2n\pi \pm \alpha \text{ radians} \qquad (167\text{-}3)$$
$$= 360n \pm \alpha \text{ degrees}$$

Therefore if $\alpha = 65°$, cos $(-65°)$, cos 295°, cos 425°, cos 785°, and so on all have the same value as cos 65°: 0.4226.

If $\tan \theta = \tan \alpha$, where θ is an angle of any size and α is an acute angle, then:

$$\text{Angle } \theta = n\pi + \alpha \text{ radians} \qquad (167\text{-}4)$$
$$= 180n + \alpha \text{ degrees}$$

Therefore if $\alpha = 65°$, tan 245°, tan 425°, tan 605°, tan 785°, tan 965°, and so on all have the same value as tan 65°: 2.145.

Example 167-1

Find the values of (a) sin 1137°, (b) cos 975°, (c) tan $(-739°)$, (d) sin $(-956°)$.

Solution

(a) OP is rotated through three complete revolutions in the counterclockwise direction and then lands in the first quadrant. The acute angle between OP and the horizontal reference line is $1137° - (3 \times 360°) = 57°$. Therefore,

$$\sin 1137° = +\sin 57° = 0.8387$$

(b) OP is rotated through two complete revolutions in the counterclockwise direction and then lands in the third quadrant, which carries a negative sign for the cosine function. The acute angle between OP and the horizontal reference line is $975° - [(2 \times 360°) + 180°] = 75°$. Therefore,

$$\cos 975° = -\cos 75° = -0.2588$$

(c) OP is rotated through two complete revolutions in a clockwise direction and then lands in the fourth quadrant, which carries a negative sign for the tangent function. The acute angle between OP and the horizontal reference line is $739° - (2 \times 360°) = 19°$. Therefore,

$$\tan (-739°) = -\tan 19° = -0.3443$$

(d) OP is rotated through two complete revolutions in the clockwise direction and then lands in the second quadrant, which carries a positive sign for the sine function. The acute angle between OP and the horizontal reference line is $956° - [(2 \times 360°) + 180°] = 56°$. Therefore,

$$\sin (-956°) = +\sin 56° = 0.8290$$

Most, but not all, scientific calculators enable the values of the functions for any angle to be obtained directly. In the example of sin (−956°),

Entry, Key	Display
956° $\boxed{\pm}$	−956°
$\boxed{\sin}$	0.8290

Therefore:

$$\sin (-956°) = 0.8290$$

Example 167-2

The equation of a sinewave voltage is $e = 2.7 \sin 2\pi ft$, where $f = 200$ kHz. What is the instantaneous value of the voltage if $t = 12$ μs?

Solution

Instantaneous voltage,

$$e = 2.7 \sin (2\pi \times 200 \times 10^3 \times 12 \times 10^{-6})^c$$
$$= 2.7 \sin (15.08^c)$$
$$= 2.7 \sin (864°)$$
$$= 1.59 \text{ V}$$

Chapter 168
The sine and cosine rules

CHAPTERS 163 THROUGH 167 HAVE ONLY BEEN CONCERNED WITH RIGHT-ANGLED triangles. However, in the analysis of ac circuits the phasor diagrams can involve general triangles. Moreover, you will need more than Pythagoras' theorem to find the resultant of two phasors that are not mutually perpendicular.

A general triangle can be precisely defined in any one of three ways. In such cases the given information is:

- One side and two angles
- Two sides and the included angle
- Three sides

In the first case the triangle is solved completely with the aid of the sine rule. For the second and third cases, the cosine rule is used.

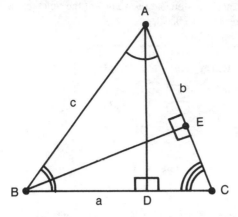

Figure 168-1

MATHEMATICAL DERIVATIONS
The sine rule

In Fig. 168-1, lines AD and BE are respectively perpendicular to BC and AC.
    In $\triangle ABD$:

$$\sin B = \frac{AD}{c} \qquad (168\text{-}1)$$

or,

$$AD = c \sin B$$

In $\triangle ADC$:

$$\sin C = \frac{AD}{b} \qquad (168\text{-}2)$$

or,

$$AD = b \sin C$$

Therefore:

$$AD = c \sin B$$
$$= b \sin C$$

and

$$\frac{c}{\sin C} = \frac{b}{\sin B} \qquad (168\text{-}3)$$

In $\triangle ABE$:

$$\sin A = \frac{BE}{c} \qquad (168\text{-}4)$$

or,

$$BE = c \sin A$$

In $\triangle BCE$:

$$\sin C = \frac{BE}{a} \qquad \text{(168-5)}$$

or,

$$BE = a \sin C$$

Therefore:

$$BE = c \sin A$$
$$= a \sin C$$

and,

$$\frac{c}{\sin C} = \frac{a}{\sin A} \qquad \text{(168-6)}$$

Combining equations 168-3 and 168-6 yields:

$$\frac{a}{\sin A} = \frac{b}{\sin B} = \frac{c}{\sin C} \qquad \text{(168-7)}$$

This important relationship is called the *sine rule* of a general triangle.

The cosine rule

In Fig. 168-2, AD is perpendicular to BC. If $BD = x$, $CD = a - x$. Then in $\triangle ABD$:

$$AD^2 = c^2 - x^2$$

and,

$$x = c \cos B$$

In $\triangle ACD$,

$$AD^2 = b^2 - (a - x)^2$$

Figure 168-2

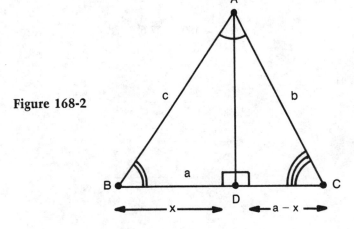

Therefore:

$$c^2 - x^2 = b^2 - (a - x)^2$$
$$= b^2 - a^2 + 2ax - x^2$$

or,

$$b^2 = a^2 + c^2 - 2ax \tag{168-8}$$
$$= a^2 + c^2 - 2ac \cos B$$

Therefore:

$$b = \sqrt{a^2 + c^2 - 2ac \cos B}$$
$$a = \sqrt{b^2 + c^2 - 2bc \cos A} \tag{168-9}$$
$$c = \sqrt{a^2 + b^2 - 2ab \cos C}$$

These relationships constitute the *cosine rule* of a general triangle. This rule will be used to find the third side of a triangle when two sides and the included angle are given. The sine rule then allows a second angle to be found.

The relationships for the cosine rule can be rearranged as:

$$\cos A = \frac{b^2 + c^2 - a^2}{2bc} \tag{168-10}$$

$$\cos B = \frac{a^2 + c^2 - b^2}{2ac} \tag{168-11}$$

$$\cos C = \frac{a^2 + b^2 - c^2}{2ab} \tag{168-12}$$

This form of the cosine rule is used to find one of the angles when the three sides are given. The sine rule then allows a second angle to be obtained. Note that in Equation 168-10, it is possible for $a^2 > b^2 + c^2$ so that the value of $\cos A$ is negative; angle A is then obtuse.

Example 168-1

In a general triangle ABC, $A = 37.6°$, $B = 78.7°$, and $c = 11.6$. Find the values of C, a, and b.

Solution

$$\text{Angle } C = 180° - 37.6° - 78.7°$$
$$= 63.7°$$

Then

$$\frac{a}{\sin A} = \frac{c}{\sin C} \tag{168-7}$$

$$\text{Side } a = \frac{c \sin A}{\sin C}$$

$$= \frac{11.6 \times \sin 37.6°}{\sin 63.7°}$$

Entry, Key	Display
11.6 $\times$	11.6
37.6° sin $\div$	7.078
63.7° sin	0.8956
=	7.895

The value of side a is 7.895.

$$\text{Side } b = \frac{c \sin B}{\sin C}$$

$$= \frac{11.6 \sin 78.7°}{\sin 63.7°}$$

$$= 12.69$$

Example 168-2

In a general triangle ABC, $b = 27.5$, $c = 47.8$, and $A = 37.2°$. Find the values of a, B, and C.

Solution

$$\text{Side } a = \sqrt{b^2 + c^2 - 2bc \cos A} \qquad (168\text{-}9)$$

$$= \sqrt{27.5^2 + 47.8^2 - 2 \times 27.5 \times 47.8 \times \cos 37.2°}$$

$$= 30.7$$

Entry, Key	Display
27.5 x^2 $+$	756.25
47.8 x^2 $-$	2284.84
$($	0
2 $\times$	2
27.5 $\times$	55
47.8 $\times$	2629
37.2° cos $)$	2094
= $\sqrt{x}$	30.7

then,

$$\frac{\sin B}{b} = \frac{\sin A}{a} \qquad (168\text{-}7)$$

Therefore:

$$\text{Angle } B = \text{inv sin} \left(\frac{b \sin A}{a} \right)$$

$$= \text{inv sin} \left(\frac{27.5 \sin 37.2°}{30.7} \right)$$

$$= 32.7°$$

and,

$$\text{Angle } C = 180° - 37.2° - 32.7°$$
$$= 110.1°$$

Example 168-3

In a general triangle ABC, $c = 23.5$, $b = 17.3$, and $a = 12.7$. Find the values of the angles C, B, and A.

Solution

$$\cos C = \frac{a^2 + b^2 - c^2}{2ab} \tag{168-12}$$

Therefore:

$$\text{Angle } C = \text{inv cos} \left(\frac{12.7^2 + 17.3^2 - 23.5^2}{2 \times 12.7 \times 17.3} \right)$$

Entry, Key	Display
12.7 x^2 $+$	161.29
17.3 x^2 $-$	299.29
23.5 x^2 $=$ $\div$	−91.67
2 $\div$	−45.835
12.7 $\div$	−3.609
17.3 $=$	−0.2086
inv cos	102.04°

Therefore angle A is obtuse and has a value of 102.04°.

Then,

$$\frac{\sin B}{b} = \frac{\sin C}{c} \tag{168-7}$$

$$\text{Angle } B = \text{inv sin} \left(\frac{b \sin C}{c} \right)$$

$$= \text{inv sin} \left(\frac{17.3 \sin 102.04°}{23.5} \right)$$

$$= 46.05°$$

$$\text{Angle } A = 180° - 102.04° - 46.05°$$
$$= 31.91°$$

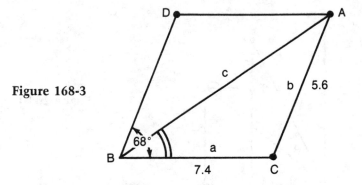

Figure 168-3

Example 168-4

In the parallelogram of Fig. 168.3, find the values of c and the angle $\angle ABC$.

Solution

$$\angle ACB = 180° - 68°$$
$$= 112° \text{ (an obtuse angle)}$$

$$\text{Side } c = \sqrt{a^2 + b^2 - 2ab \cos C} \tag{168-9}$$
$$= \sqrt{7.4^2 + 5.6^2 - 2 \times 7.4 \times 5.6 \cos 112°}$$
$$= 10.8$$

$$\angle ABC = \text{inv sin} \left(\frac{5.6 \times \sin 112°}{10.8} \right)$$
$$= 28.7°$$

This example illustrates one method of obtaining the sum of two phasors (see Chapter 64).

Chapter 169
Multiple angles

CHAPTERS 163 THROUGH 167 HAVE CONTAINED ONLY EQUATIONS THAT INVOLVE single angles, such as A and B. Now turn your attention to multiple angles that involve more than one angle; for example, $A + B$ and $A - B$ are regarded as multiple angles.

The expansions of sin $(A + B)$ and cos $(A + B)$ lead to other equations that are important in electronics and communications theory. The analysis of an amplitude-modulated wave depends on multiple angle equations. Such equations also enable us to obtain the true power in an ac circuit.

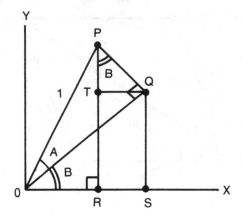

Figure 169-1

MATHEMATICAL DERIVATIONS

In Fig. 169-1, let OP be of unit length.

In $\triangle OPR$,

$$\sin (A + B) = \frac{PR}{OP} = \frac{PR}{1} = PR$$

In $\triangle OPQ$,

$$\sin A = \frac{PQ}{OP} = \frac{PQ}{1} = PQ$$

In $\triangle PTQ$,

$$\cos B = \frac{PT}{PQ} = \frac{PT}{\sin A}$$

Therefore:

$$PT = \sin A \cos B$$

In $\triangle OPQ$,

$$\cos A = \frac{OQ}{OP} = \frac{OQ}{1} = OQ$$

In $\triangle OQS$:

$$\sin B = \frac{QS}{OQ} = \frac{QS}{\cos A}$$

Therefore:

$$QS = TR = \cos A \sin B$$

Since,

$$PR = PT + TR$$

$$\sin (A + B) = \sin A \cos B + \cos A \sin B \qquad (169\text{-}1)$$

590

In △*OPR*:

$$\cos (A + B) = \frac{OR}{OP} = \frac{OR}{1} = OR$$

In △*OPQ*:

$$\cos A = \frac{OQ}{OP} = \frac{OQ}{1} = OQ$$

In △*OQS*:

$$\cos B = \frac{OS}{OQ} = \frac{OS}{\cos A}$$

Therefore:

$$OS = \cos A \cos B$$

In △*PTQ*:

$$\sin B = \frac{QT}{PQ} = \frac{QT}{\sin A}$$

Therefore:

$$QT = \sin A \sin B$$

Since,

$$OR = OS - SR = OS - QT$$

$$\cos (A + B) = \cos A \cos B - \sin A \sin B \qquad (169\text{-}2)$$

Alternative proof

In Fig. 169-2 let *OP* represent a unit vector. Resolving *OP* along the perpendicular directions *OX*, *OY*, the component along the direction *OX* = cos (*A* + *B*). The component along the direction *OY* = sin (*A* + *B*).

Resolving *OP* along the perpendicular directions *OQ*, *OR*, the component along the direction *OQ* = cos *A*. The component along the direction *OR* = sin *A*.

Resolving the components for the directions *OQ*, *OR* in the direction *OY*, the resultant along the direction *OY* = sin *A* cos *B* + cos *A* sin *B*. Therefore:

$$\sin (A + B) = \sin A \cos B + \cos A \sin B \qquad (169\text{-}1)$$

Resolving the components for the directions *OQ*, *OR* in the direction *OX*, the resultant along the direction *OX* = cos *A* cos *B* − sin *A* sin *B*. Therefore:

$$\cos (A + B) = \cos A \cos B - \sin A \sin B \qquad (169\text{-}2)$$

This vector proof is true for angles of any magnitude.

In Equation 169-1, replace +*B* by −*B*. Then,

$$\sin (A - B) = \sin A \cos (-B) + \cos A \sin (-B)$$

or:

$$\sin (A - B) = \sin A \cos B - \cos A \sin B \qquad (169\text{-}3)$$

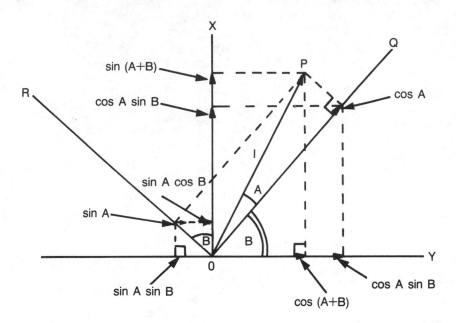

Figure 169-2

Adding Equations 169-1 and 169-3:

$$\sin (A + B) + \sin (A - B) = 2 \sin A \cos B \qquad (169\text{-}4)$$

Subtracting Equation 169-3 from 169-1:

$$\sin (A + B) - \sin (A - B) = 2 \cos A \sin B \qquad (169\text{-}5)$$

In Equation 169-2, replace $+B$ by $-B$. Then,

$$\cos (A - B) = \cos A \cos (-B) - \sin A \sin (-B)$$

or:

$$\cos (A - B) = \cos A \cos B + \sin A \sin B \qquad (169\text{-}6)$$

Adding Equations 169-2 and 169-6:

$$\cos (A + B) + \cos (A - B) = 2 \cos A \cos B \qquad (169\text{-}7)$$

Subtracting Equation 169-2 from Equation 169-6:

$$\cos (-B) - \cos (A + B) = 2 \sin A \sin B \qquad (169\text{-}8)$$

In Equation 169-1, replace B by A then,

$$\sin (A + A) = \sin A \cos A + \cos A \sin A$$

or,

$$\sin 2A = 2\sin A \cos A \qquad (169\text{-}9)$$

In Equation 169-2, replace B by A. Then,

$$\cos 2A = \cos A \cos A - \sin A \sin A \qquad (169\text{-}10)$$
$$= \cos^2 A - \sin^2 A$$
$$= \cos^2 A - (1 - \cos^2 A)$$
$$= 2 \cos^2 A - 1 \qquad (169\text{-}11)$$

or,

$$\cos 2A = \cos^2 A - \sin^2 A$$
$$= (1 - \sin^2 A) - \sin^2 A$$
$$= 1 - 2 \sin^2 A \qquad (169\text{-}12)$$

Similarly:

$$\sin 3A = \sin (A + 2A) = \sin A \cos 2A + \sin 2A \cos A$$
$$= \sin A (1 - 2 \sin^2 A) + 2 \sin A \cos A \times \cos A$$
$$= \sin A - 2 \sin^3 A + 2 \sin A (1 - \sin^2 A)$$
$$= 3 \sin A - 4 \sin^3 A \qquad (169\text{-}13)$$

$$\cos 3A = \cos (A + 2A) = \cos A \cos 2A = \sin 2A \sin A$$
$$= \cos A (2 \cos^2 A - 1) - 2 \sin A \cos A \times \sin A$$
$$= 2 \cos^3 A - \cos A - 2 \cos A (1 - \cos^2 A)$$
$$= 4 \cos^3 A - 3 \cos A \qquad (169\text{-}14)$$

Divide Equation 169-1 by 169-2. Then:

$$\frac{\sin (A + B)}{\cos (A + B)} = \frac{\sin A \cos B + \cos A \sin B}{\cos A \cos B - \sin A \sin B}$$

$$\tan (A + B) = \frac{(\sin A \cos B + \cos A \sin B)/\cos A \cos B}{(\cos A \cos B - \sin A \sin B)/\cos A \cos B}$$

$$\tan (A + B) = \frac{\tan A + \tan B}{1 - \tan A \tan B} \qquad (169\text{-}15)$$

In Equation 169-15, replace $+B$ by $-B$. Then:

$$\tan (A - B) = \frac{\tan A + \tan (-B)}{1 - \tan A \tan (-B)}$$

$$\tan (A - B) = \frac{\tan A - \tan B}{1 + \tan A \tan B} \qquad (169\text{-}16)$$

Example 169-1

In an ac circuit, the source voltage $e = 155 \sin 337t$ V, and the corresponding source current $i = 3.7 \cos 377t$ A. Calculate the value of the average power over the cycle of the source voltage.

Solution

$$\text{Instantaneous power} = e \times i$$

$$= 155 \sin 377t \times 3.7 \cos 377t$$

$$= \frac{155 \times 3.7}{2} \sin (2 \times 377t) \tag{169-9}$$

$$= 286.75 \sin 754t \text{ VAr}$$

The average value of sin $754t$ over a cycle is 0. Therefore, the average power over the cycle is zero, because the load on the source is a capacitive reactance whose value is 155 V/3.7 A = 42 Ω, rounded off.

Example 169-2

In an ac circuit, the source voltage $e = 80 \sin 1000t$ V, and the corresponding source current $i = 4 \sin (1000t - 40°)$ A. Calculate the value of the average power over the complete cycle of the source voltage.

Solution

$$\text{Instantaneous power,} \ p = e \times i = 80 \sin 1000t \times 4 \sin (1000t - 40°)$$

$$= \frac{80 \times 4}{2} [\cos 40° - \cos (2000t - 40°)] \tag{169-8}$$

$$= 160 \cos 40° - 160 \cos (2000t - 40°)$$

The average value of the term "$-160 \cos (2000t - 40°)$" is zero over the complete cycle of the source voltage. Therefore, the average or true power over the cycle is $160 \cos 40° = 122.6$ W.

Example 169-3

The equation of an amplitude-modulated wave is $e = A \sin 2\pi f_c t + B \sin 2\pi f_c t \sin 2\pi f_m t$. Analyze the wave into its separate sine and cosine components (see chapter 136).

Solution

The equation of the AM wave is:

$$e = A \sin 2\pi f_c t + B \sin 2\pi f_c t \sin 2\pi f_m t$$

$$= A \sin 2\pi f_c t + \frac{B}{2} [\cos 2\pi (f_c - f_m)t - \cos 2\pi (f_c + f_m)t] \tag{169-1}$$

The separate components are:

$A \sin 2\pi f_c t$, which is the carrier component

$\dfrac{B}{2} \cos 2\pi (f_c - f_m)t$, which is the lower sideband

$\dfrac{B}{2} \cos 2\pi (f_c + f_m)t$, which is the upper sideband.

Chapter 170
The binomial series

CONSIDER THE FOLLOWING EXPANSIONS:

$$(1 + x)^1 = 1 + x$$
$$(1 + x)^2 = 1 + 2x + x^2$$
$$(1 + x)^3 = 1 + 3x + 3x^2 + x^3$$
$$(1 + x)^4 = 1 + 4x + 6x^2 + 4x^3 + x^4$$

Clearly there is a pattern to the values of the coefficients of x, x^2, x^3, and so on. We can assume that in the expansion of $(1 + x)^5$, the coefficients of x, x^2, x^3, and x^4 are respectively $1 + 4 = 5$, $4 + 6 = 10$, $6 + 4 = 10$, $4 + 1 = 5$, so that:

$$(1 + x)^5 = 1 + 5x + 10x^2 + 10x^3 + 5x^4 + x^5$$

A general form of these expansions is given by the binomial series:

$$(1 + x)^n = 1 + \frac{nx}{1} + \frac{n(n - 1)x^2}{1 \times 2} + \frac{n(n - 1)(n - 2)x^3}{1 \times 2 \times 3}$$
$$+ \cdots + \frac{n(n - 1)(n - 2) \cdots (n - r + 1)x^r}{1 \times 2 \times 3 \times \cdots \times r} + \cdots + x^n$$

If $n = 5$,

$$(1 + x)^5 = 1 + \frac{5x}{1} + \frac{5 \times 4x^2}{1 \times 2} + \frac{5 \times 4 \times 3x^3}{1 \times 2 \times 3} + \frac{5 \times 4 \times 3 \times 2x^4}{1 \times 2 \times 3 \times 4} + \frac{5 \times 4 \times 3 \times 2 \times 1x^5}{1 \times 2 \times 3 \times 4 \times 5}$$
$$= 1 + 5x + 10x^2 + 10x^3 + 5x^4 + x^5$$

This is the same result as we previously deduced.

The terms of the binomial expansion can be simplified by using the *factorial* notation. "Factorial n" is written as $n!$ or $\angle n$ and is defined by:

$$n! = n \times (n - 1) \times (n - 2) \cdots \times 4 \times 3 \times 2 \times 1$$

As an example,

$$6! = 6 \times 5 \times 4 \times 3 \times 2 \times 1 = 720$$

The binomial expansion can then be rewritten as:

$$(1 + x)^n = 1 + \frac{nx}{1!} + \frac{n(n - 1)x^2}{2!} + \frac{n(n - 1)(n - 2)x^3}{3!}$$
$$+ \cdots + \frac{n(n - 1)(n - 2) \cdots (n - r + 1)x^r}{r!} + \cdots + x^n \qquad (170\text{-}1)$$

When the value of x is much less than 1 ($x \ll 1$), the binomial expansion can be used to simplify certain expressions. For example, $(1 + x)^6 \approx 1 + 6x$ if $x \ll 1$. The missing terms $15x^2$, $20x^3$, $15x^4$, $6x^5$, x^6 all have small values when compared to $6x$. If, for example, the value of x is 0.001, $(1 + x)^6 = (1 + 0.001)^6 \approx 1 + 0.006 = 1.006$ as opposed to the exact value of 1.006015020015006001. Incidentally, the binomial series has provided us with this exact value, which would not be available from some scientific calculators.

MATHEMATICAL DERIVATIONS

The term,

$$\frac{n(n-1)(n-2)\cdots(n-r+1)x^r}{r!}$$

is the general term of the expansion and is used to find the coefficient of x^r, where r is a positive integer.

Consider the expansion of $(1+x)^{n+1}$:

$$(1+x)^{n+1} = (1+x)(1+x)^n$$

$$= (1+x)\left[1 + \frac{nx}{1!} + \frac{n(n-1)x^2}{2!} + \cdots + \frac{n(n-1)\cdots(n-r+2)x^r}{r-1!}\;^1 \right.$$

$$\left. + \frac{n(n-1)\cdots(n-r+1)x^r}{r!} + \cdots + x^n \right]$$

The x^r term in the expansion of $(1+x)^{n+1}$ is:

$$\left[\frac{n(n-1)\cdots(n-r+2)}{r-1!} + \frac{n(n-1)\cdots(n-r+1)}{r!} \right]x^r$$

$$= \left[\frac{n(n-1)\cdots(n-r+2)}{r-1!} \times \left(1 + \frac{n-r+1}{r}\right) \right]x^r$$

$$= \frac{(n+1)(n)\cdots(n-r+2)x^r}{r!}$$

This expression is the same as would be obtained by replacing n by $n+1$ in the coefficient of the x^r term in the expansion of $(1+x)^n$. Consequently, if the series is true for $(1+x)^n$, the series must be true for $(1+x)^{n+1}$. But the series is obviously true for $n=1$ because $(1+x)^1 = 1 + x$. It follows that the series is true for $n=2$, $n=3$, and so on. The binomial theorem has, therefore, been proven if n is a positive integer. This form of proof is known as the *induction method*.

If n is not a positive integer, there is an infinite number of terms in the expansion. Under these conditions, the series is not valid unless the value of x lies between $+1$ and -1 ($1 > x > -1$).

Example 170-1

Use the binomial series to expand the following:

(a) $(1+x)^7$, (b) $(1-x)^5$, (c) $\sqrt{1+x}$, (d) $1/(1-x)$, (e) $(1+x)^{5/2}$

Solution

(a) $(1+x)^7 = 1 + 7x + \dfrac{7 \times 6}{1 \times 2}x^2 + \dfrac{7 \times 6 \times 5}{1 \times 2 \times 3}x^3 + \dfrac{7 \times 6 \times 5 \times 4}{1 \times 2 \times 3 \times 4}x^4 + \dfrac{7 \times 6 \times 5 \times 4 \times 3}{1 \times 2 \times 3 \times 4 \times 5}x^5$

$\qquad + \dfrac{7 \times 6 \times 5 \times 4 \times 3 \times 2}{1 \times 2 \times 3 \times 4 \times 5 \times 6}x^6 + \dfrac{7 \times 6 \times 5 \times 4 \times 3 \times 2 \times 1}{1 \times 2 \times 3 \times 4 \times 5 \times 6 \times 7}x^7$

$\qquad = 1 + 7x + 21x^2 + 35x^3 + 35x^4 + 21x^5 + 7x^6 + x^7$

Verification:

$$\text{If } x = 1,$$

$$(1 + 1)^7 = 2^7$$

$$= 1 + 7 + 21 + 35 + 35 + 21 + 7 + 1$$

$$= 128$$

(b) $(1 - x)^5 = 1 + 5(-x) + \dfrac{5 \times 4}{1 \times 2}(-x)^2 + \dfrac{5 \times 4 \times 3}{1 \times 2 \times 3}(-x)^3 + \dfrac{5 \times 4 \times 3 \times 2}{1 \times 2 \times 3 \times 4}(-x)^4$

$$+ \dfrac{5 \times 4 \times 3 \times 2 \times 1}{1 \times 2 \times 3 \times 4 \times 5}(-x)^5 = 1 - 5x + 10x^2 - 10x^3 + 5x^4 - x^5$$

(c) $\sqrt{1 + x} = (1 + x)^{1/2}$

$$= 1 + \frac{1}{2}x + \frac{(1/2) \times (-1/2)}{1 \times 2}x^2 + \frac{(1/2) \times (-1/2) \times (-3/2)}{1 \times 2 \times 3}x^3$$

$$+ \frac{(1/2) \times (-1/2) \times (-3/2) \times (-5/2)}{1 \times 2 \times 3 \times 4}x^4 \cdots$$

$$= 1 + \frac{1}{2}x - \frac{1}{8}x^2 + \frac{1}{16}x^3 - \frac{5}{128}x^4 \cdots$$

Note that there is an infinite number of terms in the series.
If $x \ll 1$:

$$\sqrt{1 + x} \approx 1 + \frac{1}{2}x$$

For example, if $x = 0.004$,

$$\sqrt{1 + 0.004} = (1.004)^{1/2} \approx 1 + \frac{0.004}{2}$$

$$= 1.002$$

(d) $\dfrac{1}{1 - x} = (1 - x)^{-1}$

$$= 1 + (-1)(-x) + \frac{(-1)(-2)}{1 \times 2}(-x)^2 + \frac{(-1)(-2)(-3)}{1 \times 2 \times 3}(-x)^3 + \cdots$$

$$= 1 + x + x^2 + x^3 + \cdots$$

If $x = 2$,

$$\frac{1}{1 - x} = -1,$$

which is clearly not equal to $1 + 2 + 2^2 + 2^3 + \cdots$. The expansion is not valid when $x > 1$.
If $x \ll 1$:

$$\frac{1}{1 - x} = (1 - x)^{-1} \approx 1 - x$$

For example, if $x = 0.003$,

$$\frac{1}{1 - x} = \frac{1}{1 - 0.003}$$

$$= \frac{1}{0.997} \approx 1 + x$$

$$= 1 + 0.003$$

$$= 1.003$$

(e) $(1 + x)^{5/2} = 1 + \dfrac{5x}{2} + \dfrac{(5/2)(3/2)x^2}{1 \times 2} + \dfrac{(5/2)(3/2)(1/2)x^3}{1 \times 2 \times 3} + \dfrac{(5/2)(3/2)(1/2)(-1/2)x^4}{1 \times 2 \times 3 \times 4} + \cdots$$

$$= 1 + \frac{5x}{2} + \frac{15x^2}{8} + \frac{5x^3}{16} - \frac{5x^4}{128} \cdots$$

If $x \ll 1$

$$(1 + x)^{5/2} \approx 1 + \frac{5}{2}x$$

For example, if $x = 0.005$:

$$(1 + x)^{5/2} = (1.005)^{5/2} \approx 1 + \frac{5}{2} \times 0.005$$

$$= 1.0125$$

Example 170-2

On a transmission line, the propagation constant,

$$\gamma = \alpha + j\beta$$

$$= \sqrt{(R + j\omega L)(G + j\omega C)}$$

(chapter 148). If $\omega L \gg R$ and $\omega C \gg G$, find the values of α and β.

Solution

Propagation constant,

$$\gamma = j\omega\sqrt{LC}\sqrt{\left(1 + \frac{R}{j\omega L}\right)\left(1 + \frac{G}{j\omega C}\right)}$$

$$= j\omega\sqrt{LC} \times \left(1 + \frac{R}{j\omega L}\right)^{1/2} \times \left(1 + \frac{G}{j\omega C}\right)^{1/2}$$

$$= j\omega\sqrt{LC}\left(1 + \frac{R}{2j\omega L} + \text{neglected terms}\right)\left(1 + \frac{G}{2j\omega C} + \text{neglected terms}\right) \quad (170\text{-}1)$$

using the binomial expansion.
Therefore:

$$\alpha + j\beta = j\omega\sqrt{LC}\left(1 + \frac{R}{2j\omega L} + \frac{G}{2j\omega C} + \text{neglected terms}\right)$$

$$= \frac{R}{2}\sqrt{\frac{C}{L}} + \frac{G}{2}\sqrt{\frac{L}{C}} + j\omega\sqrt{LC}$$

Equating real and imaginary parts, (see Chapter 88),

$$\text{Attenuation constant, } \alpha = \frac{R}{2}\sqrt{\frac{C}{L}} + \frac{G}{2}\sqrt{\frac{L}{C}} \text{ nepers per meter}$$

$$\text{Phase shift constant, } \beta = \omega\sqrt{LC} \text{ radians per meter.}$$

Chapter 171
The exponential and
logarithmic functions

EXPONENTIAL GROWTH AND DECAY OCCUR FREQUENTLY IN VARIOUS SCIENTIFIC subjects. For example, trees grow exponentially since their heights increase rapidly when they are initially planted. As the tree becomes taller, its height increases less rapidly; theoretically it would take an infinite time to achieve its maximum height, and it is doubtful whether a tree could live that long! However, in a limited time period the tree attains over 99 percent of its theoretical maximum height.

Newton's Law of Cooling involves exponential decay; this law states that the rate of heat loss from a hot body is directly proportional to the difference between the temperature of the body and the temperature of its surroundings. Consequently, as the temperature of the body approaches that of its surroundings, the rate of heat loss lessens. Theoretically, it would take an infinite time for the temperatures of the body and its surroundings to become equal, but the body loses virtually all of its excess heat in a limited period of time.

Radioactive substances decay exponentially, so as their intensities decline, the rate of decline is reduced. To lose all of the radioactivity would theoretically require inifinite time, but we can refer to the *half-life period,* which is the time taken for the radioactivity to fall to half its initial value. For different substances the half-life period varies from less than one microsecond to several thousand years.

In electronics the charging of a capacitor through a resistor is an example of exponential growth, while its discharge involves exponential decay (see Chapter 51). On a transmission line used in radio communications, the voltage and current waves decay exponentially as they travel along the line. The equations involving both growth and decay are explored in the mathematical derivations.

Napierian logarithms are closely related to the exponential number e (approximately 2.7183). The logarithm of a number, N, to base, a, is the exponent or power to which a must be raised to produce N. For example the logarithm of 8 to the base 2 (written as $\log_2 8$) equals 3 because $2^3 = 8$. Napierian, or natural logarithms use the base e so that $\log_e 8 = 2.08$ approximately, because $2.7183^{2.08} \approx 8$. Note that the natural logarithm is normally abbreviated ln and consequently if $\ln N = x$, $N = e^x$.

MATHEMATICAL DERIVATIONS

The exponential series is derived from the Binomial Theorem.

Consider the expansion of $\left(1 + \dfrac{1}{n}\right)^{nx}$.

The rth term of this expansion is:

$$\frac{nx(nx - 1)(nx - 2) \cdots (nx - r + 1)}{r!} \times \frac{1}{n^r}$$

As $n \to \infty$, the value of each term in parentheses in the numerator approaches nx.

Therefore:

$$\text{as } n \to \infty, \text{ the } r^{th} \text{ term} \to \frac{x^r}{r!}$$

Consequently when $n \to \infty$,

$$\left(1 + \frac{1}{n}\right)^{nx} = 1 + \frac{x}{1!} + \frac{x^2}{2!} + \cdots + \frac{x^r}{r!} + \cdots \text{ to infinity} \qquad (171\text{-}1)$$

It can be shown that this infinite series is valid for all values of x.

Equation 171-1 can be written as,

$$e^x = 1 + \frac{x}{1!} + \frac{x^2}{2!} + \cdots + \frac{x^r}{r!} + \cdots \text{ to infinity}$$

where:

$$e = \underset{n \to \infty}{\text{Limit}} \left(1 + \frac{1}{n}\right)^n \qquad (171\text{-}2)$$

If $x = 1$,

$$e^x = e^1 = e = 1 + \frac{1}{1!} + \frac{1}{2!} + \frac{1}{3!} + \cdots \text{ to infinity}$$

or

$$e = 2.71828 \cdots$$

The graphs of e^x (exponential growth) and e^{-x} (exponential decay) versus x are shown in Fig. 171-1. However the more common equation for exponential growth is $y = 1 - e^{-x}$ and the graphs of $1 - e^{-x}$ and e^{-x} versus x are plotted in Fig. 171-2 for values of x between 0 and 5 (Also see Table 171-1).

Table 171-1. Exponential growth and decay vs. x.

x	0	1	2	3	4	5
$1 - e^{-x}$	0	0.632	0.865	0.950	0.982	0.993
e^{-x}	1	0.368	0.135	0.050	0.018	0.007

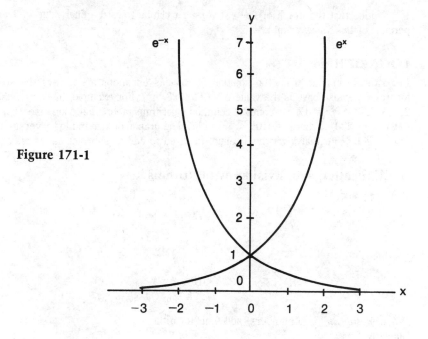

Figure 171-1

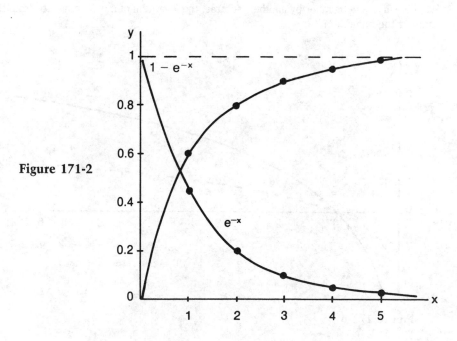

Figure 171-2

Notice that the graph of $1 - e^{-x}$ versus x climbs toward a final value of 1 and is within 1 percent of this value when $x = 5$.

LOGARITHMS

Logarithms obtained from the solution of calculus equations are always the so-called natural logarithms whose base is the value of e (2.71828. . .). For example, ln 17 = 2.833. . . because $2.71828. . .^{2.833} = 17$. By contrast common logarithms (abbreviated log) use a base of 10 so that log 17 = 1.230. . . because $10^{1.2.3.0.\cdots} = 17$. The graphs of ln x and log x versus x are shown in Fig. 171-3. Note that logarithms cannot be used to add or subtract numbers.

Multiplication and division by logarithms

If $N = a^x$ and $M = a^y$,

$$NM = a^x \times a^y$$
$$= a^{x+y}$$

Therefore:

$$\log_a (NM) = x + y \qquad (171\text{-}3)$$
$$= \log_a N + \log_a M$$

When *multiplying* two numbers, add their logarithms.
Similarly,

$$\log_a \frac{N}{M} = \log_a N - \log_a M \qquad (171\text{-}4)$$

When *dividing* one number by another, subtract the logarithm of the denominator from the logarithm of the numerator.

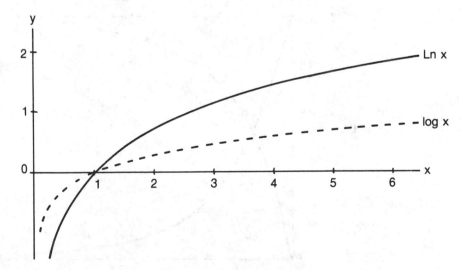

Figure 171-3

Exponents or powers

If the number M is raised to the power n.

$$M^n = (a^y)^n = a^{ny}$$

and

$$\log_a M^n = ny = n \log_a M \tag{171-5}$$

When raising a number to a certain power (exponent), *multiply* the logarithm of the number by the value of the exponent.

Also:

$$\log_a \frac{1}{M} = \log_a M^{-1} = -y = -\log_a M \tag{171-6}$$

The logarithm of a number's reciprocal is equal to the negative of the number's logarithm. Note that the logarithm of 1 to any base is equal to 0 and the logarithm of 0 to a positive base greater than 1 is negative infinity.

Example 171-1

In the equation $y = 1 - e^{-x}$, find the value of x when $y = 0.9$. In the same equation, find the value of y if $x = 6$.

Solution

When $y = 0.9$,

$$e^{-x} = 1 - 0.9 = 0.1$$

so that $e^x = 10$.
Then

$$x = \ln 10 = 2.3$$

When $x = 6$,

$$y = 1 - e^{-x} = 1 - e^{-6}$$
$$= 1 - 0.002479$$
$$= 0.997521$$

Example 171-2

In the equation $y = e^{-x}$, find the value of x when the value of $y = 7.6$.

Solution

$$\text{If } y = 7.6, \ e^x = \frac{1}{7.6} = 0.1316.$$

Then

$$x = \ln 0.1316 = -2.03$$

Example 171-3

Obtain the values of: (a) $\log_2 16$, (b) $\log_2 1/32$, (c) $\log_8 128$, (d) $\log_{25} 1/125$.

Solution

(a)
$$\log_2 16 = \log_2 2^4$$
$$= 4$$

(b)
$$\log_2 1/32 = -\log_2 32$$
$$= -\log_2 2^5$$
$$= -5$$

(c)
$$\log_8 128 = \log_8 (64 \times 2) \qquad\qquad (171\text{-}3)$$
$$= \log_8 64 + \log_8 2$$
$$= \log_8 8^2 + \log_8 8^{1/3}$$
$$= 2 + 1/3 = 2\frac{1}{3}$$
$$= \frac{7}{3}$$

(d)
$$\log_{25} 1/125 = -\log_{25} 125$$
$$= -\log_{25} (5 \times 25)$$
$$= -\log_{25} 25^{1/2} - \log_{25} 25$$
$$= -\frac{1}{2} - 1$$
$$= -\frac{3}{2}$$

Chapter 172
Introduction to differential calculus

DIFFERENTIAL CALCULUS IS INVOLVED WITH THE RATE OF CHANGE BETWEEN ONE variable and another. For example, if one variable, y, is a function of another variable, x ($y = f(x)$), by how much will y increase (or decrease) if the value of x is increased by some small amount? For functions in general, the answer will depend on the initial value of x. This fact is illustrated graphically in Fig. 172-1. When the curve of $y = f(x)$ is drawn, the rate of change of y with respect to x is proportional to the slope of the curve at a particular point, P, corresponding to the initial value of x. If the curve is very steep at that point, a small increase in x will produce a proportionately large increase in y. By contrast, if the curve is fairly flat, the increase in y will be relatively small.

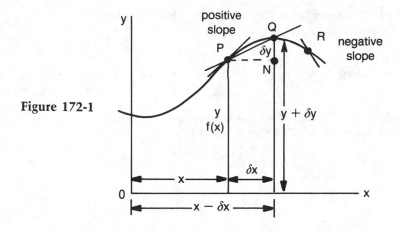

Figure 172-1

For any graph other than a straight line, the value of the slope, or the *rate of change,* varies from one point to another. If the slope is to be measured by considering the ratio of a change in y to a corresponding change in x, it is vital that these changes be as small as possible; otherwise the ratio will not provide a true value of the curve's slope at the chosen point.

The small changes in the values of x and y are usually denoted by δx and δy (alternatively Δx, Δy) with the Greek lowercase letter, δ, or capital letter, Δ, (delta) signifying "a small change in. . .". Note that if δx is small, $(\delta x)^2$ will be even smaller and its value can be neglected in comparison with the value of δx. Higher powers of δx such as $(\delta x)^3$ and $(\delta x)^4$ are smaller still.

In Fig. 172-1 the rate of change is determined by the ratio of the changes, δy, δx in y and x or by the value of $\delta y/\delta x$. To find the slope at the point $P(x, y)$, we select an adjacent point Q, whose coordinates are $(x + \delta x, y + \delta y)$. The value of the ratio $\delta y/\delta x$ is equal to the tangent function of the angle QPN and therefore to the slope of the line PQ. If Q is moved closer and closer to P, the value of $\delta y/\delta x$ approximates more and more to the slope of the curve. If δy and δx both tend to be zero we have the condition for $\delta y/\delta x$ to equal the slope of the tangent line at point P.

Although both δy and δx become zero as Q reaches P, the ratio $\delta y/\delta x$ approaches some finite value, which is denoted by dy/dx. In other words, dy/dx is the limit of the ratio $\delta y/\delta x$ as δy and $\delta x \to 0$. In equation form:

$$\frac{dy}{dx} = \operatorname*{Limit}_{\delta x, \delta y \to 0} \frac{\delta y}{\delta x}$$

The quantity dy/dx is equal to the rate of change of y with respect to x; this is, in turn, equal to the value of the slope at point P. Provided the graph of y versus x is not a straight line; the value of the slope depends on the position of the point, P, and therefore dy/dx is a function of x. To find the numerical value of the slope at a particular point, the value of x for that point must be substituted in the expression for dy/dx.

It was stated that y is a function of x, as denoted by $f(x)$. The quantity dy/dx is also a function of x and is denoted by $f'(x)$, which is the result of differentiating y with respect to x. The function $f'(x)$ is called the *first derivative,* or *first differential coefficient,* of $f(x)$. As an equation,

$$\frac{dy}{dx} = \frac{df(x)}{dx} = f'(x)$$

MATHEMATICAL DERIVATIONS

Calculation of $dy/dx = f'(x)$,

The quantity $f'(x)$ is defined as the limit of $\delta y/\delta x$ when δx, $\delta y \to 0$.

At the point P,

$$y = f(x) \qquad (172\text{-}1)$$

At the point Q, x increases to $x + \delta x$, and as a result, y increases to $y + \delta y$. Therefore,

$$y + \delta y = f(x + \delta x) \qquad (172\text{-}2)$$

This condition must be fulfilled in order for point Q to be on the curve.

Subtracting Equation 172-2 from Equation 172-1:

$$\delta y = f(x + \delta x) - f(x)$$

and

$$\frac{\delta y}{\delta x} = \frac{f(x + \delta x) - f(x)}{\delta x} \qquad (172\text{-}3)$$

Then the first derivative,

$$f'(x) = \frac{dy}{dx} \qquad (172\text{-}4)$$

$$= \underset{\delta x, \delta y \to 0}{\text{Limit}} \frac{\delta y}{\delta x}$$

$$= \underset{\delta x, \delta y \to 0}{\text{Limit}} \frac{f(x + \delta x) - f(x)}{\delta x}$$

If this procedure is repeated for point R in Fig. 172-1, the increase from x to $x + \delta x$ will be accompanied by a decrease from y to $y - \delta y$. It follows that the slope at point R is negative.

In the particular case of $y = x^2$, the graph of y versus x is a parabola, which is illustrated in Fig. 172-2. At points $P(1,1)$ and $Q(3,9)$, tangent lines have been estimated and their slopes have been calculated. At point P, for which $x = 1$, the slope is 2, while at point Q, where $x = 3$, the slope is 6. These results are compatible with the equation $dy/dx = 2x$.

Substituting $f(x) = y = x^2$ in Equation 172-4:

$$f'(x) = \frac{dy}{dx} = \underset{\delta x, \delta y \to 0}{\text{Limit}} \frac{(x + \delta x)^2 - x^2}{\delta x}$$

$$= \underset{\delta x, \delta y \to 0}{\text{Limit}} \frac{x^2 + 2x\delta x + (\delta x)^2 - x^2}{\delta x}$$

$$= \underset{\delta x, \delta y \to 0}{\text{Limit}} \ 2x + \delta x$$

$$= 2x$$

This proves our deduction that the slope of $y = x^2$ at any point is equal to $2x$. For example, at the point $(3,9)$, the slope is $2 \times 3 = 6$.

Derivative of x^n

Let $f(x) = x^n$. Then,

$$\frac{\delta y}{\delta x} = \frac{(x + \delta x)^n - x^n}{\delta x}$$

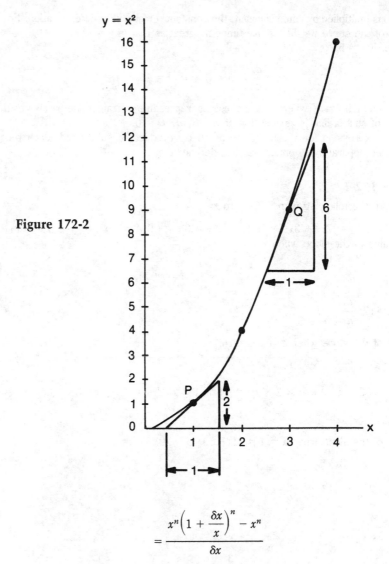

Figure 172-2

$$= \frac{x^n\left(1 + \dfrac{\delta x}{x}\right)^n - x^n}{\delta x}$$

Expand $[1 + (\delta x/x)]^n$ by the binomial series, where n and x can have any value because $-1 < \delta x/x < 1$ (see Chapter 170). Then,

$$\frac{\delta y}{\delta x} = \frac{x^n\left[1 + n\dfrac{\delta x}{x} + \dfrac{n(n-1)}{2!}\left(\dfrac{\delta x}{x}\right)^2 + \cdots\right] - x^n}{\delta x}$$

$$= nx^{n-1} + \frac{n(n-1)x^{n-2}}{2!} \cdot \delta x + \cdots$$

In the limit when δx, $\delta y \to 0$,

$$\frac{dy}{dx} = nx^{n-1} \tag{172-5}$$

If $f(x)$ is multiplied by some constant, the constant remains after differentiation. For example the slope of the curve $y = 5x^3$ is five times as great as that of $y = x^3$. Therefore, if $y = 5x^3$:

$$\frac{dy}{dx} = 5 \times 3 \times x^{3-1} = 15x^2$$

The equation $y = K$ where K is a constant, represents a straight line parallel to the x axis. The slope of such a line is zero so that if $y = K$, $dy/dx = 0$.

If the expression for y contains terms that are added or subtracted, each term can be differentiated separately to obtain the first derivative.

Example 172-1

Differentiate the following with respect to x:

(a) $y = -17$, (b) $y = 7x^3 + 3x^2 - 5x + 6$, (c) $y = 3\sqrt{x}$, (d) $y = 11/x^2$, (e) $y = 5/x^{3/2}$. In each case find the value of the slope when $x = 1$.

Solution

(a) If $y = -17$, $\dfrac{dy}{dx} = 0$.

The value of the slope when $x = 1$ is zero.

(b) If $y = 7x^3 + 3x^2 - 5x + 6$,

$$\frac{dy}{dx} = 7 \times 3 \times x^{3-1} + 3 \times 2 \times x^{2-1} - 5 \times 1 \times x^{1-1} + 0$$

$$= 21x^2 + 6x - 5$$

The value of the slope when $x = 1$ is $21 + 6 - 5 = 22$

(c) If $y = 3\sqrt{x} = 3x^{1/2}$,

$$\frac{dy}{dx} = 3 \times \frac{1}{2} \times x^{1/2-1} = \frac{3}{2}x^{-1/2}$$

$$= \frac{3}{2\sqrt{x}}.$$

The value of the slope when $x = 1$ is $\dfrac{3}{2\sqrt{1}} = 1.5$

(d) If $y = \dfrac{11}{x^2} = 11x^{-2}$,

$$\frac{dy}{dx} = 11 \times (-2) \times x^{-2-1} = -22x^{-3}$$

$$= \frac{-22}{x^3}.$$

The value of the slope when $x = 1$ is $-22/1^3 = -22$

(e) If $y = \dfrac{5}{x^{3/2}} = 5x^{-3/2}$

$$\frac{dy}{dx} = 5 \times \left(\frac{-3}{2}\right) \times x^{-3/2-1}$$

$$= \frac{-15}{2}x^{-5/2}$$

The value of the slope when $x = 1$ is $-15/2 \times (1)^{-5/2} = -7.5$

Chapter 173
Differentiation of a function of a function

IF, FOR EXAMPLE, $y = f(x) = (3x^2 - 7x)^3$, WE COULD, WITH SOME DIFFICULTY, EXPAND the expression and then differentiate each term separately. However the process would be very tedious, and it is preferable to regard y as equal to u^3 where $u = 3x^2 - 7x$. In other words, y is a function of u where u is a function of x. We can, therefore, say that y is a function of a function of x; mathematically this could be shown as $y = F(f(x))$.

Since $\dfrac{\delta y}{\delta x} = \dfrac{\delta y}{\delta u} \times \dfrac{\delta u}{\delta x}$,

$$\frac{dy}{dx} = \frac{dy}{du} \times \frac{du}{dx} \tag{173-1}$$

in the limit as δy, δx, $\delta u \to 0$.
Therefore:

$$y = u^3, \frac{dy}{du} = 3u^2 = 3(3x^2 - 7x)^2$$

$$u = 3x^2 - 7x, \frac{du}{dx} = 6x - 7$$

Consequently,

$$\frac{dy}{dx} = \frac{dy}{du} \times \frac{du}{dx}$$
$$= 3(3x^2 - 7x)^2(6x - 7)$$

If the expression for y had been expanded:

$$y = (3x^2 - 7x)^3$$
$$= (9x^4 - 42x^3 + 49x^2)(3x^2 - 7x)$$
$$= 27x^6 - 126x^5 + 147x^4 - 63x^5 + 294x^4 - 343x^3$$
$$= 27x^6 - 189x^5 + 441x^4 - 343x^3$$

$$\frac{dy}{dx} = 162x^5 - 945x^4 + 1764x^3 - 1029x^2$$

Comparing this result with:

$$\frac{dy}{dx} = 3(3x^2 - 7x)^2 \times (6x - 7)$$

$$= 3(9x^4 - 42x^3 + 49x^2) \times (6x - 7)$$

$$= 3(54x^5 - 252x^4 + 294x^3 - 63x^4 + 294x^3 \times 343x^2)$$

$$= 3(54x^3 - 315x^4 + 588x^3 - 343x^2)$$

$$= 162x^5 - 945x^4 + 1764x^3 - 1029x^2$$

The two methods produce the same result, but it is much easier and quicker to obtain $3(3x^2 - 7x)^2 \times (6x - 7)$, rather than $162x^5 - 945x^4 + 1764x^3 - 1029x^2$. Moreover, if $y = (3x^2 - 7x)^{20}$, expansion is virtually impossible, but by using the function of the function, $dy/dx = 20(3x^2 - 7x)^{19} \times (6x - 7)$.

Note that a function such as $y = \sqrt{3x^2 - 7x} = (3x^2 - 7x)^{1/2}$ cannot be expanded, but by using the function of a function method,

$$\frac{dy}{dx} = \frac{1}{2}(3x^2 - 7x)^{1/2-1} \times (6x - 7)$$

$$= \frac{6x - 7}{2\sqrt{3x^2 - 7x}}$$

MATHEMATICAL DERIVATIONS

The principle of differentiating the function of a function can be extended to any number of functions. For example, if y is a function of u, u is a function of v and v is a function of x,

$$\frac{\delta y}{\delta x} = \frac{\delta y}{\delta u} \times \frac{\delta u}{\delta v} \times \frac{\delta v}{\delta x} \tag{173-2}$$

then

$$\frac{dy}{dx} = \frac{dy}{du} \times \frac{du}{dv} \times \frac{dv}{dx}$$

in the limit as δy, δx, δu, $\delta v \to 0$.
As an example, let $y = 7(1 + 4\sqrt{5x^3 - 8x})^4$. Then,

$$y = 7u^4$$

$$u = 1 + 4\sqrt{v}$$

$$v = 5x^3 - 8x$$

$$\frac{dy}{du} = 28u^3$$

$$\frac{du}{dv} = 2v^{-1/2}$$

$$\frac{dv}{dx} = 15x^2 - 8$$

Therefore:

$$\frac{dy}{dx} = 28u^3 \times 2v^{-1/2} \times (15x^2 - 8)$$

$$= 56\frac{(1 + 4\sqrt{5x^3 - 8x})^3 \times (15x^2 - 8)}{\sqrt{5x^3 - 8x}}$$

Example 173-1

Differentiate the following with respect to x.

(a) $5(3x^2 + 2)^8$, (b) $8\sqrt{x^3 - 3x}$, (c) $\dfrac{6}{(5x^3 + 6x - 2)^3}$, (d) $4(5x^4 - 4x^2)^{1/3}$.

Solution

(a) If $y = 5(3x^2 + 2)^8$,

$$\frac{dy}{dx} = 5 \times 8 \times (3x^2 + 2)^{8-1} \times 6x$$

$$= 240x \times (3x^2 + 2)^7.$$

(b) If $y = 8\sqrt{x^3 - 3x} = 8(x^3 - 3x)^{1/2}$,

$$\frac{dy}{dx} = 8 \times \frac{1}{2} \times (x^3 - 3x)^{1/2-1} \times (3x^2 - 3)$$

$$= \frac{12(x^2 - 1)}{\sqrt{x^3 - 3x}}$$

(c) If $y = \dfrac{6}{(5x^3 + 6x - 2)^3} = 6 \times (5x^3 + 6x - 2)^{-3}$,

$$\frac{dy}{dx} = 6 \times (-3) \times (5x^3 + 6x - 2)^{-4} \times (15x^2 + 6)$$

$$= \frac{-18(15x^2 + 6)}{(5x^3 + 6x - 2)^4}.$$

(d) If $y = 4(5x^4 - 4x^2)^{1/3}$,

$$\frac{dy}{dx} = 4 \times \frac{1}{3} \times (5x^4 - 4x^2)^{1/3-1} \times (20x^3 - 8x)$$

$$= \frac{16(5x^3 - 2x)}{3(5x^4 - 4x^2)^{2/3}}$$

Chapter 174
Differentiation of a product

IF YOU WERE ASKED TO OBTAIN THE FIRST DERIVATIVE OF $(2x + 7)(3x^2 + 4)$, YOU could multiply out the expression and then differentiate each term separately.

The procedure would be as follows:

$$y = (2x + 7)(3x^2 + 4)$$
$$= 6x^3 + 21x^2 + 8x + 28$$

Then,

$$\frac{dy}{dx} = 18x^2 + 42x + 8.$$

The procedure in this case is relatively simple, but if we need to obtain the derivative of $(2x + 7)^6(3x^2 + 4)^9$, multiplication would be extremely tedious and the first derivative would contain more than twenty terms. However, we already know how to differentiate $(2x + 7)^6$ and $(3x^2 + 4)^9$ by using the function of a function process, as outlined in Chapter 173. We therefore require a formula for differentiating the product of these two functions of x.

MATHEMATICAL DERIVATIONS

Consider $y = f(x) = uv$, where u and v are both functions of x. Then,

$$\frac{\delta y}{\delta x} = \frac{(u + \delta u)(v + \delta v) - uv}{\delta x}$$

$$= \frac{uv + u\delta v + v\delta u + \delta u\delta v - uv}{\delta x}$$

$$= u\frac{\delta v}{\delta x} + v\frac{\delta u}{\delta x} + \frac{\delta u\delta v}{\delta x}$$

As δx approaches zero, so also will δu and δv. The term $\delta u\delta v/\delta x$ is then infinitesimally small and can be neglected.

In the limits as δy, δx, δu, $\delta v \to 0$,

$$\frac{dy}{dx} = u\frac{dv}{dx} + v\frac{du}{dx} \tag{174-1}$$

If $y = (2x + 7)(3x^2 + 4)$, let $u = 2x + 7$ and $v = 3x^2 + 4$. Then,

$$\frac{dy}{dx} = \underbrace{(2x + 7)}_{} \times 6x + \underbrace{(3x^2 + 4)}_{} \times 2$$
$$u \times \frac{dv}{dx} \qquad + \qquad v \times \frac{du}{dx}$$

$$= 12x^2 + 42x + 6x^2 + 8$$

$$= 18x^2 + 42x + 8.$$

The result is the same as the one we previously obtained.

If $y = (2x + 7)^6(3x^2 + 4)^9$, let $u = (2x + 7)^6$ and $v = (3x^2 + 4)^9$. Then,

$$\frac{dy}{dx} = \underbrace{(2x + 7)^6}_{u} \times \underbrace{9 \times (3x^2 + 4)^8 \times 6x}_{\dfrac{dv}{dx}}$$

$$+ \underbrace{(3x^2 + 4)^9}_{v} \times \underbrace{6 \times (2x + 7)^5 \times 2}_{\dfrac{du}{dx}}$$

$$= 54x(2x + 7)^6(3x^2 + 4)^8 + 12(3x^2 + 4)^9(2x + 7)^5$$

Example 174-1

Differentiate the following with respect to x:

(a) $\sqrt{3x^2 + 7x} \times (8x^3 - 6)^{-2}$, (b) $\dfrac{7x^{3/2}}{2x^3 - 5x^2}$.

Solution

(a) $y = \sqrt{3x^2 + 7x} \times (8x^3 - 6)^{-2}$

Let $u = \sqrt{3x^2 + 7x} = (3x^2 + 7x)^{1/2}$ and $v = (8x^3 - 6)^{-2}$. Then,

$$\frac{dy}{dx} = \underbrace{(3x^2 + 7x)^{1/2}}_{u} \times \underbrace{(-2) \times (8x^3 - 6)^{-3} \times 24x^2}_{\dfrac{dv}{dx}}$$

$$+ \underbrace{(8x^3 - 6)^{-2}}_{v} \times \underbrace{\left(\frac{1}{2}\right) \times (3x^2 + 7x)^{-1/2} \times (6x + 7)}_{\dfrac{du}{dx}}$$

$$= \frac{-48x^2\sqrt{3x^2 + 7x}}{(8x^3 - 6)^3} + \frac{6x + 7}{2(8x^3 - 6)^2\sqrt{3x^2 + 7x}}.$$

(b) $y = \dfrac{7x^{3/2}}{(2x^3 - 5x^2)^2} = 7x^{3/2} \times (2x^3 - 5x^2)^{-2}$

Let $u = 7x^{3/2}$ and $v = (2x^3 - 5x^2)^{-2}$. Then

$$\frac{dy}{dx} = \underbrace{7x^{3/2}}_{u} \times \underbrace{(-2) \times (2x^3 - 5x^2)^{-3} \times (6x^2 - 10x)}_{\dfrac{dv}{dx}}$$

$$+ \underbrace{(2x^3 - 5x^2)^{-2}}_{v} \times \underbrace{7 \times 3/2 \times x^{1/2}}_{\dfrac{du}{dx}}$$

$$= \frac{-14x^{3/2}(6x^2 - 10x)}{(2x^3 - 5x^2)^3} + \frac{21x^{1/2}}{2(2x^3 - 5x^2)^2}$$

Example 174-2

The expression for the power dissipated in a load is:

$$P_L = \frac{E^2 R_L}{(R_i + R_L)^2}$$

where E and R_i are constants. Differentiate P_L with respect to R_L.

Solution

Power, $P_L = \dfrac{E^2 R_L}{(R_i + R_L)^2}$

$$= E^2 R_L \times (R_i + R_L)^{-2}$$

Let $u = E^2 R_L$ and $v = (R_i + R_L)^{-2}$. Then,

$$\frac{dP_L}{dR_L} = E^2 R_L \times (-2) \times (R_i + R_L)^{-3} + (R_i + R_L)^{-2} \times E^2$$

$$= -\frac{2E^2 R_L}{(R_i + R_L)^3} + \frac{E^2}{(R_i + R_L)^2}$$

$$= \frac{E^2(R_i - R_L)}{(R_i + R_L)^3}$$

Chapter 175
Differentiation of a quotient

IN THE SOLUTION TO EXAMPLE 174-2, WE DIFFERENTIATED THE FUNCTION:

$$f(R_L) = P_L = \frac{E^2 R_L}{(R_i + R_L)^2}$$

by rearranging $f(R_L)$ as the product $E^2 R_L \times (R_i + R_L)^{-2}$. However, as originally expressed, $f(R_L)$ is a quotient, and therefore we need to know the formula for differentiating all such quotients.

MATHEMATICAL DERIVATIONS

Let $y = u/v$, where u and v are both functions of x. Then,

$$\frac{\delta y}{\delta x} = \frac{\dfrac{u + \delta u}{v + \delta v} - \dfrac{u}{v}}{\delta x}$$

$$= \frac{\dfrac{u}{v}\left[\dfrac{1 + \dfrac{\delta u}{u}}{1 + \dfrac{\delta v}{v}}\right] - \dfrac{u}{v}}{\delta x}$$

$$= \frac{\dfrac{u}{v}\left(1 + \dfrac{\delta u}{u}\right)\left(1 + \dfrac{\delta v}{v}\right)^{-1} - \dfrac{u}{v}}{\delta x}$$

Expanding $(1 + \delta v/v)^{-1}$ by the binomial series and neglecting $(\delta v/v)^2$ as well as all other higher-order terms, then

$$\frac{\delta y}{\delta x} = \frac{\dfrac{u}{v}\left(1 + \dfrac{\delta u}{u}\right)\left(1 - \dfrac{\delta v}{v} + \text{neglected terms}\right) - \dfrac{u}{v}}{\delta x}$$

$$\frac{\delta y}{\delta x} = \frac{\dfrac{u}{v}\left(1 + \dfrac{\delta u}{u} - \dfrac{\delta v}{v} - \dfrac{\delta u \delta v}{uv}\right) - \dfrac{\delta u}{v}}{x}$$

$$= \frac{\dfrac{\delta u}{v} - \dfrac{u\delta v}{v^2} - \dfrac{\delta u \delta v}{v^2}}{\delta x}$$

As δy, δx, δu, $\delta v \to 0$, $\delta u \delta v / v^2 \to 0$ and therefore:

$$\frac{dy}{dx} = \frac{v\dfrac{du}{dx} - u\dfrac{dv}{dx}}{v^2} \tag{175-1}$$

This is the fomula for obtaining the first derivative of a quotient.

Example 175-1

Differentiate the following with respect to x.

(a) $3x^2/(x^3 + 7)$, (b) $(5x - 6)/(x^4 - 2x^2)$, (c) $(2x^3 - 3x)/\sqrt{x^2 + 7}$.

Solution

(a) $y = \dfrac{3x^2}{x^3 + 7}$

Let $u = 3x^2$ and $v = x^3 + 7$. Then,

$$\frac{dy}{dx} = \frac{(x^3 + 7)6x - 3x^2 \times 3x^2}{(x^3 + 7)^2}$$

$$= \frac{-3x^4 + 42x}{(x^3 + 7)^2} \tag{175-1}$$

(b) $y = \dfrac{5x - 6}{x^4 - 2x^2}$

Let $u = 5x - 6$ and $v = x^4 - 2x^2$. Then,

$$\frac{dy}{dx} = \frac{(x^4 - 2x^2) \times 5 - (5x - 6) \times (4x^3 - 4x)}{(x^4 - 2x^2)^2}$$

$$= \frac{-15x^4 + 24x^3 + 10x^2 - 24x}{(x^4 - 2x^2)^2}$$

(c) $y = \dfrac{2x^3 - 3x}{\sqrt{x^2 + 7}}$

Let $u = 2x^3 - 3x$ and $v = \sqrt{x^2 + 7} = (x^2 + 7)^{1/2}$. Then,

$$\frac{dy}{dx} = \frac{(x^2 + 7)^{1/2} \times (6x^2 - 3) - (2x^3 - 3x) \times 1/2 \times (x^2 + 7)^{-1/2} \times 2x}{x^2 + 7} \qquad (175\text{-}1)$$

$$= \frac{4x^4 + 42x^2 - 21}{(x^2 + 7)^{3/2}}$$

Example 175-2

If

$$P_L = \frac{E^2 R_L}{(R_i + R_L)^2},$$

differentiate P_L with respect to R_L. (Regard E and R_i as constants.)

Solution

$$\text{Power, } P_L = \frac{E^2 R_L}{(R_i + R_L)^2}$$

Let $u = E^2 R_L$ and $v = (R_i + R_L)^2$ Then

$$\frac{dP_L}{dR_L} = \frac{(R_i + R_L)^2 \times E^2 - E^2 R_L \times 2 \times (R_i + R_L)}{(R_i + R_L)^4} \qquad (175\text{-}1)$$

$$= \frac{E^2 \times (R_i^2 - R_L^2)}{(R_i + R_L)^4}$$

$$= \frac{E^2(R_i - R_L)}{(R_i + R_L)^3}$$

In this case differentiation by the quotient formula is simpler than using the product method (Example 174-2).

Chapter 176
Differentiation of the trigonometrical functions

THE VOLTAGE INDUCED IN A COIL IS DIRECTLY PROPORTIONAL TO THE RATE OF change of the current flowing through the coil. If the waveform of the current in the coil is an alternating sine wave, we need to be capable of differentiating the sine function in order to determine the expression for the voltage across the coil. Only in this way can we prove the *eLi* relationship, namely that for an inductor L, the voltage, e, leads the current, i, by 90°. As the result of differentiation, we can further prove the inductive reactance equation:

$$X_L = 2\pi f L \ \Omega$$

(See Chapter 67.)

In an ac circuit the current associated with a capacitor is directly proportional to the rate of change of the voltage across the capacitor. Again we need to be capable of differentiating the sine function to determine the expression for the capacitor's current. We can then prove the *iCe* relationship, namely that for a capacitor C, the current, i, leads the voltage, e, by 90°. As the result of differentiation, we can further prove the capacitive reactance equation:

$$X_C = \frac{1}{2\pi f C} \ \Omega$$

(See Chapter 68.)

Once the sine and cosine functions have been differentiated, the first derivatives of other trigonometrical functions can be obtained through the use of the function-of-a-function, product, and quotient formulas.

MATHEMATICAL DERIVATIONS

Let $y = \sin x$. Then,

$$\frac{\delta y}{\delta x} = \frac{\sin(x + \delta x) - \sin x}{\delta x} \tag{169-1}$$

$$= \frac{\sin x \cos \delta x + \cos x \sin \delta x - \sin x}{\delta x}$$

As δy and $\delta x \rightarrow 0$, $\cos \delta x \rightarrow 1$ and $\sin \delta x \rightarrow x$. $\tag{165-6}$

Therefore:

$$\frac{dy}{dx} = \frac{\sin x \times 1 + \cos x \times \delta x - \sin x}{\delta x} \tag{176-1}$$

$$= \cos x$$

This result can be verified by referring to Fig. 176-1.

When $x = 0°$, $y = \sin x = 0$, $dy/dx = \cos x = 1$, and the slope of $y = \sin x$ is a maximum in a positive direction. When $x = 90° = \pi^c/2$, $y = \sin x = 1$, $dy/dx = \cos x = 0$, and the slope of $y = \sin x$ is zero. When $x = 180° = \pi^c$, $y = \sin x = 0$, $dy/dx = \cos x = -1$, and the slope of $y = \sin x$ is a maximum in the negative direction.

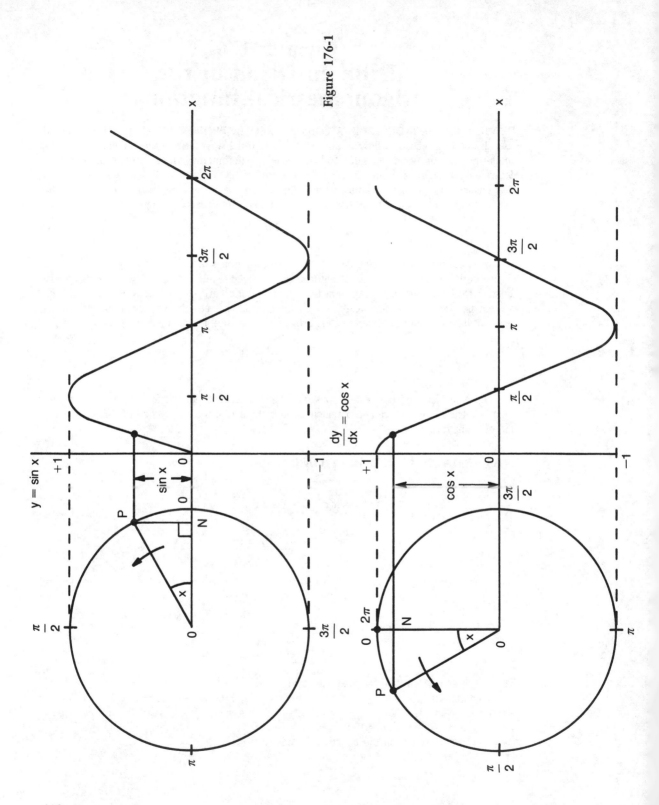

Figure 176-1

Table 176-1.
First derivatives of
the trigonometric functions.

y	dy/dx
$\sin x$	$\cos x$
$\cos x$	$-\sin x$
$\tan x$	$\sec^2 x$
$\cot x$	$-\csc^2 x$
$\sec x$	$\sec x \times \tan x$
$\text{cosec } x$	$-\csc x \times \cot x$

Let $y = \cos x = \sin (\pi/2 - x)$. Then,

$$\frac{dy}{dx} = \cos \left(\frac{\pi}{2} - x \right) \times (-1) \qquad (173\text{-}1)$$

$$= -\sin x \qquad (176\text{-}2)$$

Let $y = \tan x = \sin x/\cos x$. This can be considered as a quotient. Therefore:

$$\frac{dy}{dx} = \frac{\cos x \cos x - \sin x \, (-\sin x)}{\cos x} \qquad (175\text{-}1)$$

$$= \frac{\cos^2 x + \sin^2 x}{\cos^2 x}$$

$$= \frac{1}{\cos^2 x} = \sec^2 x \qquad (164\text{-}9), \ (166\text{-}5)$$

The first derivatives of the six trigonometrical functions are listed in Table 176-1.

Differentiation of the inverse trigonometrical functions

If $y = \sin^{-1} x$, $x = \sin y$. Then,

$$\frac{dx}{dy} = \cos y$$

Therefore:

$$\frac{dy}{dx} = \frac{1}{\cos y} \qquad (176\text{-}4)$$

$$= \frac{1}{\sqrt{1 - \sin^2 y}}$$

$$= \frac{1}{\sqrt{1 - x^2}}$$

Likewise, if $y = \cos^{-1} x$, $x = \cos y = \sin [(\pi/2) - y]$, and

$$\frac{dy}{dx} = \frac{1}{\sqrt{1 - x^2}} \qquad (176\text{-}5)$$

If $y = \tan^{-1}x$, $x = \tan y$. Then,

$$\frac{dx}{dy} = \sec^2 y \qquad (176\text{-}3)$$

$$= 1 + \tan^2 y \qquad (166\text{-}14)$$
$$= 1 + x^2$$

Therefore:

$$\frac{dy}{dx} = \frac{1}{1 + x^2} \qquad (176\text{-}6)$$

Example 176-1

Differentiate the following with respect to x:
(a) 5 sin 3x, (b) 7 cos 6x, (c) 3 tan 2x.

Solution

(a) If $y = 5 \sin 3x$, $y = \sin u$ where $u = 3x$ so that y is a function of a function. Therefore:

$$\frac{dy}{dx} = (5 \cos 3x) \times 3 \qquad (176\text{-}1)$$
$$= 15 \cos 3x$$

(b) If $y = 7 \cos 6x$,

$$\frac{dy}{dx} = (-7 \sin 6x) \times 6 \qquad (176\text{-}2)$$
$$= -42 \sin 6x$$

(c) If $y = 3 \tan 2x$,

$$\frac{dy}{dx} = (3 \sec^2 2x) \times 2 \qquad (176\text{-}3)$$
$$= 6 \sec^2 2x$$

Example 176-2

The alternating current flowing through an inductor is represented by $i = I_{peak} \sin \omega t$. If the inductor is connected directly across an ac source, determine the expression for the source voltage, e.

Solution

$$\text{Source voltage, } e = L\frac{di}{dt} \qquad (176\text{-}1)$$
$$= L\frac{d(I_{peak} \sin \omega t)}{dt}$$
$$= \omega L\, I_{peak} \cos \omega t$$
$$= \omega L\, I_{peak} \sin \left(\omega t + \frac{\pi}{2} \right)$$
$$= E_{peak} \sin \left(\omega t + \frac{\pi}{2} \right)$$

Therefore e leads i by $\pi/2$ radians or 90°. Moreover,

$$E_{\text{peak}} = \omega L\, I_{\text{peak}}$$

$$\frac{E_{\text{peak}}}{I_{\text{peak}}} = \frac{E_{\text{rms}}}{I_{\text{rms}}} = X_L = \omega L = 2\pi f L \ \Omega$$

The equation $X_L = 2\pi f L \ \Omega$ is the formula for obtaining the value of the inductive reactance, X_L.

Example 176-3

The voltage applied across a capacitor is represented by $e = E_{\text{peak}} \sin \omega t$. Determine the expression for the current that is alternately charging and discharging the capacitor.

Solution

$$\text{Source voltage, } e = E_{\text{peak}} \sin \omega t$$

Since the charge, $q = Ce$,

$$\frac{dq}{dt} = C\frac{de}{dt}$$

The current is equivalent to the rate of change of charge. Therefore:

$$\text{Current, } i = \frac{dq}{dt} = C\frac{d(E_{\text{peak}} \sin \omega t)}{dt} \tag{176-1}$$

$$= \omega C\, E_{\text{peak}} \cos \omega t$$

$$= I_{\text{peak}} \sin\left(\omega t + \frac{\pi}{2}\right)$$

Therefore i leads e by $\pi/2$ radians or 90°. Moreover:

$$I_{\text{peak}} = \omega C\, E_{\text{peak}}$$

$$\frac{E_{\text{peak}}}{I_{\text{peak}}} = \frac{E_{\text{rms}}}{I_{\text{rms}}} = X_C = \frac{1}{\omega C} = \frac{1}{2\pi f C} \ \Omega$$

Equation $X_C = 1/(2\pi f C) \ \Omega$ is the formula for obtaining the value of the capacitive reactance X_C.

Chapter 177
Differentiation of the
exponential and logarithmic functions

T HE APPLICATIONS OF THE EXPONENTIAL AND LOGARITHMIC FUNCTIONS TO electronics and communications were explored in Chapter 171. You are now going to obtain the first derivatives of these quantities. This will lead to series expansions of the sine and cosine functions.

Mathematical Derivations

The first derivative of the exponential function can be found by expressing the function in terms of its series. Each term of the series is differentiated in turn, and the resulting derivatives are then added.

If $y = e^x$, then

$$y = 1 + \frac{x}{1!} + \frac{x^2}{2!} + \frac{x^3}{3!} + \frac{x^4}{4!} + \cdots \tag{171-2}$$

$$\frac{dy}{dx} = 0 + 1 + \frac{2x}{2!} + \frac{3x^2}{3!} + \frac{4x^3}{4!} + \cdots$$

$$= 1 + \frac{x}{1!} + \frac{x^2}{2!} + \frac{x^3}{3!} + \cdots$$

or,

$$\frac{dy}{dx} = e^x$$

Therefore:

$$\frac{d(e^x)}{dx} = e^x$$

This result can be used to find the first derivative of $\ln x$.

If $y = \ln x$, then,

$$x = e^y \tag{177-1}$$

$$\frac{dx}{dy} = \frac{d(e^y)}{dy}$$

$$= e^y$$

Therefore:

$$\frac{dy}{dx} = \frac{1}{e^y}$$

$$= \frac{1}{x}$$

so that,

$$\frac{d(\ln x)}{dx} = \frac{1}{x} \tag{177-2}$$

Differentiation of a phasor

If phasor $z = R \cos \theta + jR \sin \theta$, $\qquad$ (87-6, 87-7)

$$\frac{dz}{d\theta} = -R \sin \theta + jR \cos \theta \qquad \text{(176-1, 176-2)}$$

then,

$$\frac{dz}{d\theta} = jz \tag{177-3}$$

Differentiation of a phasor with respect to its angle results in the phasor being multiplied by the operator j.

Comparing Equation 177-3 with $d(e^x)/dx = e^x$,

$$\text{Phasor, } z = Re^{j\theta} \tag{177-4}$$

This leads to,

$$e^{j\theta} = \cos\theta + j\sin\theta \tag{177-5}$$

and,

$$e^{-j\theta} = \cos(-\theta) + j\sin(-\theta) \tag{177-6}$$
$$= \cos\theta - j\sin\theta$$

Therefore:

$$\cos\theta + j\sin\theta = 1 + \frac{j\theta}{1!} + \frac{(j\theta)^2}{2^1} + \frac{(j\theta)^3}{3!} + \frac{(j\theta)^4}{4!} + \cdots$$

$$= 1 + \frac{j\theta}{1!} - \frac{\theta^2}{2!} - \frac{j\theta^3}{3!} + \frac{\theta^4}{4!} + \cdots$$

Equating real and imaginary parts (see Chapter 88),

$$\sin\theta = \theta - \frac{\theta^3}{3!} + \frac{\theta^5}{5!} - \frac{\theta^7}{7!} + \cdots \tag{177-7}$$

and,

$$\cos\theta = 1 - \frac{\theta^2}{2!} + \frac{\theta^4}{4!} - \frac{\theta^6}{6!} + \cdots \tag{177-8}$$

In these series expansions of $\sin\theta$ and $\cos\theta$, the angle is measured in radians. For example if $\theta = 30° = 0.5236^c$,

$$\sin 30° = \sin 0.5236^c$$

$$= 0.5236 - \frac{0.5236^3}{6} + \frac{0.5236^5}{120} - \cdots$$

$$= 0.5236 - 0.0239 + 0.0003 - \cdots \text{ rounded off}$$

$$= 0.5$$

$$\cos 30° = \cos 0.5236° = \frac{0.5236^2}{2} + \frac{0.5236^4}{24} - \cdots$$

$$= 1 - 0.1371 + 0.0031 - \cdots \text{ rounded off}$$

$$= 0.8660$$

Adding Equations 177-5 and 177-6,

$$2\cos\theta = e^{j\theta} + e^{-j\theta}$$

or,

$$\cos\theta = \frac{e^{j\theta} + e^{-j\theta}}{2} \tag{177-9}$$

Subtracting Equation 177-6 from 177-5.

$$j2 \sin \theta = e^{j\theta} - e^{-j\theta}$$

or,

$$\sin \theta = \frac{e^{j\theta} - e^{-j\theta}}{j2} \qquad (177\text{-}10)$$

It follows that:

$$\tan \theta = -j \times \frac{e^{j\theta} - e^{-j\theta}}{e^{j\theta} + e^{-j\theta}} \qquad (177\text{-}11)$$

$$\sec \theta = \frac{2}{e^{j\theta} + e^{-j\theta}} \qquad (177\text{-}12)$$

$$\csc \theta = \frac{j2}{e^{j\theta} - e^{-j\theta}} \qquad (177\text{-}13)$$

$$\cot \theta = j \frac{e^{j\theta} + e^{-j\theta}}{e^{j\theta} - e^{-j\theta}} \qquad (177\text{-}14)$$

Example 177-1

Differentiate the following with respect to x:
(a) $y = 5e^{3x}$, (b) $y = 4 \ln (5x)$, (c) $y = \ln (\sin x)$, (d) $y = \ln (\tan \sqrt{x^2 + 1})$.

Solution

(a) If $y = 5e^{2x}$, $y = e^u$, where $u = 3x$. Therefore y is a function of a function of x. Then:

$$\frac{dy}{dx} = 5e^{3x} \times 3 \qquad (177\text{-}1)$$

$$= 15e^{3x}$$

(b) If $y = 4 \ln (5x)$,

$$\frac{dy}{dx} = \frac{4}{5x} \times 5 \qquad (177\text{-}2)$$

$$= \frac{4}{x}$$

(c) If $y = \ln (\sin x)$,

$$\frac{dy}{dx} = \frac{1}{\sin x} \times \cos x \qquad (176\text{-}1,\ 177\text{-}2)$$

$$= \cot x$$

(d) This is a function (ln) of a function (tan) of a function ($\sqrt{\ }$) of a function ($x^2 + 1$), which is, in turn, a function of x.
Therefore, if $y = \ln (\tan \sqrt{x^2 + 1})$,

$$\frac{dy}{dx} = \frac{1}{\tan \sqrt{x^2 + 1}} \times \sec^2 \sqrt{x^2 + 1} \times \frac{1}{2} \times (x^2 + 1)^{-1/2} \times 2x \qquad (176\text{-}3,\ 177\text{-}2)$$

$$= \frac{x}{\sqrt{x^2 + 1} \times \sin \sqrt{x^2 + 1} \times \cos \sqrt{x^2 + 1}}.$$

Chapter 178
The hyperbolic functions
and their derivatives

THE FUNCTIONS $(e^\theta - e^{-\theta})/2$ AND $(e^\theta + e^{-\theta})/2$ HAVE PROPERTIES THAT ARE analogous to those of sin θ and cos θ. These functions are called, respectively, the *hyperbolic sine* and the *hyperbolic cosine* of θ. Therefore,

$$\sinh \theta = \frac{1}{2}(e^\theta - e^{-\theta})$$

and,

$$\cosh \theta = \frac{1}{2}(e^\theta + e^{-\theta})$$

It follows that

$$\cosh^2 \theta - \sinh^2 \theta = \frac{1}{2}(e^\theta + e^{-\theta})^2 - \frac{1}{2}(e^\theta - e^{-\theta})^2$$

$$= 1$$

(The hyperbolic sine, sinh, is pronounced "shine," while the hyperbolic cosine is pronounced "kosh.")

The trigonometrical functions sine and cosine are called *circular functions* because they are connected with the geometry of the circle. The coordinates of any point, P, on the circle $x^2 + y^2 = a^2$ (Fig. 178-1) can be expressed in the form ($a \cos \theta$, $a \sin \theta$) since $a^2 \cos^2 \theta + a^2 \sin^2 \theta = a^2$.

The functions cosh θ, sinh θ are connected in a similar way with the geometry of a rectangular hyperbola, and this is the origin of the term *hyperbolic functions*. Figure 178-2 shows the rectangular hyperbola $x^2 - y^2 = a^2$. The coordinates of any point, P, on this hyperbola can be expressed in the form ($a \cosh \theta$, $a \sinh \theta$) because $a^2 \cosh^2 \theta - a^2 \sinh^2 \theta = a^2$. For comparison purposes, the shaded areas in Figs. 178-1 and 178-2 can both be expressed by $a^2 \theta/2$.

Figure 178-1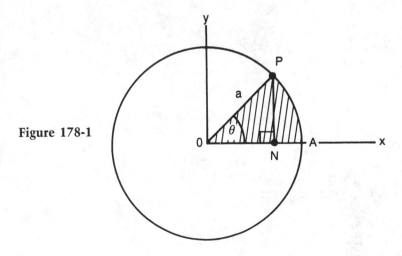

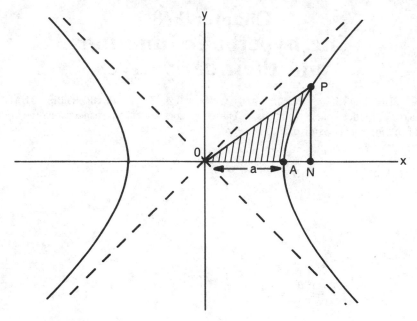

Figure 178-2

The hyperbolic functions appear in the equations of transmission line theory. (See Example 178-2.)

MATHEMATICAL DERIVATIONS
Relationships between circular and hyperbolic functions

$$\sinh(-\theta) = \frac{e^{-\theta} - e^{\theta}}{2} \qquad (178\text{-}1)$$

$$= -\sinh\theta$$

$$\cosh(-\theta) = \frac{e^{-\theta} + e^{\theta}}{2} \qquad (178\text{-}2)$$

$$= \cosh\theta$$

Moreover,

$$\cosh\theta + \sinh\theta = e^{\theta} \qquad (178\text{-}3)$$

and

$$\cosh\theta - \sinh\theta = e^{-\theta} \qquad (178\text{-}4)$$

Since $e^{\theta} = 1 + \dfrac{\theta}{1!} + \dfrac{\theta^2}{2!} + \dfrac{\theta^3}{3!} + \dfrac{\theta^4}{4!} + \cdots$ $\qquad (171\text{-}2)$

and $e^{-\theta} = 1 - \dfrac{\theta}{1!} + \dfrac{\theta^2}{2!} - \dfrac{\theta^3}{3!} + \dfrac{\theta^4}{4!} - \cdots,$

626

$$\cosh \theta = 1 + \frac{\theta^2}{2!} + \frac{\theta^4}{4!} + \cdots \qquad (178\text{-}5)$$

$$\sinh \theta = \theta + \frac{\theta^3}{3!} + \frac{\theta^5}{5!} + \cdots \qquad (178\text{-}6)$$

From Equations 177-5 and 178-3:

$$e^{j\theta} = \cos \theta + j \sin \theta$$
$$= \cosh j\theta + \sinh j\theta$$

Equating the "real" and "imaginary" parts by using Equations 178-5 and 178-6,

$$\cosh j\theta = \cos \theta \qquad (178\text{-}7)$$
$$\sinh j\theta = j \sin \theta$$

The following conversion rules apply:

Circular $\rightarrow$ Hyperbolic	Hyperbolic $\rightarrow$ Circular
$\sin \theta = -j \sinh j\theta$	$\sinh \theta = -j \sin j\theta$
$\cos \theta = \cosh j\theta$	$\cosh \theta = \cos j\theta$
$\sin j\theta = j \sinh \theta$	$\sinh j\theta = j \sin \theta$
$\cos j\theta = \cosh \theta$	$\cosh j\theta = \cos \theta$

The hyperbolic tangent is defined by:

$$\tanh \theta = \frac{\sinh \theta}{\cosh \theta} \qquad (178\text{-}8)$$

$$= \frac{e^{\theta} - e^{-\theta}}{e^{\theta} + e^{-\theta}}$$

$$= \frac{e^{2\theta} - 1}{e^{2\theta} + 1}$$

(*Tanh* is pronounced "than").
Moreover,

$$\tanh \theta = \frac{\sinh \theta}{\cosh \theta} = \frac{j \sin j\theta}{\cos j\theta} \qquad (178\text{-}9)$$

$$= -j \tan j\theta$$

and,

$$\tan \theta = -j \tanh j\theta \qquad (178\text{-}10)$$

The graphs of the hyperbolic functions are shown in Fig. 178-3.

Note that: 1) Sinh θ lies between $-\infty$ and $+\infty$. 2) Cosh θ is always greater or equal to 1. 3) Tanh θ lies between $+1$ and -1. 4) The hyperbolic functions, unlike the circular functions, are not periodic.

The remaining hyperbolic functions are defined as follows:

$$\text{csch } \theta = \frac{1}{\sinh \theta} \qquad (178\text{-}11)$$

$$= \frac{2}{e^{\theta} - e^{-\theta}} = j \csc j\theta$$

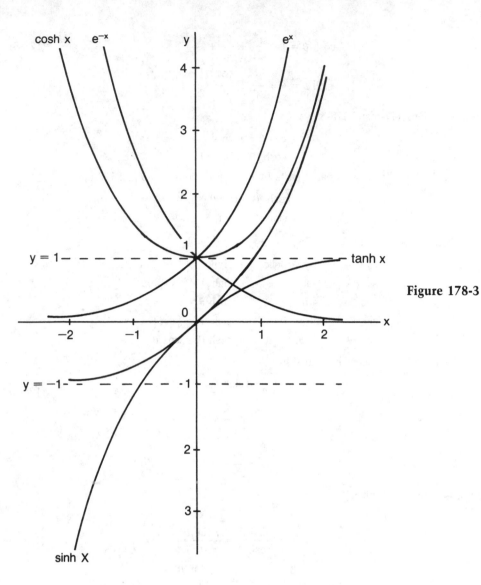

Figure 178-3

(*csch* is pronounced "co-shek").

$$\text{sech } \theta = \frac{1}{\cosh \theta} \qquad (178\text{-}12)$$

$$= \frac{2}{e^{\theta} + e^{-\theta}}$$

$$= \sec j\theta$$

("sech" is pronounced as "shek").

$$\coth \theta = \frac{1}{\tanh \theta} \qquad (178\text{-}13)$$

$$= \frac{e^{2\theta} + 1}{e^{2\theta} - 1}$$

$$= j \cot j\theta$$

(*coth* is pronounced as "koth").

Hyperbolic identities

$$\sinh (\theta + \phi) = \frac{1}{2} \left[e^{\theta + \phi} - e^{-\theta - \phi} \right] \tag{178-14}$$

$$= \frac{1}{4} \left[2e^{\theta} e^{\phi} - \frac{2}{e^{\theta} e^{\phi}} \right]$$

$$= \frac{1}{4} \left[\left(e^{\theta} - \frac{1}{e^{\theta}} \right) \left(e^{\phi} + \frac{1}{e^{\phi}} \right) + \left(e^{\theta} + \frac{1}{e^{\theta}} \right) \left(e^{\phi} - \frac{1}{e^{\phi}} \right) \right]$$

$$= \sinh \theta \cosh \phi + \cosh \theta \sinh \phi$$

This result suggests that the relationships connecting hyperbolic functions are closely related to those connecting circular functions. In fact, hyperbolic identities can be written down using Osborn's Rule:

In any formula connecting the circular functions of general angles, replace each circular function by the corresponding hyperbolic function and change the sign of every product (or implied product) of two sines.

Therefore

$$\sinh (\theta - \phi) = \sinh \theta \cosh \phi - \cosh \theta \sinh \phi \tag{178-15}$$

$$\cosh (\theta + \phi) = \cosh \theta \cosh \phi + \sinh \theta \sinh \phi \tag{178-16}$$

$$\cosh (\theta - \phi) = \cosh \theta \cosh \phi - \sinh \theta \sinh \phi \tag{178-17}$$

$$\sinh (\theta + \phi) + \sinh (\theta - \phi) = 2 \sinh \theta \cosh \phi \tag{178-18}$$

$$\sinh (\theta + \phi) - \sinh (\theta - \phi) = 2 \cosh \theta \sinh \phi \tag{178-19}$$

$$\cosh (\theta + \phi) + \cosh (\theta - \phi) = 2 \cosh \theta \cosh \phi \tag{178-20}$$

$$\cosh (\theta + \phi) - \cosh (\theta - \phi) = 2 \sinh \theta \sinh \phi \tag{178-21}$$

$$\sinh 2\theta = 2 \sinh \theta \cosh \theta \tag{178-22}$$

$$\cosh 2\phi = \cosh^2 \theta + \sinh^2 \theta \tag{178-23}$$

$$= 2 \cosh^2 \theta - 1$$

$$= 2 \sinh^2 \theta + 1$$

$$\tanh (\theta + \phi) = \frac{\tanh \theta + \tanh \phi}{1 + \tanh \theta \times \tanh \phi} \tag{178-24}$$

$$\tanh 2\theta = \frac{2 \tanh \theta}{1 + \tanh^2 \theta} \tag{178-25}$$

629

Derivatives of Hyperbolic Functions

The six hyperbolic function derivatives are presented here:

If $y = \sinh \theta = (e^\theta - e^{-\theta})/2$,

$$\frac{dy}{dx} = \frac{d(\sinh \theta)}{d\theta}$$

$$= \frac{1}{2}(e^\theta + e^{-\theta})$$

$$= \cosh \theta$$

(178-26)

If $y = \cosh \theta = (e^\theta + e^{-\theta})/2$,

$$\frac{dy}{dx} = \frac{d(\cosh \theta)}{d\theta}$$

$$= \frac{1}{2}(e^\theta - e^{-\theta})$$

$$= \sinh \theta$$

(178-27)

If $y = \tanh \theta = \sinh \theta/\cosh \theta$

$$\frac{dy}{d\theta} = \frac{d(\tanh \theta)}{d\theta}$$

$$= \frac{\cosh^2 \theta - \sinh^2 \theta}{\cosh^2 \theta}$$

$$= \frac{1}{\cosh^2 \theta}$$

$$= \operatorname{sech}^2 \theta$$

(178-28)

If $y = \operatorname{csch} \theta = 1/\sinh \theta = (\sinh \theta)^{-1}$,

$$\frac{dy}{d\theta} = \frac{d(\operatorname{csch} \theta)}{d\theta}$$

$$= \frac{-\cosh \theta}{\sinh^2 \theta}$$

$$= -\operatorname{csch} \theta \coth \theta$$

(178-29)

If $y = \operatorname{sech} \theta = 1/\cosh \theta = (\cosh \theta)^{-1}$,

$$\frac{dy}{d\theta} = \frac{d(\operatorname{sech} \theta)}{d\theta}$$

$$= \frac{-\sinh \theta}{\cosh^2 \theta}$$

$$= -\operatorname{sech} \theta \tanh \theta$$

(178-30)

If $y = \coth \theta = \cosh \theta/\sinh \theta$,

$$\frac{dy}{d\theta} = \frac{d(\coth \theta)}{d\theta}$$

(178-31)

$$= \frac{\sinh^2 \theta - \cosh^2 \theta}{\sinh^2 \theta}$$

$$= -\text{csch}^2 \theta$$

Example 178-1

Differentiate the following with respect to θ:

(a) $y = \sinh^2 3\theta$, (b) $y = \ln \sinh \theta$, (c) $y = \ln \tanh \theta$, (d) $y = \text{sech}^2 2\theta$.

Solution

(a) If $y = \sinh^2 3\theta$,

$$\frac{dy}{d\theta} = 2 \sinh 3\theta \cosh 3\theta \times 3$$

$$= 3 \sinh 6\theta$$

(b) If $y = \ln \sinh \theta$,

$$\frac{dy}{d\theta} = \frac{\cosh \theta}{\sinh \theta}$$

$$= \coth \theta$$

(c) If $y = \ln \tanh \theta$,

$$\frac{dy}{d\theta} = \frac{\text{sech}^2 \theta}{\tanh \theta} = \frac{1}{\sinh \theta \cosh \theta}$$

$$= 2 \text{csch} 2\theta$$

(d) If $y = \text{sech}^2 2\theta$.

$$\frac{dy}{d\theta} = 2 \text{sech} 2\theta \times (-\text{sech} 2\theta \tanh 2\theta) \times 2$$

$$= -4 \sinh 2\theta \text{sech}^3 2\theta$$

Example 178-2

The current on a transmission line is given by $I = A \cosh \gamma x + B \sinh \gamma x$. If the voltage $E = [1/(G + j\omega C)] \times dI/dx$, express E as a function of γx.

Solution

$$\text{Voltage, } E = \frac{1}{G + j\omega C} \times \frac{dI}{dx}$$

$$= \frac{1}{G + j\omega C} \times \frac{d}{dx} (A \cosh \gamma x + B \sinh \gamma x)$$

$$= \frac{\gamma}{G + j\omega C} (A \sinh \gamma x + b \cosh \gamma x)$$

Chapter 179
Applications of derivatives: maxima, minima, and inflexion conditions

BECAUSE dy/dx IS THE SLOPE OF $y = f(x)$ AT ANY POINT, dy/dx CAN BE USED TO find the points where y is a maximum or a minimum. At the maximum and minimum points, the curve of $y = f(x)$ is instantaneously horizontal so that the slope is 0. The values of x for which y is a maximum or a minimum are the solutions to the equation $dy/dx = 0$.

For the position of a maximum, the slope is positive before the position and negative afterwards (Fig. 179-1); therefore, the rate of change of the slope is negative. This means a negative value for the second derivative, which is written as d^2y/dx^2, rather than $d(dy/dx)/dx$. The same notation—namely d^3y/dx^3, d^4y/dx^4 and so on—is used to express further derivatives.

For the position of a minimum, the *slope* is negative before the position and negative after (Fig. 179-2). Therefore, the rate of change of the slope is positive, and the value of d^2y/dx^2 is positive.

At a point where $d^2y/dx^2 = 0$, the rate of change of the slope is momentarily zero, and we have a stationary tangent! The simplest case of this condition is at a point of *inflexion* where the curve crosses its tangent (Fig. 179-3). It is further necessary that d^2y/dx^2 changes its sign as x increases through its value at the point of inflexion.

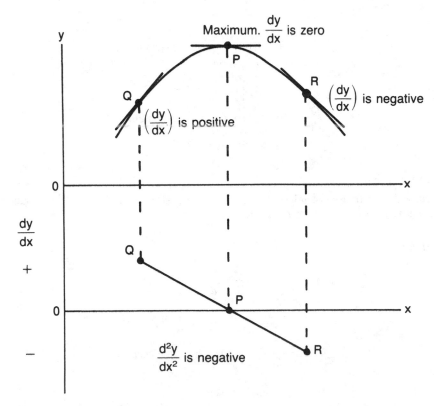

Figure 179-1.

632

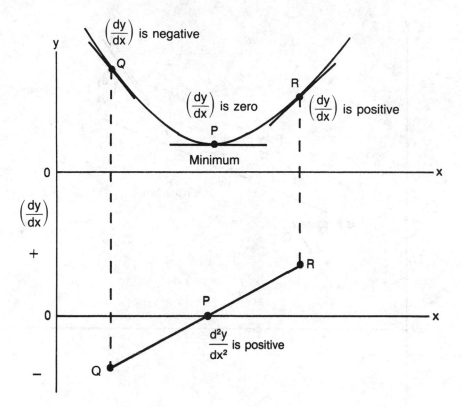

Figure 179-2

Example 179-1
Examine the curve $y = x^4 - 5x^2 + 4$ for maxima, minima, and inflexion points.

Solution
If $y = x^4 - 5x^2 + 4$,

$$\frac{dy}{dx} = 4x^3 - 10x$$

$$\frac{d^2y}{dx^2} = 12x^2 - 10.$$

Maxima or minima occur when,

$$\frac{dy}{dx} = 4x^3 - 10x = 0.$$

Solutions are $x = 0$, $x = \pm\sqrt{5/2}$

If $x = 0$,

$$\frac{d^2y}{dx^2} = -10 \text{ and is negative.}$$

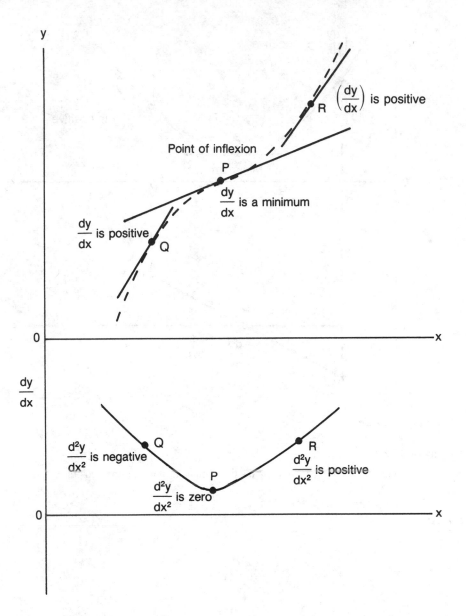

Figure 179-3

Therefore a maximum occurs at point (0,4).

If $x = \pm\sqrt{\dfrac{5}{2}}$,

$$\frac{d^2y}{dx^2} = 12 \times \frac{5}{2} - 10$$

$$= 20 \text{ and is positive.}$$

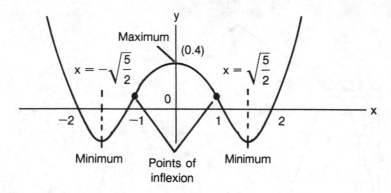

Figure 179-4

Therefore minima occur at the points $(+\sqrt{5/2},\ -9/4)$ and $(-\sqrt{5/2},\ -9/4)$.

When $d^2y/dx^2 = 0$,

$$12x^2 - 10 = 0$$

$$x = \pm\sqrt{\frac{5}{6}}$$

Since d^2y/dx^2 changes sign as x increases through each of the values $\pm\sqrt{5/6}$, there are points of inflexion at $(+\sqrt{5/6},\ 19/36)$ and $(-\sqrt{5/6},\ 19/36)$. The curve for $y = x^4 - 5x^2 + 4$ is shown in Fig. 179-4.

Example 179-2

The equation for the maximum power in a load is:

$$P_L = \frac{E^2 R_L}{(R_i - R_L)^2}$$

where E and R_i are constants. Find the condition for maximum power in the load.

Solution

$$\text{If } P_L = \frac{E^2 R_L}{(R_i + R_L)^2},$$

$$\frac{dP_L}{dR_L} = E^2 \times \left[\frac{(R_i + R_L)^2 - R_L \times 2(R_i + R_L)}{(R_i + R_L)^4}\right]$$

$$= E^2 \times \frac{R_i - R_L}{(R_i + R_L)^3}$$

$$\frac{d^2P}{dR_L{}^2} = E^2 \times \left[\frac{-(R_i + R_L)^3 - (R_i - R_L) \times 3(R_i + R_L)^2}{(R_i + R_L)^6}\right]$$

For a maximum or minimum value of P_L,

$$\frac{dP_L}{dR_L} = E^2 \times \frac{(R_i - R_L)}{(R_i + R_L)^3} = 0$$

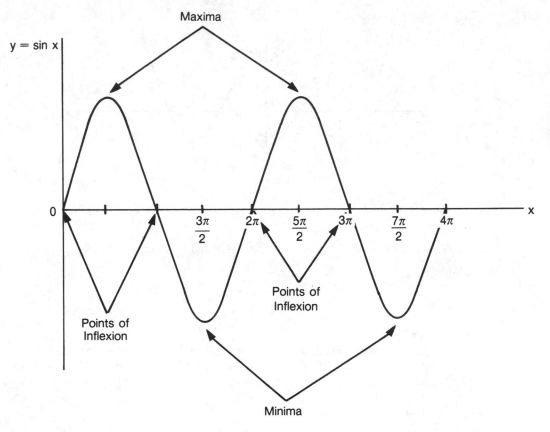

Figure 179-5

This yields,

$$R_L = R_i$$

When $R_L = R_i$, d^2P/dR_L^2 is negative and therefore P_L is at its maximum value.

Example 179-3

Examine the curve $y = \sin x$ for points of inflexion.

Solution

If $y = \sin x$,

$$\frac{dy}{dx} = \cos x \qquad\qquad (176\text{-}1)$$

$$\frac{d^2y}{dx^2} = -\sin x = -y \qquad\qquad (176\text{-}2)$$

Therefore d^2y/dx^2 changes sign whenever the curve crosses the x axis. These are points of inflexion since the values of d^2y/dx^2 at these points are zero (Fig. 179-5). Maxima occur at $x = 2n\pi + \pi/2$ radians, where n is any whole number. For these points $dy/dx = 0$ and d^2y/dx^2 is negative. Minima exist at $x = 2n\pi - \pi/2$ radians. For these points, $dy/dx = 0$, but d^2y/dx^2 is positive.

Chapter 180
Introduction to integral calculus

INTEGRATION IS THE REVERSE OF DIFFERENTIATION. THEREFORE IF $dy/dx = f'(x)$, y is the "integral" of $f'(x)$ with respect to x. Mathematically this is written as:

$$y = \int f'(x)\, dx$$

There is not a complete set of rules for integration. In fact, for certain integrals, the results are unknown. The process of integration depends on recalling the various results of differentiation.

Before proceeding further, examine in detail one interpretation of the integration symbol, $\int$. Referring to Fig. 180-1, the area $OMPN$ under the curve $y = f(x)$, and bounded by the axes and the ordinate NP, is a function of x. The area cannot be found at once, but its first derivative can. If x and y are respectively increased to $x + \delta x$ and $y + \delta y$, the area A increases by a small amount δA, which is equal to the area $PP'N'N$. Regarding this incremental area to be an approximate rectangle, $\delta A = y\delta x$, we can take the sum (the symbol $\int$ is a form of the letter S standing for "summation") of all the incremental areas to obtain the area A. In the limit as $\delta x \to 0$, the approximation disappears and,

$$\text{Area, } A = \int y\, dx = \int f(x)\, dx$$

Therefore the area underneath the curve up to the point $P = (x, y)$ is found by integration.

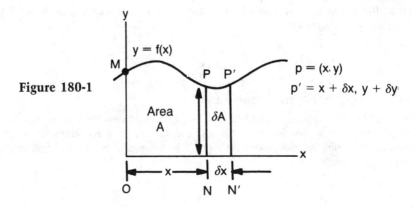

Figure 180-1

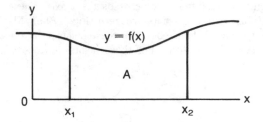

Figure 180-2

However, A does not have an exact numerical value because no precise limits have been set for the x coordinates. Therefore, $\int f(x)\, dx$ is known as an *indefinite integral*.

Sometimes the area is required between two coordinates, such as x_1 and x_2 in Fig. 180-2. This is found by calculating the area up to x_2 and subtracting from the result the area up to x_1. If $A = \int f(x)\, dx$, the notation used for the area under consideration is:

$$[A]_{x_1}^{x_2}$$

which is the value of A when $x = x_2$ minus the value of A when $x = x_1$. This *definite* integral is written as,

$$\int_{x_1}^{x_2} f(x)\, dx$$

MATHEMATICAL DERIVATIONS

Integral of x^n

Consider the indefinite integral of x^n, namely

$$y = \int x^n\, dx$$

From a knowledge of differentiation, the first derivative of $x^{n+1} = (n + 1)x^n$. Therefore:

$$y = \int x^n\, dx = \left(\frac{1}{n + 1}\right)x^{n+1} + K \qquad (180\text{-}1)$$

For indefinite integrals a constant, K, is always included in the answer because the first derivative of any constant is 0. With definite integrals, the constant of integration disappears as a result of the subtraction process.

Equation 180-1 is valid for all values of n except $n = -1$. In this case you will recall that if $y = \ln x$, $dy/dx = 1/x$. Therefore:

$$y = \int x^{-1}\, dx = \int \frac{1}{x}\, dx = \ln x + K \qquad (180\text{-}2)$$

Integrals resulting in natural logarithms

If the expression $\ln f(x)$ is differentiated, the result is $f'(x)/f(x)$ and any expression of this form can be integrated directly as $\ln f(x)$. In other words, if the numerator of an expression is the first derivative of the denominator, then the integral of the expression is the natural logarithm of the denominator. For example,

$$y = \int \tan x \, dx = \int \frac{\sin x}{\cos x} \, dx \qquad (180\text{-}3)$$

$$= -\int \frac{(-\sin x)}{\cos x} \, dx$$

$$= -\ln \cos x + K$$

$$= \ln \sec x + K$$

Integration by partial fractions

When the denominator of an expression can be factored, the integral can often be found by rearranging the expression as partial fractions. For example,

$$y = \int \frac{x + 1}{x^2 + 5x + 6} \, dx$$

$$= \int \frac{x + 1}{(x + 3)(x + 2)} \, dx$$

$$= \int \frac{2}{x + 3} \, dx - \int \frac{1}{x + 2} \, dx$$

$$= 2 \ln (x + 3) - \ln (x + 2) + K$$

Integration by parts

This method of integration is obtained from the rule for differentiating a product. From Equation 174-1, if $y = uv$:

$$\frac{dy}{dx} = \frac{d}{dx}(uv) = u\frac{dv}{dx} + v\frac{du}{dx}$$

then,

$$uv = \int \left(u\frac{dv}{dx} \right) dx + \int \left(v\frac{du}{dx} \right) dx$$

or,

$$\int \left(u\frac{dv}{dx} \right) dx = uv - \int \left(v\frac{du}{dx} \right) dx \qquad (180\text{-}4)$$

This equation represents the method of integration by parts. For example, in the integral $\int x \ln x \, dx$, let $u = \ln x$ and $dv/dx = x$ so that $v = x^2/2$. Then:

$$\int x \ln x \, dx = \frac{1}{2}x^2 \ln x - \int \frac{1}{2}x^2 \left(\frac{1}{x} \right) dx$$

$$= \frac{1}{2}x^2 \ln x - \int \frac{1}{2}x \, dx$$

$$= \frac{1}{2}x^2 \ln x - \frac{1}{4}x^2 + K$$

As another example, let $u = \ln x$ and $dv/dx = 1$ in the integral $\int \ln x \, dx$. Then:

$$\int \ln x \, dx = x \ln x - \int x\left(\frac{1}{x}\right) dx \qquad (180\text{-}5)$$

$$= x \ln x - \int 1 \, dx$$

$$= x \ln x - x + K$$

Trigonometrical transformations

Certain products and powers of the sine and cosine functions can sometimes be integrated by converting to the forms $\sin nx$ and $\cos nx$. For example since $\cos 2x = 1 - 2 \sin^2 x$ (formula 169-12),

$$\int \sin^2 x \, dx = \int \frac{1}{2}(1 - \cos 2x) \, dx$$

$$= \int \frac{1}{2} \, dx - \int \frac{1}{2} \cos 2x \, dx$$

$$= \frac{x}{2} - \frac{\sin 2x}{4} + K$$

Integration by substitution

The process of integration can frequently be simplified by making a substitution, either algebraic or trigonometric. It is important to realize that the dx term must be converted into terms of the new variable. As an example, consider the integral of $x(4 + 2x^2)^6$ and notice that the derivative of x^2 is $2x$. Then let,

$$u = 4 + 2x^2$$

so that

$$du = 4x \, dx$$

or,

$$x \, dx = \frac{du}{4}$$

then,

$$\int x(4 + 2x^2)^6 \, dx = \int \frac{1}{4} u^6 \, du$$

$$= \frac{u^7}{28} + K$$

$$= \frac{(4 + 2x^2)^7}{28} + K$$

If we integrate from $x = 1$ to $x = 2$, the limits must be changed when we made the substitution in the definite integral.

When $x = 1$, $u = 4 + 2 \times 1^2 = 6$.

When $x = 2$, $u = 4 + 2 \times 2^2 = 12$.

Then,

$$\int_{1}^{2} x(4 + 2x^2)^6 \, dx = \int_{6}^{12} \frac{1}{4} u^6 \, du$$

$$= \left[\frac{u^7}{28} \right]_{6}^{12}$$

$$= \left[\frac{12^7}{28} - \frac{6^7}{28} \right]$$

$$= 1.27 \times 10^6, \text{ rounded off.}$$

Recommendations for trigonometrical and hyperbolic substitutions

If the expression to be integrated contains:

1. $\sqrt{a^2 - x^2}$, substitute $x = a \sin \theta$ or $x = a \cos \theta$
2. $a^2 + x^2$, substitute $x = a \tan \theta$
3. $\sqrt{x^2 - a^2}$, substitute $x = a \cosh \theta$
4. $\sqrt{a^2 + x^2}$, substitute $x = a \sinh \theta$
5. $a^2 - x^2$, substitute $x = a \tanh \theta$

As an example, consider the integral

$$\int \frac{1}{\sqrt{a^2 - x^2}} \, dx$$

Substituting $x = a \sin \theta$, $dx = a \cos \theta \, d\theta$ and therefore:

$$\int \frac{1}{\sqrt{a^2 - x^2}} \, dx = \int \frac{a \cos \theta \, d\theta}{\sqrt{a^2 - a^2 \sin^2 \theta}} \qquad (180\text{-}6)$$

$$= \int d\theta$$

$$= \theta + K$$

$$= \sin^{-1} \left(\frac{x}{a} \right) + K$$

This is known as a *standard integral*, which allows similar expressions to be integrated. For example,

$$\int \frac{5}{\sqrt{4 - x^2}} \, dx = 5 \sin^{-1} \frac{x}{2} + K$$

A list of selected standard integrals is shown in Table 180-1.

Example 180-1

Find the area of the circle $x^2 + y^2 = a^2$.

Solution

Let us consider the area, A, of one quadrant. Then

$$A = \int_{0}^{a} \sqrt{a^2 - x^2} \, dx$$

Table 180-1. Common derivatives and integrals.

y	$\dfrac{dy}{dx}$	$\displaystyle\int y\,dx$
C	0	Cx
$a\,f(x)$	$a\dfrac{d}{dx}f(x)$	$a\displaystyle\int f(x)\,dx$
x^n	nx^{n-1}	$\dfrac{1}{n+1}x^{n+1}$ when $n \neq -1$
x^{-1}	$-x^{-2}$	$\log x$
e^{ax}	ae^{ax}	$\dfrac{1}{a}e^{ax}$
xe^x	$e^x(x+1)$	$e^x(x-1)$
a^x	$a^x \log_e a$	$a^x(\log_e a)$
$\log_e x$	x^{-1}	$x(\log x - 1)$
$\log_a x$	$\dfrac{1}{x}\log_a e$	$x \log_a\left(\dfrac{x}{e}\right)$
uv	$v\,du + u\,dv$	$u\displaystyle\int v\,dx - \int\left(\int v\,dx\right)du$
$\dfrac{u}{v}$	$\dfrac{v\,du - u\,dv}{v^2}$	$u\displaystyle\int \dfrac{1}{v}dx - \int\left(\int \dfrac{1}{v}dx\right)du$
$\sin ax$	$a\cos ax$	$-\dfrac{1}{a}\cos ax$
$\cos ax$	$-a\sin ax$	$+\dfrac{1}{a}\sin ax$
$\tan ax$	$a\sec^2 ax$	$-\dfrac{1}{a}\log_e \cos ax$
$\sin^{-1}\dfrac{x}{a}$	$\dfrac{1}{\sqrt{a^2 - x^2}}$	$x\sin^{-1}\dfrac{x}{a} + \sqrt{a^2 - x^2}$
$\cos^{-1}\dfrac{x}{a}$	$\dfrac{-1}{\sqrt{a^2 - x^2}}$	$x\cos^{-1}\dfrac{x}{a} - \sqrt{a^2 - x^2}$
$\tan^{-1}\dfrac{x}{a}$	$\dfrac{a}{a^2 + x^2}$	$x\tan^{-1}\dfrac{x}{a} - \dfrac{1}{a}\log\sqrt{a^2 + x^2}$
$\sinh ax$	$a\cosh ax$	$\dfrac{1}{a}\cosh ax$
$\cosh ax$	$a\sinh ax$	$\dfrac{1}{a}\sinh ax$
$\tanh ax$	$a\,\text{sech}^2 ax$	$\dfrac{1}{a}\log_e \cosh ax$

y	$\dfrac{dy}{dx}$	$\displaystyle\int y\,dx$
$\sinh^{-1}\dfrac{x}{a}$	$\dfrac{1}{\sqrt{a^2+x^2}}$	$x\sinh^{-1}\dfrac{x}{a}-\sqrt{x^2+a^2}$
$\cosh^{-1}\dfrac{x}{a}$	$\dfrac{1}{\sqrt{x^2-a^2}}$	$x\cosh^{-1}\dfrac{x}{a}-\sqrt{x^2-a^2}$
$\tanh^{-1}\dfrac{x}{a}$	$\dfrac{a}{a^2-x^2}$	$x\tanh^{-1}\dfrac{x}{a}+a\log_e\sqrt{a^2-x^2}$
$\dfrac{1}{\sin x}$	$-\dfrac{\cos x}{\sin^2 x}$	$\log_e\tan\dfrac{x}{2}$
$\dfrac{1}{\cos x}$	$+\dfrac{\sin x}{\cos^2 x}$	$\log_e\tan\left(\dfrac{x}{2}+\dfrac{n}{4}\right)$ $=\log_e(\tan x+\sec x)$
$\sin^2 ax$	$a\sin 2ax$	$\left(\dfrac{x}{2}-\dfrac{\sin ax\cos ax}{2a}\right)=\left(\dfrac{x}{2}-\dfrac{\sin 2ax}{4a}\right)$
$\cos^2 ax$	$-a\sin 2ax$	$\left(\dfrac{x}{2}+\dfrac{\sin ax\cos ax}{2a}\right)=\left(\dfrac{x}{2}+\dfrac{\sin 2ax}{4a}\right)$
$e^{ax}\sin\beta x$		$\dfrac{\alpha\,e^{ax}\sin\beta x-\beta e^{ax}\cos\beta x}{\alpha^2+\beta^2}$
$e^{ax}\cos\beta x$		$\dfrac{\alpha\,e^{ax}\cos\beta x+\beta e^{ax}\sin\beta x}{\alpha^2+\beta^2}$
$e^{ax}\sinh\beta x$		$\dfrac{\alpha\,e^{ax}\sinh\beta x-\beta e^{ax}\cosh\beta x}{\alpha^2-\beta^2}$
$e^{ax}\cosh\beta x$		$\dfrac{\alpha\,e^{ax}\cosh\beta x-\beta e^{ax}\sinh\beta x}{\alpha^2-\beta^2}$
$\sinh ax\sin\beta x$		$\dfrac{\alpha\cosh ax\sin\beta x-\beta\sinh ax\cos\beta x}{\alpha^2+\beta^2}$
$\cosh ax\sin\beta x$		$\dfrac{\alpha\sinh ax\sin\beta x-\beta\cosh ax\cos\beta x}{\alpha^2+\beta^2}$
$\sinh ax\cos\beta x$		$\dfrac{\alpha\cosh ax\cos\beta x+\beta\sinh ax\sin\beta x}{\alpha^2+\beta^2}$
$\cosh ax\cos\beta x$		$\dfrac{\alpha\sinh ax\cos\beta x+\beta\cosh ax\sin\beta x}{\alpha^2+\beta^2}$

Note: In this table "ln" is shown as "$\log_e$". The constant of integration is omitted from the expressions for $\int y\,dx$. The integrals of dy/dx are expressions for y.

Let $x = a \sin \theta$ so that $dx = a \cos \theta \, d\theta$. When $x = 0$, $\theta = 0$ and when $x = a$, $\theta = \pi^c/2$. Therefore:

$$A = \int_0^{\pi/2} a^2 \cos^2 \theta \, d\theta$$

$$= \int_0^{\pi/2} \frac{a^2}{2} \times (1 + \cos 2\theta) \, d\theta \qquad (169\text{-}11)$$

$$= \frac{a^2}{2} \left[\theta + \frac{\sin 2\theta}{2} \right]_0^{\pi/2} \qquad (176\text{-}1)$$

$$= \frac{a^2}{2} \left(\frac{\pi}{2} - 0 + 0 - 0 \right)$$

$$= \frac{\pi a^2}{4}$$

This is the area of one quadrant of the circle. The total area of the circle is:

$$4 \times \frac{\pi a^2}{4} = \pi a^2$$

Example 180-2

Find the mean height of the curve $y = A \sin \theta$ between the limits $\theta = 0$ and $\theta = \pi$.

Solution

The mean height is determined by finding the area beneath the curve between the given limits and then dividing this area by the length (π) of the base. Therefore:

$$\text{Mean height} = \frac{\displaystyle\int_0^{\pi} A \sin \theta \, d\theta}{\pi}$$

$$= \left[\frac{-A \cos \theta}{\pi} \right]_0^{\pi}$$

$$= \frac{A}{\pi} [-(-1) - (-1)]$$

$$= \frac{2A}{\pi}$$

Therefore, $2A/\pi$ is the average value of the curve $y = A \sin \theta$ between the limits 0 and π.

Example 180-3

Integrate the following with respect to x:
(a) $7x^{5/2}$, (b) $x\sqrt{x + 1}$, (c) $\sqrt{x/(1 - x)}$, (d) $x \sin x$, (e) $x^2/(1 + x^3)$, (f) $e^{\sin x} \cos x$, (g) xe^x, (h) $\sqrt{a^2 + x^2}$.

Solution

(a)
$$7x^{5/2}\,dx = \frac{7}{\frac{5}{2}+1}x^{5/2+1} + K \tag{180-1}$$
$$= 2x^{7/2} + K$$

(b) Let $x + 1 = z^2$ so that $dx = 2z\,dz$.
Then,

$$\int x\sqrt{x+1}\,dx = \int (z^2 - 1) \times z \times 2z\,dz$$

$$= \int 2z^4 - 2z^2\,dz$$

$$= \frac{2}{5}z^5 - \frac{2}{3}z^3 + K$$

$$= \frac{2}{5}(x+1)^{5/2} - \frac{2}{3}(x+1)^{3/2} + K$$

(c) Let $x = \sin^2\theta$ so that $dx = 2\sin\theta\cos\theta\,d\theta$

$$\int \sqrt{\frac{x}{1-x}}\,dx = \int \frac{\sin\theta \times 2\sin\theta\cos\theta}{\cos\theta}\,d\theta$$

$$= \int 2\sin^2\theta\,d\theta = \int 1 - \cos 2\theta\,d\theta$$

$$= \theta - \frac{\sin 2\theta}{2} + K$$

$$= \theta - \sin\theta\cos\theta + K$$

$$= \sin^{-1}\sqrt{x} - \sqrt{x - x^2} + K$$

(d)
$$\int x\sin x\,dx = -x\cos x + \int 1 \times \cos x\,dx \tag{180-4}$$
$$= -\cos x + \sin x + K$$

(e)
$$\int \frac{x^2}{1+x^3} = \frac{1}{3}\int \frac{3x^2}{1+x^3}\,dx = \frac{1}{3}\ln(1+x^3) + K$$

(f) Let $y = \sin x$ so that $dy = \cos x\,dx$. Then

$$\int e^{\sin x}\cos x\,dx = \int e^y\,dy = e^y + K = e^{\sin x} + K$$

(g)
$$\int x\,e^x\,dx = x\,e^x - \int 1(e^x)\,dx \tag{180-4}$$
$$= xe^x - e^x + K$$

(h) Let $x = a\sinh\theta$ so that $dx = a\cosh\theta\,d\theta$

645

$$\int \sqrt{a^2 + x^2}\, dx = \int \sqrt{a^2 + a^2 \sinh^2 \theta} \times a \cosh \theta\, d\theta$$

$$= \int a^2 \cosh^2 \theta\, d\theta$$

$$= \int \frac{a^2 (\cosh 2\theta + 1)}{2}\, d\theta \qquad\qquad (178\text{-}22)$$

$$= \frac{a^2}{4} \sinh 2\theta + \frac{a^2 \theta}{2} + K \qquad\qquad (178\text{-}25)$$

$$= \frac{a^2}{2} \sinh \theta \cosh \theta + \frac{a^2 \theta}{2} + K \qquad\qquad (178\text{-}21)$$

$$= \frac{1}{2} x \sqrt{a^2 + x^2} + \frac{1}{2} a^2 \sinh^{-1} \frac{x}{a} + K$$

7
PART

Digital principles

Chapter 181
Number systems

A *NUMBER SYSTEM* IS ANY SET OF SYMBOLS OR CHARACTERS USED TO ENU-
merate objects and perform mathematical computations such as addition, subtraction, multiplication, and division. All number systems are related to each other by symbols or characters
commonly referred to as *digits*. Modern number systems have certain digits in common; however, these systems do not all use the same number of digits, as shown in Table 181-1.

The most commonly used system is the Hindu-Arabic system, which uses the digits 0, 1, 2,
3, 4, 5, 6, 7, 8, and 9. There are a total of ten digits, so you have a decimal or base-10 system.
Because most measurements are made with this system, it will be used as the basis for a discussion of other number systems.

Number systems in ancient times were used primarily to take measurements and keep
records because mathematical computations using the Greek, Roman, and Egyptian number systems were extremely difficult. The lack of an adequate number system was probably a major
factor in hampering scientific development in these early civilizations. Obviously, mathematical
computations were difficult with Roman numerals where, for example, MCMLXXXIX is equivalent to 1989 in the decimal system.

The acceptance of two basic concepts has greatly simplified mathematical computations and
has led to the development of modern number systems. These two concepts are: The use of zero
to signify the absence of a unit, and the principle of *positional value*.

The principle of positional value consists of assigning to a digit a value that depends on its
position within a given number. For example, the digit 6 has a different value in each of the
decimal numbers 876, 867, and 687. In the first number, 876, the digit 6 has its base value of 6.
In the second number, 867, the digit 6 has a value of 60 (6×10 or 6×10^1). In the third number,
687, the value of the digit 6 is 600 (6×100, or 6×10^2).

Table 181-1. Comparison of number systems.

Decimal	Binary	Octal	Hexadecimal	Duodecimal
0	00000	0	0	0
1	00001	1	1	1
2	00010	2	2	2
3	00011	3	3	3
4	00100	4	4	4
5	00101	5	5	5
6	00110	6	6	6
7	00111	7	7	7
8	01000	10	8	8
9	01001	11	9	9
10	01010	12	A	t
11	01011	13	B	e
12	01100	14	C	10
13	01101	15	D	11
14	01110	16	E	12
15	01111	17	F	13
16	10000	20	10	14
17	10001	21	11	15
18	10010	22	12	16

Sometimes a position within a given number does not have a value. However, if this position is totally disregarded, there is no way to distinguish between two different numbers, such as 706 and 76. The 0 is therefore used to signify that a particular position within a certain number has no value assigned.

The use of 0 and positional value has greatly simplified counting and mathematical computations. Consequently, these concepts are used in all modern number systems.

MATHEMATICAL DERIVATIONS

Positional notation

The standard shorthand form of writing numbers is known as *positional notation*. The value of a particular digit depends not only on its basic value but also on its position within a number. For example, the decimal number 2365.74 is the standard shorthand form of the quantity two thousand three hundred sixty-five, seven-tenths, four-hundredths. Expressing this number in its general form:

$$2365.74 = (2 \times 10^3) + (3 \times 10^2) + (6 \times 10^1) + (5 \times 10^0) + (7 \times 10^{-1}) + (4 \times 10^{-2})$$

In this number, the 2 carries the most weight of all the digits and is called the *most significant digit* (MSD). By contrast, the 4 carries the least weight and is referred to as the *least significant digit* (LSD).

A number can be expressed with positional notation in any system. The general form for expressing a number is:

$$N = (d_n \times r^n) + \cdots + (d_2 \times r^2) + (d_1 \times r^1) + (d_0 \times r^0) + (d_{-1} \times r^{-1}) + (d_{-2} \times r^{-2})$$
$$+ \cdots + (d_{-n} \times r^{-n}) \quad (184\text{-}1)$$

where: N = the number expressed in a positional notation form

r = the base which is raised in turn to a series of exponents

d = the digits of the number system.

A base point, such as a decimal point, is not required in the general form because at the position of the point, the exponent changes from positive to negative. In the shorthand form the base point is between the $d_0 \times r^0$ and $d_{-1} \times r^{-1}$ values.

The base

Every number system has a base with a certain value. The hexadecimal, duodecimal, decimal, octal, and binary systems have bases whose values are respectively 16, 12, 10, 8, and 2. The division between integers (whole numbers) and fractions is recognized by the position of the base point. In addition:

1. The base of a number system is equal to the number of the different characters used to indicate all the various magnitudes that a digit might represent. For example, the decimal system, with its base of 10, has ten digits whose magnitudes are 0 through 9.
2. The value of the base is always one unit greater than the largest-value character in the system. This follows from the fact that the base is equal to the number of the characters, while the characters themselves start from zero. As an example, the highest value digit in the decimal system is 9; therefore, the base is 9 + 1 = 10.
3. The positional notation does not, by itself, indicate the value of the base. The symbol "123.41" could represent a number written in a system that has a base value of five (4 + 1) or more. To avoid confusion, numbers written in systems other than the decimal

system, have the base denoted by a subscript, such as 123.41_8. The base subscript is always written as a decimal number.

4. Any number can easily be multiplied or divided by the base of its number system. When multiplying a number by its base, move the base point one digit to the right of its former position. For example, $123.41_8 \times 8 = 1234.1_8$. To divide a number by its base, move the base point one digit to the left of its former position so that $123.41_8 \div 8 = 12.341_8$.

5. In any number system the symbol "10" always equals the value of the base. This follows from the fact that the value of the base is one unit greater than the highest-value character.

Counting

In any system using positional notation, the rules for counting are the same, and are independent of the base. With the octal system as an example, the rules are:

1. Start from zero and then add 1 to the least significant digit until the series of all the basic characters in sequence is complete. Such a series is known as a cycle which, for the octal system, would be 0, 1, 2, 3, 4, 5, 6, 7.

2. Because 7 is the highest-value character in the octal system, the next number requires two digits. Begin the start of the two-digit numbers with 0 as the least significant digit and place a 1 to the left of the 0. Therefore the series becomes 0, 1, 2, 3, 4, 5, 6, 7, 10, 11, 12, 13, 14, 15, 16, 17, 20, 21.

3. When a digit reaches its maximum value, replace it with a 0 and then add 1 to the next more significant digit. Consequently, the series is:

    . . . 16,17,20,21 . . . 26,27,30,31 . . . 36,37,40,41
    . . . 46,47,50,51 . . . 56,57,60,61 . . . 66,67,70,71
    . . . 76,77

4. When two or more consecutive digits reach the maximum value, replace them with 0's and add 1 to the next more significant digit. The series continues as:

    . . . 76,77,100,101 . . . 176,177,200,201 . . . 276,277,300,301
    . . . 376,377,400,401 . . . 476,477,500,501 . . . 576,577,600,601
    . . . 676,677,700,701 . . . 767,777,1000,1001 . . .

5. Note that, in any number system, the maximum whole number to be expressed by N digits is given by,

$$\text{Maximum number} = r^N - 1 \qquad (184\text{-}2)$$

where r = base value. For example, $777_8 = 8^3 - 1 = 511_{10}$

Example 181-1

Express (a) 375_{16}, (b) 483_{12}, (c) 726_8, (d) 11010_2 as numbers to the base 10.

Solution

(a)
$$\text{Decimal number} = (3 \times 16^2) + (7 \times 16^1) + (5 \times 16^0) \qquad (184\text{-}1)$$
$$= 768 + 112 + 5$$
$$= 885$$

(b)
$$\text{Decimal number} = (4 \times 12^2) + (8 \times 12^1) + (3 \times 12^0) \qquad (184\text{-}1)$$
$$= 576 + 96 + 3$$
$$= 675$$

(c)
$$\text{Decimal number} = (7 \times 8^2) + (2 \times 8^1) + (6 \times 8^0) \qquad (184\text{-}1)$$
$$= 448 + 16 + 6$$
$$= 470$$

(d)
$$\text{Decimal number} = (1 \times 2^4) + (1 \times 2^3) + (0 \times 2^2) + \qquad (184\text{-}1)$$
$$(1 \times 2^1) + (0 \times 2^0)$$
$$= 16 + 8 + 0 + 2 + 0$$
$$= 26$$

Chapter 182
The binary number system

THE BINARY NUMBER SYSTEM HAS A BASE OF 2 AND IS USED IN VIRTUALLY ALL digital circuits, as well as in computers. Because the base is 2, there will only be two digits: 0 and 1. This is an enormous advantage because it is relatively easy to design electronic circuits having only two possible states. For example, a bipolar transistor can either be in the saturation mode or in the cut off mode, and these two states can then correspond to two different output voltages from the circuit containing the transistor. Owing, for example, to temperature variations, such voltages are bound to fluctuate to a certain extent, so the two states will each correspond to a limited voltage range. This is illustrated in Fig. 182-1, where the 0 state is from 0 V to 1 V, and the 1 state exists between 9 V and 12 V; this means that the range between 1 V and 9 V is not used. The situation as described is sometimes referred to as *positive logic*. By contrast, *negative logic* means that the voltage levels are interchanged so that the range of 9 V to 12 V represents the 0 state, while the 0V to 1V range indicates the 1 state. These chapters are only concerned with positive logic.

The simplicity of the binary system is its advantage over other systems. For example, with a decimal system it would be necessary to design circuits capable of working with ten different voltage ranges, each of which would correspond to a digit between 0 and 9.

Binary numbers use positional notation, so that the value of a particular digit depends not only on the digit's value but also on its position within the number.

Using Equation 181-1,

$$\text{Binary Number } \underbrace{11010}_{\substack{\text{whole} \\ \text{number}}} \cdot \underbrace{011}_{\text{fraction}}{}_2 = (1 \times 2^4) + (1 \times 2^3) + (0 \times 2^2)$$
$$+ (1 \times 2^1) + (0 \times 2^0) + (0 \times 2^{-1})$$
$$+ (1 \times 2^{-2}) + (1 \times 2^{-3})$$

base
point

$$= 16 + 8 + 2 + 0.25 + 0.125$$
$$= 26.375_{10}$$

As an abbreviation, binary digits are referred to as *bits*. In our example the most significant bit (MSB) is the leftmost 1, which has a value of $(1 \times 2^4) = 16$, whereas the least significant bit (LSB) is the rightmost 1, which has a value of only $(1 \times 2^{-3}) = 0.125$.

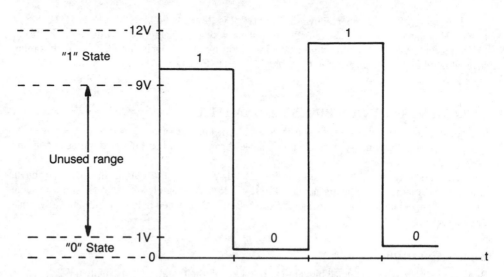

Figure 182-1

In a binary number, the leftmost 1 (Fig. 182-2) is referred to as the MSB because it is multiplied by the highest coefficient. Once the MSB has been determined, all positions to the left of the MSB have no significance even though they might be occupied by zeros; this must be true because in a given number, any position unoccupied or occupied by a zero does not have a value assigned.

Whether or not a value is assigned, all bit positions to the right of the MSB must be occupied by a 1 or a 0 (so that one number can be distinguished from another). The bit position at the extreme right of a given number is always considered to be occupied by the LSB even though it might contain a zero to indicate that no value has been assigned to this position.

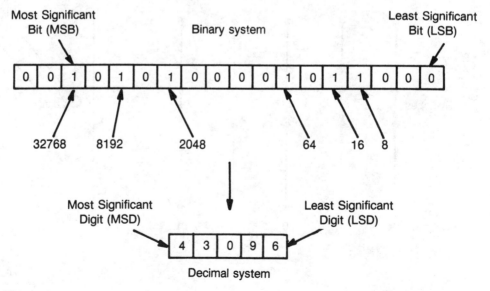

Figure 182-2

The value of the number shown in Fig. 182-2 is $32768 + 8192 + 2048 + 64 + 16 + 8 = 43096$. This illustrates the disadvantage of the binary system, namely the large number (16) of bits required to be equivalent to a decimal number having only 5 significant digits. The maximum decimal number to be represented by N bits is $2^N - 1$ (Equation 181-2). For example, $1111_2 = 2^4 - 1 = 15_{10}$.

REPRESENTATION OF BINARY QUANTITIES

If a device has only two possible operating states, a number of such devices can be used to represent a binary quantity. A single-pole, single-throw switch has only two states: open and closed.

If an open switch is in the 0 state, while a closed switch is in the 1 state, the switches shown in Fig. 182-3A indicate the binary number $11001_2 = 25_{10}$. Alternately we can represent the same number by punching holes in a paper tape (Fig. 182-3B). A punched hole then indicates a 1, while the absence of a hole is a 0.

Binary Counting

Beginning with a zero count, all bits in Fig. 182-4 are in the 0 state. First count: the 2^0 position changes from 0 to 1. Second count: the 2^0 position reverts back to 0, while the 2^1 position changes from 0 to 1. Third count: the 2^0 position changes to 1, and the 2^1 position remains at 1. Fourth count: the 2^0 and 2^1 position are 0, and the 2^2 position changes from 0 to 1. Observing the pattern, the 2^0 position changes from 0 to 1 or 1 to 0 with every count. In the general 2^N position, the bit will stay at 0 for 2^N counts and then change to 1 for the next 2^N counts.

MATHEMATICAL DERIVATIONS
Binary-to-decimal conversion

The binary-to-decimal conversion can be achieved by using Equation 181-1. For example,

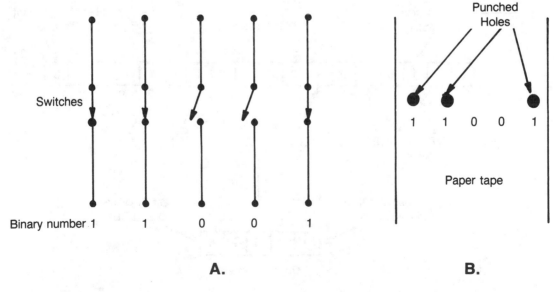

Figure 182-3

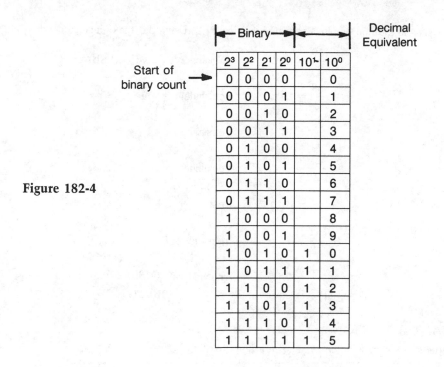

Figure 182-4

$$10011_2 = (1 \times 2^4) + (0 \times 2^3) + (0 \times 2^2) + (1 \times 2^1) + (1 \times 2^0)$$
$$= 16 + 2 + 1 = 19_{10}$$

Alternatively, the conversion can be carried out using a calculator. First locate the MSB and then proceed as follows:

Entry, Key	Display
1 ⊗ 2 ⊕	2
0 ⊗ 2 ⊕	4
0 ⊗ 2 ⊕	8
1 ⊗ 2 ⊕	18
1 =	19

This method is sometimes referred to as the *double-dabble*.

Decimal-to-binary conversions

One method of decimal-to-binary conversion is to reverse the procedure described for binary-to-decimal conversion. For example, the nearest power of two to 119_{10} is 64 (2^6). The remainder is $119 - 64 = 55$, for which the nearest power of two is 32. This leaves a new remainder of $55 - 32 = 23$. Repeating the process, $23 - 16 = 7$, $7 - 4 = 3$, $3 - 2 = 1$. Then,

$$119_{10} = 2^6 + 2^5 + 2^4 + 0 + 2^2 + 2^1 + 2^0$$
$$= 1 \quad 1 \quad 1 \quad 0 \quad 1 \quad 1 \quad 1$$
$$= 1110111_2$$

A second method employs repeated division by 2. Using the same decimal number 119_{10},

$$\frac{119}{2} = 59 \qquad \text{remainder} \qquad 1 \quad \text{(LSB)}$$

$$\frac{59}{2} = 29 \qquad \text{remainder} \qquad 1$$

$$\frac{29}{2} = 14 \qquad \text{remainder} \qquad 1$$

$$\frac{14}{2} = 7 \qquad \text{remainder} \qquad 0$$

$$\frac{7}{2} = 3 \qquad \text{remainder} \qquad 1$$

$$\frac{3}{2} = 1 \qquad \text{remainder} \qquad 0$$

$$\frac{1}{2} = 0 \qquad \text{remainder} \ 1 \ \text{(MSB)}$$

Then, 119_{10} is equal to $\qquad\qquad 1010111_2$

BINARY ADDITION

The rules of binary addition are shown in Fig. 182-5,

where $A + B = C$. Therefore:

$$0 + 0 = 0$$
$$0 + 1 = 1$$
$$1 + 0 = 1$$
$$1 + 1 = 10$$

$1 + 1 = 10$ means that when 1 is added to 1, the result is 0 with a 1 carried over to the next column. As an example add 1011_2 (11_{10}) and 11011_2 (27_{10}):

$$\begin{array}{r} 1011_2 = 11_{10} \\ +11011_2 = 27_{10} \\ \hline \text{Answer} = \ 100110_2 = 38_{10} \end{array}$$

In the second column from the right you were faced with adding three 1's, in which case $1 + 1 + 1 = 10 + 1 = 11$.

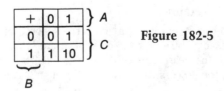

Figure 182-5

BINARY SUBTRACTION

In Fig. 182-5, $A = C - B$ so that,

$$0 - 0 = 0$$
$$1 - 0 = 1$$
$$1 - 1 = 0$$
$$10 - 1 = 1$$

$10 - 1 = 1$ means that when 1 is subtracted from 10, 1 is borrowed from the next column to the left. As an example, subtract 10010_2 (18_{10}) from 11101_2 (29_{10}):

$$11101_2 = 29_{10}$$
$$10010_2 = 18_{10}$$
$$\text{Answer} = 01011_2 = 11_{10}$$

BINARY MULTIPLICATION

The rules of binary multiplication are illustrated in Fig. 182-6, where $A \times B = C$. Therefore:

$$0 \times 0 = 0$$
$$0 \times 1 = 0$$
$$1 \times 0 = 0$$
$$1 \times 1 = 1$$

As an example multiply 10111 (23_{10}) by 1111_2 (15_{10}):

$$10111_2 = 23_{10}$$
$$1111_2 = 15_{10}$$
$$10111$$
$$10111$$
$$10111$$
$$10111$$
$$\text{Answer} = 101011001_2 = 23_{10} \times 15_{10} = 345_{10}$$

Note that in the third column from the right you had to add four 1s which is equal to 100_2 so that 1 is carried forward to the fifth column. In that column we added five 1s; this equals 101_2 so that 1 appears in the fifth column and 1 is carried forward to the seventh column.

BINARY DIVISION

Using Fig. 182-6, binary division will be defined by $A = C/B$. Therefore:

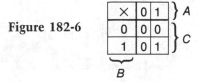

Figure 182-6

$$0 \div 0 = 0$$
$$0 \div 1 = 0$$
$$1 \div 0 = 0$$
$$1 \div 1 = 1$$

Using the "long-division" method, divide 101011001_2 (345_{10}) by 10111_2 (23_{10}):

$$
\begin{array}{r}
1111 \\
10111\overline{)101011001} \\
\underline{10111} \\
101000 \\
\underline{10111} \\
100010 \\
\underline{10111} \\
010111 \\
\underline{10111} \\
0
\end{array}
$$

Therefore, $\dfrac{101011001_2}{10111_2} = 1111_2$

or, $\dfrac{345_{10}}{23_{10}} = 15_{10}$

Example 182-1

Convert (a) 111001101_2 to a decimal number, and (b) 377_{10} to a binary number.

Solution

(a) $\qquad 111001101_2 = (1 \times 2^8) + (1 \times 2^7) + (1 \times 2^6) + (0 \times 2^5)$
$$+ (0 \times 2^4) + (1 \times 2^3) + (1 \times 2^2) + (0 \times 2^1) + (1 \times 2^0)$$
$$= 256 + 128 + 64 + 8 + 4 + 1$$
$$= 461_{10} \qquad\qquad\qquad (181\text{-}1)$$

Find the answer to (b) using the double-dabble method.

(b) $\qquad\qquad \dfrac{377}{2} = 188$, remainder 1 (LSB)

$$\dfrac{188}{2} = 94, \text{ remainder } 0$$

$$\dfrac{94}{2} = 47, \text{ remainder } 0$$

$$\dfrac{47}{2} = 23, \text{ remainder } 1$$

$$\dfrac{23}{2} = 11, \text{ remainder } 1$$

$$\dfrac{11}{2} = 5, \text{ remainder } 1$$

$$\frac{5}{2} = 2, \text{ remainder } 1$$

$$\frac{2}{2} = 1, \text{ remainder } 0$$

$$\frac{1}{2} = 0, \text{ remainder } 1 \text{ (MSB)}$$

Therefore, $377_{10} = 101111001_2$

Example 182-2

What are the binary values of:
(a) $111001_2 + 1011011_2$, (b) $1011101_2 - 100111_2$, (c) $1011101_2 \times 110011_2$, (d) $1100111_2 \div 1011_2$?

Solution

(a)
$$
\begin{array}{r}
111001_2 = 57_{10} \\
+\ 1011011_2 = 91_{10} \\
\hline
10010100_2 = 148_{10}
\end{array}
$$

Therefore, $111001_2 + 1011011_2 = 10010100_2$

(b)
$$
\begin{array}{r}
1011101_2 = 93_{10} \\
-\ 100111_2 = 39_{10} \\
\hline
110110_2 = 54_{10}
\end{array}
$$

Therefore, $1011101_2 - 100111_2 = 110110_2$

(c)
$$
\begin{array}{r}
1011101_2 = 93_{10} \\
\times\ 110011_2 = 51_{10} \\
\hline
1011101 \\
1011101 \\
1011101 \\
1011101 \\
\hline
1001010000111_2 = 93_{10} \times 51_{10} = 4743_{10}
\end{array}
$$

Therefore, $1011101_2 \times 110011_2 = 1001010000111_2$

(d)
$$
\begin{array}{r}
1001_2 \\
1011_2 \overline{)1100111_2} \\
1011 \\
\hline
0001111 \\
1011 \\
\hline
0100_2, \text{ remainder}
\end{array}
$$

In decimal form,

$$\frac{103_{10}}{11_{10}} = 9_{10} + \text{ remainder } 4_{10}$$

Therefore $1100111_2 \div 1011_2 = 1001_2$, remainder 0100_2

Chapter 183
Binary-coded decimal system

THE BINARY-CODED DECIMAL (BCD) IS NOT A TRUE NUMBER SYSTEM, BUT IT IS used in digital readout meters and in other situations where speed is not important. Compared with the binary number system, BCD has the advantage of simpler conversions to and from decimal numbers, but the disadvantage of requiring a greater number of bits. Moreover, certain complications arise when BCD numbers are added. For these reasons, BCD is not applicable to high-speed computers.

BCD makes use of groups of binary bits to represent a decimal number. Since there are ten digits in the decimal system, only ten groups are required, each containing four binary bits:

Decimal Digit	BCD Groups
0	0000
1	0001
2	0010
3	0011
4	0100
5	0101
6	0110
7	0111
8	1000
9	1001

Because a group of four binary bits has a maximum decimal value of 15, there will be six forbidden groups in BCD: 1010, 1011, 1100, 1101, 1110, and 1111.

To convert a decimal number into a BCD number, the appropriate binary group is substituted for each decimal digit. As an example,

$$381_{10} = \underset{3}{0011} \quad \underset{8}{1000} \quad \underset{1}{0001} = 001110000001 \ \text{(BCD)}$$

In true binary,

$$381_{10} = 101111101_2$$

BCD requires 12 bits, while the true binary system only needs 9 bits.

To convert a BCD number to a decimal number, separate the binary bits into groups of four. Then write down the decimal digit corresponding to each group. For example,

$$\text{BCD } 0100100110010011 = \underset{4}{0100} \quad \underset{9}{1001} \quad \underset{9}{1001} \quad \underset{3}{0011}$$

$$= 4993_{10}$$

MATHEMATICAL DERIVATIONS
BCD addition

Consider the following addition:

Decimal	BCD		
132	0001	0011	0010
+216	+0010	0100	1000
348	3	4	8

By applying the normal rules of binary addition, the correct answer of 348 was obtained. However, in this example there was no "carry-over" in the decimal addition.

When the sum in a particular BCD group exceeds decimal 9, problems arise as illustrated in the following example:

Decimal	BCD		
362	0011	0110	0010
+456	+0100	0101	0110
818	0111	1011	1000

The middle group "1011" is forbidden when operating in BCD. The difficulty is solved by adding 0110 (decimal 6) to the middle group. The result is:

Decimal	BCD		
362	0111	1011	1000
+456	+	0110	
818	1000	0001	1000

Whenever a forbidden group is observed, 0110 must be added to that group in order to obtain the correct BCD sum. The extra requirement increases the complexity of the electronic circuitry and slows down the addition process.

Example 183-1

Convert (a) 5728_{10} into its BCD equivalent, (b) convert BCD 1001011001000011 into its equivalent decimal number, (c) add 786_{10} and 559_{10} using BCD.

Solution

(a)

$$\begin{array}{ccccc} & 5 & 7 & 2 & 8 \\ 5728_{10} = & 0101 & 0111 & 0010 & 1000 \end{array}$$

$$= 0101011100101000 \text{ BCD}$$

(b)

$$\begin{array}{ccccc} \text{BCD } 1001011001000011 = & 1001 & 0110 & 0100 & 0011 \\ & 9 & 6 & 4 & 3 \end{array}$$

$$= 9643_{10}$$

(c)

Decimal		BCD	
786_{10}	0111	1000	0110
$+\ 559_{10}$	+0101	0101	1001
1345_{10}	1100	1101	1111	Three forbidden groups.
	+0110	0110	0110	Add "0110" to each group.
0001	0011	0100	0101
1	3	4	5_{10}

∞

Chapter 184
The octal number system

IT IS VERY SIMPLE TO CONVERT BETWEEN OCTAL (BASE 8) AND BINARY (BASE 2) numbers. Consequently, when a system such as a computer is involved with large quantities of binary numbers, each consisting of many bits, it is more convenient for the operators to program the information using octal numbers. However remember that the computer's digital circuits operate only with the binary system.

Although you can convert between octal and decimal numbers directly, it is preferable to use the binary system as an intermediate step (Fig. 184-1). The same principle applies to hexadecimal-to-decimal conversions

MATHEMATICAL DERIVATIONS
Direct octal-to-decimal conversion

To achieve this conversion, use Equation 181-1. For example,

$$235.7_8 = (2 \times 8^2) + (3 \times 8^1) + (5 \times 8^0) + (7 \times 8^{-1})$$
$$= 157.875_{10}$$

Direct decimal-to-octal conversion

To convert a decimal whole number into its equivalent octal number, use the repeated division method with 8 as the division factor. For example, 437_{10} is equivalent to:

$$\frac{437}{8} = 54, \text{ remainder 5 (LSD)}$$

$$\frac{54}{8} = 6, \text{ remainder 6}$$

$$\frac{6}{8} = 0, \text{ remainder 6 (MSD)}$$

Therefore, $437_{10} = 665_8$

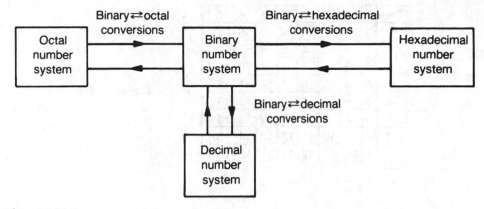

Figure 184-1

Octal-to-binary conversion

Because the base of the binary system is 2 and the base of the octal system is $8 = 2^3$, we can convert from octal whole numbers to binary whole numbers by changing each octal digit into its three-bit binary equivalent. As an example,

$$1426_8 = \underbrace{1}_{001} \quad \underbrace{4}_{100} \quad \underbrace{2}_{010} \quad \underbrace{6}_{110}$$

$$= 1100010110_2$$

Note that the two 0's to the left of the MSB are discarded.

Binary-to-octal conversion

To convert from binary whole numbers to octal whole numbers, reverse the octal-to-binary procedure. Starting at the binary point and moving to the left, mark off the bits in groups of 3, then express each group in its octal equivalent. If the number of bits is not exactly divisible by 3, complete the last group by adding one or two 0's to the left of the MSB. For example,

$$11001011_2 = \underbrace{011}_{3} \quad \underbrace{001}_{1} \quad \underbrace{011}_{3_8} \cdot$$

with labels "added '0'" above the first group and "binary point" above the last.

Therefore, $11001011_2 = 313_8$

OCTAL ADDITION

The results of octal addition are shown in Fig. 184-2. When an addition carry is made, the amount of the carry is 8. For example, when you add 7_8 and 6_8, the sum is 13 in base 10, but 13_{10} in base 8 is $(1 \times 8^1) + (5 \times 8^0) = 15_8$. This result is shown in Fig. 184-2, where $A + B = C$. As another example,

$$
\begin{aligned}
1743_8 &= 995_{10} \\
+ \quad 652_8 &= 426_{10} \\
\hline
\text{Answer} = \quad 2615_8 &= 1421_{10}
\end{aligned}
$$

Therefore, $1743_8 + 652_8 = 2615_8$

+	0	1	2	3	4	5	6	7
0	0	1	2	3	4	5	6	7
1	1	2	3	4	5	6	7	10
2	2	3	4	5	6	7	10	11
3	3	4	5	6	7	10	11	12
4	4	5	6	7	10	11	12	13
5	5	6	7	10	11	12	13	14
6	6	7	10	11	12	13	14	15
7	7	10	11	12	13	14	15	16

A (top row), B (left column), C (body)

Figure 184-2

OCTAL SUBTRACTION

Because $A = C - B$, Fig. 184-2 can be used to determine the results of subtraction so that if $C = 13_8$ and $B = 6_8$, $A = 5_8$. As another example,

$$
\begin{aligned}
3752_8 &= 2026_{10} \\
-\ 674_8 &= -444_{10} \\
\hline
\text{Answer} =\ 3056_8 &= 1582_{10}
\end{aligned}
$$

Therefore, $3752_8 - 674_8 = 3056_8$

When a subtraction borrow is made, the amount of the borrow is 8.

OCTAL MULTIPLICATION

The results of octal multiplication are shown in Fig. 184-3, where $A \times B = C$. When, for example, 6_8 is multiplied by 7_8, the product would be 42_{10}, which is equal to $(5 \times 8^1) + (2 \times 8^0) = 52_8$. As another example, multiply 473_8 by 67_8:

$$
\begin{aligned}
473_8 &= 315_{10} \\
\times\ 67_8 &=\ 55_{10} \\
\hline
4235 & \\
3542\ \ & \\
\hline
\text{Answer} =\ 41655_8 &= 17325_{10}
\end{aligned}
$$

×	0	1	2	3	4	5	6	7
0	0	0	0	0	0	0	0	0
1	0	1	2	3	4	5	6	7
2	0	2	4	6	10	12	14	16
3	0	3	6	11	14	17	22	25
4	0	4	10	14	20	24	30	34
5	0	5	12	17	24	31	36	43
6	0	6	14	22	30	36	44	52
7	0	7	16	25	34	43	52	61

A (top row), B (left column), C (body)

Figure 184-3

Therefore, $473_8 \times 67_8 = 41655_8 = 17325_{10}$

OCTAL DIVISION

Using Fig. 184-3, octal division is defined by $A = C/B$. When, for example, 234_8 is divided by 6_8,

$$
\begin{array}{r}
32_8 \\
6_8\overline{)234_3} \\
22 \\
\hline
14 \\
14 \\
\hline
0
\end{array}
$$

Therefore, $234_8 \div 6_8 = 32_8$

Example 184-1

Convert (a) 673_8 into its equivalent binary and decimal numbers, (b) 11011001_2 into its equivalent octal number, and (c) 5693_{10} into its equivalent octal number.

Solution

(a) $\qquad 673_8 = \underbrace{110}_{6} \quad \underbrace{111}_{7} \quad \underbrace{011}_{3} \qquad = 110111011_2$

$\qquad 673_8 = (6 \times 8^2) + (7 \times 8^1) + (3 \times 8^0) = 443_{10}$

(b) $\qquad\qquad 11011001_2 = \underbrace{011}_{3} \; \underbrace{011}_{3} \; \underbrace{001}_{1} = 331_8$

(c) $\qquad\qquad\qquad \dfrac{5693_{10}}{8} = 711, \text{ remainder } 5$

$$\dfrac{711}{8} = 88, \text{ remainder } 7$$

$$\dfrac{88}{8} = 11, \text{ remainder } 0$$

$$\dfrac{11}{8} = 1, \text{ remainder } 3$$

$$\dfrac{1}{8} = 0, \text{ remainder } 1$$

Answer: $5693_{10} = 13075_8$

Check:

$$
\begin{aligned}
13075_8 &= 001011000111101 \\
&= 1011000111101_2 \\
&= 5693_{10}, \text{ using the double-dabble method.}
\end{aligned}
$$

Example 184-2

What are the octal values of (a) $1672_8 + 354_8$, (b) $13875_8 - 6437_8$, (c) $132_8 \times 65_8$, (d) $765_8 \div 4_8$?

Solution

(a)
$$1672_8 = 954_{10}$$
$$+\ 354_8 = 236_{10}$$
$$\text{Answer} = \overline{2246_8 = 1190_{10}}$$

Therefore, $1672_8 + 354_8 = 2246_8 = 1190_{10}$

(b)
$$13675_8 = 6077_{10}$$
$$-\ 6437_8 = 3359_{10}$$
$$\text{Answer} = \overline{5236_8 = 2718_{10}}$$

Therefore, $13675_8 - 6437_8 = 5236_8 = 2718_{10}$

(c)
$$132_8 = 90_{10}$$
$$\underline{65_8 = 53_{10}}$$
$$702$$
$$\underline{1034}$$
$$\text{Answer} = 11242_8 = 4770_{10}$$

Therefore, $132_8 \times 65_8 = 11242_8$

(d)
$$
\begin{array}{r}
175_8 \\
4_8\overline{)765_8} \\
4 \\
\overline{36} \\
34 \\
\overline{25} \\
24 \\
\overline{1,\ \text{remainder}}
\end{array}
$$

Therefore,

$$\frac{765_8}{4_8} = 175_8 \text{ with a remainder of } 1_8.$$

Check:

$$\frac{765_8}{4_8} = \frac{501_{10}}{4_{10}} = 125_{10} \ (\text{or } 175_8)$$

with a remainder of $1_{10}(=1_8)$.

Chapter 185
The hexadecimal number system

THE HEXADECIMAL NUMBER SYSTEM (BASE 16) REQUIRES 16 CHARACTERS: THE digits 0,1,2,3,4,5,6,7,8,9, followed by the letters A,B,C,D,E,F. As shown in Fig. 181-1, the letters A through F are equivalent to the decimal values 10 through 15.

Although you can convert between hexadecimal and decimal numbers directly, it is possible to use the binary system as an intermediate step. (See Fig. 184-1.)

Hexadecimal numbers are frequently used in conjunction with computer memories and with various forms of programming. During such operations it is necessary to add or subtract hexadecimal numbers.

MATHEMATICAL DERIVATIONS
Direct hexadecimal-to-decimal conversion

To obtain a hexadecimal-to-decimal conversion, use Equation 181-1. For example,

$$1B6_{16} = (1 \times 16^2) + (11 \times 16^1) + (6 \times 16^0)$$
$$= 256 + 176 + 6$$
$$= 438_{10}$$

Direct decimal-to-hexadecimal conversion

A decimal whole number is converted into its equivalent hexadecimal number using the repeated division method with 16 as the division factor. For example, 5831_{10} is equivalent to:

$$\frac{5831}{16} = 364, \text{ remainder } 7$$

$$\frac{364}{16} = 22, \text{ remainder } 12(C)$$

$$\frac{22}{16} = 1, \text{ remainder } 6$$

$$\frac{1}{16} = 0, \text{ remainder } 1$$

Therefore, $5831_{10} = 16C7_{16}$.

Hexadecimal-to-binary conversion

Because the base of the binary system is 2 and the base of the hexadecimal system is $16 = 2^4$, you can convert from hexadecimal whole numbers to binary whole numbers by changing each hexadecimal digit into its four-bit binary equivalent. As an example,

$$3B7D_{16} = \underbrace{3}_{0011} \quad \underbrace{B}_{1011} \quad \underbrace{7}_{0111} \quad \underbrace{D}_{1101}$$
$$= 11101101111101_2$$
$$= 15229_{10} \text{ by the "double-dabble" method.}$$

It is clear that the hexadecimal system can represent large numbers more efficiently than the binary system can.

Binary-to-hexadecimal conversion

To convert from binary whole numbers to hexadecimal whole numbers, reverse the hexadecimal-to-binary procedure. Starting at the binary point and moving to the left, mark off the bits in groups of four, then express each group in its hexadecimal equivalent. If the number of bits is not exactly divisible by 4, the last group can be completed by adding one, two, or three 0's to the left of the MSB. For example,

$$11010111010011_2 = \underbrace{0011}_{3} \quad \underbrace{0101}_{5} \quad \underbrace{1101}_{D} \quad \underbrace{0011}_{3}$$

Therefore,

$$11010111010011_2 = 35D3_{16}$$
$$= 13779_{10}.$$

HEXADECIMAL ADDITION

The results of hexadecimal addition are shown in Fig. 185-1. When an addition carry is made, the amount of the carry is 16. For example when you add D_{16} and E_{16}, the sum is $13 + 14 = 27_{10}$, which in base 16 is $(1 \times 16^1) + (11 \times 16^0) = 1B_{16}$. This result is shown in Fig. 185-1, where $X + Y = Z$. As an example,

+	0	1	2	3	4	5	6	7	8	9	A	B	C	D	E	F
0	0	1	2	3	4	5	6	7	8	9	A	B	C	D	E	F
1	1	2	3	4	5	6	7	8	9	A	B	C	D	E	F	10
2	2	3	4	5	6	7	8	9	A	B	C	D	E	F	10	11
3	3	4	5	6	7	8	9	A	B	C	D	E	F	10	11	12
4	4	5	6	7	8	9	A	B	C	D	E	F	10	11	12	13
5	5	6	7	8	9	A	B	C	D	E	F	10	11	12	13	14
6	6	7	8	9	A	B	C	D	E	F	10	11	12	13	14	15
7	7	8	9	A	B	C	D	E	F	10	11	12	13	14	15	16
8	8	9	A	B	C	D	E	F	10	11	12	13	14	15	16	17
9	9	A	B	C	D	E	F	10	11	12	13	14	15	16	17	18
A	A	B	C	D	E	F	10	11	12	13	14	15	16	17	18	19
B	B	C	D	E	F	10	11	12	13	14	15	16	17	18	19	1A
C	C	D	E	F	10	11	12	13	14	15	16	17	18	19	1A	1B
D	D	E	F	10	11	12	13	14	15	16	17	18	19	1A	1B	1C
E	E	F	10	11	12	13	14	15	16	17	18	19	1A	1B	1C	1D
F	F	10	11	12	13	14	15	16	17	18	19	1A	1B	1C	1D	1E

X

Z

Y

$$3DB_{16} = 987_{10}$$
$$+ \ 7A_{16} = \ 122_{10}$$
$$\text{Answer} = \ 455_{16} = 1109_{10}$$

Therefore, $3DB_{16} + 7A_{16} = 455_{16}$.

HEXADECIMAL SUBTRACTION

Because $X = Z - Y$, Fig. 185-1 can be used to determine the results of subtraction so that if $Z = ID_{16}$ and $Y = E_{16}$, $X = F_{16}$. When a subtraction borrow is made, the amount of the borrow is 16. As a further example,

$$4C8_{16} = 1224_{10}$$
$$- \ D9_{16} = \ 217_{10}$$
$$\text{Answer} = \ 3EF_{16} = 1007_{10}$$

Therefore, $4C8_{16} - D9_{16} = 3EF_{16}$.

HEXADECIMAL MULTIPLICATION

The results of hexadecimal multiplication are shown in Fig. 185-2, in which $X \times Y = Z$. When for example, you multiply A_{16} by D_{16}, the product is $10_{10} \times 13_{10} = 130_{10} = (8 \times 16^1) + (2 \times 10^0) = 82_{16}$. As another example, multiply $7C3_{16}$ by $A6_{16}$:

×	0	1	2	3	4	5	6	7	8	9	A	B	C	D	E	F
0	0	0	0	0	0	0	0	0	0	0	0	0	0	0	0	0
1	0	1	2	3	4	5	6	7	8	9	A	B	C	D	E	F
2	0	2	4	6	8	A	C	E	10	12	14	16	18	1A	1C	1E
3	0	3	6	9	C	F	12	15	18	1B	1E	21	24	27	2A	2D
4	0	4	8	C	10	14	18	1C	20	24	28	2C	30	34	38	3C
5	0	5	A	F	14	19	1E	23	28	2D	32	37	3C	41	46	4B
6	0	6	C	12	18	1E	24	2A	30	36	3C	42	48	4E	54	5A
7	0	7	E	15	1C	23	2A	31	38	3F	46	4D	54	5B	62	69
8	0	8	10	18	20	28	30	38	40	48	50	58	60	68	70	78
9	0	9	12	1B	24	2D	36	3F	48	51	5A	63	6C	75	7E	87
A	0	A	14	1E	28	32	3C	46	50	5A	64	6E	78	82	8C	96
B	0	B	16	21	2C	37	42	4D	58	63	6E	79	84	8F	9A	A5
C	0	C	18	24	30	3C	48	54	60	6C	78	84	90	9C	A8	B4
D	0	D	1A	27	34	41	4E	5B	68	75	82	8F	9C	A9	B6	C3
E	0	E	1C	2A	38	46	54	62	70	7E	8C	9A	A8	B6	C4	D2
F	0	F	1E	2D	3C	4B	5A	69	78	87	96	A5	B4	C3	D2	E1

Figure 185-2

$$\begin{array}{r} 7C3_{16} = 1987_{10} \\ \times \quad A6_{16} = \underline{166_{10}} \\ \hline 2E92 \\ \underline{4D9E} \end{array}$$

$$\text{Answer} = \underline{50872_{16}} = 329842_{10}$$

Therefore, $7C3_{16} \times A6_{16} = 50872_{16}$.

HEXADECIMAL DIVISION

Using Fig. 185-2, hexadecimal division is defined by $X = Z/Y$. When, for example, $A8_{16}$ (168_{10}) is divided by E_{16} (14_{10}), the result is C_{16} (12_{10}). As another example, divide $3F1D8_{16}$ by $B8_{16}$:

$$\begin{array}{r} 57D_{16} \\ B8_{16} \overline{)3F1D8_{16}} \\ \underline{398} \\ 59D \\ \underline{508} \\ 958 \\ \underline{958} \\ 0 \end{array}$$

Therefore, $3F1D8_{16} \div B8_{16} = 57D_{16}$.

Example 185-1

Convert (a) $A6B_{16}$ into its equivalent binary and decimal numbers, (b) 11100101111_2 into its equivalent hexadecimal number, and (c) 4376_{10} into its equivalent hexadecimal value.

Solution

(a)
$$A6B_{16} = \underbrace{A(10)}_{1010} \quad \underbrace{6}_{0110} \quad \underbrace{B(11)}_{1011}$$
$$= 101001101011_2$$
$$= 2667_{10}$$

Alternatively,

$$A6B_{16} = (10 \times 16^2) + (6 \times 16^1) + (11 \times 16^0)$$
$$= 2667_{10}$$

(b)
$$11100101111_2 = \underbrace{0111}_{8} \quad \underbrace{0010}_{2} \quad \underbrace{1111}_{F}$$
$$= 82F_{16}$$

(c)
$$\frac{4376}{16} = 273, \text{ remainder } 8$$

$$\frac{273}{16} = 17, \text{ remainder } 1$$

$$\frac{17}{16} = 1, \text{ remainder } 1$$

$$\frac{1}{16} = 0, \text{ remainder } 1$$

Therefore, $4376_{10} = 1118_{16}$.

Example 185-2

What are the hexadecimal values of (a) $9CD_{16} + 47E_{16}$, (b) $17AB_{16} - C6F_{16}$, (c) $7A5_{16} \times 83_{16}$, (d) $109C8_{16} \div B4_{16}$?

Solution

(a)
$$9CD_{16} = 2509_{10}$$
$$+ 47E_{16} = 1151_{10}$$
$$\text{Answer} = \quad E4C_{16} = 3660_{10}$$

Therefore, $9CD_{16} + 47E_{16} = E4C_{16}$

(b)
$$17AB_{16} = 6059_{10}$$
$$- \quad C6F_{16} = 2671_{10}$$
$$\text{Answer} = \quad D3C_{16} = 3388_{10}$$

Therefore, $17AB_{16} - C6F_{16} = D3C_{16}$.

(c)
$$7A5_{16} = \quad 1957_{10}$$
$$\times \quad 83_{16} = \quad 131_{10}$$
$$16EF$$
$$3D28$$
$$\text{Answer} = \quad 3E96F_{16} = 256367_{10}$$

Therefore, $7A5_{16} \times 83_{16} = 3E96F_{16}$.

(d)
$$\begin{array}{r} 17A_{16} \\ B4_{16}\overline{)109C8_{16}} \\ \underline{B4} \\ 55C \\ \underline{4EC} \\ 708 \\ \underline{708} \\ 0 \end{array}$$

Therefore, $109C8_{16} \div B4_{16} = 17A_{16}$.

Chapter 186
The duodecimal number system

THE MAIN USE OF THE DUODECIMAL NUMBER SYSTEM IS FOR ERROR DETECTION and correction in certain digital equipment. Because the base of the system is 12, it requires 12 characters: the digits $0,1,2,3,4,5,6,7,8,9$, followed by the letters t and e, which respectively stand for "*t*en" and "*e*leven."

MATHEMATICAL DERIVATIONS
Duodecimal-to-decimal conversion

To achieve a duodecimal-to-decimal conversion, use Equation 181-1. For example,

$$2t8_{12} = (2 \times 12^2) + (10 \times 12^1) + (8 \times 12^0)$$
$$= 288 + 120 + 8$$
$$= 416_{10}$$

Decimal-to-duodecimal conversion

A decimal whole number is converted into its equivalent duodecimal number by using the repeated division method with 12 as the division factor. For example, 5831_{10} is equivalent to:

$$\frac{5831}{12} = 485, \text{ remainder } 11(e)$$

$$\frac{485}{12} = 40, \text{ remainder } 5$$

$$\frac{40}{12} = 3, \text{ remainder } 4$$

$$\frac{3}{12} = 0, \text{ remainder } 3$$

Therefore, $5831_{10} = 345e_{12}$.

Binary $\rightleftarrows$ duodecimal, octal $\rightleftarrows$ duodecimal and hexadecimal $\rightleftarrows$ duodecimal conversions can be conveniently carried out by using the decimal system as the intermediate step. However a second method involves the use of repeated division and is as follows:

(1) Consider a number N in a nondecimal system whose base is r_1. It is required to convert N to its equivalent number in another nondecimal system, whose base is r_2.
(2) Convert the base r_2 to its equivalent value in the base r_1.
(3) Carry out repeated division of the number N using the base r_1.
(4) Convert all remainders to the base r_2. As an example, convert 2767_8 to its equivalent number in base 12. Because $12 = 14_8$, the repeated division process is

$$\frac{2767_8}{14_8} = 177_8, \text{ remainder } 3_8 = 3_{12}$$

$$\frac{177_8}{14_8} = 12_8, \text{ remainder } 7_8 = 7_{12}$$

$$\frac{12_8}{14_8} = 0, \text{ remainder } 12_8 = t_{12}$$

Therefore, $2767_8 = t73_{12}$.

A third method consists of the following steps:

1. Perform all arithmetic operations in the desired base.
2. Express the base of the original number in terms of the desired base.
3. Multiply the number obtained in step 2 by the MSD of the original number and add the product to the digit immediately on the right of the MSD.
4. Repeat step 3 as many times as there are digits in the original number. The final sum is the required answer.

For example, convert 173_{12} to base 8.

Step 1. All arithmetic operations will be carried out in base 8.
Step 2. $12 = 14_8$
Step 3.

$$
\begin{array}{lll}
[\ 1_8 & 7_8 & 3_8]_{12} \\
\times 14_8 & \times\ 14_8 & \times 344_8 \\
\hline
14_8 & 23_8 & 347_8 = \text{Answer} \\
& \times\ 14_8 & \\
& \overline{\ 114} & \\
& \ \ 23 & \\
& \overline{344_8} &
\end{array}
$$

Therefore, $173_{12} = 347_8$.

DUODECIMAL ADDITION

The results of duodecimal addition are shown in Fig. 186-1. When an addition carry is made, the amount of the carry is 12. For example, when you add t_{12} and e_{12}, the result is $10_{10} + 11_{10} = 21_{10}$, which in base 12 is $(1 \times 12^1) + (9 \times 12^0) = 19_{12}$. This is shown in Fig. 186-1, where $A + B = C$.

Figure 186-1

+	0	1	2	3	4	5	6	7	8	9	t	e
0	0	1	2	3	4	5	6	7	8	9	t	e
1	1	2	3	4	5	6	7	8	9	t	e	10
2	2	3	4	5	6	7	8	9	t	e	10	11
3	3	4	5	6	7	8	9	t	e	10	11	12
4	4	5	6	7	8	9	t	e	10	11	12	13
5	5	6	7	8	9	t	e	10	11	12	13	14
6	6	7	8	9	t	e	10	11	12	13	14	15
7	7	8	9	t	e	10	11	12	13	14	15	16
8	8	9	t	e	10	11	12	13	14	15	16	17
9	9	t	e	10	11	12	13	14	15	16	17	18
t	t	e	10	11	12	13	14	15	16	17	18	19
e	e	10	11	12	13	14	15	16	17	18	19	1t

A (top row bracket), B (bottom bracket), C (right bracket)

As another example,

$$8te2_{12} = 15398_{10}$$
$$+ \ 9e4_{12} = \ \underline{1432_{10}}$$
$$\text{Answer} = \ 98t6_{12} = 16830_{10}$$

Therefore, $8te2_{12} + 9e4_{12} = 98t6_{12}$.

DUODECIMAL SUBTRACTION

Because $A = C - B$, Fig. 186-1 can be used to determine the results of the subtraction so that if $C = 19_{12}$ and $B = e_{12}$, $A = t_{12}$. When a subtraction borrow is made, the amount of the borrow is 12. As a further example,

$$t64_{12} = 1516_{10}$$
$$-2e9_{12} = \ \underline{429_{10}}$$
$$\text{Answer} = \ 767_{12} = 1087_{10}$$

Therefore, $t64_{12} - 2e9_{12} = 767_{12}$.

DUODECIMAL MULTIPLICATION

The results of duodecimal multiplication are shown in Fig. 186-2, where $A \times B = C$. When, for example, 9_{12} is multiplied by 5_{12}, the product is 45_{10}, which is equal to $(3 \times 12^1) + (9 \times 12^0) = 39_{12}$. As another example, multiply $3e7_{12}$ by $t4_{12}$:

$$3e7_{12} = \ 571_{10}$$
$$\times \ \ t4_{12} = \ 124_{10}$$
$$13t4$$
$$\underline{337t}$$
$$\text{Answer} = \ 34e84_{12} = 70804_{10}$$

Therefore, $3e7_{12} \times t4_{12} = 34e84_{12}$.

×	0	1	2	3	4	5	6	7	8	9	t	e
0	0	0	0	0	0	0	0	0	0	0	0	0
1	0	1	2	3	4	5	6	7	8	9	t	e
2	0	2	4	6	8	t	10	12	14	16	18	1t
3	0	3	6	9	10	13	16	19	20	23	26	29
4	0	4	8	10	14	18	20	24	28	30	34	38
5	0	5	t	13	18	21	26	2e	34	39	42	47
6	0	6	10	16	20	26	30	36	40	46	50	56
7	0	7	12	19	24	2e	36	41	48	53	5t	65
8	0	8	14	20	28	34	40	48	54	60	68	74
9	0	9	16	23	30	39	46	53	60	69	76	83
t	0	t	18	26	34	42	50	5t	68	76	84	92
e	0	e	1t	29	38	47	56	65	74	83	92	t1

A, B, C

Figure 186-2

DUODECIMAL DIVISION

Using Fig. 186-2, duodecimal division will be defined by $A = C/B$. When, for example, 446_{12} is divided by 6_{12}:

$$
\begin{array}{r}
89_{12} \\
6_{12}\overline{)446_{12}} \\
40 \\
\hline
46 \\
46 \\
\hline
0
\end{array}
$$

Therefore, $446_{12} \div 6_{12} = 89_{12}$.

As another example, divide $1te74_{12}$ by $5e_{12}$:

$$
\begin{array}{r}
3t7_{12} \\
5e_{12}\overline{)1te75_{12}} \\
159 \\
\hline
527 \\
4e2 \\
\hline
355 \\
355 \\
\hline
0
\end{array}
$$

Therefore, $1te75_{12} \div 5e_{12} = 3t7_{12}$.

Example 186-1

Convert (a) $5t8_{12}$ into its equivalent decimal and binary numbers, and (b) 7834_{10} into its equivalent duodecimal number.

Solution

(a)
$$5t8_{12} = 5 \times 12^2 + 10 \times 12^1 + 8 \times 12^0 \qquad (181\text{-}1)$$
$$= 848_{10}$$

But,

$$848 = (1 \times 2^9) + (1 \times 2^8) + (0 \times 2^7) + (1 \times 2^6) + (0 \times 2^5)$$
$$+ (1 \times 2^4) + (0 \times 2^3) + (0 \times 2^2) + (0 \times 2^1) + (0 \times 2^0)$$
$$= 1101010000_2$$

Therefore, $5t8_{12} = 1101010000_2$.

(b) By the repeated division method,

$$\frac{7834_{10}}{12} = 652, \text{ remainder } 10 = t_{12}$$

$$\frac{652}{12} = 54, \text{ remainder } 4_{12}$$

$$\frac{54}{12} = 4, \text{ remainder } 6_{12}$$

$$\frac{4}{12} = 0, \text{ remainder } 4_{12}$$

Therefore, $7834_{10} = 464t_{12}$.

Example 186-2

What are the duodecimal values of (a) $8t3e_{12} + 748_{12}$, (b) $t4e6_{12} - 387t_{12}$, (c) $2e7_{12} \times 4t_{12}$, (d) $19546_{12} \div 5e_{12}$?

Solution

(a)
$$
\begin{aligned}
8t3e_{12} &= 15311_{10} \\
+\ 748_{12} &= \underline{\quad 1064_{10}} \\
\text{Answer} = \quad 9587_{12} &= 16375_{10}
\end{aligned}
$$

Therefore, $8t3e_{12} + 748_{12} = 9587_{12}$.

(b)
$$
\begin{aligned}
t4e6_{12} &= 17994_{10} \\
-387t_{12} &= \underline{\quad 6430_{10}} \\
\text{Answer} = \quad 6838_{12} &= 11564_{10}
\end{aligned}
$$

Therefore, $t4e6_{12} - 387t_{12} = 6838_{12}$.

(c)
$$
\begin{aligned}
2e7_{12} &= \quad 427_{10} \\
\times \quad 4t_{12} &= \underline{\quad 58_{10}} \\
257t& \\
\underline{et4\quad}& \\
\text{Answer} = \quad 123et_{12} &= 24766_{10}
\end{aligned}
$$

Therefore, $2e7_{12} \times 4t_{12} = 123et_{12}$.

(d)
$$
\begin{array}{r}
376_{12} \\
5e_{12}\overline{)19546_{12}} \\
\underline{159\quad} \\
384 \\
\underline{355} \\
2e6 \\
\underline{2e6} \\
0
\end{array}
$$

Therefore, $19546_{12} \div 5e_{12} = 376_{12}$.

Chapter 187
The arithmetic of complements

A COMPUTER MIGHT BE DESIGNED TO PERFORM ARITHMETIC OPERATIONS BY using addition only. Such a computer must employ some method for identifying and manipulating both positive and negative numbers. This is normally accomplished by using complement arithmetic. When you wish to subtract B from A, the result, $A - B$, can be written $A + (-B)$. If you regard $(-B)$ as the complement of B, then you can carry out the subtraction process by adding A to the complement of B.

In any number system, the *complement* of a number is defined as the difference between the number and the next higher power of the base. Therefore,

$$C = r^D - N \qquad \text{(187-1)}$$

where: N = the number

C = the complement of the number

D = the number of digits in the number

r = the base of the number system

For example, if the base is 10,

2_{10} is the complement of 8_{10}, $(2 = 10^1 - 8)$

26_{10} is the complement of 74_{10}, $(26 = 10^2 - 74)$

744_{10} is the complement of 256_{10}, $(744 = 10^3 - 256)$

Similarly in base 8,

2_8 is the complement of 6_8, $(10_8 - 6_8 = 2^8)$

4_8 is the complement of 74_8, $(100_8 - 74_8 = 4_8)$

522_8 is the complement of 256_8, $(1000_8 - 256_8 = 522_8)$

Consider the example of subtracting 26_{10} from 83_{10}. The simple subtraction is $83_{10} - 26_{10} = 57_{10}$. However, the complement of $26_{10} = 10^2 - 26_{10} = 74_{10}$, so that by addition,

$$
\begin{array}{r}
83_{10} \\
+\ 74_{10} \\
\hline
157_{10}
\end{array}
$$

If the final carry-over of 1 is ignored, the addition process using complement-arithmetic produces the same answer as simple subtraction. This must be true because when you add 83_{10} and $10_{10}^2 - 26_{10}$, the answer is $83_{10} + (10_{10}^2 - 26_{10}) = 10_{10}^2 + (83_{10} - 26_{10})$, or 100 plus the result of subtracting 26_{10} from 83_{10}.

The use of binary complements is important since it means that the same circuitry of a digital computer can be used for both addition and subtraction. Consequently the total amount of circuitry required is reduced.

MATHEMATICAL DERIVATIONS

In Chapters 182 through 186, the binary bits have been used to represent the magnitude of a number. However, since you are now considering positive and negative numbers, you need some method of indicating the sign of a number. The normal convention is to use 0 as the positive sign bit, which is positioned immediately to the left of the MSB. In a similar way a negative number is proceeded by a 1 bit. As examples,

$$0100111 = +39$$
$$1010010 = -18$$

If you wish to use addition in order to subtract 18 from 39, the $+39$ will remain in the binary form as shown, but the -18 must be converted into some form of complement.

2's-complement

From Equation 187-1, the complement of $1001_2 = 2^4 - 1001_2 = 16_{10} - 9_{10} = 7_{10} = 0111_2$. This is equivalent to interchanging 0 and 1 in the original number and then adding 1 to the LSB. The result is called the *2's-complement form*.

For example, the process of converting -37_{10} into its 2's-complement form is as follows:

$$37_{10} = \quad 100101_2$$
$$011010 \quad \text{(interchanging 0 and 1)}$$
$$+ \quad\quad 1 \quad \text{(Add 1 to the LSB)}$$
$$011011 \quad \text{(2's-complement form)}$$

When you include the sign bit, $-37_{10} = 1011011$ in its 2's-complement form.

To convert from the 2's-complement form back to the true binary value, repeat the 2's-complement process. Therefore 011011 becomes $100100 + 1$ or 100101, which is the original binary value. Because $+37 = 0100101$ and -37 in its 2's-complement form is 1011011, the complete operation when performed on an entire number, including the sign bit, changes a positive number to a negative number and vice versa.

Combining two signed numbers

Now use the 2's-complement system to add together two signed numbers. If both numbers are positive such as $+7_{10}$ and $+5_{10}$, neither number is expressed in its 2's-complement form, and the procedure is:

$$+7 \rightarrow \overset{+}{0} \quad 0111$$
$$+ \quad +5 \rightarrow \overset{+}{0} \quad \underline{0101}$$
$$\text{Sum} = \quad +12 \rightarrow \overset{+}{0} \quad 1100$$

When one number is positive $(+7)$ and the other is a smaller negative number (-5), the -5 must be in 2's-complement form, which is 11011. When you add $+7$ and -5, it is equivalent to subtracting 5 from 7, and you anticipate an answer of $+2$. The binary process is

$$+7 \rightarrow \overset{+}{0} \quad 0111$$
$$+ \ (-5) \rightarrow \overset{-}{1} \quad \underline{1011} \ \text{(2's-complement form)}$$
$$\text{Sum} = \quad +2 \ \cancel{1} \overset{+}{0} \quad 0010$$
$$\uparrow$$
$$\text{discard}$$

Although the sign bits are involved in the addition process, any carry-over from the sign bits is disregarded. Note that the same number (4) of bits is used for the magnitude of each number; this is essential if the 2's-complement form is to be used.

Now add -7 and $+5$. This is equivalent to subtracting 7 from 5 so look for an answer of -2. The binary process is,

$$+5 \rightarrow \overset{+}{0} \quad 0101$$

$$+ \, (-7) \rightarrow \overset{-}{1} \quad \underline{1001} \text{ (2's-complement form)}$$

$$\text{Answer} \rightarrow \overset{-}{1} \quad 1110 \text{ (2's-complement form)}$$

$$\text{Sum} = -2 \quad \overset{-}{1} \quad 0010 \text{ (true binary sum)}$$

Because the answer is negative, it is in its 2's-complement form and the true binary sum is $-(0010_2) = -2_{10}$.

Finally add -7 and -5 for a sum of -12. Both numbers will be expressed in their 2's-complement form and therefore:

$$(-7) \rightarrow \overset{-}{1} \quad 1001 \text{ (2's-complement form)}$$

$$+ \, (-5) \rightarrow \overset{-}{1} \quad \underline{1011} \text{ (2's-complement form)}$$

$$\text{Answer} \rightarrow \cancel{1} \, \overset{-}{1} \quad 0100 \text{ (2's-complement form}$$

$$\underset{\text{discard}}{\uparrow}$$

$$\text{Sum} = -12 \rightarrow \overset{-}{1} \quad 1100 \text{ (true binary sum)}$$

The carry-over from the sign bits is disregarded. Because the answer is negative, it is in its 2's-complement form, which must be converted to its true binary sum.

In each of the four combinations [$(+7) + (+5) = +12$, $(+7) + (-5) = +2$, $(-7) + (+5) = -2$ and $(-7) + (-5) = -12$], the magnitude of the sum was accommodated within the four bits available. However, if you add $(+10)$ to $(+8)$,

$$(+10) \rightarrow \overset{+}{0} \quad 1010$$

$$(+8) \rightarrow \overset{+}{0} \quad \underline{1000}$$

$$\text{Answer} \rightarrow \overset{-}{1} \quad 0010$$

the answer is wrong because it contains a negative sign bit, while the sign of the sum is clearly positive. Because the sum of $+18$ is not accommodated within the available four bits, there is an overflow into the sign bits and the answer is then wrong. Overflow is detected by the circuitry of the computer, and the necessary correction is made.

Sometimes digital circuitry must multiply two signed numbers. Since digital circuitry uses the 2's-complement form, both numbers are first converted into their true binary forms and then the multiplication is carried out. If the product of the two signed numbers is positive, the 0 sign bit is given to the true binary answer. However, if the product is negative, the true binary answer is converted to its 2's-complement form and then given a 1 sign bit.

Example 187-1

What are the complements of (a) 213_{10}, (b) 61_8, (c) 1011_2?

Solution

(a) Complement $= 10^3 - 213_{10} = 787_{10}$

(b) Complement $= 8^2 - 61_8 = 100_8 - 61_8 = 17_8$

(c) 2's-complement $= 0100$
$$\begin{array}{r} 0100 \\ +1 \\ \hline 0101 \end{array}$$

Example 187-2

Carry out the following operations in binary, using the 2's-complement form when necessary:
(a) $(+8) + (+3)$, (b) $(+8) + (-3)$, (c) $(+3) + (-8)$, (d) $(-8) + (-3)$.

Solution

(a)
$$
\begin{array}{rl}
(+8) \rightarrow & \overset{+}{0}\,1000_2 \\
\underline{(+3) \rightarrow} & \underline{\overset{+}{0}\,0011_2} \\
\text{Sum} = +11 \rightarrow & 0\,1011_2
\end{array}
$$

(b)
$$
\begin{array}{rl}
(+8) \rightarrow & \overset{+}{0}\,1000 \\[4pt]
\underline{+(-3) \rightarrow} & \underline{\overset{-}{1}\,1101}\ \text{(2's-complement form)} \\
\text{Sum} = \quad +5\ \not{1}\ & \overset{+}{0}\,0101 \\
& \quad \uparrow \\
& \text{discard}
\end{array}
$$

(c)
$$
\begin{array}{rl}
(+3) \rightarrow & \overset{+}{0}\,0011 \\[4pt]
\underline{-(+8) \rightarrow} & \underline{\overset{-}{1}\,1000}\ \text{(2's-complement form)} \\
\text{Answer} \rightarrow & \overset{-}{1}\,1011\ \text{(2's-complement form)} \\
\text{Sum} = -5 \rightarrow & \overset{-}{1}\,0101\ \text{(true binary form)}
\end{array}
$$

(d)
$$
\begin{array}{rl}
(-8) \rightarrow & \overset{-}{1}\,1000\ \text{(2's-complement form)} \\
\underline{+(-3) \rightarrow} & \underline{\overset{-}{1}\,1101}\ \text{(2's-complement form)} \\
\text{Answer} \rightarrow \not{1}\ & \overset{-}{1}\,0101\ \text{(2's-complement form)} \\
& \quad \uparrow \\
& \text{discard} \\
\text{Sum} = -11 \rightarrow & \overset{-}{1}\,1011\ \text{(true binary form)}
\end{array}
$$

Chapter 188
Introduction to boolean algebra

BOOLEAN ALGEBRA WAS DEVELOPED BY THE ENGLISH LOGICIAN AND mathematician George Boole (1815–1864). In the spring of 1847, Boole wrote a pamphlet entitled *A Mathematical Analysis of Logic*. This was followed in 1854 by his more exhaustive treatise, *An Investigation of the Laws of Thought*. It is this later work that forms the basis for our present-day mathematical theories used for the analysis of logical processes.

Although conceived in the nineteenth century, little practical application was found for Boole's work until 1938, when it was discovered that his algebra could be adapted to the analysis of switching circuits. With the advent of modern computers, Boolean algebra has become an important subject in the understanding of complex digital circuitry.

CLASSES AND ELEMENTS

In our world it is logical to visualize two divisions: all things of interest are in one division, and all things not of interest in the discussion are in the other division. These two divisions make up a set or a class that is designated as the *universal class*. All things contained in the universal class are referred to as *elements* or *variables*. You can also visualize another class, which contains no elements; such a set is designated as the *null class*.

In a particular discussion, certain elements of the universal class can be grouped together to form combinations, which are known as *subclasses*. Each subclass of the universal class is dependent on its elements and the possible states (stable, unstable, or both) that these elements might have.

Boolean algebra is limited to the use of elements that only possess two possible logic states, both of which are stable. If you have two elements, A and B, their possible states can be designated in a number of ways, as shown in Table 188-1.

Notice that in Boolean algebra there are only two numbers: 0 and 1. Moreover, this form of algebra does not contain such concepts as squares, square roots, reciprocals, and logarithms.

If you have two elements, each of which has two states, there are 2^2, or 4, possible subclasses. If the states of A and B are true or false and you use the connective word AND, the four subclasses are:

A true AND B false

A true AND B true

A false AND B true

A false AND B false

**Table 188-1. Possible states
for *A* and *B*.**

Logic 1	Logic 0
True	False
Yes	No
High	Low
+10 V	0 V
ON	OFF
Closed Switch	Open Switch

However, if the connective word OR is used, there are four additional subclasses:

$$A \text{ true OR } B \text{ false}$$
$$A \text{ true OR } B \text{ true}$$
$$A \text{ false OR } B \text{ true}$$
$$A \text{ false OR } B \text{ false}$$

MATHEMATICAL DERIVATIONS

Venn diagrams

A *Venn diagram* is a topographical picture of logic. Such a diagram is composed of the universal class, which is divided into subclasses. The number of these subclasses depends on the number of elements.

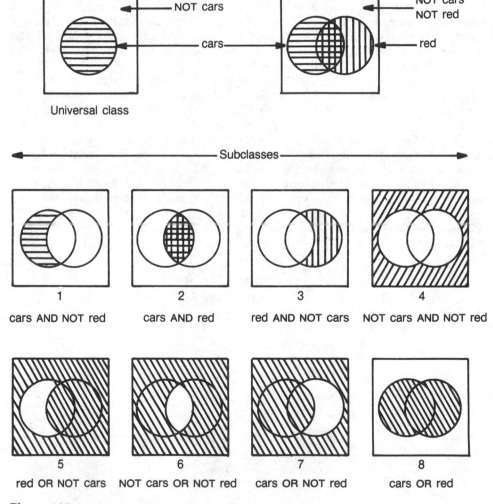

Figure 188-1

As an example, let A equal *cars* and B equal the color *red*. With the connective word AND, the four subclasses are:

Cars AND *Not red*

Cars AND *red*

Red AND *Not cars*

Not cars AND *Not red*

With the connective word OR, the four additional subclasses are:

Red OR *Not cars*

Not cars OR *Not red*

Cars OR *Not red*

Cars OR *red*

All these subclasses are shown in Fig. 188-1. The shaded area in each diagram represents the particular subclass.

Symbols in Boolean algebra

The symbols used in Boolean algebra are common in other branches of mathematics, but in certain cases their meaning is slightly different. The principal symbols are explained in Table 188-2.

Table 188-2. Boolean algebra symbols.

Symbol	Meaning
=	As in conventional mathematics the equal sign represents the relationship of equivalence between the expressions which are so connected.
· or ×	These symbols indicate the logic product which is also known as the AND operation. Frequently this operation is indicated without the use of a symbol so that $AB = A \times B = A \cdot B$
+	The plus sign indicates the logic sum which is also known as the OR operation.
−	The overbar signifies logical complementation or inversion which is known as the NOT operation.
( ), [ ], { }	The grouping symbols mean that all the contained terms must be treated as a unit.

**Table 188-3. Truth table
for A and B.**

Input Elements		Output
A	*B*	*f(A, B)*
0	0	?
0	1	?
1	0	?
1	1	?

It follows that logic circuits have only three basic operations: AND, OR, and NOT.

Truth tables

For any Boolean operation, there is a corresponding truth table, which shows the various outputs of the operation for each possible way the states of the input elements can be assigned. In a truth table, the states are designated as 0 and 1; if there are only two input elements, the corresponding truth table is as shown in Table 188-3.

The operation's output will be a function of *A* and *B* and is therefore designated as *f(A,B)*. Note that the top-to-bottom sequence of the input elements is the same as that of binary counting.

Example 188-1

Design a truth table for three input elements designated *A*, *B*, and *C*.

Solution

Input Elements			Output
A	*B*	*C*	*f(A,B,C)*
0	0	0	?
0	0	1	?
0	1	0	?
0	1	1	?
1	0	0	?
1	0	1	?
1	1	0	?
1	1	1	?

Chapter 189
The OR operation

THE VENN DIAGRAM IN FIG. 189-1A HAS TWO ELEMENTS, OR VARIABLES, WHICH are designated A and B. The shaded area represents the subclass, which is $A + B$ in Boolean notation. The corresponding equation in Boolean algebra is expressed as:

$$f(A,B) = A + B \qquad (189\text{-}1)$$

This expression is called an OR *operation,* since it represents the last of the four OR subclasses illustrated in the Venn diagrams of Fig. 188-1. The equation $f(A,B) = A + B$ is read as either "$f(A,B)$ equals A plus B" or "$f(A,B)$ equals A OR B".

Figure 189-1B illustrates the truth table of the OR operation. When A and B are each 0, the output is also 0. If either A or B takes the value of 1, the output $f(A,B)$ equals 1. However, if both A and B are equal to 1, $f(A,B)$ is 1, not 2 as in binary addition, because $A + B$ represents the *logic* sum and in Boolean algebra, only two numbers exist: 0 and 1.

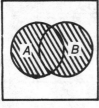

Venn Diagram
for A OR B

A.

Input Elements		OR Output
A	B	$f(A,B) = A + B$
0	0	0
0	1	1
1	0	1
1	1	1

Truth table for
OR operation

B.

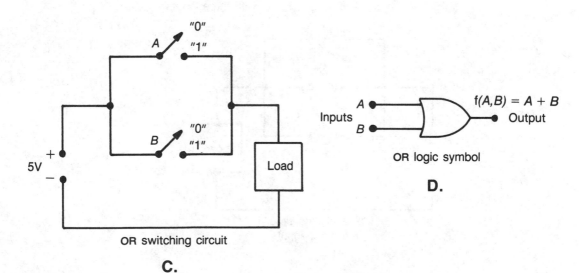

OR switching circuit

C.

OR logic symbol

D.

Figure 189-1

685

Figure 189-1C shows the switching circuit that represents the OR operation. It consists of two switches in parallel. The circuit is ON (1) if either switch A or switch B is in the CLOSED (1) state. The circuit is also ON (1) if both switch A and switch B are CLOSED (1), but it is OFF (0) if, and only if, both switch A and switch B, are OPEN (0). Note that the high and low values applied to the load are 5 V and 0 V respectively.

THE OR GATE

The OR gate is a digital circuit that has two or more inputs represented by the logic voltage levels 0 and 1. For example, the logic 1 level (high) might be +5 V, while the logic 0 level (low) is only +0.5 V. The output of the gate is the logic sum of the inputs and has a voltage level (either 1 or 0), that is the result of carrying out the OR operation on the inputs. The logic symbol for the OR gate is shown in Fig. 189-1D.

If there are only two inputs, A and B, the OR gate operates in such a way that its output, A OR B, is low (0) if, and only if, both A and B are in the 0 state. By contrast, if either A or B or both are in the 1 state, the gate's output is high and in the 1 state.

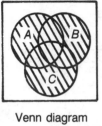

Venn diagram
for A OR B OR C

A.

Input elements			OR Output
A	B	C	$f(A,B,C) = A + B + C$
0	0	0	0
0	0	1	1
0	1	0	1
0	1	1	1
1	0	0	1
1	0	1	1
1	1	0	1
1	1	1	1

Truth table for
OR operation
with three inputs

B.

OR switching circuit
with three inputs

C.

Figure 189-2

With more than two inputs, the same principles apply. In all cases when one or more of the inputs is in the 1 state, the output is high. The outputs will be low only when all of the inputs are in the 0 state.

MATHEMATICAL DERIVATIONS

To summarize the results of the OR operation:

1. For the OR operation, the output is in the 0 state if, and only if, all of the inputs are in the 0 state. Therefore,

$$f(A,B,C\ldots) = A + B + C + \cdots = 0 \qquad (189\text{-}2)$$

$$\text{if } A = B = C = \cdots = 0$$

2. For the OR operation, the output is in the 1 state when any one of the inputs is in the 1 state. Therefore,

$$f(A,B,C\ldots) = A + B + C + \cdots = 1 \qquad (189\text{-}3)$$

$$\text{if } A \text{ or } B \text{ or } C \text{ or } \cdots = 1$$

3. For the OR operation, the output is in the 1 state if more than one of the inputs is in the 1 state. For example,

$$f(A,B,C\ldots) = 1 + 0 + 1 + 1 + 0 + 0 + 1 \cdots = 1.$$

Example 189-1

An OR gate has three inputs. Derive the corresponding Venn diagram, truth table, and switching circuit.

Solution

The solution is illustrated in Fig. 189-2.

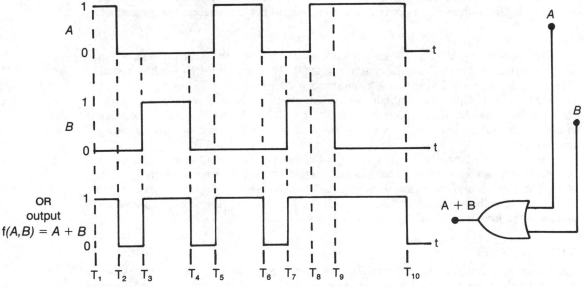

Figure 189-3

687

Example 189-2

Figure 189-3 shows an OR gate with two inputs, A and B, whose logic voltage levels are given. Determine the output's logic voltage level as it varies with time.

Solution

The solution is illustrated in Fig. 189-3. The OR outputs are:

Between times T_1 and T_2: $A = 1$, $B = 0$, $A + B = 1$.

Between times T_2 and T_3: $A = 0$, $B = 0$, $A + B = 0$.

Between times T_3 and T_4: $A = 0$, $B = 1$, $A + B = 1$.

Between times T_4 and T_5: $A = 0$, $B = 0$, $A + B = 0$.

Between times T_5 and T_6: $A = 1$, $B = 0$, $A + B = 1$.

Between times T_6 and T_7: $A = 0$, $B = 0$, $A + B = 0$.

Between times T_7 and T_8: $A = 0$, $B = 1$, $A + B = 1$.

Between times T_8 and T_9: $A = 1$, $B = 1$, $A + B = 1$.

Between times T_9 and T_{10}: $A = 1$, $B = 0$, $A + B = 1$.

Time t_{10} and beyond: $A = 0$, $B = 0$, $A + B = 0$.

Chapter 190
The AND operation

THE VENN DIAGRAM IN FIG. 190-1A HAS TWO ELEMENTS, OR VARIABLES, WHICH ARE designated A and B. The shaded area represents the subclass, which is AB or $A \cdot B$ or $A \times B$ in Boolean notation. The corresponding equation in Boolean algebra is expressed as:

$$f(A,B) = AB = A \cdot B = A \times B \qquad (190\text{-}1)$$

This expression is called an AND operation because it represents the second of the four AND subclasses illustrated in the Venn diagrams of Fig. 188-1. The equation $f(A,B) = AB$ is read as "$f(A,B)$ equals A AND B," which is the logic product of A and B.

Figure 190-1B depicts the truth table for the AND operation. When A and B are each 0, the output is also 0. If either A or B takes the value of 0, $f(A,B)$ is 0. However, if both A and B are in the 1 state, $f(A,B)$ is 1. The rules for the AND operation are the same as those for binary multiplication.

Figure 190-1C shows the switching circuit that represents the AND operation. It consists of two switches in series so that the circuit is OFF (0) if either switch A or switch B is in the OPEN (0) state. The circuit is also OFF (0) if both switch A and switch B are OPEN (0). The circuit is ON (1) if, and only if, both switches A and B are CLOSED (1). Note that the high and low logic values applied to the load are 5 V and 0 V, respectively.

THE AND GATE

The AND gate is a digital circuit that has two or more inputs represented by the logic voltage levels 0 and 1. For example, the logic 1 level (high) might be +5 V, while the logic 0 level (low) is only

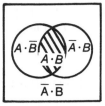

Venn diagram
for *A* AND *B*

A.

Input elements		AND Output
A	*B*	f(A,B) = A · B
0	0	0
0	1	0
1	0	0
1	1	1

Truth table for
AND operation

B.

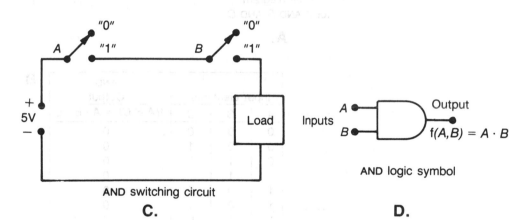

AND switching circuit

C.

AND logic symbol

D.

Figure 190-1

+0.5 V. The output of the gate is the logic product of the inputs and has a voltage level (either 1 or 0) that is the result of carrying out the AND operation on the inputs. The logic symbol for the AND gate is shown in Fig. 190-1D.

If there are only two inputs, *A* and *B*, the AND gate operates in such a way that its output, *A* AND *B*, is high (1) if, and only if, *A* and *B* are in the 1 state. By contrast, if either *A* or *B* or both are in the 0 state, the gate's output is low and in the 0 state.

With more than two inputs, the same principles apply. In all cases, when one or more of the inputs is in the 0 state, the output is low. The output will be high only when all the inputs are in the 1 state.

MATHEMATICAL DERIVATIONS

To summarize the results of the AND operation:

1. For the AND operation, the output is in the 1 state if, and only if, all inputs are in the 1 state. Therefore,

$$f(A,B,C \ldots) = ABC \cdots = 1 \text{ if } A = B = C \cdots = 1 \qquad (190\text{-}2)$$

Between times T_7 and T_8: $A = 0$, $B = 1$, $AB = 0$.

Between times T_8 and T_9: $A = 1$, $B = 1$, $AB = 1$.

Between times T_9 and T_{10}: $A = 1$, $B = 0$, $AB = 0$.

Time T_{10} and beyond: $A = 0$, $B = 0$, $AB = 0$.

These AND outputs should be compared with the OR outputs of Example 188-2. For the AND gate, the output is high only when all of the inputs are high. By contrast, the output from an OR gate is high when any input is high.

Chapter 191
The NOT operation

THE VENN DIAGRAM IN FIG. 191-1A HAS ONE INPUT ELEMENT, OR VARIABLE, which is designated as A and is represented by the clear area. The shaded area represents the Boolean output, which is indicated by $\overline{A}$ or A', and is the *complement*, or inverse, of A.

The corresponding equation is:

$$f(A) = \overline{A} = A' \tag{191-1}$$

This equation is read as "Output, $f(A)$ is NOT A." Notice that, unlike the AND and OR operations, the NOT operation can be performed on a single input element.

Because the logic value of $\overline{A}$ is opposite to the logic value of A, it follows that: (a) if $A = 0$, $\overline{A} = 1$ because NOT 0 is 1; and (b) if $A = 1$, $\overline{A} = 0$ because NOT 1 is 0. The results of this reasoning are shown in the truth table of Fig. 191-1B.

The NOT switching circuit is shown in Fig. 191-1C. The requirement of a NOT circuit is that a signal injected at the input produces the complement, or inversion, of this signal at the output. When switch A is closed and is in state 1, the relay opens the circuit to the load and the load circuit is, therefore, OFF (0). However, when switch A is open and is in state 0, the relay closes the load circuit, which is then in the ON (1) condition. Note that the high and low logic values applied to the load are 5 V and 0 V, respectively.

The logic symbol for the NOT circuit is shown in Fig. 191-1D. The presence of the small circle, or *bubble*, on a logic symbol is always an indication of inversion. The input and output symbols mean that the logic levels are interchanged, and consequently an input level of 1 corresponds to an output level of 0 and vice versa.

MATHEMATICAL DERIVATIONS

The NOT operation is summarized as follows:

$$f(A) = \overline{A} = 1 \text{ when } A = 0.$$
$$f(A) = \overline{A} = 0 \text{ when } A = 1.$$

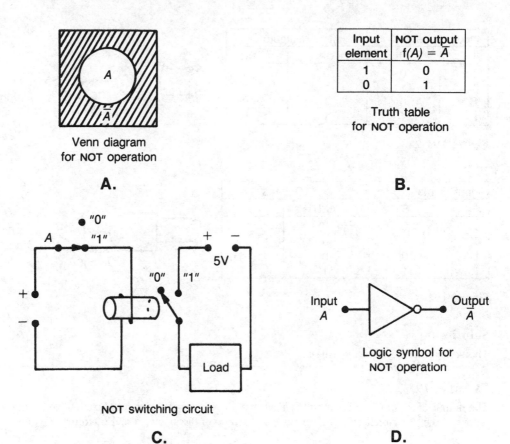

Venn diagram
for NOT operation

A.

Input element	NOT output $f(A) = \bar{A}$
1	0
0	1

Truth table
for NOT operation

B.

NOT switching circuit

C.

Logic symbol for
NOT operation

D.

Figure 191-1

The OR, AND, and NOT circuits provide the three basic Boolean operations. Comparing the rules for these operations:

OR	AND	NOT
$0 + 0 = 0$	$0 \cdot 0 = 0$	$\bar{0} = 1$
$0 + 1 = 1$	$0 \cdot 1 = 0$	$\bar{1} = 0$
$1 + 0 = 1$	$1 \cdot 0 = 0$	
$1 + 1 = 1$	$1 \cdot 1 = 1$	

Example 191-1

The output of an OR gate is fed to the input of a NOT circuit. Determine the truth table for the output of the NOT circuit with the various combinations of the inputs, A and B, to the OR gate.

693

Input elements		Outputs	
A	B	OR	NOT
0	0	0	1
0	1	1	0
1	0	1	0
1	1	1	0

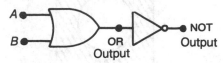

Figure 191-2

Input elements		Outputs	
A	B	AND	NOT
0	0	0	1
0	1	0	1
1	0	0	1
1	1	1	0

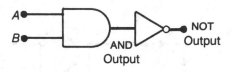

Figure 191-3

Solution

The solution is shown in Fig. 191-2.

Example 191-2

The output of an AND gate is fed to the input of a NOT circuit. Determine the truth table for the output of the NOT circuit with the various combinations of the inputs, A and B, to the AND gate.

Solution

The solution is shown in Fig. 191-3.

Chapter 192
The NOR operation

IF THERE ARE TWO INPUT ELEMENTS, A AND B, EACH OF WHICH HAS TWO alternative stable states, 0 and 1, there will be a total of 16 possible truth tables. Assuming that the 0,0,0,0 and 1,1,1,1 outputs have no use, there are 14 remaining tables that can be arranged in 7 pairs. In each pair, the outputs of one table will be the complements of the outputs from the other table. It follows that there must exist a truth table that is the complement or inversion of the table for the OR operation (Chapter 189). This new truth table will refer to the NOR operation, which is a combination of the OR and NOT operations.

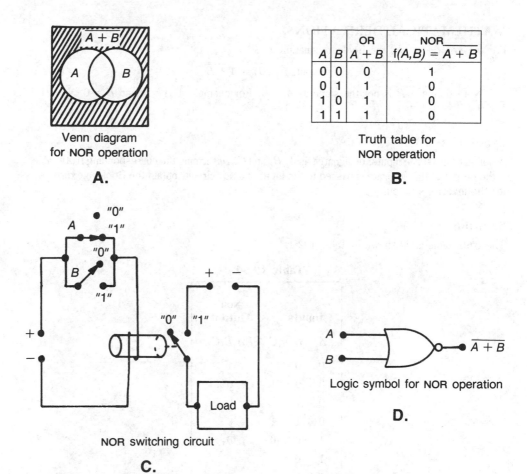

		OR	NOR
A	B	A + B	$f(A,B) = \overline{A + B}$
0	0	0	1
0	1	1	0
1	0	1	0
1	1	1	0

Venn diagram
for NOR operation

A.

Truth table for
NOR operation

B.

NOR switching circuit

C.

Logic symbol for NOR operation

D.

Figure 192-1

In Fig. 192-1A, the unshaded area of the Venn diagram represents the subclass A or B; the shaded area is the subclass A or B, when negated or complemented. The Boolean expression for A or B, when negated or inverted is $f(A,B) = \overline{A \text{ or } B} = \overline{A + B}$. DeMorgan's Theorems (Chapter 198) will show that $\overline{A \text{ or } B} = \overline{A} \cdot \overline{B}$.

The truth table for the NOR operation is shown in Fig. 192-1B. This table shows that if either A or B or both A and B is in the 1 state, then the output $f(A,B)$ is in the 0 state. When both A and B are in the 0 state, the output $f(A,B)$ is in the 1 state.

The NOR equivalent switching arrangement is the result of combining the OR switching circuit and the NOT switching circuit (Fig. 192-1C). If either switch A or switch B, or both, is in the closed (1) position, the load circuit is inactive or OFF (0). However, if both of the switches A and B are open (0), the load circuit is active or ON (1).

The logic symbol for the NOR gate is shown in Fig. 192-1D. It is a combination of the OR gate and the bubble, which indicates inversion. The NOR gate is commonly used in digital circuitry and is equivalent to an OR gate followed by an inverter.

MATHEMATICAL DERIVATIONS

For the NOR operation, the Boolean equation is:

$$\text{Output, } f(A,B) = \overline{A + B} \tag{192-1}$$

Note that $\overline{A + B}$ is not the same as $\overline{A} + \overline{B}$. For example, if $A = 1$ and $B = 0$, $\overline{A + B} = \overline{1 + 0} = \overline{1} = 0$, but $\overline{A} + \overline{B} = \overline{1} + \overline{0} = 0 + 1 = 1$.

Example 192-1

A NOR gate has three inputs, designated as A, B, and C. Determine the corresponding truth table. If the output of the NOR gate is passed to the input of a NOT circuit, obtain the Boolean expression for the inverter's output.

Solution

The truth table is as shown in Table 192-1.

Table 192-1

Inputs			NOR Output
A	B	C	$f(A,B,C)$
0	0	0	1
0	0	1	0
0	1	0	0
0	1	1	0
1	0	0	0
1	0	1	0
1	1	0	0
1	1	1	0

$$\text{NOR gate output} = \overline{A + B + C} \tag{192-1}$$

$$\text{Inverter output} = \overline{\overline{A + B + C}} = A + B + C.$$

Example 192-2

Figure 192-2 shows the logic voltage levels of two inputs A and B to a NOR gate. Determine the logic voltage levels of the gate's output.

Solution

The output of the NOR gate is shown in Fig. 192-2. When the input elements are both in the 0 state, the output is in the 1 state. Under all other conditions, the output is in the 0 state.

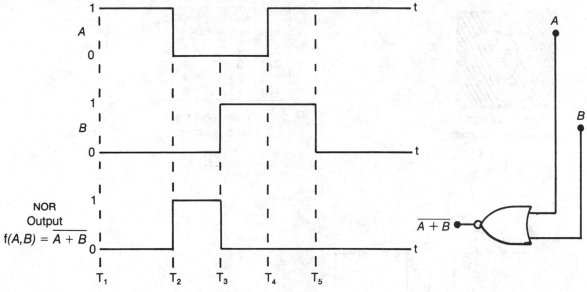

NOR
Output
$f(A,B) = \overline{A + B}$

$\overline{A + B}$

T_1 T_2 T_3 T_4 T_5

Figure 192-2

Chapter 193
The NAND operation

IN FIG. 193-1A THE UNSHADED AREA OF THE VENN DIAGRAM REPRESENTS THE subclass A AND B. The shaded area is the subclass A AND B when negated or complemented; this subclass, $\overline{A}$ AND $\overline{B}$, is shown in the Venn diagrams of Fig. 188-1. The Boolean equation for A AND B, when negated or inverted, is $f(A,B) = \overline{A \cdot B}$, which is equal to $\overline{A} + \overline{B}$. (See DeMorgan's Theorem in Chapter 198.)

The truth table for the NAND operation is shown in Fig. 193-1B. The table shows that if both A and B are in the 1 state, the output is in the 0 state. For all other possible combinations of the A and B states, the output is in the 1 state.

The equivalent NAND switching arrangement (Fig. 193-1C) is a combination of AND and NOT circuits. If either or both of the switches A and B are OPEN (0), the load circuit is ON (1). However if both of the switches A and B are CLOSED (1), the load circuit is OFF (0).

The logic symbol for the NAND gate is shown in Fig. 193-1D. There are two input elements, A and B, into an AND gate whose output, $A \cdot B$, is fed into a NOT circuit; the final output, $f(A,B)$, is then $\overline{A \cdot B}$.

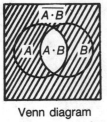

Venn diagram
for NAND operation

A.

A	B	AND $A \cdot B$	NAND $f(A,B) = \overline{A \cdot B}$
0	0	0	1
0	1	0	1
1	0	0	1
1	1	1	0

Truth table for
NAND operation

B.

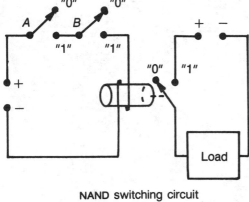

NAND switching circuit

C.

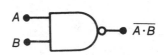

Logic symbol for NAND operation

D.

Figure 193-1

MATHEMATICAL DERIVATIONS

For the NAND operation, the Boolean equation is:

$$\text{Output, } f(A,B) = \overline{A \cdot B} \tag{193-1}$$

Note that the entire group, $A \cdot B$, is complemented, and this is not the same as the logic product of the two elements when they are complemented separately $(\overline{A} \cdot \overline{B})$. For example if $A = 1$ and $B = 0$, $\overline{A \cdot B} = \overline{0} = 1$, but $\overline{A} \cdot \overline{B} = 0 \cdot 1 = 0$.

Example 193-1

A NAND gate has three inputs, designated as A, B, and C. Determine the corresponding truth table. If the NAND gate is followed by an inverter, obtain the Boolean expression for the inverter's output.

Solution

The truth table is shown in Table 193-1.

Table 193-1

Inputs			NAND Output
A	B	C	$f(A,B,C)$
0	0	0	1
0	0	1	1
0	1	0	1
0	1	1	1
1	0	0	1
1	0	1	1
1	1	0	1
1	1	1	0

$$\text{NAND gate output} = \overline{A \cdot B \cdot C} \qquad (193\text{-}1)$$
$$\text{Inverter output} = \overline{\overline{A \cdot B \cdot C}} = A \cdot B \cdot C$$

Example 193-2

Figure 193-2 shows the logic voltage levels of the input elements A and B to a NAND gate. Obtain the output waveform from the gate.

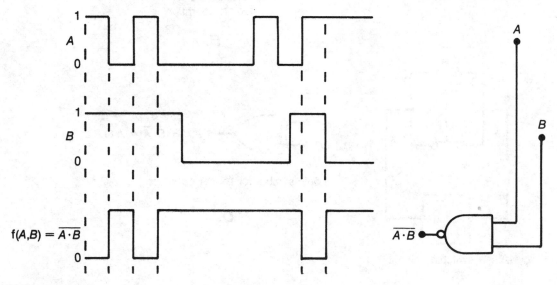

Figure 193-2

699

Solution

The solution is shown in Fig. 193-2. The NAND output is in the 0 state only when both of the inputs are in their 1 state. Under all other conditions, the NAND output is in the 1 state.

Chapter 194
The exclusive-OR operation

THE EXCLUSIVE-OR (ABBREVIATED EX-OR) OPERATION IS ACTUALLY A SPECIAL application of the OR operation (Chapter 189) and appears quite frequently in digital circuitry. In the EX-OR operation, either of the inputs A or B must be in the 1 state in order for the output, $f(A,B)$ to be in the 1 state; however, if both A and B are in the 1 state or in the 0 state at the same time, the output is in the 0 state.

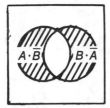

Venn diagram
for EX-OR operation

A.

Input elements		EX-OR output
A	B	$f(A,B) = A \cdot \overline{B} + B \cdot \overline{A}$
0	0	0
0	1	1
1	0	1
1	1	0

Truth table for
EX-OR operation

B.

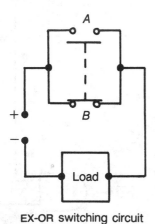

EX-OR switching circuit

C.

$A \bullet$
$B \bullet$
$A \cdot \overline{B} + B \cdot \overline{A} = A \oplus B$

Logic symbol for EX-OR operation

D.

Figure 194-1

The Venn diagram for the EX-OR operation is shown in Fig. 194-1A. This diagram does not appear in any of those shown in Fig. 188-1 and is, therefore, a new subclass.

The Boolean equation for the EX-OR operation is:

$$f(A,B) = \overline{A} \cdot B + A \cdot \overline{B} \tag{194-1}$$

If
$$A = 0,\ B = 0,\ f(A,B) = 1 \cdot 0 + 0 \cdot 1 = 0$$
$$A = 0,\ B = 1,\ f(A,B) = 1 \cdot 1 + 0 \cdot 0 = 1$$
$$A = 1,\ B = 0,\ f(A,B) = 0 \cdot 0 + 1 \cdot 1 = 1$$
$$A = 1,\ B = 1,\ f(A,B) = 0 \cdot 1 + 1 \cdot 0 = 0$$

These results are shown in the truth table of Fig. 194-1B. Note that the EX-OR gate always has only two inputs. There are no EX-OR gates with three or more inputs.

In the equivalent switching circuit (Fig. 194-1C) for an EX-OR gate, the two switches are mechanically linked together so that one or the other, but not both, can be closed at any particular time.

The EX-OR gate actually consists of inverters, AND gates, and an OR gate. The required combination of these logic circuits is shown in Fig. 194-2. However, the EX-OR gate is used sufficiently often to be given its own symbol (Fig. 194-1D).

MATHEMATICAL DERIVATIONS

For the EX-OR operation,

$$f(A,B) = \overline{A} \cdot B + A \cdot \overline{B}$$
$$= A \oplus B \tag{194-2}$$

The $\oplus$ symbol is used to represent the EX-OR gate operation.

To summarize the EX-OR gate operation:

1. The output is only in the 1 state when the two input elements are in different states.
2. If the two input elements are in the same state, the output is in the 0 state.

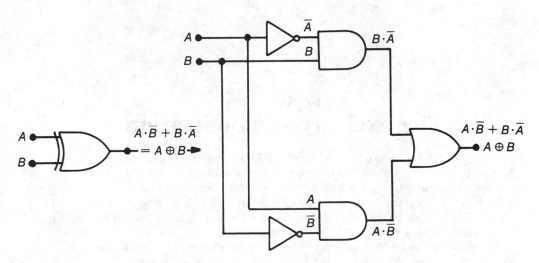

Figure 194-2

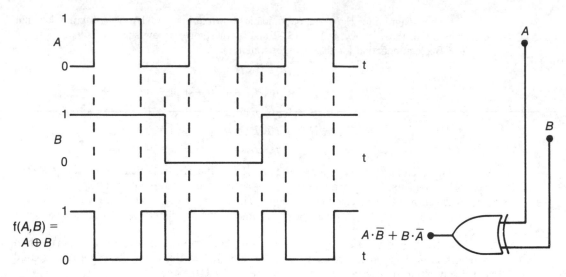

Figure 194-3

Example 194-1

Figure 194-3 shows the logic voltage levels for the two inputs to an EX-OR gate. Obtain the waveform for the logic output.

Solution

The solution appears in Fig. 194-3. Note that when B is in the 1 state, the output waveform is the inverse of the A waveform, but when B is in the 0 state, the A waveform and output waveform are the same.

Chapter 195
The exclusive-NOR operation

THE EXCLUSIVE-NOR (ABBREVIATED EX-NOR) OPERATION IS THE LEAST COMMON of all the operations discussed. It is merely the inverse of the EX-OR operation, consequently, an EX-NOR circuit can be formed by following an EX-OR gate with an inverter.

In the EX-NOR operation both of the input elements, A and B, must be in the same state (either 0 or 1) in order for the output, $f(A,B)$, to be in the 1 state; however, if A and B are in opposite states at the same time, the output is in the 0 state.

The Venn diagram for the EX-NOR operation is shown in Fig. 195-1A. This diagram does not appear in any of those shown in Fig. 188-1, and is therefore a new subclass.

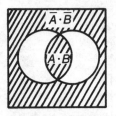

Venn diagram
for EX-NOR operation

A.

Input elements		EX-NOR output
A	B	$f(A,B) = A \cdot B + \overline{A} \cdot \overline{B}$
0	0	1
0	1	0
1	0	0
1	1	1

Truth table for
EX-NOR operation

B.

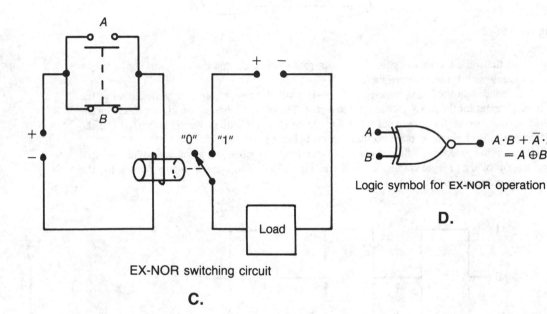

EX-NOR switching circuit

C.

$A \cdot B + \overline{A} \cdot \overline{B}$
$= A \oplus B$

Logic symbol for EX-NOR operation

D.

Figure 195-1

The Boolean equation for the EX-NOR operation is:

$$f(A,B) = A \cdot B + \overline{A} \cdot \overline{B} \qquad (195\text{-}1)$$

Note that this logic expression is not the same as $A \cdot B + \overline{A \cdot B}$, which must always equal 1.

If

$$A = 0, B = 0, f(A,B) = 0 \cdot 0 + 1 \cdot 1 = 1$$
$$A = 0, B = 1, f(A,B) = 0 \cdot 1 + 1 \cdot 0 = 0$$
$$A = 1, B = 0, f(A,B) = 1 \cdot 0 + 0 \cdot 1 = 0$$
$$A = 1, B = 1, f(A,B) = 1 \cdot 1 + 0 \cdot 0 = 1$$

These results are shown in the truth table of Fig. 195-1B.

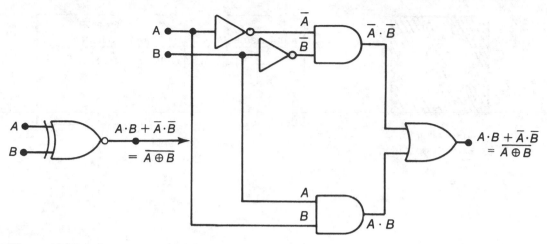

Figure 195-2

Note that, like the EX-OR gate, the EX-NOR gate always has only two inputs. There are no EX-NOR gates with three or more inputs.

The equivalent switching circuit (Fig. 195-1C) for the EX-NOR gate is a combination of the switching circuit for the EX-OR gate and the inverter. The two switches, A and B, are mechanically linked so that one or the other, but not both, can be closed. If either A or B is CLOSED (1), the switch in the load circuit is open so that the load circuit is OFF (0).

The EX-NOR gate is actually composed of inverters, AND gates, and an OR gate. The required combination of these logic circuits is shown in Fig. 195-2. The corresponding symbol for the EX-NOR gate appears in Fig. 195-1D.

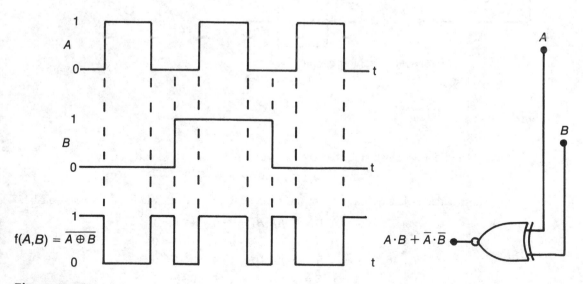

Figure 195-3

MATHEMATICAL DERIVATIONS

For the EX-NOR operation,

$$f(A,B) = A \cdot B + \overline{A} \cdot \overline{B}$$

This can be abbreviated to:

$$f(A,B) = \overline{A \oplus B} \qquad (195\text{-}2)$$

To summarize the EX-NOR operation,

1. The output is only in the 1 state when the two input elements are in the same state.
2. If the two input elements are in opposite states, the output is in the 0 state.

Example 195-1

Figure 195-3 shows the logic voltage levels for the two inputs to an EX-NOR gate. Obtain the waveform for the logic output.

Solution

The solution appears in Fig. 195-3. Note that when B is in the 0 state, the output waveform is the inverse of the A waveform; however, when B is in the 1 state, the A waveform and the output waveform are the same.

Chapter 196
Combinational logic circuits

IN CHAPTERS 189 THROUGH 195, YOU HAVE LOOKED AT INDIVIDUAL LOGIC operations, each with its own Venn diagram, Boolean equation, and truth table. You will now investigate the result of combining inverters with the various gates to produce combination logic circuits. In each such circuit, you will derive an expression for the Boolean output and then obtain its truth table. However, the output expression might not be in its simplest form, so that you will need the axioms, laws, and theorems of Boolean algebra, as outlined in Chapters 197 and 198.

Simplification is important because you might be given the required expression for the output and then be asked to design the corresponding logic circuit. If the expression can be simplified, you can achieve the most economical solution by creating the necessary circuit with the minimum number of gates.

In the mathematical derivations, you will summarize the individual logic operations and look at alternative symbols for these operations. In one set of alternative symbols, the input elements and the output are inverted, while the AND and OR operations are interchanged. The inversions are signified by adding bubbles to the input and output lines. Because both inputs and outputs have been inverted, you can regard these new symbols as DeMorgan equivalents of the standard symbols (Chapter 198).

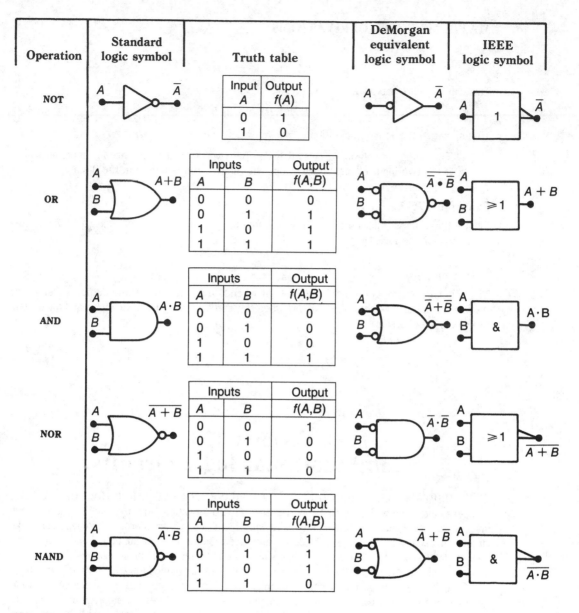

Operation	Standard logic symbol	Truth table		DeMorgan equivalent logic symbol	IEEE logic symbol

NOT

Input A	Output f(A)
0	1
1	0

OR

Inputs		Output
A	B	f(A,B)
0	0	0
0	1	1
1	0	1
1	1	1

AND

Inputs		Output
A	B	f(A,B)
0	0	0
0	1	0
1	0	0
1	1	1

NOR

Inputs		Output
A	B	f(A,B)
0	0	1
0	1	0
1	0	0
1	1	0

NAND

Inputs		Output
A	B	f(A,B)
0	0	1
0	1	1
1	0	1
1	1	0

Figure 196-1

MATHEMATICAL DERIVATIONS

The most common logic operations are summarized in Fig. 196-1. For a particular operation, the standard and DeMorgan equivalent symbols show the same truth table. As a result, the following relationship must exist:

$$\text{NOR} \quad \overline{A} \cdot \overline{B} = \overline{A + B} \qquad (196\text{-}1)$$

$$\text{NAND} \quad \overline{A \cdot B} = \overline{A} + \overline{B} \qquad (196\text{-}2)$$

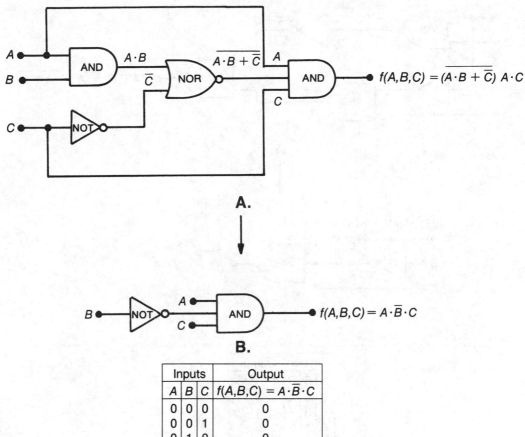

Figure 196-2

These results will be demonstrated by DeMorgan's Theorem (Chapter 198).

The rectangular symbols were first approved in 1984 and are now used by some digital IC manufacturers and the military. The signs "1, &, ⩾ 1" respectively indicate the NOT, AND, and OR operations, while the right-angled triangle "◁" represents inversion.

Consider the combinational logic circuits of Fig. 196-2A. The Boolean expression for the output is:

$$f(A,B,C) = \overline{(A \cdot B + \overline{C})} \cdot A \cdot C \qquad (196\text{-}1)$$
$$= A \cdot C \cdot \overline{\overline{C}} \cdot \overline{A \cdot B}$$

But $\overline{\overline{C}} = C$ and $C \cdot C = C$. Therefore,

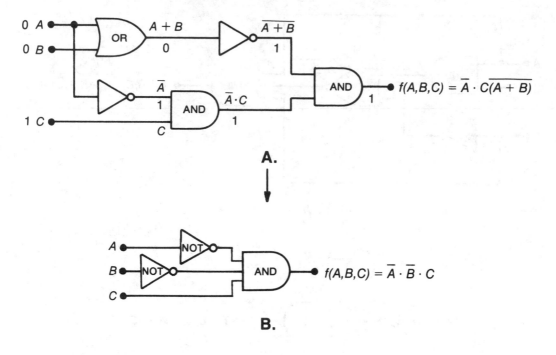

A.

B.

Inputs			Output
A	B	C	f(A,B,C)
0	0	0	0
0	0	1	1
0	1	0	0
0	1	1	0
1	0	0	0
1	0	1	0
1	1	0	0
1	1	1	0

C.

Figure 196-3

$$f(A,B,C) = A \cdot C \cdot (\overline{A} + \overline{B}) \qquad (196\text{-}2)$$
$$= A \cdot \overline{B} \cdot C$$

because $A \cdot \overline{A} = 0$.

Consequently, the logic circuit with its three gates and one inverter can be simplified to the circuit of Fig. 196-2B. The corresponding truth table is shown in Fig. 196-2C.

Example 196-1

Obtain the Boolean expression for the output of the combinational logic circuit of Fig. 196-3A. Derive the corresponding truth table.

Solution

$$\text{Output, } f(A,B,C) = \overline{A} \cdot C(\overline{A + B}) \tag{196-2}$$
$$= \overline{A} \cdot C \cdot \overline{A} \cdot \overline{B}$$
$$= \overline{A} \cdot \overline{B} \cdot C$$

because $\overline{A} \cdot \overline{A} = \overline{A}$.

The simplified logic circuit and its corresponding truth table are respectively shown in Fig. 196-3B and C. Note that the truth table can be derived by applying, in turn, each sequence of the three inputs to the original logic circuit. (See Fig. 196-3A for the inputs $A = 0$, $B = 0$, $C = 1$.)

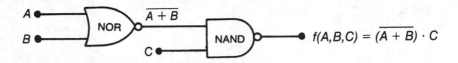

$$f(A,B,C) = \overline{(A + B)} \cdot C$$

A.

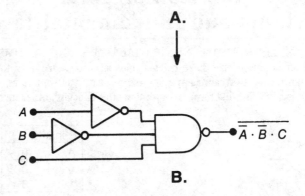

$$\overline{\overline{A} \cdot \overline{B} \cdot C}$$

B.

Inputs			Output
A	B	C	f(A,B,C)
0	0	0	1
0	0	1	0
0	1	0	1
0	1	1	1
1	0	0	1
1	0	1	1
1	1	0	1
1	1	1	1

C.

Figure 196-4

Example 196-2

Develop a logic circuit for which the Boolean output expression is $f(A,B,C) = \overline{(\overline{A + B}) \cdot C}$. Derive the corresponding truth table.

Solution

The NOR gate provides an output of $\overline{A + B}$ (see Fig. 196-1). The NOR gate output is then one input to a NAND gate, whose other input is C. The required logic circuit is shown in Fig. 196-4A.

$$\text{Note that:} \qquad f(A,B,C) = \overline{(\overline{A + B}) \cdot C} \qquad (196\text{-}2)$$
$$= \overline{\overline{A} \cdot \overline{B} \cdot C}$$

The alternative logic circuit is shown in Fig. 196-4B, while the corresponding truth table appears in Fig. 196-4C.

Chapter 197
Boolean algebra:
axioms and fundamental laws

THIS CHAPTER LISTS THE FUNDAMENTAL LAWS AND AXIOMS OF BOOLEAN algebra. You should memorize these relationships since they are used to simplify logic expressions. Although these laws can change any expression, it is sometimes difficult to decide whether the result is in its simplest form. To overcome this difficulty, other methods of simplifica-

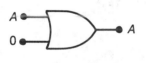

Inputs		Output
A	0	*f(A,0)*
0	0	0
1	0	1

$$A + 0 \;=\; A$$

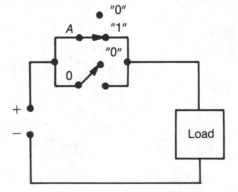

Figure 197-1

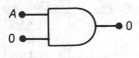

Inputs		Output
A	0	f(A,0)
0	0	0
1	0	0

$$A \cdot 0 = 0$$

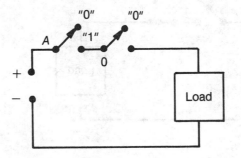

Figure 197-2

tion—the Karnaugh map and the Harvard Chart—are discussed in Chapters 199 and 200. For each axiom and law there is a corresponding switching circuit, logic diagram, and truth table.

MATHEMATICAL DERIVATIONS
Axioms

$$A + 0 = A \qquad (197\text{-}1)$$

$$A \cdot 0 = 0 \qquad (197\text{-}2)$$

$$A + 1 = 1 \qquad (197\text{-}3)$$

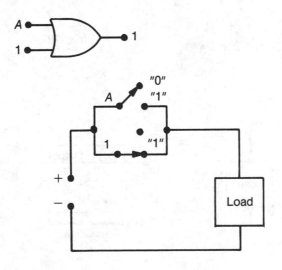

Inputs		Output
A	1	f(A,1)
0	1	1
1	1	1

$$A + 1 = 1$$

Figure 197-3

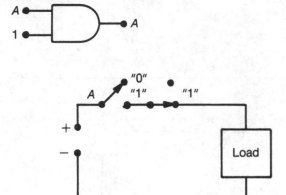

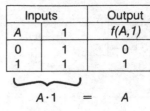

Inputs		Output
A	1	f(A,1)
0	1	0
1	1	1

$$A \cdot 1 = A$$

Figure 197-4

$$A \cdot 1 = A \qquad (197\text{-}4)$$

These axioms are true whether A is 0 or 1. They are illustrated in Figs. 197-1, 2, 3, and 4.

Laws

Identity Law

$$A = A \qquad \{\text{Fig. 197-5}\} \quad (197\text{-}5)$$

Complementary Law

(a) $\qquad\qquad A \cdot \overline{A} = 0 \qquad\qquad\qquad (197\text{-}6)$

$\qquad\qquad\qquad\qquad\qquad\qquad \{\text{Fig. 197-6}\}$

(b) $\qquad\qquad A + \overline{A} = 1 \qquad\qquad\qquad (197\text{-}7)$

Idempotent Law

(a) $\qquad\qquad A \cdot A = A \qquad\qquad\qquad (197\text{-}8)$

$\qquad\qquad\qquad\qquad\qquad\qquad \{\text{Fig. 197-7}\}$

(b) $\qquad\qquad A + A = A \qquad\qquad\qquad (197\text{-}9)$

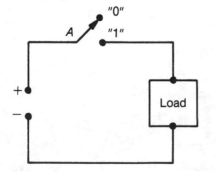

Input	Output
A	A
0	0
1	1

$$A = A$$

Figure 197-5

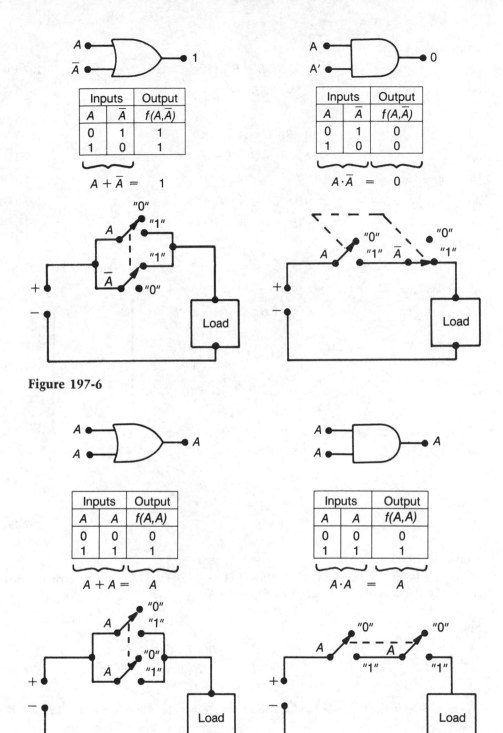

Figure 197-6

Figure 197-7

Commutative Law

(a)
$$A \cdot B = B \cdot A \qquad (197\text{-}10)$$
(b)
$$A + B = B + A \qquad (197\text{-}11)$$

{Fig. 197-8}

Associative Law

(a)
$$(A \cdot B) \cdot C = A \cdot (B \cdot C) \qquad (197\text{-}12)$$
(b)
$$(A + B) + C = A + (B + C) \qquad (197\text{-}13)$$

{Fig. 197-9}

Distributive Law

(a)
$$A \cdot (B + C) = (A \cdot B) + (A \cdot C) \qquad (197\text{-}14)$$
(b)
$$A + (B \cdot C) = (A + B) \cdot (A + C) \qquad (197\text{-}15)$$

{Fig. 197-10}

Double Negation Law

$$\overline{\overline{A}} = A \qquad \text{\{Fig. 197-11\}} \quad (197\text{-}16)$$

Absorption Law

(a)
$$A \cdot (A + B) = A \cdot A + A \cdot B$$
$$= A + A \cdot B$$
$$= A \cdot (1 + B)$$
$$= A \cdot 1 = A$$

Fig. 197-12 \quad (197-17)

(b)
$$A + (A \cdot B) = A(1 + B)$$
$$= A \cdot 1$$
$$= A$$

(197-18)

(c)
$$A + \overline{A} \cdot B = A + A \cdot B + \overline{A} \cdot B$$
$$= A \cdot \overline{A} + A \cdot A + A \cdot B + A \cdot B$$
$$= (A + \overline{A}) \cdot (A + B)$$
$$= 1 \cdot (A + B)$$
$$= A + B \qquad (197\text{-}19)$$

Example 197-1

Simplify the Boolean equation $f(A, B, C, D) = A \cdot C + A \cdot D + B \cdot C + B \cdot D$. Derive the logic circuits for the original and the simplified expressions.

Solution

$$f(A, B, C, D) = A \cdot C + A \cdot D + B \cdot C + B \cdot D$$
$$= A \cdot (C + D) + B \cdot (C + D) \qquad (197\text{-}14)$$
$$= (A + B) \cdot (C + D)$$

The logic circuits for the original and simplified expressions are shown in Fig. 197-13A and B.

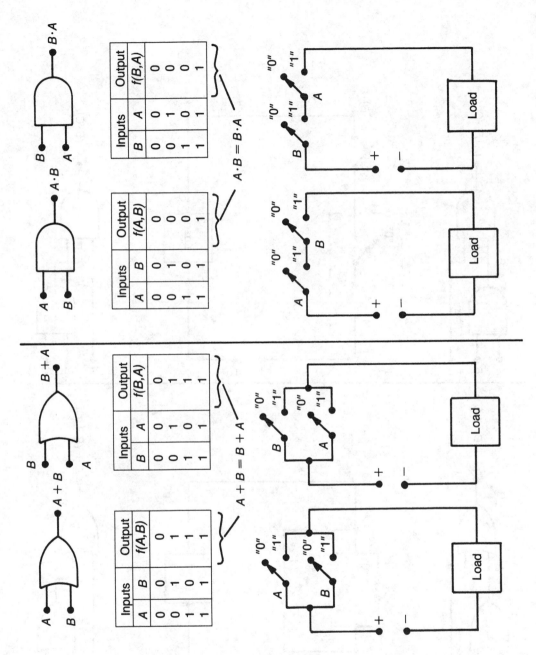

Figure 197-8

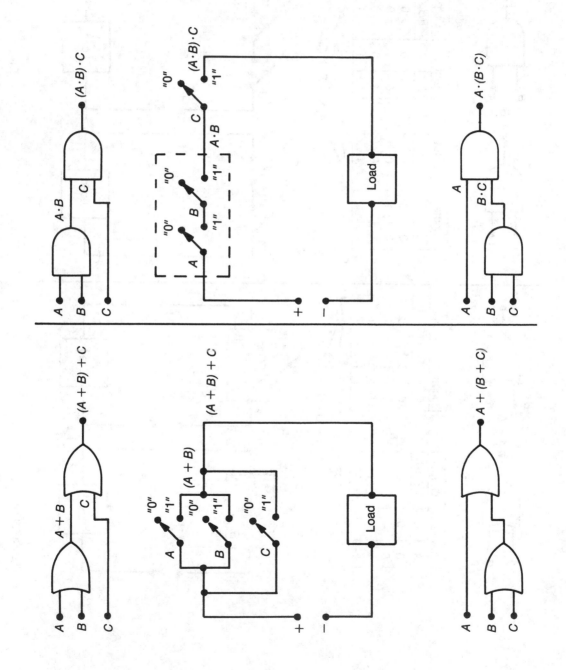

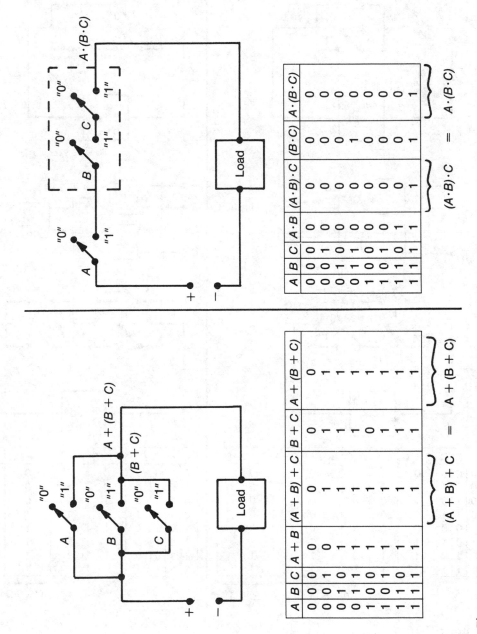

Figure 197-9

717

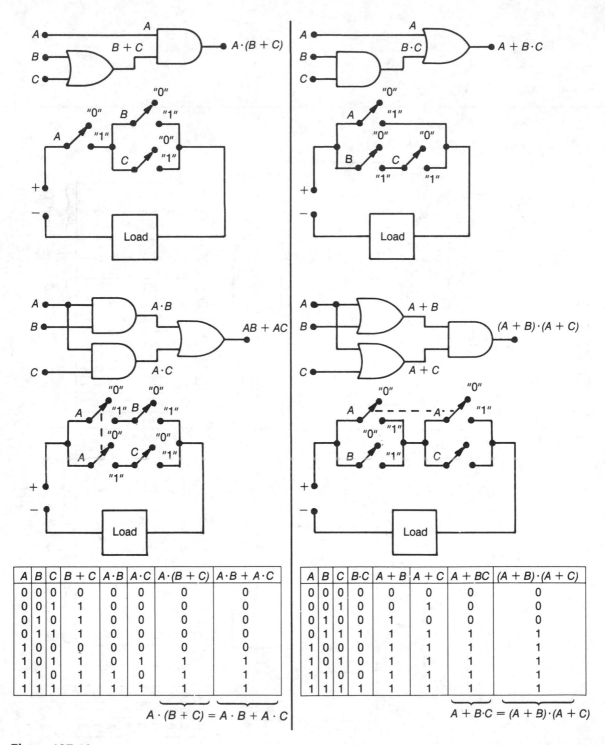

A	B	C	B+C	A·B	A·C	A·(B+C)	A·B + A·C
0	0	0	0	0	0	0	0
0	0	1	1	0	0	0	0
0	1	0	1	0	0	0	0
0	1	1	1	0	0	0	0
1	0	0	0	0	0	0	0
1	0	1	1	0	1	1	1
1	1	0	1	1	0	1	1
1	1	1	1	1	1	1	1

$$A \cdot (B + C) = A \cdot B + A \cdot C$$

A	B	C	B·C	A+B	A+C	A+BC	(A+B)·(A+C)
0	0	0	0	0	0	0	0
0	0	1	0	0	1	0	0
0	1	0	0	1	0	0	0
0	1	1	1	1	1	1	1
1	0	0	0	1	1	1	1
1	0	1	0	1	1	1	1
1	1	0	0	1	1	1	1
1	1	1	1	1	1	1	1

$$A + B \cdot C = (A + B) \cdot (A + C)$$

Figure 197-10

718

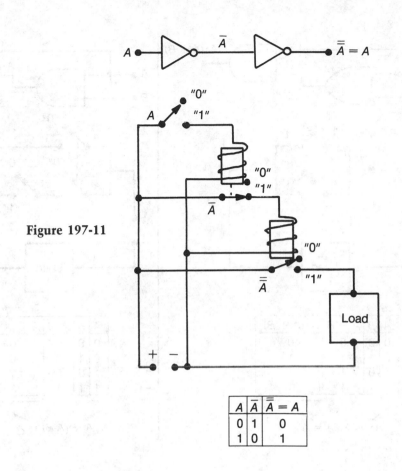

Figure 197-11

A	$\overline{A}$	$\overline{\overline{A}} = A$
0	1	0
1	0	1

Example 197-2

Simplify the Boolean equation $f(A,B,C,D) = A \cdot B \cdot C + A \cdot B \cdot \overline{D} + A \cdot \overline{C} + \overline{A} \cdot \overline{B} \cdot \overline{C} \cdot \overline{D} + \overline{A} \cdot C$. Derive the logic circuits for the original and simplified expressions.

Solution

The logic circuit for the original equation appears in Fig. 197-14A.

$$f(A,B,C,D) = A \cdot B \cdot C + A \cdot B \cdot \overline{D} + A \cdot \overline{C} + \overline{A} \cdot \overline{B} \cdot \overline{C} \cdot \overline{D} + \overline{A} \cdot C$$

$$= A \cdot B \cdot C + A \cdot \overline{C} + \overline{A} \cdot \overline{B} \cdot \overline{C} \cdot \overline{D} + \overline{A} \cdot C + A \cdot B \cdot \overline{D} \quad (197\text{-}11)$$

$$= A \cdot (B \cdot C + \overline{C}) + \overline{A} \cdot (\overline{B} \cdot \overline{C} \cdot \overline{D} + C) + A \cdot B \cdot \overline{D} \quad (197\text{-}14)$$

$$= A \cdot (B + \overline{C}) + \overline{A} \cdot (\overline{B} \cdot \overline{D} + C) + A \cdot B \cdot \overline{D} \quad (197\text{-}6)$$

$$= A \cdot B + A \cdot \overline{C} + \overline{A} \cdot \overline{B} \cdot \overline{D} + \overline{A} \cdot C + A \cdot B \cdot \overline{D} \quad (197\text{-}14)$$

$$= (A \cdot B + A \cdot B \cdot \overline{D}) + A \cdot \overline{C} + \overline{A} \cdot C + \overline{A} \cdot \overline{B} \cdot \overline{D}$$

$$= A \cdot B + A \cdot \overline{C} + \overline{A} \cdot C + \overline{A} \cdot \overline{B} \cdot \overline{D} \quad (197\text{-}18)$$

This form does, in fact, produce the simplest logic circuit (Fig. 197-14B). However, the simplification process for the equation can be carried one step further by factoring so that,

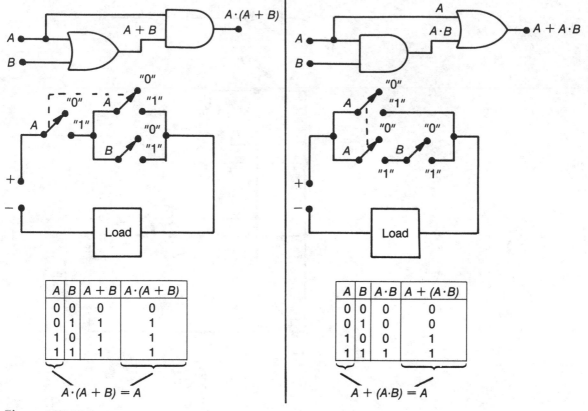

Figure 197-12

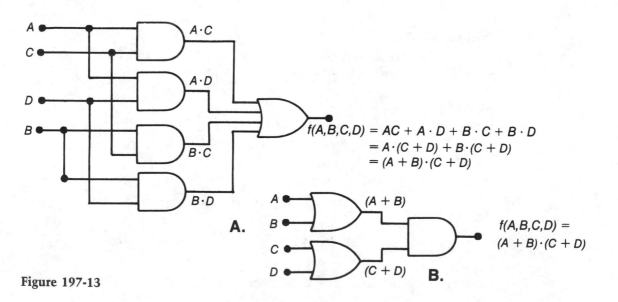

Figure 197-13

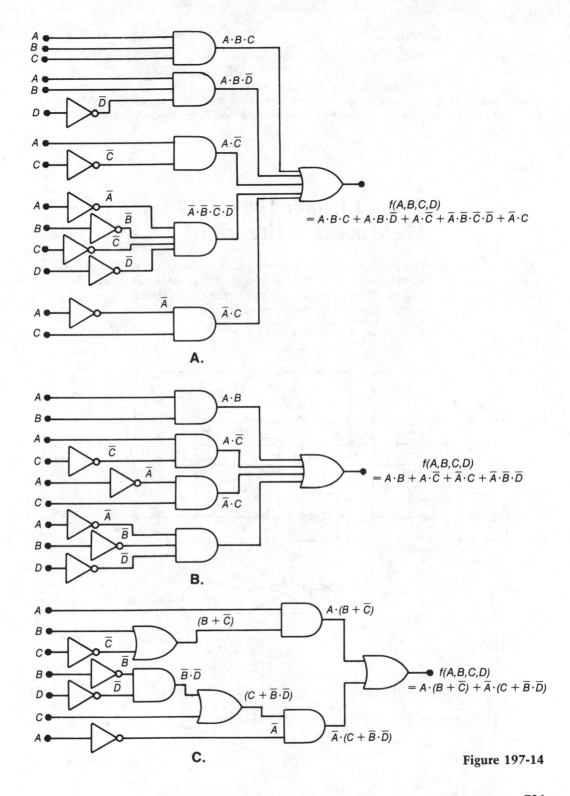

A.

$$f(A,B,C,D)$$
$$= A \cdot B \cdot C + A \cdot B \cdot \overline{D} + A \cdot \overline{C} + \overline{A} \cdot \overline{B} \cdot \overline{C} \cdot \overline{D} + \overline{A} \cdot C$$

B.

$$f(A,B,C,D)$$
$$= A \cdot B + A \cdot \overline{C} + \overline{A} \cdot C + \overline{A} \cdot \overline{B} \cdot \overline{D}$$

C.

$$f(A,B,C,D)$$
$$= A \cdot (B + \overline{C}) + \overline{A} \cdot (C + \overline{B} \cdot \overline{D})$$

Figure 197-14

$$f(A,B,C,D) = A \cdot (B + \overline{C}) + \overline{A} \cdot (C + \overline{B} \cdot \overline{D}) \qquad (197\text{-}14)$$

The logic circuit for this equation requires six gates, as does the logic circuit for the original equation. The simplest logic circuit of Fig. 197-14B needs only five gates.

This example of simplification shows that the process is rather difficult at first with no positive indication that the equation for the simplest logic circuit has been reached. Skill in making the correct choice can only be acquired by repeated use of the Boolean laws and axioms.

Chapter 198
DeMorgan's theorems

A FAMOUS MATHEMATICIAN, AUGUSTUS DeMORGAN (1806–1871), STATED TWO theorems that relate to Boolean algebra. These theorems are very useful when you need to simplify expressions containing the sum or product of two or more elements that are complemented (inverted). In equation form, the two theorems are:

$$\overline{A} \cdot \overline{B} \cdot \overline{C} \cdots = \overline{A + B + C +} \qquad (198\text{-}1)$$

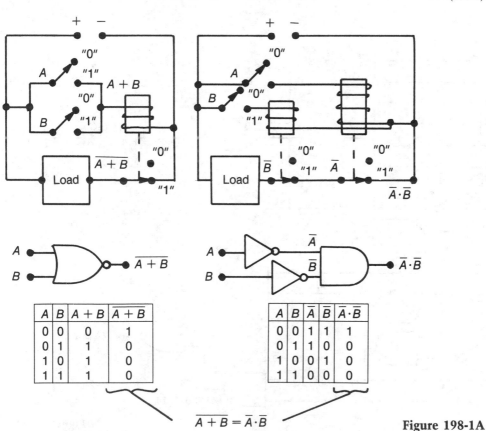

Figure 198-1A

and,

$$\overline{A} + \overline{B} + \overline{C} + \cdots = \overline{A \cdot B \cdot C} \tag{198-2}$$

The equivalent switching circuits, logic diagrams, and truth tables for the DeMorgan theorems with two input elements are shown in Fig. 198-1A and B.

Equations 198-1 and 198-2 contain only single elements—A, B, C, and so on. However each of these elements might be an expression containing a number of variables. As examples:

$$\begin{aligned} f(A,B,C) &= \overline{(A + B) \cdot \overline{C}} \\ &= \overline{A + B} + \overline{\overline{C}} \\ &= \overline{A} \cdot \overline{B} + C \end{aligned}$$

and

$$\begin{aligned} f(A,B,C,D) &= \overline{A \cdot \overline{B} \cdot C + D} \\ &= \overline{A \cdot \overline{B} \cdot C} \cdot \overline{D} \\ &= (\overline{A} + \overline{\overline{B}} + \overline{C}) \cdot \overline{D} \\ &= (\overline{A} + B + \overline{C}) \cdot \overline{D} \end{aligned}$$

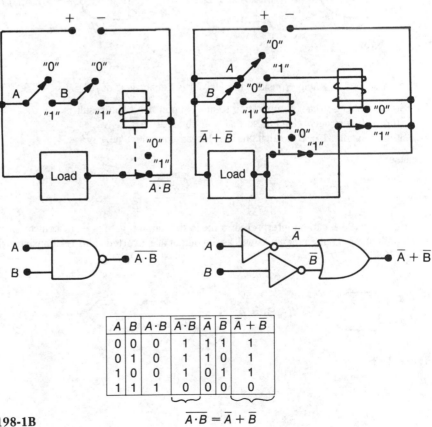

A	B	A·B	$\overline{A \cdot B}$	$\overline{A}$	$\overline{B}$	$\overline{A} + \overline{B}$
0	0	0	1	1	1	1
0	1	0	1	1	0	1
1	0	0	1	0	1	1
1	1	1	0	0	0	0

$$\overline{A \cdot B} = \overline{A} + \overline{B}$$

Figure 198-1B

723

There is a step-by-step approach for obtaining the DeMorgan equivalent of a Boolean expression:

Step 1. Interchange AND operator symbols ($\cdot$) with OR operator symbols ($+$).
Step 2. Invert all the elements so they are complemented.
Step 3. Invert (complement) the results of steps 1 and 2 to obtain the DeMorgan equivalent.

For example, the DeMorgan equivalent of $f(A,B) = A + B$ is:

Step 1. Change the OR operation to an AND operation. The result is $A \cdot B$.
Step 2. Invert each element to obtain $\overline{A} \cdot \overline{B}$.
Step 3. Invert $\overline{A} \cdot \overline{B}$ so that the DeMorgan equivalent of $A + B$ is $\overline{\overline{A} \cdot \overline{B}}$.

Because

$$A + B = \overline{\overline{A} \cdot \overline{B}} \tag{198-3}$$

$$\overline{A + B} = \overline{\overline{\overline{A} \cdot \overline{B}}} = \overline{A} \cdot \overline{B} \tag{198-1}$$

In terms of logic circuits, Equation 198-2 means that an OR gate is equivalent to an AND gate with both of the input elements and the output inverted. This relationship is illustrated in Fig. 198-2.

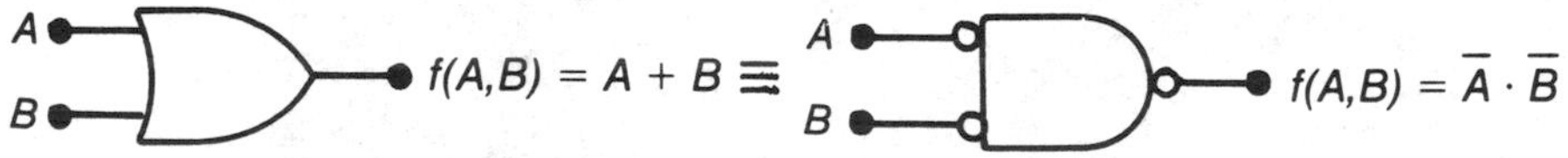

$$f(A,B) = A + B \equiv \qquad f(A,B) = \overline{A} \cdot \overline{B}$$

Figure 198-2

As a second example, consider the DeMorgan equivalent of $f(A,B) = A \cdot B$.

Step 1. Change the AND operation to an OR operation. The result is $A + B$.
Step 2. Invert each element so that you obtain $\overline{A} + \overline{B}$.
Step 3. Invert $\overline{A} + \overline{B}$ so that the DeMorgan equivalent of $A \cdot B$ is $\overline{\overline{A} + \overline{B}}$.

Because

$$A \cdot B = \overline{\overline{A} + \overline{B}}, \tag{198-4}$$

$$\overline{A \cdot B} = \overline{\overline{\overline{A} + \overline{B}}} = \overline{A} + \overline{B} \tag{198-2}$$

Equation 198-3 can be interpreted to mean that an AND gate is equivalent to an OR gate in which both of the input elements as well as the output are inverted. This relationship is illustrated in Fig. 198-3.

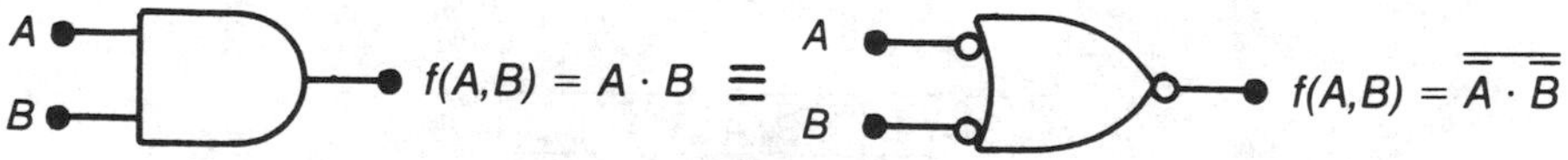

$$f(A,B) = A \cdot B \equiv \qquad f(A,B) = \overline{\overline{A} \cdot \overline{B}}$$

Figure 198-3

MATHEMATICAL DERIVATIONS

To summarize the equations for the DeMorgan theorems with two input elements:

$$\overline{A} \cdot \overline{B} = \overline{A + B} \qquad (198\text{-}5)$$

or,

$$A + B = \overline{\overline{A} \cdot \overline{B}} \qquad (198\text{-}6)$$

$$\overline{A} + \overline{B} = \overline{A \cdot B} \qquad (198\text{-}7)$$

or,

$$A \cdot B = \overline{\overline{A} + \overline{B}} \qquad (198\text{-}8)$$

Equations 198-5 and 198-7 are illustrated in the Venn diagrams of Fig. 198-4A and B. The shaded area of Fig. 198-4A represents $\overline{A} \cdot \overline{B}$ as well as $\overline{A + B}$ and means that NOT A AND NOT B is equivalent to NOT (A OR B). (See the Venn diagram in Fig. 188-1.) Similarly, the shaded area of Fig. 198-4B represents $\overline{A} + \overline{B}$ as well as $\overline{A \cdot B}$ and means NOT A OR NOT B is equivalent to NOT (A AND B). (See the Venn diagram in Fig. 188-1.)

Example 198-1

Simplify the Boolean expressions (a) $f(A,B,C,D) = \overline{(\overline{A} + B) \cdot (C + \overline{D})}$, (b) $f(A,B,C) = A \cdot B \cdot C + A \cdot B\overline{(\overline{A} \cdot C)}$, and (c) $f(A,B,C,D) = \overline{A} \cdot B\overline{(A \cdot C \cdot D)} + \overline{A} \cdot \overline{B} \cdot C \cdot \overline{D} + A \cdot B \cdot \overline{C}$.

Solution

(a)
$$
\begin{aligned}
f(A,B,C,D) &= \overline{(\overline{A} + B) \cdot (C + \overline{D})} \\
&= \overline{\overline{A} + B} + \overline{C + \overline{D}} \\
&= \overline{\overline{A}} \cdot \overline{B} + \overline{C} \cdot \overline{\overline{D}} \\
&= A \cdot \overline{B} + \overline{C} \cdot D
\end{aligned}
$$

This example illustrates one of the principles of simplification. The inversion overbar is broken by interchanging the AND and OR operations. This process is repeated until you are left with only inverted elements.

(b)
$$
\begin{aligned}
f(A,B,C) &= A \cdot B \cdot C + A \cdot \overline{B}\,\overline{(\overline{A} \cdot \overline{C})} \\
&= A \cdot B \cdot C + A \cdot \overline{B}(\overline{\overline{A}} + \overline{\overline{C}}) \qquad (198\text{-}7)
\end{aligned}
$$

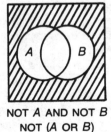

NOT A AND NOT B
NOT (A OR B)

A.

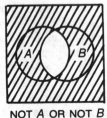

NOT A OR NOT B
NOT (A AND B)

B.

Figure 198-4

$$= A \cdot B \cdot C + A \cdot A \cdot \overline{B} + A \cdot \overline{B} \cdot C \qquad \text{(197-16)}$$
$$= A \cdot B \cdot C + A \cdot \overline{B} + A \cdot \overline{B} \cdot C \qquad \text{(197-8)}$$
$$= A \cdot C \cdot (B + \overline{B}) + A \cdot \overline{B}$$
$$= A \cdot C + A \cdot \overline{B} \qquad \text{(197-7)}$$
$$= A \cdot (\overline{B} + C)$$

The logic circuit for the original expression requires four gates and three inverters, while the simplified expression only needs two gates and one inverter.

(c)
$$f(A,B,C,D) = \overline{A} \cdot B(\overline{\overline{A} \cdot C \cdot D}) + \overline{A} \cdot \overline{B} \cdot C \cdot \overline{D} + A \cdot B \cdot \overline{C}$$
$$- \overline{A} \cdot B \cdot (\overline{\overline{A}} + \overline{C} + \overline{D}) + \overline{A} \cdot \overline{B} \cdot C \cdot \overline{D} + A \cdot B \cdot \overline{C} \qquad \text{(198-7)}$$
$$= A \cdot \overline{A} \cdot B + \overline{A} \cdot B \cdot \overline{C} + \overline{A} \cdot B \cdot \overline{D} + \overline{A} \cdot \overline{B} \cdot C \cdot \overline{D} + A \cdot B \cdot \overline{C}$$
$$= \overline{A} \cdot B \cdot \overline{C} + \overline{A} \cdot B \cdot \overline{D} + \overline{A} \cdot \overline{B} \cdot C \cdot \overline{D} + A \cdot B \cdot \overline{C} \qquad \text{(197-6)}$$
$$= B \cdot \overline{C}(A + \overline{A}) + \overline{A} \cdot \overline{D}(B + \overline{B} \cdot C)$$
$$= B \cdot \overline{C} + \overline{A} \cdot \overline{D} \cdot (B + C) \qquad \text{(197-19)}$$

The original expression requires five inverters and five gates, while the simplified result only needs three inverters and four gates.

Chapter 199
Karnaugh maps

RATHER THAN EMPLOY THE BOOLEAN LAWS AND AXIOMS TO SIMPLIFY LOGIC expressions, you can use a Karnaugh map. Such a map provides a very quick and easy way of finding the simplest form of a logic equation. As shown in Fig. 199-1, Karnaugh maps can be readily constructed for two, three, or four elements, but they become more unwieldly if five or six elements are involved.

Because each element has two possible states, such as A and B, the number of squares required is 2^N, where N is the number of variables. Consequently for four, five, six, seven, and eight elements, the corresponding Karnaugh maps respectively contain 16, 32, 64, 128, and 256 squares. If the number of elements exceed six, the Karnaugh maps become too unwieldy and other methods of simplification should be used.

An exploded view of a four-element Karnaugh map is shown in Fig. 199-2. Note the division of the map into labeled columns and rows. Entries into the map are placed in these columns and rows in accordance with the logic terms contained in the given Boolean expression.

In Fig. 199-2, the square in the upper left corner of the Karnaugh map contains the elements A, B, $\overline{C}$, and D; the adjacent lower square represents the elements A, B, $\overline{C}$, and $\overline{D}$; the next lower square contains A, $\overline{B}$, $\overline{C}$, and D; while the lowest lefthand square has the elements A, $\overline{B}$, $\overline{C}$, and $\overline{D}$. The other 12 squares in the map are similarly identified. Note that the elements $A \cdot \overline{C}$ are contained in each of the four terms just identified. Since the elements $B \cdot \overline{B}$, and $D \cdot \overline{D}$ are also contained in the same four squares, these will disappear when the four squares are added

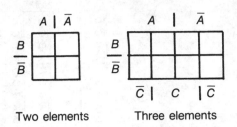

Two elements   Three elements

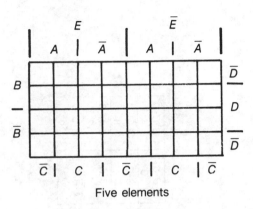

Four elements   Five elements

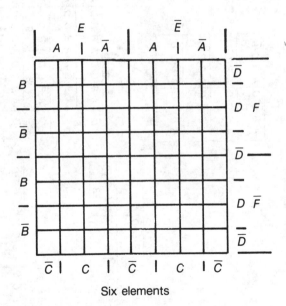

Six elements

Figure 199-1

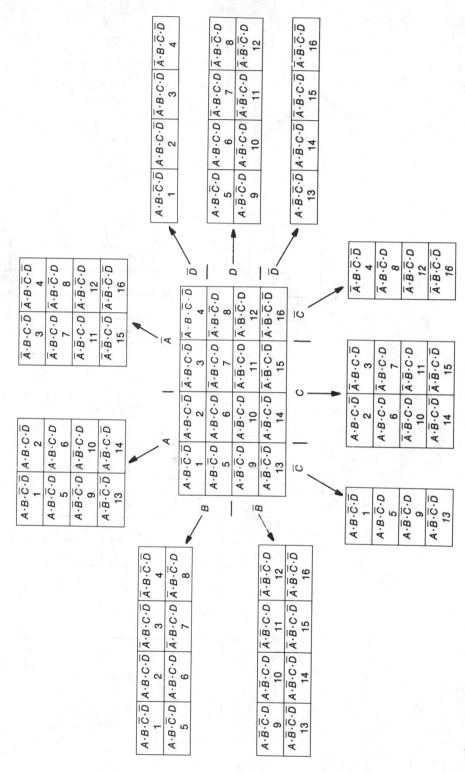

Figure 199-2

together. In other words, the left vertical column represents the term $A \cdot \overline{C}$. This can be proven as follows:

$$\begin{aligned}
A \cdot \overline{C} &= A \cdot \overline{C} \cdot 1 \\
&= A \cdot \overline{C} \cdot (B + \overline{B}) \text{ since } B + \overline{B} = 1 \qquad\qquad (197\text{-}7) \\
&= A \cdot B \cdot \overline{C} \cdot (D + \overline{D}) + A \cdot \overline{B} \cdot \overline{C} \cdot (D + \overline{D}) \\
&= A \cdot B \cdot \overline{C} \cdot D + A \cdot B \cdot \overline{C} \cdot \overline{D} + A \cdot \overline{B} \cdot \overline{C} \cdot D + A \cdot \overline{B} \cdot \overline{C} \cdot \overline{D} \qquad (199\text{-}1)
\end{aligned}$$

These terms have the same sequence as that of the four squares.

Note that a two-element term is represented in Fig. 199-2 by four squares. Further study of the diagram reveals that a one-element term is represented by eight squares, while a three-element term is formed from two squares.

MATHEMATICAL DERIVATIONS

Now use the Karnaugh map to simplify the logic equation:

$$f(A,B,C,D) = A \cdot B \cdot C + A \cdot B \cdot \overline{D} + A \cdot \overline{C} + \overline{A} \cdot \overline{B} \cdot \overline{C} \cdot \overline{D} + \overline{A} \cdot C.$$

Because there are four elements, you will need the Karnaugh map that contains 16 squares (Fig. 199-3). The step-by-step procedure is then as follows:

Step 1. For the purpose of identification, label the squares from 1 to 16.

Step 2. On a term-by-term basis, plot the logic expression on the map. In other words, place a 1 in each of the squares that are required to represent a particular term.

The term $A \cdot B \cdot C$ is represented by squares 2 and 6, since $A \cdot B \cdot C \cdot D + A \cdot B \cdot C \cdot \overline{D} = A \cdot B \cdot C(D + \overline{D}) = A \cdot B \cdot C$.

The term $A \cdot B \cdot \overline{D}$ is identified by squares 1 and 2, since $A \cdot B \cdot \overline{C} \cdot \overline{D} + A \cdot B \cdot C \cdot \overline{D} = A \cdot B \cdot \overline{D}(C + \overline{C}) = A \cdot B \cdot \overline{D}$. Since a 1 already exists in square 2, it is only necessary to insert a 1 in square 1.

The term $A \cdot \overline{C}$ requires the squares 1, 5, 9, and 13 since Equation 199-1 has already shown that $A \cdot \overline{C} = A \cdot B \cdot \overline{C} \cdot D + A \cdot B \cdot \overline{C} \cdot \overline{D} + A \cdot \overline{B} \cdot \overline{C} \cdot D + A \cdot \overline{B} \cdot \overline{C} \cdot \overline{D}$.

The term $\overline{A} \cdot \overline{B} \cdot \overline{C} \cdot \overline{D}$ corresponds to square 16.

The term $\overline{A} \cdot C$ is the result of combining the squares 3, 7, 11, and 15 as follows:

$$\begin{aligned}
\overline{A} \cdot B \cdot C \cdot \overline{D} + \overline{A} \cdot B \cdot C \cdot D + \overline{A} \cdot \overline{B} \cdot C \cdot D + \overline{A} \cdot \overline{B} \cdot C \cdot \overline{D} \\
= \overline{A} \cdot B \cdot C(\overline{D} + D) + \overline{A} \cdot \overline{B} \cdot C(D + \overline{D}) \\
= \overline{A} \cdot B \cdot C + \overline{A} \cdot \overline{B} \cdot C \\
= \overline{A} \cdot C(B + \overline{B}) \\
= \overline{A} \cdot C
\end{aligned}$$

Step 3. Obtain the simplified logic equation by using Fig. 199-3 and observing the following rules:

(a) If 1's are located in adjacent squares or at opposite ends of any row or column, one of the elements can be dropped.

(b) Two of the elements can be dropped if: (1) any row or column of squares is filled with 1's; (2) any block of four squares is filled with 1's; (3) the four end squares of any adjacent rows or columns are filled with 1's (4) the four squares of a corner are filled with 1's.

(c) Three of the elements can be dropped if: (1) two adjacent rows or columns are filled with 1's; (2) the top and bottom rows or the right and left columns are completely filled with 1's.

(d) To reduce the original logic equation to its simplest form, all of the 1's must be included in the final expression. A particular 1 can be used more than once, but look for the largest possible combinations to form groups of eight, four, and two squares.

Note that, although the Karnaugh map is shown as a table, it is considered to be a cylinder, so squares 13 and 16 are contiguous. In addition, the top of the table can be folded back and down, while the bottom of the table is folded back and up, so the "$\overline{D}$" squares become adjacent.

Using Rule (b) in Step 3, squares 1, 5, 9, and 13 are combined to yield $A \cdot \overline{C}$. Squares 3, 7, 11, and 15 are grouped to yield $\overline{A} \cdot C$. Squares 1, 2, 5, and 6 when taken together, are equivalent to $A \cdot B$. Finally Rule (a) in Step 3 is used to group squares 15 and 16 and yields $\overline{A} \cdot \overline{B} \cdot \overline{D}$. All of these groups are shown in Fig. 199-3.

The Karnaugh map has produced the following simplification:

$$f(A,B,C,D) = A \cdot B \cdot C + A \cdot B \cdot \overline{D} + A \cdot \overline{C} + \overline{A} \cdot \overline{B} \cdot \overline{C} \cdot \overline{D} + \overline{A} \cdot C$$
$$= A \cdot B + A \cdot \overline{C} + \overline{A} \cdot C + \overline{A} \cdot \overline{B} \cdot \overline{D}$$

The original logic equation requires six gates, while the simplified expression requires five gates.

Alternatively, if squares 13 and 16 are combined by Rule (a),

$$f(A,B,C,D) = A \cdot B + A \cdot \overline{C} + \overline{A} \cdot C + \overline{B} \cdot \overline{C} \cdot \overline{D}$$

A single Boolean expression can, therefore, be represented by more than one simplified form.

A Karnaugh map also provides a convenient means of finding the complement of a logic

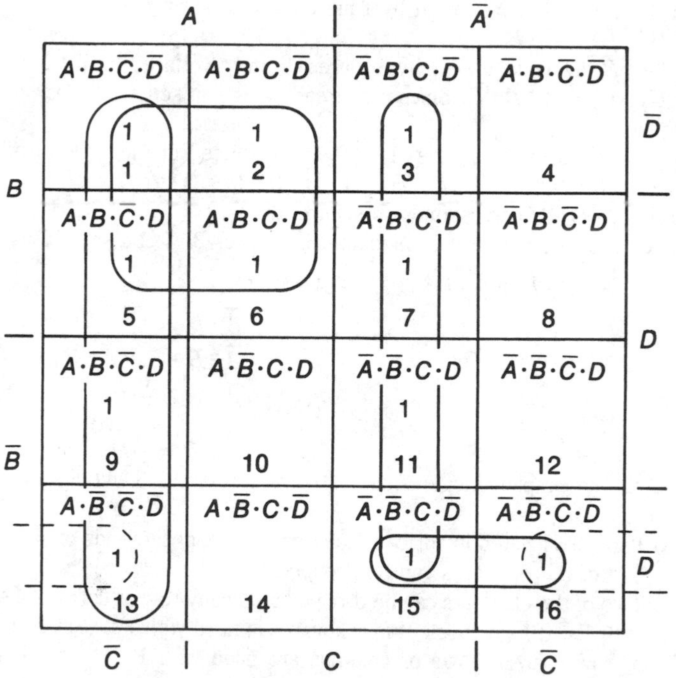

Figure 199-3

730

expression. This is done by plotting the logic terms on a map; then, on a second map, placing 1's in all the squares that did not contain 1's in the original map.

As an example,

$$f(A,B,C) = A \cdot B \cdot C$$

simplify
$$\overline{f(A,B,C)} = \overline{A \cdot B \cdot C}.$$

Fig. 199-4A shows the original map, while the complement map appears in Fig. 199-4B. As illustrated, squares 3, 4, 7, and 8 are grouped to produce $\overline{A}$. Similarly, squares 5, 6, 7, and 8 are combined to form $\overline{B}$, while squares 1, 5, 4, and 8 are equivalent to $\overline{C}$. Therefore,

$$f(A,B,C) = \overline{A \cdot B \cdot C}$$
$$= \overline{A} + \overline{B} + \overline{C}$$

This result is a DeMorgan's theorem, which appears in Equation 198-2.

Example 199-1

Use a Karnaugh map to simplify the logic equation:

$$f(A,B,C,D) = A \cdot \overline{B} \cdot \overline{C} + \overline{A} \cdot \overline{B} \cdot C \cdot D + A \cdot C \cdot \overline{D} + \overline{A} \cdot C \cdot \overline{D} + \overline{C} \cdot \overline{D}$$

A.

	A		$\overline{A}$	
	$A \cdot B \cdot \overline{C}$	$A \cdot B \cdot C$	$\overline{A} \cdot B \cdot C$	$\overline{A} \cdot B \cdot \overline{C}$
B	1	2	3	4
$\overline{B}$	$A \cdot \overline{B} \cdot \overline{C}$	$A \cdot \overline{B} \cdot C$	$\overline{A} \cdot \overline{B} \cdot C$	$\overline{A} \cdot \overline{B} \cdot \overline{C}$
	5	6	7	8
	$\overline{C}$	C		$\overline{C}$

Karnaugh map for $f(A,B,C) = A \cdot B \cdot C$

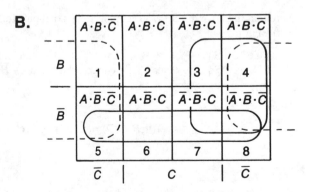

B.

Karnaugh map for $\overline{f(A,B,C)} = \overline{A \cdot B \cdot C}$

Figure 199-4

731

	A		Ā	
	$A \cdot B \cdot \overline{C} \cdot \overline{D}$	$A \cdot B \cdot C \cdot \overline{D}$	$\overline{A} \cdot B \cdot C \cdot \overline{D}$	$\overline{A} \cdot B \cdot \overline{C} \cdot \overline{D}$	$\overline{D}$
B	1	2	3	4	
	$A \cdot B \cdot \overline{C} \cdot D$	$A \cdot B \cdot C \cdot D$	$\overline{A} \cdot B \cdot C \cdot D$	$\overline{A} \cdot B \cdot \overline{C} \cdot D$	
	5	6	7	8	D
	$A \cdot \overline{B} \cdot \overline{C} \cdot D$	$A \cdot \overline{B} \cdot C \cdot D$	$\overline{A} \cdot \overline{B} \cdot C \cdot D$	$\overline{A} \cdot \overline{B} \cdot \overline{C} \cdot D$	
$\overline{B}$	9	10	11	12	
	$A \cdot \overline{B} \cdot \overline{C} \cdot \overline{D}$	$A \cdot \overline{B} \cdot C \cdot \overline{D}$	$\overline{A} \cdot \overline{B} \cdot C \cdot \overline{D}$	$\overline{A} \cdot \overline{B} \cdot \overline{C} \cdot \overline{D}$	$\overline{D}$
	13	14	15	16	
	$\overline{C}$	C		$\overline{C}$	

Figure 199-5

Solution

$$f(A,B,C,D) = A \cdot \overline{B} \cdot \overline{C} \cdot (D + \overline{D}) + \overline{A} \cdot \overline{B} \cdot C \cdot D + A \cdot C \cdot \overline{D} \cdot (B + \overline{B})$$
$$+ \overline{A} \cdot C \cdot \overline{D} \cdot (B + \overline{B}) + \overline{C} \cdot \overline{D} \cdot (\overline{A} + A) \cdot (B + \overline{B})$$
$$= A \cdot \overline{B} \cdot \overline{C} \cdot D + A \cdot \overline{B} \cdot \overline{C} \cdot \overline{D} + \overline{A} \cdot \overline{B} \cdot C \cdot D + A \cdot B \cdot C \cdot \overline{D}$$
$$+ A \cdot \overline{B} \cdot C \cdot \overline{D} + \overline{A} \cdot B \cdot C \cdot \overline{D} + \overline{A} \cdot \overline{B} \cdot C \cdot \overline{D} + A \cdot B \cdot \overline{C} \cdot \overline{D}$$
$$+ \overline{A} \cdot B \cdot \overline{C} \cdot \overline{D} + A \cdot \overline{B} \cdot \overline{C} \cdot \overline{D} + \overline{A} \cdot \overline{B} \cdot \overline{C} \cdot \overline{D}$$

Note that the term $A \cdot \overline{B} \cdot \overline{C} \cdot \overline{D}$ is repeated so that only ten 1's will be plotted in the four-element Karnaugh map of Fig. 199-5.

Rule (c). Squares 1, 2, 3, 4, 13, 14, 15, 16 are combined to yield $\overline{D}$.
Rule (a) Squares 9 and 13 are combined to yield $A \cdot \overline{B} \cdot \overline{C}$. Squares 11 and 15 are combined to yield $\overline{A} \cdot \overline{B} \cdot C$.

Therefore:

$$f(A,B,C,D) = \overline{D} + A \cdot \overline{B} \cdot \overline{C} + \overline{A} \cdot \overline{B} \cdot C.$$

Chapter 200
Harvard chart

THE HARVARD CHART IS ANOTHER TECHNIQUE FOR SIMPLIFYING BOOLEAN expressions. With expressions containing four elements or less, the Karnaugh map is superior to the Harvard chart. When five or more elements are involved, however, the Harvard chart has distinct advantages.

As an example, consider the logic equation:

$$f(A,B,C) = A \cdot B \cdot \overline{C} + A \cdot B \cdot C + \overline{A} \cdot B \cdot C + \overline{A} \cdot B \cdot \overline{C} + A \cdot \overline{B} \cdot C$$

$$= A \cdot B \cdot (C + \overline{C}) + \overline{A} \cdot B \cdot (C + \overline{C}) + A \cdot \overline{B} \cdot C \qquad (197\text{-}14)$$

$$= A \cdot B + \overline{A} \cdot B + A \cdot \overline{B} \cdot C \qquad (197\text{-}7)$$

$$= B \cdot (A + \overline{A}) + A \cdot \overline{B} \cdot C \qquad (197\text{-}14)$$

$$= B + \overline{B} \cdot A \cdot C \qquad (197\text{-}7)$$

$$= B + A \cdot C \qquad (197\text{-}19)$$

We will use the Harvard chart to obtain the same simplified result in the Mathematical Derivations.

MATHEMATICAL DERIVATIONS

A Harvard chart for three elements is shown in Fig. 200-1. The steps for obtaining the simplified result are as follows:

Step 1. Draw lines through all the rows whose terms are not contained in the expression being simplified. In our example, Step 1 will apply to rows 1, 2, and 5.

Step 2. Starting with the leftmost column (column 1), cross out all terms that are the same as those already lined out in accordance with Step 1. Since $\overline{A}$ was lined out in rows 1, and 2, and A was lined out in row 5, all terms must be lined out in column 1.

Step 3. In column 2, only $\overline{B}$ was lined out in accordance with step 1. Therefore all B's are circled, since B is part of the final simplified result.

Figure 200-1

Columns							
1	2	3	4	5	6	7	Rows
A	B	C	$A \cdot B$	$A \cdot C$	$B \cdot C$	$A \cdot B \cdot C$	
~~$\overline{A}$~~	~~$\overline{B}$~~	~~$\overline{C}$~~	~~$\overline{A} \cdot \overline{B}$~~	~~$\overline{A} \cdot \overline{C}$~~	~~$\overline{B} \cdot \overline{C}$~~	~~$\overline{A} \cdot \overline{B} \cdot \overline{C}$~~	1
~~$\overline{A}$~~	~~$\overline{B}$~~	~~C~~	~~$\overline{A} \cdot \overline{B}$~~	~~$\overline{A} \cdot C$~~	~~$\overline{B} \cdot C$~~	~~$\overline{A} \cdot \overline{B} \cdot C$~~	2
~~$\overline{A}$~~	(B)	~~$\overline{C}$~~	~~$\overline{A} \cdot B$~~	~~$\overline{A} \cdot \overline{C}$~~	~~$B \cdot \overline{C}$~~	~~$\overline{A} \cdot B \cdot \overline{C}$~~	3
~~$\overline{A}$~~	(B)	~~C~~	~~$\overline{A} \cdot B$~~	~~$\overline{A} \cdot C$~~	~~$B \cdot C$~~	~~$\overline{A} \cdot B \cdot C$~~	4
~~A~~	~~$\overline{B}$~~	~~$\overline{C}$~~	~~$A \cdot \overline{B}$~~	~~$A \cdot \overline{C}$~~	~~$\overline{B} \cdot \overline{C}$~~	~~$A \cdot \overline{B} \cdot \overline{C}$~~	5
~~A~~	~~$\overline{B}$~~	~~C~~	~~$A \cdot \overline{B}$~~	(A·C)	~~$\overline{B} \cdot C$~~	~~$A \cdot \overline{B} \cdot C$~~	6
~~A~~	~~B~~	~~$\overline{C}$~~	~~$A \cdot B$~~	~~$A \cdot \overline{C}$~~	~~$B \cdot \overline{C}$~~	~~$A \cdot B \cdot \overline{C}$~~	7
~~A~~	~~B~~	~~C~~	~~$A \cdot B$~~	(A·C)	~~$B \cdot C$~~	~~$A \cdot B \cdot C$~~	8

Step 4. By moving to the right of the chart, line out those items containing B in all of the rows that possess a circled B.

Step 5. Repeat Steps 3 and 4 for columns 3 and 4; no circled terms are discovered in these columns. However, in column 5, the term $A \cdot C$ is circled, and step 4 is repeated. All terms in columns 6 and 7 have now been lined out so that the simplification procedure has been concluded.

From the Harvard chart, $f(A,B,C) = B + A \cdot C$, which is the same simplified result you obtained from using the Boolean laws and axioms.

Example 200-1

By using a Harvard chart, simplify the logic equation:

$$f(A,B,C) = \overline{A} \cdot \overline{B} \cdot \overline{C} + \overline{A} \cdot B \cdot C + A \cdot B \cdot C + A \cdot \overline{B} \cdot \overline{C} + A \cdot \overline{B} \cdot C.$$

Solution

Start by simplifying the equation through the use of the Boolean laws and axioms.

$$\begin{aligned} f(A,B,C) &= \overline{A} \cdot \overline{B} \cdot \overline{C} + \overline{A} \cdot B \cdot C + A \cdot B \cdot C + A \cdot \overline{B} \cdot \overline{C} + A \cdot \overline{B} \cdot C \\ &= B \cdot C \cdot (\overline{A} + A) + A \cdot \overline{B} \cdot (\overline{C} + C) + \overline{A} \cdot \overline{B} \cdot \overline{C} \\ &= B \cdot C \cdot 1 + A \cdot \overline{B} \cdot 1 + \overline{A} \cdot \overline{B} \cdot \overline{C} && (197\text{-}7) \\ &= B \cdot C + \overline{B} \cdot (A + \overline{A} \cdot \overline{C}) \\ &= B \cdot C + \overline{B} \cdot (A + \overline{C}) && (197\text{-}19) \end{aligned}$$

From the Harvard Chart of Fig. 200-2, the simplified result is:

$$f(A,B,C) = A \cdot \overline{B} + A \cdot C + \overline{B} \cdot \overline{C} + B \cdot C$$

At first glance this does not appear to be the same result because there is an additional $A \cdot C$ term. However,

Columns							
1	2	3	4	5	6	7	Rows
A	B	C	$A \cdot B$	$A \cdot C$	$B \cdot C$	$A \cdot B \cdot C$	
$\overline{A}$	$\overline{B}$	$\overline{C}$	$\overline{A} \cdot \overline{B}$	$\overline{A} \cdot \overline{C}$	$\overline{B} \cdot \overline{C}$	$\overline{A} \cdot \overline{B} \cdot \overline{C}$	1
$\overline{A}$	$\overline{B}$	C	$\overline{A} \cdot \overline{B}$	$\overline{A} \cdot C$	$\overline{B} \cdot C$	$\overline{A} \cdot \overline{B} \cdot C$	2
$\overline{A}$	B	$\overline{C}$	$\overline{A} \cdot B$	$\overline{A} \cdot \overline{C}$	$B \cdot \overline{C}$	$\overline{A} \cdot B \cdot \overline{C}$	3
$\overline{A}$	B	C	$\overline{A} \cdot B$	$\overline{A} \cdot C$	$B \cdot C$	$\overline{A} \cdot B \cdot C$	4
A	$\overline{B}$	$\overline{C}$	$A \cdot \overline{B}$	$A \cdot \overline{C}$	$\overline{B} \cdot \overline{C}$	$A \cdot \overline{B} \cdot \overline{C}$	5
A	$\overline{B}$	C	$A \cdot \overline{B}$	$A \cdot C$	$\overline{B} \cdot C$	$A \cdot \overline{B} \cdot C$	6
A	B	$\overline{C}$	$A \cdot B$	$A \cdot \overline{C}$	$B \cdot \overline{C}$	$A \cdot B \cdot \overline{C}$	7
A	B	C	$A \cdot B$	$A \cdot C$	$B \cdot C$	$A \cdot B \cdot C$	8

Figure 200-2

$$f(A,B,C) = A \cdot \overline{B} + A \cdot C + \overline{B} \cdot \overline{C} + B \cdot C$$
$$= A \cdot \overline{B} + C \cdot B(B + \overline{B} \cdot A) + \overline{B} \cdot \overline{C} \qquad \text{(197-8, 19)}$$
$$= A \cdot \overline{B} + C \cdot B \cdot (B + A) + \overline{B} \cdot \overline{C} \qquad \text{(197-19)}$$
$$= A \cdot \overline{B} + B \cdot C + \overline{B} \cdot \overline{C} \qquad \text{(197-17)}$$
$$= B \cdot C + \overline{B} \cdot (A + \overline{C})$$

the $A \cdot C$ term has been absorbed, and the two results are now the same.

Appendix A
Elements of a remote-control AM station

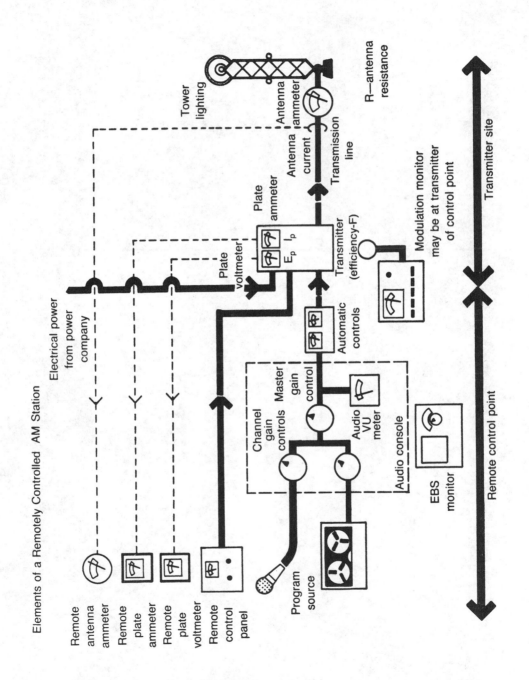

Elements of a Remotely Controlled AM Station

Tower lighting

Antenna ammeter

Antenna current

Transmission line

R—antenna resistance

Transmitter site

Plate ammeter

Plate voltmeter

E_p I_p

Transmitter (efficiency-F)

Modulation monitor may be at transmitter of control point

Electrical power from power company

Automatic controls

Master gain control

Channel gain controls

Audio VU meter

Audio console

Remote control point

EBS monitor

Program source

Remote antenna ammeter

Remote plate ammeter

Remote plate voltmeter

Remote control panel

Appendix B
AM station license

FCC Form 352

UNITED STATES OF AMERICA
FEDERAL COMMUNICATIONS COMMISSION

File No.: BL-13,040

Call Sign: W S L W

STANDARD BROADCAST STATION LICENSE

Subject to the provisions of the Communications Act of 1934, subsequent Acts, and Treaties, and Commission Rules made thereunder, and further subject to conditions set forth in this license, [1]the LICENSEE
REGIONAL RADIO, INC.
is hereby authorized to use and operate the radio transmitting apparatus hereinafter described for the purpose of broadcasting for the term ending 3 a.m. Local Time October 1, 1972
The licensee shall use and operate said apparatus only in accordance with the following terms:

1. On a frequency of 1310 kHz.
2. With nominal power of   -   watts nighttime and 5 kilo watts day time,
   with antenna input power of - watts - directional
   antenna nighttime.................................. [-   current   -   amperes
                                                               -        resistance -   ohms
   and antenna input power of 5 kilo watts non directional
   antenna daytime................................... [antenna current   8.90 amperes
                                                          antenna resistance 63.0 ohms
3. Hours of operation: Daytime as follows:
   Jan. 7:30am to 5:30pm; Feb. 7.15am to 6:00pm;
   Mar. 6:30am to 6:30pm; Apr. 5:45am to 7:00pm;
   May 5:15am to 7:30pm; June 5:00am to 7:45pm;
   July 5:15am to 7:45pm; Aug. 5:30am to 7:15pm;
   Sep. 6:00am to 6:30pm; Oct. 6:30am to 5:45pm;
   Nov. 7:00am to 5:15pm; Dec. 7:30am to 5:00pm;
   Eastern Standard Time (non-advanced)

   Transmitter may be operated by remote control from 73 East Main Street, White Sulphur Springs, West Virginia

4. With the station located at: White Sulphur Springs, West Virginia
5. With the main studio located at:
   73 East Main Street
   White Sulphur Springs, West Virginia
6. The apparatus herein authorized to be used and operated is located at: North Latitude: 37° 48' 34.5"
   Rural area 0.75 mi. North of White Sulphur Springs, West Virginia   West Longitude: 80° 17'   59"

7. Transmitter(s): BAUER, FB-5V

(or other transmitter currently listed in the Commission's "Radio Equipment List, Part B, Aural Broadcast Equipment" for the power herein authorized).**

8. Obstruction marking specifications in accordance with the following paragraphs of FCC Form 715: 1, 3, 11, and 21
9. Conditions:
   **ANTENNA: 190' (193' overall height) uniform cross section, guyed, series excited vertical radiator.
   Ground system consists of 120 equally spaced, buried copper radials 106 to 190 feet in length plus 120 interspaced radials 50 to 106 feet in length.
   The Commission reserves the right during said license period of terminating this license or making effective any changes or modification of this license which may be necessary to comply with any decision of the Commission rendered as a result of any hearing held under the rules of the Commission prior to the commencement of the license period or any decision rendered as a result of any such hearing which has been designated but not held, prior to the commencement of this license period.
   This license is issued on the licensee's representation that the statements contained in licensee's application are true and that the undertakings therein contained so far as they are consistent herewith, will be carried out in good faith. The license shall, during the term of this license, render such broadcasting service as will serve public interest, convenience, or necessity to the full extent of the privileges herein conferred.
   This license shall not vest in the licensee any right to operate nor any right in the use of the frequency designated in the license beyond the term hereof, nor in any other manner than authorized herein. Neither the license nor the right granted hereunder shall be assigned or otherwise transferred in violation of the Communications Act of 1934. This license is subject to the right of use or control by the Government of the United States conferred by Section 606 of the Communications Act of 1934.

[1]This license consists of this page and pages _____ —

Dated: NOVEMBER 4, 1971

FEDERAL
COMMUNICATIONS
COMMISSION

739

Appendix C
Elements of a directional AM station

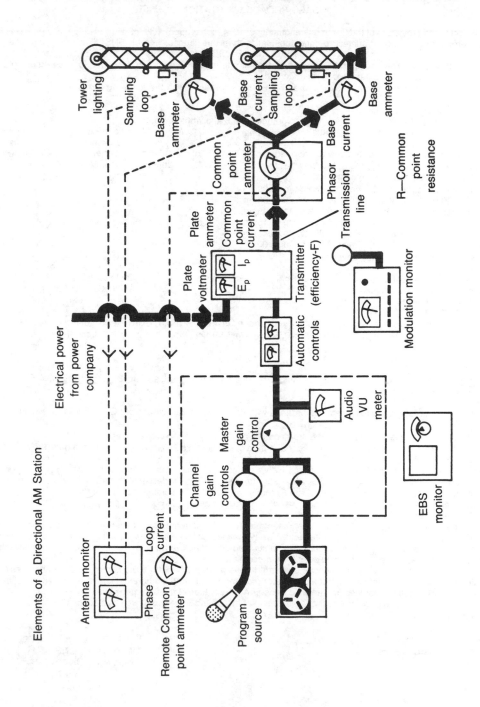

Elements of a Directional AM Station

Appendix D
Directional AM station license

FCC Form 352

STANDARD BROADCAST STATION LICENSE
MAIN AND AUXILIARY TRANSMITTERS

Call Sign: K X X O

Subject to the provisions of the Communications Act of 1934, subsequent Acts, and Treaties, and Commission Rules made thereunder, and further subject to conditions set forth in this license, [1]the LICENSEE

SAN ANTONIO BROADCASTING, INC.

is hereby authorized to use and operate the radio transmitting apparatus hereinafter described for the purpose of broadcasting for the term ending 3 a.m. Local Time JUNE 1, 1977

The licensee shall use and operate said apparatus only in accordance with the following terms:

1. On a frequency of 1300 kHz.
2. With nominal power of 1 kilo watts nighttime and 5 kilo watts daytime,
   with antenna input power of 1.08 kilowatts - directional antenna nighttime................................
   and antenna input power of 5.4 kilo watts directional antenna daytime................................

	Common Point	current	3.93	amperes
	Common Point	resistance	70	ohms
	Common Point	current	8.79	amperes
	Common Point	resistance	70	ohms

3. Hours of operation: Unlimited Time.

   Average hours of sunrise and sunset:

Jan.	7:30 am	to	5:30 pm;	Feb.	7.15 am	to	6:00 pm;
Mar.	6:30 am	to	6:30 pm;	Apr.	6:00 am	to	7:00 pm;
May	5:15 am	to	7:30 pm;	June	5:00 am	to	7:45 pm;
July	5:15 am	to	7:45 pm;	Aug.	5:45 am	to	7:15 pm;
Sep.	6:00 am	to	6:30 pm;	Oct.	6:30 am	to	5:45 pm;
Nov.	7:00 am	to	5:15 pm;	Dec.	7:30 am	to	5:15 pm;

   Transmitters may be operated by remote control from 2805 East Skelly Drive, Tulsa, Oklahoma.

   Central Standard Time (Non-Advanced).

4. With the station located at: Tulsa, Oklahoma

5. With the main studio located at:
   2805 East Skelly Drive
   Tulsa, Oklahoma

6. The apparatus herein authorized to be used and operated is located at:
   8601 South Harvard
   Tulsa, Oklahoma

   North Latitude: 36° 02' 19"
   West Longitude: 95° 56' 07"

7. Transmitter(s): COLLINS, 820E-1 (Main)

   WESTERN ELECTRIC, 405-B2 (Auxiliary)

   (or other transmitter currently listed in the Commission's "Radio Equipment List, Part B, Aural Broadcast Equipment" for the power herein authorized).

8. Obstruction marking specifications in accordance with the following paragraphs of FCC Form 715:**
9. Conditions: (See Page 1A.)

   **TOWERS 1, 2, & 4: Paragraphs 1, 3, 12 & 21. Beacons and all obstruction lights
   shall be flashed, with flashing of towers synchronized so that
   at any instant two towers are lighted and one tower is not.

   Tower 3: Paragraph 1.

The Commission reserves the right during said license period of terminating this license or making effective any changes or modification of this license which may be necessary to comply with any decision of the Commission rendered as a result of any hearing held under the rules of the Commission prior to the commencement of the license period or any decision rendered as a result of any such hearing which has been designated but not held, prior to the commencement of this license period.

This license is issued on the licensee's representation that the statements contained in licensee's application are true and that the undertakings therein contained so far as they are consistent herewith, will be carried out in good faith. The license shall, during the term of this license, render such broadcasting service as will serve public interest, convenience, or necessity to the full extent of the privileges herein conferred.

This license shall not vest in the licensee any right to operate nor any right in the use of the frequency designated in the license beyond the term hereof, nor in any other manner than authorized herein. Neither the license nor the right granted hereunder shall be assigned or otherwise transferred in violation of the Communications Act of 1934. This license is subject to the right of use or control by the Government of the United States conferred by Section 606 of the Communications Act of 1934.

[1]This license consists of this page and pages 1a, 2, 3, & 4.

Dated: DECEMBER 17, 1975

FEDERAL
COMMUNICATIONS
COMMISSION

1. DESCRIPTION OF DIRECTIONAL ANTENNA SYSTEM

No. and Type of Elements:	Four, triangular cross-section, guyed, series-excited vertical towers. Two towers used daytime, three used nighttime. A communications type antenna is side-mounted near the top of the W (No. 4) tower.
Height above Insulators:	284′ (135°)
Overall Height:	288′ (Towers 1 and 3); 290′ (Tower 4); 289′ (Tower 2)
Spacing Orientation:	West to West Center Tower, 273.5′ (130°) - Day; West to East Center Tower, 547′ (260°) East Center to East Tower, 547′ (260°) - Night Line of towers bears 72° true.
Non-Directional Antenna:	None used.

Ground System consists of 240 - 43′ buried copper wire radials equally spaced about each tower; 120 - 43′ to 284′ buried copper wire radials alternately spaced. All radials bonded together by a copper wire at a radius of 43′ from each tower. Copper strap between transmitter ground and bond straps at each tower.

2. THEORETICAL SPECIFICATIONS

		E(No.1)	EC(No.2)	WC(No.3)	W(No.4)
Phasing:	Night	0°	−9.44°	-	0°
	Day	-	-	0°	−52°
Field Ratio:	Night	0.85	1.36	-	0.65
	Day	-	-	1.0	0.80

3. CREATING SPECIFICATIONS

Phase Indication*	Night	17°	0°	-	8°
	Day	-	-	0°	57°
Antenna Base Current Ratio:	Night	0.608	1.0	-	0.473
	Day	-	-	1.0	0.818
Antenna Monitor Sample Current Ratio	Night	56	100	-	40
	Day	-	-	100	80

*As indicated by Potomac Instruments Antenna Monitor. PM-112

Appendix E
Elements of an FM station

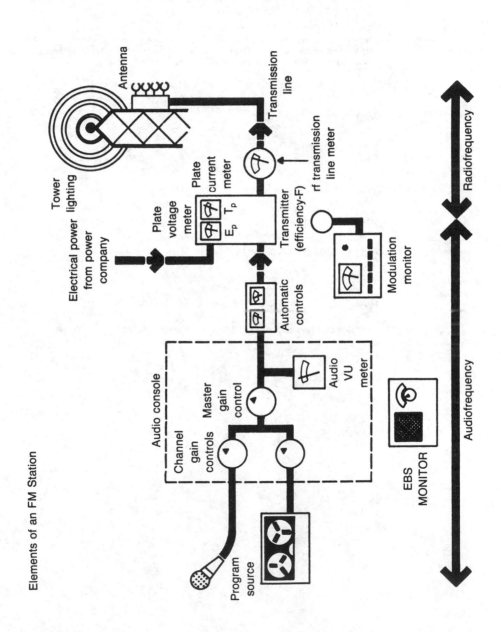

Elements of an FM Station

Appendix F
FM station license

FCC Form 352-A
United States of America
FEDERAL COMMUNICATIONS COMMISSION

File No.: BRH-2019

Call Sign: W F Y N-FM

FM BROADCAST STATION LICENSE

Subject to the provisions of the Communications Act of 1934, as amended, treaties, and Commission Rules, and further subject to conditions set forth in this license, [1]the LICENSEE

FLORIDA KEYS BROADCASTING CORPORATION

is hereby authorized to use and operate the radio transmitting apparatus hereinafter described for the purpose of broadcasting for the term ending 3 a.m. Local Time FEBRUARY 1, 1979

The licensee shall use and operate said apparatus only in accordance with the following terms:

1. Frequency (MHz)........: 92.5
2. Transmitter output power..: 10 kilowatts
3. Effective radiated power..: 25 kilowatts (Horiz.) & 23.5 kilowatts (Vert.)
4. Antenna height above
   average terrain (feet)...: 135' (Horiz.) & 130' (Vert.)
5. Hours of operation.......: Unlimited
6. Station location..........: Key West, Florida
7. Main studio location......:
   Fifth Avenue Stock Island
   Key West, Florida
8. Remote Control point.....:

9. Antenna & supporting structure: North Latitude 24° 34' 01"
   West Longitude 81° 44' 54"

ANTENNA: COLLINS, 37M-5/300-C-5, Five-sections (Horiz. & Vert.), FM antenna side-mounted near the top of the north tower of WKIZ(AM) directional array. Overall height above ground 155 feet.

10. Transmitter location......:

    Fifth Avenue Stock Island
    Key West, Florida

11. Transmitter(s)..........: COLLINS, 830-F-1A

12. Obstruction marking specifications in accordance with the following paragraphs of FCC Form 715: 1, 3, 11 & 21.
13. Conditions:

The Commission reserves the right during said license period of terminating this license or making effective any changes or modification of this license which may be necessary to comply with any decision of the Commission rendered as a result of any hearing held under the rules of the Commission prior to the commencement of the license period or any decision rendered as a result of any such hearing which has been designated but not held, prior to the commencement of this license period.

This license is issued on the licensee's representation that the statements contained in licensee's application are true and that the undertakings therein contained so far as they are consistent herewith, will be carried out in good faith. The license shall, during the term of this license, render such broadcasting service as will serve public interest, convenience, or necessity to the full extent of the privileges herein conferred.

This license shall not vest in the licensee any right to operate nor any right in the use of the frequency designated in the license beyond the term hereof, nor in any other manner than authorized herein. Neither the license nor the right granted hereunder shall be assigned or otherwise transferred in violation of the Communications Act of 1934. This license is subject to the right of use or control by the Government of the United States conferred by Section 606 of the Communications Act of 1934.

[1]This license consists of this page and pages —

Dated: January 28, 1976

Federal
Communications
Commission

Appendix G
FCC emission designations

In 1987 the FCC introduced new emission designations into its rules and regulations. Each type of emission is now designated by three symbols of which the first and third symbols are letters and the second or middle symbol is a number. The meanings of these symbols are:

First Symbol (letter)   Type of Modulation.
Second Symbol (number)   Nature of Signal.
Third Symbol (letter)   Type of Information.

FIRST SYMBOL

A   Amplitude Modulation. Double Sideband. Full Carrier.
C   Vestigial Sideband.
F   Frequency Modulation.
G   Phase Modulation.
H   Single Sideband. Full Carrier.
J   Single Sideband. Suppressed Carrier.
K   Pulse Amplitude Modulation.
L   Pulse Width (Duration) Modulation.
M   Pulse Position Modulation.
N   Unmodulated Carrier.
P   Unmodulated Pulse Sequence.
R   Single Sideband. Reduced Carrier.

SECOND SYMBOL

0   Absence of Any Modulation.
1   Telegraphy On-Off Keying without the use of a Modulating AF Tone.
2   Telegraphy by On-Off Keying of a Modulating AF Tone, or by the On-Off Keying of the Modulated Emission.
3   Analog Voice Communication.

THIRD SYMBOL

A   Telegraphy (aural reception).
B   Telegraphy (reception by automatic machine).
C   Facsimile.
D   Telemetry, Data Transmission.
E   Telephony.
F   Television (video signal).
N   No Information.

EXAMPLES OF EMISSIONS

A1A   Telegraphy by On-Off Keying. Previously Designated as A1.
A3C   AM Facsimile. Previously Designated as A4.
A3E   Amplitude Modulated, Double Sideband, Telephony. Previously Designated as A3.
C3F   Vestigial Sideband Transmission for Television's Video Signal. Previously Designated as A5C.
F1B   Frequency Shift Keying (FSK). Previously Designated as F1.
F3C   FM Facsimile. Previously Designated as F4.

F3E   FM Telephony. Previously Designated as F3.

G3E   Phase Modulated (PM) Telephony. Previously Designated as F3.

H3E   Single Sideband, Full Carrier. Previously Designated as A3H.

J3E   Single Sideband, Suppressed Carrier (SSSC). Previously Designated as A3J.

NON   Unmodulated Carrier. No Information. Previously Designated as A0.

R3E   Single Sideband, Reduced Carrier. Previously Designated as A3A.

Note that sometimes a designation is preceeded by a number that represents the allowed bandwidth in kilohertz.

Appendix H
FCC tolerances and standards

Carrier frequency

Standard AM broadcast stations	±20 Hz
Commercial FM broadcast stations	±2 kHz
Television broadcast stations—aural and visual transmitters	±1 kHz

Non-commercial educational FM broadcast stations:

(1) Licensed for power of more than 10 watts	±2 kHz
(2) Licensed for power of 10 watts or less	±3 kHz

Studio transmitter link (STL)	0.005%
International broadcast stations	0.0015%

Public Safety Radio Services

Frequency range · All mobile stations

	All fixed and base stations	Over 3 W	3 W or less
MHz	Percent	Percent	Percent
Below 25	0.01	0.01	0.02
25 to 50	.002	.002	.005
50 to 450[1]	.0005	.0005	.005
450 to 470[2,3]	.00025	.0005	.0005
470 to 512	.00025	.0005	.0005
806 to 820	.00015	.00025	.00025
351 to 888	.00015	.00025	.00025
250 to 1,427[2]			
1,427 to 1,435[4]	.03	.03	.03
Above 1,435[2]			

[1]Stations authorized for operation on or before Dec. 1, 1961, in the frequency band 73.0–74.6 MHz may operate with a frequency tolerance of 0.005 percent.

[2]Radiolocation equipment using pulse modulation shall meet the following frequency tolerance: the frequency at which maximum emission occurs shall be within the authorized frequency band and shall not be closer than $1.5/T$ MHz to the upper and lower limits of the authorized frequency band where T is the pulse duration in microseconds. For other radiolocation equipment, tolerances will be specified in the station authorization.

(3)Operational fixed stations controlling mobile relay stations, through the use of the associated mobile frequency, may operate with a frequency tolerance of 0.0005 percent.

(4)For fixed stations with power above 200 watts, the frequency tolerance is 0.01 percent if the necessary bandwidth of the emission does not exceed 3 kHz. For fixed station transmitters with a power of 200 watts or less and using time division multiplex, the frequency tolerance can be increased to 0.05 percent.

Power
Transmitters of standard AM and FM commercial broadcast stations . . . . 10% below and
5% above

Aural and visual transmitters of TV broadcast stations . . . . 20% below and
10% above

Current
All currents 5%

Modulation—Standard AM Broadcast
Minimum modulation on *average modulation peaks* 85%
Maximum modulation on *positive modulation peaks* 125%
Maximum modulation on *negative modulation peaks* 100%
Maximum carrier shift allowed 5%

Temperature for Master-Oscillator Crystals
X-cut and Y-cut crystals . . . . ±0.1° C
Low temperature-coefficient crystals . . . . ±1.0° C

Final rf stage: Plate Voltage and Plate Current Meters
Accuracy at full-scale reading 2%
Maximum permissible full-scale reading: 5 times minimum normal reading

Meters: Recording Antenna Current
Accuracy at full-scale reading 2%
Maximum permissible full-scale reading for the scale of current-squared meters . . . .
. . . . 3 times minimum normal reading
Portion of scale used for accuracy with current-squared meters . . . . Upper two-thirds

FCC STANDARDS

Standard AM Broadcast
Band . . . . 535-1605 kHz
Channel width . . . . 10 kHz

FM Commercial Broadcast
Band . . . . 88-108 MHz
Channel width . . . . 200 kHz
Transmitted AF range (main channel) . . . . 50 to 15000 Hz
100% modulation (deviation ratio = 5) . . . . ±75 kHz swing
Time constant for pre-emphasis and de-emphasis . . . . 75 microseconds

Television Broadcast

Bands

Channels 2 through 4 . . . . .54 to 72 MHz
Channels 5 and 6 . . . . .76 to 88 MHz
Channels 7 through 13 . . . . .174 to 216 MHz
Channels 14 through 83 . . . . .470 to 890 MHz
Channel width . . . . .6 MHz
Field frequency . . . . .60 Hz
Frame frequency . . . . .30 Hz
Lines per frame 525
Horizontal scanning frequency . . . . .15750 Hz
Aspect ratio . . . . .4 to 3
Visual bandwidth . . . . .4.9 MHz
Frequency separation between aural carrier frequency and channel upper limit . . . . . . .0.25 MHz
Frequency separation between visual carrier frequency (below) and aural carrier frequency . 4.5 MHz
Frequency separation between visual carrier frequency (below) and chrominance sub-carrier frequency
. . . . 3.579545 MHz ±10 Hz
Reference white level . . . . .12.5% of peak carrier level (±2.5%)
Reference black level . . . . .70% of peak carrier level
Blanking level in a monochrome TV signal . . . . .75% of peak
carrier level (±2.5%)

100% modulation for the aural FM transmission . . . . ±25 kHz swing
Transmitted AF range (main channel) . . . . .50 Hz to 15 kHz
Deviation ratio 1.667

Public Safety Radio Services

Maximum audio frequency . . . . .3 kHz
A1A emission—maximum bandwidth . . . . .0.25 kHz
A3E emission—maximum bandwidth . . . . .8 kHz
Minimum modulation on average modulation peaks 70%
Maximum modulation on negative modulation peaks 100%

F3E emission

Frequency band (MHz)	Authorized bandwidth (kHz)	Frequency deviation (kHz)
25 to 50	20	5
50 to 150	*20	*5
150 to 450	20	5
450 to 470	20	5
470 to 512	20	5
806 to 821	20	5
831 to 866	20	5

In each frequency band the deviation ratio is 1.667

*Stations authorized for operation on or before Dec. 1, 1961, in the frequency band 73.0-74.6 MHz may continue to operate with a bandwidth of 40 kHz and a deviation of 15 kHz.

Harmonic Attenuation

The mean power of emissions shall be attenuated below the mean power output of the transmitter in accordance with the following schedule:

(1) On any frequency removed from the assigned frequency by more than 50% up to and including 100% of the authorized bandwidth: at least 25 decibels.

(2) On any frequency removed from the assigned frequency by more than 250% of the authorized bandwidth: at least 35 decibels.

(3) On any frequency removed from the assigned frequency by more than 250 percent of the authorized bandwidth: at least 43 plus 10 log (mean output power in watts) decibels or 80 decibels, whichever is the lesser attenuation.

Index

Other Bestsellers of Related Interest

ENGINEERING DESIGN: Reliability, Maintainability, and Testability—James V. Jones

Today's economy demands low-cost, reliable products. To meet that demand, design engineers must be able to coordinate their efforts with those of end users, test technicians, manufacturing and support personnel, and other engineers. In this comprehensive guide, management and logistics expert Jim Jones presents a total, field-tested plan that will help you keep your customers satisfied and increase your chances for success. 334 pages, 188 illustrations, Book No. 3151.

AUTOCAD: Methods and Macros—Jeff Guenther, Ed Ocoboc, and Anne Wayman

The ability to generate fully shaded, 3-D images of almost any object is an exciting, practical application of the powerful graphic techniques available in today's CAD software. AutoCAD has become the bestselling PC CAD software in the industry. This comprehensive single-source handbook is designed to be both an introduction to AutoCAD for the newcomer and a reliable reference guide for the professional user. 296 pages, 112 illustrations, Book No. 2989.

THE C⁴ HANDBOOK: CAD, CAM, CAE, CIM—Carl Machover

Increase your productivity and diversity with this collection of articles by the international industry experts, detailing what you can expect from the latest advances in computer aided design and manufacturing technology. Machover has created an invaluable guide to identifying equipment requirements, justifying investments, defining and selecting appropriate systems, and training staff to use the systems effectively. 449 pages, 166 illustrations, Book No. 3098.

COMPUTER TOOLS, MODELS AND TECHNIQUES FOR PROJECT MANAGEMENT—Dr. Adedeji B. Badiru and Dr. Gary E. Whitehouse

Badiru and whitehouse provide you with practical, down-to-earth guidance on the use of project management tools, models and techniques. You'll find this book filled with helpful tips and advice. You'll also discover ways to use your current computer hardware and software resources to more effectively enhance project management functions. 320 pages, 112 illustrations, Book No. 3200.

HUMAN FACTORS IN INDUSTRIAL DESIGN: The Designer's Companion—John H. Burgess

This book represents a nonscientific introduction to human-factor considerations directed specifically to the everyday needs of industrial designers, emphasizing the importance of creating products and equipment that are both safe and useful to their human users. Burgess explains the methods employed by human-factors specialists in determining the significance3 of human dimensions, capabilities, and limitations in designing everything from small hand tools to large, complex systems. 224 pages, illustrated, Book No. 3356.

VersaCAD® TUTORIAL: A Practical Approach—Carol Buehrens

Includes all version of VersaCAD, including the recently released 5.0! This book presents an introduction to VersaCAD and basic drafting principles before leading you step-by-step through the more advanced and creative options available. Among the features: Geometric configurations—Properties of linestyles, pens, and level—Plotting—Table overlays—Dimensions, textures, sizing—Modifying, copying, grouping. 312 pages, 439 illustrations, Book No. 3003.

PROJECT MANAGEMENT: Strategic Design and Implementation—T.H. Allegri, P.E.

T.H. Allegri illustrates the many ways that various disciplines have been integrated in order to produce advancements in the technology of manufacturing. Digesting dozens of references, Allegri covers: CIM, CAD, CAE, management information systems, robotics, graphics and simulation, automated materials handling systems, quality control and product life cycle, flexible manufacturing systems, and more. 400 pages, Book No. 2746.

ENGINEERING DESIGN: Reliability, Maintainability, and Testability—James V. Jones

Today's economy demands low-cost, reliable products. To meet that demand, design engineers must be able to coordinate their efforts with those of end users, test technicians, manufacturing and support personnel, and other engineers. In this comprehensive guide, management and logistics expert Jim Jones presents a total, field-tested plan that will help you keep your customers satisfied and increase your changes for success. 334 pages, 188 illustrations, Book No. 3151.